HW Gaines
723 J. Edwards

INTERNAL
COMBUSTION ENGINES

INTERNAL COMBUSTION ENGINES

BY

LESTER C. LICHTY
*Professor of Mechanical Engineering,
School of Engineering, Yale University*

FIFTH EDITION
THIRTEENTH IMPRESSION

McGRAW-HILL BOOK COMPANY, Inc.
NEW YORK AND LONDON
1939

COPYRIGHT, 1915, 1923, 1929, 1933, 1939, BY THE
MCGRAW-HILL BOOK COMPANY, INC.

PRINTED IN THE UNITED STATES OF AMERICA

*All rights reserved. This book, or
parts thereof, may not be reproduced
in any form without permission of
the publishers.*

THE MAPLE PRESS COMPANY, YORK, PA.

PREFACE

This volume is more than a revision of the book by the same title which was first written by the late Robert L. Streeter in 1915. The second edition appeared in 1923, the principal changes being made in four chapters prepared by E. E. Adams, T. E. Butterfield, G. A. Goodenough, and L. H. Pomeroy. The undersigned assumed full responsibility for the third and fourth editions in 1929 and 1933 and now assumes full authorship for this edition. The reader will recognize, in Chaps. XVII and XIX, only a very small amount of the material that appeared in the first edition, for which credit is hereby given.

The same general procedure is followed in this edition as in the previous one. The theoretical principles are developed, and their application to the analysis, operation, and design of the internal-combustion engine and its parts is well illustrated.

The analysis of the combustion and other thermodynamic processes is very much simplified by the direct use of the general energy equation with the chemical energy concept. This eliminates the indirect method used in the previous edition, which involved a fictitious combustion process usually misunderstood. The use of the combustion charts of Hersey, Eberhardt, and Hottel has made the more advanced study of the combustion process a comparatively simple matter.

The air-standard analysis is treated in a separate chapter and provides a simple approach for those who prefer to study principally parts of the text other than those dealing with the thermodynamics of the actual engine process. The actual process is treated extensively in the following chapter both with charts and with typical thermodynamic computations which show the limitations of the charts as well as their indispensability.

Much has been learned in the last few years about fuels and their characteristics, which has necessitated the rewriting of the chapters dealing with fuels, detonation, and fuel injection. A short chapter dealing with spark ignition has been added. Better knowledge of the combustion process has resulted in a more extensive treatment of combustion-chamber design. Research and analysis have provided better methods for treating heat losses. The air cooling of engines is dealt with in a fundamental manner. The theory of lubrication is given more attention, and more information on materials is included.

The foregoing indicates the extent of the major changes but is no measure of the changes in details, which necessitated rewriting to obtain a homogeneous treatment. The design of internal-combustion engines involves materials, mechanisms, stresses, moments, sections, vibration, flame-front areas, combustion shock, heat transfer, etc. General design is treated in the last chapter, but many other design features are dealt with in other parts of the book.

More material has been included than can be dealt with satisfactorily in the usual undergraduate course, considerable latitude being thus provided for the instructor to choose his material according to the type of course he prefers to teach. The author believes the student's understanding of the subject is enhanced by the solution of problems. Some of the suggested problems have similar parts which may be solved by different students to eliminate repetition. Those who prefer the discursive rather than the more technical type of course should focus attention on the principles involved and their application in the illustrative examples and should reduce problem work to the desired minimum.

Obviously, much information has been drawn from the literature dealing with the subject, for which credit is given throughout the book. The author is also greatly indebted to those users of the previous edition who offered valuable suggestions for this edition. The author acknowledges the valuable assistance of J. L. Meriam, who critically examined and checked the manuscript.

L. C. LICHTY.

NEW HAVEN, CONN.,
　August, 1939.

CONTENTS

	PAGE
PREFACE TO THE FIFTH EDITION	v

CHAPTER

		PAGE
I.	THE INTERNAL-COMBUSTION-ENGINE PROCESS	1
II.	THERMODYNAMICS FOR ENGINE ANALYSIS	10
III.	THE COMBUSTION PROCESS	32
IV.	AIR-STANDARD CYCLE ANALYSIS	55
V.	INTERNAL-COMBUSTION-ENGINE-PROCESS ANALYSIS	68
VI.	DEVIATIONS FROM IDEAL PROCESSES	97
VII.	LIQUID AND GASEOUS FUELS	118
VIII.	DETONATION AND KNOCK TESTING	165
IX.	CARBURETION AND FUEL INJECTION	204
X.	MANIFOLDS AND MIXTURE DISTRIBUTION	270
XI.	VALVES AND VALVE MECHANISMS	303
XII.	IGNITION OF THE CHARGE	345
XIII.	COMBUSTION-CHAMBER AND CYLINDER-HEAD DESIGN	355
XIV.	ENGINE LUBRICATION	398
XV.	ENGINE COOLING	428
XVI.	ENGINE PERFORMANCE	448
XVII.	MECHANICS OF PRINCIPAL MOVING PARTS	480
XVIII.	ENGINE VIBRATION AND BALANCE	499
XIX.	ENGINE DESIGN	525
APPENDIX		581
INDEX		591

INTERNAL-COMBUSTION ENGINES

CHAPTER I

THE INTERNAL-COMBUSTION-ENGINE PROCESS

The principal source of energy from which work may be obtained is the natural supply of coal and oil. Liberation of the chemical energy associated with these fuels and the transformation of some of this energy into work are accomplished by two general methods:
1. The heat-engine process.
2. The internal-combustion-engine process.

Both methods liberate chemical energy by the combustion process; this process results in a large increase in the internal energy of the substances, which undergo reaction or change in atomic and molecular arrangement. The heat-engine process transfers most of this internal-energy increase to a working medium, usually water, which is permitted to exert its expansive force on an engine piston or turbine blades and produces work. The reacting substances are the working medium for the internal-combustion-engine process and exert a force directly on the engine piston.

The heat-engine process is used in engines ranging in size from toy steam engines to reciprocating engines developing several hundred horsepower and to steam turbines developing well over 100,000 hp. Internal-combustion engines range in size from the model aircraft engines to large engines developing nearly 10,000 hp.

The internal-combustion engine is inherently well adapted for transportation purposes, and it is in the automotive field that remarkable development has taken place. The application of this engine in the aviation industry promises to do for the world what the automobile has done for this country.

While the foregoing application of the internal-combustion engine has seemed spectacular to the general public, there are numerous other applications that are very important to various industries. Much of the power developed for the oil and gas industry is by means of liquid-fuel or gas engines. Other important applications are found in the steel and marine industries, and in numerous industrial plants.

Vacuum-principle Engines.—In the earliest engines, such as Street's in 1794 (Fig. 1), air was pumped by hand into the cylinder, causing the piston to rise a portion of the stroke. Liquid fuel was then poured in and ignited by the hot end of the cylinder located in the furnace. The combustion and expansion

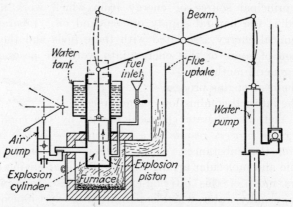

Fig. 1.—Street's engine (1794). (*From "Evolution of the Internal-combustion Engine" by Edward Butler, Charles Griffin & Co., Ltd., 1912.*)

of gases forced the piston upward and did work not only in elevating the heavy piston but in pumping water to a higher elevation.

The upper end of the cylinder was surrounded by a water jacket for cooling the expanded gases and producing a partial vacuum which, combined with gravity, caused the piston to return to the bottom of its stroke and pump water from a well or reservoir.

From about 1860 to 1880, the vacuum principle of obtaining work was developed to the fullest extent. The free-piston engine, such as Otto and Langen's (Fig. 2), forced a piston upward during the combustion and expansion process. No work was done during the expansion process except to elevate the piston. The gases were then cooled, and the vacuum as well as the weight of the piston was used to perform work. On the vacuum stroke the piston rod was connected to the work mechanism, whereas on the expansion stroke a "freewheeling" mechanism permitted free motion of the piston.

The early engines were hand-charged and operated at the rate of a few cycles per minute, usually less than ten. Now, some of the modern high-speed racing engines complete as high as 4000 cycles min.$^{-1}$ per cylinder. This is one indication of the extensive development of the internal-combustion engine.

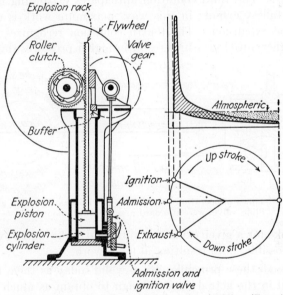

Fig. 2.—Otto and Langen's "Free-piston" engine (1866). (*From "Evolution of the Internal-combustion Engine" by Edward Butler, Charles Griffin & Co., Ltd., 1912.*)

Early Fundamental Analysis.—All the early engines were provided with a combustible charge which was ignited at about atmospheric pressure. However, as early as 1838, Barnett recognized the advantage of compression of the charge before combustion and incorporated this process in his patent of that date.

It was not until 1862 that the fundamental principles underlying the economical operation of the internal-combustion engine were set forth by Beau de Rochas, a Frenchman. His paper listed the conditions under which maximum economy could be obtained, namely:

1. Smallest possible surface-volume relation for the cylinder.
2. Most rapid expansion process possible.
3. Maximum possible expansion.
4. Maximum possible pressure at beginning of expansion process.

The first two conditions are designed to reduce the heat loss through the cylinder walls to a minimum, thus maintaining the inherent availability of the energy in the combustion products. The third condition anticipates expanding the gases to the fullest extent; in this way, maximum work is obtained from the expansion. The fourth condition recognizes the fact that higher initial pressures result in higher pressures throughout

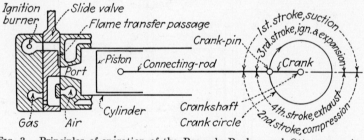

Fig. 3.—Principles of operation of the Beau de Rochas and Otto gas engine. Slide valves were standard practice until about 1890.

the stroke for a given expansion ratio and make higher expansion ratios possible, both resulting in more work.

Although these principles are as valid today as then, they are modified in the actual engine design to obtain as much of their benefits as is economically worth while. When the gain by following any fundamental principle toward its limit in the actual machine is offset by the effort or cost to do so, then the desirable limit of application of that principle has been passed.

Beau de Rochas also set forth the series of operations (Fig. 3) by which the internal-combustion engine should be operated:

1. Suction during the outward stroke of the piston.
2. Compression during the inward stroke of the piston.
3. Ignition of the charge at inward dead center, followed by expansion during the next outward stroke of the piston.
4. Exhaust during the next inward stroke of the piston.

This cycle of engine events is practically the same as is used today, the timing of the various events being modified to attain the higher speeds and outputs.

Four- and Two-stroke Cycles.—Four strokes are required to complete the engine cycle described by Beau de Rochas, from which has come the term, "four-stroke cycle." Work is obtained during only one of the four strokes The desirability of having one working stroke in every revolution led to the development of the "two-stroke cycle." Barnett, an Englishman, in 1838 described the mechanism for supplying a charge to the cylinder by means of separate pumps, the fresh charge displacing the products of the previous charge which escaped through ports

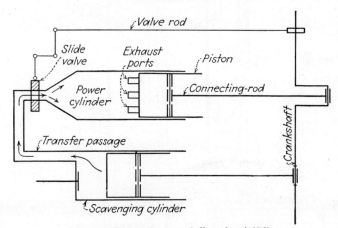

FIG. 4.—Clerk's "two-cycle" engine (1878).

in the cylinder wall. Clerk, also an Englishman, deserves most of the credit for the early development of this cycle (Fig. 4) which is desirable for certain types of engines. However, the four-stroke engine predominates in the field of internal-combustion engines.

The Otto Engine.—The first successful engine embodying the principles of Beau de Rochas was built in 1876 by Otto, a German, from which came the term "Otto Cycle," often used in place of the term "spark-ignition cycle."

This country followed the lead of the early European inventors and manufacturers and produced engines of the Otto type of large weight and small output. However, Langley's early experiments on "mechanical flight" (1887–1896) indicated the need of a lightweight internal-combustion engine. Charles M. Manly, Langley's chief assistant (1897–1903) was unable to

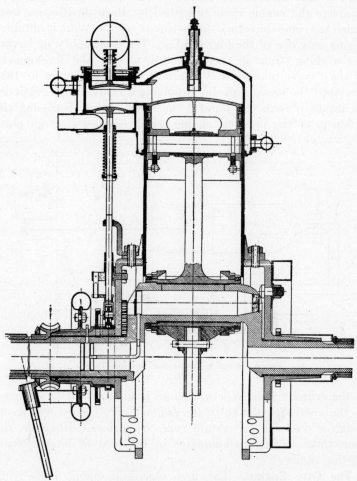

Fig. 5.—Cross-section of Manly radial engine.

obtain such an engine for the Langley airdrome from either European or American manufacturers and finally developed a light radial engine (Fig. 5) in 1901 which might well be called the forerunner of the present radial aircraft engine. This engine[1] had five water-cooled cylinders of 5-in. bore and 5½-in. stroke, weighed 184.47 lb., and developed 52.4 hp. at 950 r.p.m. for 10 consecutive hours.[2] This was a remarkable performance at that time.

The Diesel Engine.—The work of compressing a medium results in an increase in the internal energy and temperature of the medium unless unusually large heat transfer occurs during the process. In 1892, Diesel, a German, proposed compression of air alone until sufficiently high temperatures were reached to ignite the fuel which was to be injected at the end of the compression stroke. He proposed to coordinate the rate of fuel injection with the piston movement so that the heat of combustion would be liberated at a constant maximum temperature. Thus, the original process was designed to approach the Carnot engine cycle as closely as possible. However, economic and other considerations resulted in an approximately constant-pressure combustion process for the large slow-speed Diesel engine.

The modern high-speed Diesel engine has a combustion process that lies between that of the Otto engine and the slow-speed Diesel engine. The process may be assumed to be one of constant-volume combustion until some pressure limit is reached and then one of constant pressure for the remainder of the process. The terms "Diesel cycle" and "compression-ignition cycle" are used synonymously.

The principal differences between the Otto and Diesel engines are:

1. Fuel is mixed with air *after* compression in the Diesel and *before* compression in the Otto engine.

2. Much higher compression is used in the Diesel than in the Otto engine.

3. The Otto engine requires some source of ignition, but the Diesel requires none.

4. The Otto engine has approximately constant-volume combustion, but the Diesel may have approximately constant

---

[1] Langley Memoir on Mechanical Flight; *Smithsonian Contributions to Knowledge*, Vol. XXVII, Part II, p. 234 (1911). The reader will find a most interesting account of engine development in this reference.

[2] Except for a 10-min. stop to renew the lubricating oil and change the ignition batteries.

pressure combustion, or a combination of constant-volume and constant-pressure combustion.

These differences exist in the two-stroke as well as in the four-stroke cycles.

Heat-engine Cycle vs. Internal-combustion-engine Process.— A study of all types of heat engines results in the conclusion that certain elements, used either independently or in combination, are required for transforming heat into work. These are:

1. A source of heat.
2. A medium to absorb heat from the source and carry it to the engine.
3. An engine in which the medium expands and does work.
4. A refrigerator to which the medium and some of the energy remaining in the medium are rejected after expansion.

The steam engine requires a boiler in which energy is liberated and transferred to the medium, which flows to the engine and expands therein and does work. After the expansion process the steam engine rejects the medium to a condenser in which the steam rejects heat to the cooling water. The steam is condensed to water, which is pumped back into the boiler and used again. Thus, the medium in the heat-engine cycle undergoes a series of processes which return it to the original condition.

In the internal-combustion-engine process the energy liberation occurs in the engine cylinder. The liberated chemical energy appears as internal energy of the products of combustion which do work on the engine piston during the expansion process. The products are rejected from the engine during the exhaust stroke and are not returned to the original condition of fuel and air. Hence, the medium does not undergo a cycle of processes although the engine undergoes a cycle of events.

In the chapter on Air-standard Cycle Analysis, it is shown that in the Carnot cycle, which is the most efficient heat-engine cycle, all the heat is added to the medium at the highest possible temperature, some is transformed into work, and the balance is rejected from the medium at the lowest possible temperature. The temperature attained by the medium in the internal-combustion-engine process is much higher than in the steam or even the mercury-steam cycle, since with the heat-engine cycle all the heat must be transferred through metal walls to the medium. This limits the upper temperature to the maximum that the walls can withstand. Thus, on the high-temperature side the internal-combustion-engine process has a great advantage.

The heat-engine cycle expands to and rejects heat at a low temperature which is nearly atmospheric. The internal-combustion-engine process releases the products of combustion at a fairly high temperature. Thus, on the low-temperature side the heat-engine cycle has a great advantage.

A knowledge of the operating conditions in both cases is required to determine the more efficient process. However, something other than thermal efficiency is usually the deciding factor in choosing the type of power plant for a given service.

Analysis of the Process.—The principal considerations that must be dealt with in any of the cycles or processes to be used are as follows:

1. Pressure, volume, and temperature changes of the medium.
2. The performance of work on or by the medium.
3. The combustion process by whatever path is used.
4. The kinetic energy and flow work of the entering mixture and leaving products, if appreciable.
5. The loss of heat from or addition of heat to the medium.

The thermodynamic principles involved and their applications will be dealt with in the following four chapters as well as in various other places in this book.

CHAPTER II

THERMODYNAMICS FOR ENGINE ANALYSIS

Media and Coordinates.—Air is the principal constituent in the media supplied internal-combustion engines. It is in the gaseous state at atmospheric conditions. The fuel flowing to the engine is in the liquid state until leaving the carburetor jet or fuel-injection nozzle. It may either be partly or completely vaporized or be superheated before ignition occurs in the engine cylinder. Also, the fuel may be in the gaseous state as is the case with natural and manufactured gases. Thus, all the states of media ranging from liquid to gaseous are encountered in the internal-combustion-engine process.

The state of a unit quantity of a medium is defined by the elementary coordinates, pressure P, volume V, and temperature T. Either pressure or temperature defines the state of a saturated liquid or vapor. The volume is also required for defining the state of a mixture of saturated liquid and vapor. Any two of these coordinates define the superheated vapor and gaseous states, so that in general only two of the coordinates are required to define the state of a given quantity of medium.

Other factors or properties such as entropy S, internal energy E, chemical energy C, and enthalpy H, which are point functions, may be used to define the state of the medium. Thus, T and S, as well as H and S, are used as coordinates.

Characteristic Equations.—The elementary coordinates P, V, and T are related through the characteristic equation which may be of the form

$$V = f(P, T). \qquad (1)$$

The actual equation is somewhat involved for other than the perfect gaseous state, and, as a result, charts and tables of properties have been developed for most common media. Beyond such tabulated data, the characteristic equation for perfect gases may be used with little error in regions where the super-

THERMODYNAMICS FOR ENGINE ANALYSIS

heated-vapor data indicate that the enthalpy and temperature lines are nearly parallel.

The characteristic equation for perfect gases is

$$PV = MRT \tag{2}$$

where P = abs. pr., lb. ft.$^{-2}$,
V = vol., ft.^{3},
M = mols of medium, wt. $\div$ molecular wt. of medium,
R = universal gas const., 1545 ft.-lb. mol^{-1} °R.$^{-1}$
T = abs. temp., °Rankine = (°F. + 459.6).*

Though all actual gases deviate from perfect gas behavior, and particularly at the high pressures encountered in the internal-combustion-engine process, the perfect gas relationship is considered satisfactory for analysis purposes.

Example.—Determine the volume of 1 mol of gas at 14.7 lb. in.$^{-2}$ abs. pressure and a temperature of 62°F.

From the characteristic equation for gases,

$$V = \frac{MRT}{P} = \frac{1 \times 1545 \times 522}{14.7 \times 144} = 381 \text{ ft.}^3 \text{ mol}^{-1}.$$

The volume per pound of the gas may be obtained by dividing the volume per mol by the molecular weight. Thus, for methane (CH$_4$) with a molecular weight of 16,

$$V = {}^{381}\!/\!_{16} = 23.8 \text{ ft.}^3 \text{ lb.}^{-1}$$

Various conditions in the gaseous state may be related to each other by combining the characteristic equations for each condition. Thus, for conditions indicated by subscripts 1, 2, 3, etc.,

$$\frac{P_1 V_1}{M_1 T_1} = \frac{P_2 V_2}{M_2 T_2} = \frac{P_3 V_3}{M_3 T_3} = \cdots = R. \tag{3}$$

With no change in mols Eq. (3) becomes

$$\frac{P_1 V_1}{T_1} = \frac{P_2 V_2}{T_2} = \cdots = \text{const.} \tag{4}$$

Example.—Ten in.^{3} of a mixture of gases undergo chemical reaction while the volume decreases to 9 in.^{3} The final pressure is five times the initial pressure. Final mols are 10 per cent greater than initial mols. Initial temperature is 540°F. Determine the final temperature.

* Usually 460 is used, the difference being 0.1 per cent.

12   INTERNAL-COMBUSTION ENGINES

From Eq. (3), $\dfrac{P_1 V_1}{M_1 T_1} = \dfrac{P_2 V_2}{M_2 T_2}.$

Substituting, $\dfrac{1 \times 10}{1 \times 1000} = \dfrac{5 \times 9}{1.1 \times T_2},$

or $T_2 = 4090°\text{R}.$

Mixtures of Media.—Each medium in a mixture of media in the gaseous or vapor state occupies the total volume, is assumed to be at the same temperature as the other media, and exerts a pressure that is termed its "partial" pressure. Thus, if in the gaseous state,

$$V = \frac{M_1 RT}{P_1} = \frac{M_2 RT}{P_2} = \cdots = \frac{MRT}{P} \qquad (1)[1]$$

from which $\dfrac{M_1}{P_1} = \dfrac{M_2}{P_2} = \dfrac{M}{P} \qquad (2)$

where P_1 and P_2 = the partial pressures for the constituents M_1 and M_2, respectively.

Dalton's law states that the sum of the partial pressures equals the total pressure of the mixture. This follows from Eq. (1), for

$$M = M_1 + M_2 + \cdots + M_n \qquad (3)$$
and $$P = P_1 + P_2 + \cdots + P_n. \qquad (4)$$

Example.—Carbon monoxide and carbon dioxide are to be mixed in the ratio of 1:3 by weight, respectively. The total pressure is 14.7 lb. in.$^{-2}$ abs., and the temperature is 40°F. Determine the total volume per pound of mixture and the partial pressure of each constituent.

One pound of mixture will contain 0.25 lb. of CO and 0.75 lb. of CO_2.

$$\text{Mols of CO} = \frac{0.25}{28} = 0.00893$$

$$\text{Mols of } CO_2 = \frac{0.75}{44} = 0.01704$$

$$\text{Mols of mixture} = 0.02597$$

$$\text{Total } V = \frac{MRT}{P} = \frac{0.02597 \times 1545 \times 500}{14.7 \times 144} = 9.48 \text{ ft.}^3 \text{ lb.}^{-1}$$

For CO_2, $P_{CO_2} V = M_{CO_2} RT.$
Also, $PV = MRT,$

---

[1] If superheated vapor, such as water vapor, the superheated tables for the substance should be used if available.

or, $\quad P_{CO_2} = P\dfrac{M_{CO_2}}{M} = 14.7 \times \dfrac{0.01704}{0.02597} = 9.65$ lb. in.$^{-2}$

$$P_{CO} = P - P_{CO_2} = 14.7 - 9.65 = 5.05 \text{ lb. in.}^{-2}$$

The pressure of a saturated vapor is a function of the temperature. Consequently, the temperature of a mixture of a gas or gases and a saturated vapor fixes the partial pressure exerted by the vapor.

Also, $\quad\quad P_{gases} = P - P_{vapor}\quad\quad\quad\quad(5)$

Example.—A mixture of air and saturated water vapor at a temperature of 100°F. is flowing to the carburetor of an engine. Determine the volume per pound of mixture and the pounds of water vapor per pound of air.

At 100°F., $\quad P_{wv} = 0.95$ lb. in.$^{-2}$ abs.,*
and $\quad\quad\quad V_{wv} = 350.4$ ft.^{3} lb.$^{-1}$
$\quad\quad\quad\quad P_{air} = P - P_{wv} = 14.7 - 0.95 = 13.75$ lb. in.$^{-2}$ abs.

Then, $\quad\quad V_{air} = \dfrac{M_{air}RT}{P_{air}} = \dfrac{1545 \times 560}{13.75 \times 144} = 436.9$ ft.^{3} mol.$^{-1}$

$V_{air} \div$ mol. wt. $= 436.9 \div 28.95 = 15.09$ ft.^{3} lb.$^{-1}$
$\quad\quad M_{wv} = 15.09 \div 350.4 = 0.0431$ lb. wv lb.$^{-1}$ of air.

$$V_{mix} = \dfrac{V}{M} = \dfrac{15.09}{1.0431} = 14.47 \text{ ft.}^3 \text{ lb.}^{-1} \text{ of mixture.}$$

If one of the constituents is part liquid and part saturated vapor, the volume of the saturated vapor is the volume of the gaseous constituents, since the liquid occupies a volume alone.

The vapor pressure of liquids containing mutually soluble substances, such as the various hydrocarbons in petroleum fuels, is less than the sum of the partial pressures each vapor would exert if alone and in the presence of its own liquid at the same temperature. Data and relationships other than those for substances alone are required when dealing with mixtures containing mutually soluble substances.

The properties of mixtures, which are functions of the state only, are the sum of those properties of the individual constituents in mixture condition.

Thus, $\quad\quad E_{mix} = E_1 + E_2 + \cdots + E_n.\quad\quad\quad(6)$

* All properties of water are obtained from "Thermodynamic Properties of Steam" by J. H. Keenan and F. G. Keyes, John Wiley & Sons, Inc., New York, 1936.

Energy Terms.—The following forms of energy are usually encountered in engineering processes:

1. *Mechanical work* W is evaluated as the product of force and distance through which the force acts. It is also the product of pressure and change in volume, $P\,\Delta V$, if the process is reversible.[1]

Thus,
$$W = \int_{V_1}^{V_2} P\,dV. \tag{1}$$

The area beneath the path of a reversible process plotted on the

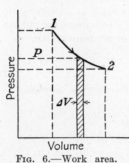

Fig. 6.—Work area.

P-V diagram (Fig. 6) represents the mechanical work done by or on the medium if acting against a mechanism. It may be evaluated by mathematical or graphical integration.

In some cases the path may be represented very nearly by the equation

$$P_1 V_1^n = P_2 V_2^n, \tag{2}$$

from which the $\int P\,dV$ may be evaluated. When $V_2 > V_1$, the work is out, that is, done by the medium on the mechanism, and vice versa. The work area is treated as a positive value regardless of the sign of the integral.

Mechanical work values are *absolute* values of energy actually being transferred from a mechanism to a medium, or vice versa.

Example.—A medium at an initial pressure of 100 lb. in.$^{-2}$ abs. and a volume of 1 ft.^{3} expands and does work on a piston which permits the volume to increase to 3 ft.^{3} The path of the process is described by the relation $PV^2 = $ const. Determine the work done by the medium.

$$\text{Work} = \int_{V_1}^{V_2} P\,dV = P_1 V_1^n \int_{V_1}^{V_2} \frac{dV}{V^n}$$
$$= \frac{P_2 V_2 - P_1 V_1}{1-n}. \tag{3}$$

Solving for the final pressure results in,

$$P_2 = P_1 \left(\frac{V_1}{V_2}\right)^n = 100\left(\frac{1}{3}\right)^2 = 11.11 \text{ lb. in.}^{-2} \text{ abs.}$$

Then, $\quad \text{Work} = \dfrac{144(11.11 \times 3 - 100 \times 1)}{1-2} = 9600$ ft.-lb.

The work is done by the medium since $V_2 > V_1$.

[1] See the section dealing with reversibility.

2. *Flow work PV* is energy entering or leaving a process owing to the flow of the fluids to and from the process. It is equivalent to the gross work required of a piston mechanism in moving the medium past a given section in the pipe. In this case $P \Delta V$ becomes PV, the product of the pressure and volume of the medium at the given section (Fig. 7). Flow work is not energy in the medium but energy transferred into and out of the process by the flowing medium. It represents an *absolute* value of energy.

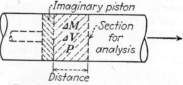

Fig. 7.—Evaluation of flow work.

Example.—Determine the flow work per mol of exhaust products flowing through an exhaust line. The pressure is 16 lb. in.$^{-2}$ abs., and the temperature is 2000°R at the section where the flow work is to be evaluated.

$$\text{Flow work} = PV = MRT$$
$$= 1 \times 1.986 \times 2000 = 3972 \text{ B.t.u.}$$

where $R = {}^{1545}\!\!/\!_{778} = 1.986$ B.t.u. mol^{-1} °R.$^{-1}$

3. *Heat Q* is energy transferred to or from media owing to temperature difference only. For reversible processes,[1] $T \Delta S$ represents the heat transferred which may be evaluated as the area beneath the path representing the process on the T-S diagram (Fig. 8).

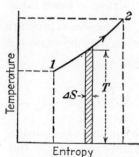

Fig. 8.—Heat-transfer area.

Thus, $$Q = \int_{S_1}^{S_2} T \, dS. \qquad (4)$$

When $S_2 > S_1$ the heat transfer will be into the medium, and vice versa. The heat transfer is treated as a positive value although the integral is negative with $S_2 < S_1$.

Heat transfer may also be evaluated from the specific heat c and the temperature change.

Thus, $$Q = \int_{T_1}^{T_2} Mc \, dT \qquad (5)$$

where $c =$ ordinarily some function T (Table II, Appendix) for a constant-pressure or -volume change. Again, the heat

---
[1] See the section dealing with reversibility.

transfer is treated as a positive value although the integral is negative with $T_2 < T_1$.

Example.—One mol of O_2 is heated at constant pressure from 1000 to 5400°R. Determine the heat added.

For O_2 (Table II, Appendix),

$$c_p = 11.515 - \frac{172}{\sqrt{T}} + \frac{1530}{T}, \qquad (540 - 5000)$$

and $$c_p = (c_p)_{540}^{5000} + \frac{0.05(T - 4000)}{1000}. \qquad (5000 - 9000)$$

$$Q_1 = \int_{1000}^{5000} c_p\, dT = \left[11.515(T_2 - T_1) - 344(T_2^{1/2} - T_1^{1/2}) \right.$$
$$\left. + 1530 \log_e \frac{T_2}{T_1} \right]_{1000}^{5000} = 34{,}455 \text{ B.t.u.}$$

$$Q_2 = \int_{5000}^{5400} c_p\, dT = \left[11.515(T_2 - T_1) - 344(T_2^{1/2} - T_1^{1/2}) + 1530 \log_e \frac{T_2}{T_1} \right.$$
$$\left. + \frac{0.025(T_2^2 - T_1^2)}{1000} - \frac{200(T_2 - T_1)}{1000} \right]_{5000}^{5400} = 4830 \text{ B.t.u.}$$

$$Q = Q_1 + Q_2 = 39{,}285 \text{ B.t.u.}$$

This should equal ΔH for the same range as indicated by Table III. Thus,

$$\Delta H = \Delta E + \Delta RT = (32{,}607 - 2539) + (10{,}724 - 1986) = 38{,}806 \text{ B.t.u.}$$

The difference between heat transfer at constant pressure and constant volume is equivalent to the work done by or on the medium during the constant-pressure change.

For the same temperature change,

$$c_p\, \Delta T - c_v\, \Delta T = P\, \Delta V = R\, \Delta T,$$
or $$c_p - c_v = R. \tag{6}$$

The specific heat of monatomic gases is a constant, as indicated by the Kinetic Theory of Gases. It amounts to $1.5R$ for constant-volume and $2.5R$ for constant-pressure processes.

Heat values represent absolute values of energy since heat is transferred from one medium to another.

4. *Potential energy PE* is the energy of position of the medium and is evaluated as the product of weight of medium and elevation above an assumed datum plane. This energy is usually so small in the internal-combustion-engine process that it is considered negligible. Potential energy is a *relative* value, being referred to some datum plane such as the earth's surface.

5. Kinetic energy KE is energy of the medium due to its motion. It is equivalent to one-half the product of the mass of medium and its velocity squared.

Thus,
$$KE = \frac{1}{2}\left(\frac{wt}{g} \times w^2\right). \tag{7}$$

Kinetic energy is a *relative* value, usually being referred to the velocity of the surface of the earth as representing zero velocity.

Example.—The velocity of exhaust gases in an exhaust manifold is 200 ft. sec.$^{-1}$ The apparent molecular weight of the gases is 28.5. Determine the kinetic energy per mol of gases.

$$KE = \frac{1}{2}\left(\frac{28.5}{32.2} \times \overline{200^2}\right) = 17,700 \text{ ft.-lb.}$$

This represents much less than 0.1 per cent of the energy liberated in the reaction producing the exhaust gases.

6. Internal energy E is the energy in the medium due to the motion and position of the molecules. It does not include the energy associated with the arrangement of the atoms in the molecule. Internal energy may be evaluated by the amount of heat transfer required to change the condition of a medium while held at constant volume.

Thus,
$$\Delta Q = M c_v \Delta T = \Delta E, \tag{8}$$
or, for media in the gaseous state,
$$E_2 - E_1 = \int_{T_1}^{T_2} c_v \, dT. \tag{9}$$

Internal-energy values (Table III, Appendix) are relative values since they are evaluated either above some datum plane or from specific heat equations which are valid approximations only for specified temperature ranges.

7. Enthalpy H is the term applied to the sum of the internal energy and the PV term which represents flow work if the medium is flowing.

Thus, Enthalpy $= H = E + PV,$ (10)
or, for media in the gaseous state,
$$H_2 - H_1 = \int_{T_1}^{T_2} c_p \, dT. \tag{11}$$

Enthalpy values are particularly useful in steady-flow processes and also in nonflow constant-pressure processes in which case the heat transfer is equivalent to the change in enthalpy. Since internal energy is a *relative* value, enthalpy is also *relative* although the PV part is *absolute* in value. Enthalpy values of various substances in the gaseous state are obtained by adding RT to the proper internal-energy values.

8. *Chemical energy C* is energy stored in the electron and bonding arrangement of the various atoms of a molecule. Combination or dissociation of atoms and molecules changes the foregoing arrangement and results in the liberation or absorption of energy. Chemical energy is evaluated[1] from the heats of reaction and the internal energy or enthalpy values of the media before and after reaction. Since chemical energy depends upon internal energy or enthalpy values, it is a *relative* value and can be used only with the tables of energy values from which it was determined. Its use simplifies the analysis of combustion processes.

9. *Electrical energy EE* is measured in watt-seconds or multiples thereof and represents an absolute value of energy being transferred into or out of a process by means of an electrical circuit.

Fundamental Difference of Energy Terms.—All the forms of energy evaluated in the foregoing except mechanical work, heat, and electrical energy are functions of the *condition, position*, or *motion* of the medium. All that is required to evaluate such forms of energy for a given quantity of medium is a knowledge of the coordinates, elevation, and velocity of the medium.

Work, heat transfer, and electrical energy are functions of the *path of the process* and cannot be evaluated by a knowledge only of the conditions at one point of the process. The plotting of the path of the process is possible only when the medium is in a state of equilibrium throughout the process.

Some of the forms of energy, W, PV, Q, and EE, are *absolute* in value, whereas the others, E, H, PE, KE, and C, are *relative* and in some cases may be based upon different reference planes. These differences must be considered in using the energy values for any process. For processes in which the media do not change to other media, differences of corresponding entering- and

---
[1] See Chemical Energy in the chapter on The Combustion Process.

leaving-energy values eliminate the reference-plane values and result in absolute-energy changes.[1]

Relations between the various energy units are given in Table I (Appendix).

Tables of Energy Values.—Tables of properties of media usually contain values for the coordinates P, V, and T, as well as values for E, H, and S. Media in the gaseous state are usually assumed to behave as perfect gases, under which condition E and H are functions of the temperature only. Thus, only T, E, PV, and S values are given in Tables III and IV (Appendix) for gases.

Any process that starts with a medium in the gaseous region and ends with the medium in the vapor region must ordinarily be solved in two parts. The first part deals with the medium from the initial condition to the upper temperature limit of the vapor data, and the second part deals with the medium from the latter condition to the final condition of the process. Data are available in the case of water up to 1600°F., at which temperature enthalpy does not vary much with pressure except at the higher pressures.

Example.—H$_2$O at 300 lb. in.$^{-2}$ abs. pressure and 5000°R. flows through an apparatus and leaves at a pressure of 100 lb. in.$^{-2}$ abs. and a temperature of 500°F. Determine the change in enthalpy.

At 5000°R., $E = 39{,}885$ B.t.u. (Table III, Appendix).
At 2060°R., $E = 10{,}820$ B.t.u.
            $\Delta E = 29{,}065$ B.t.u. mol.$^{-1}$
            $\Delta RT = 1.986(5000 - 2060) = 5839$ B.t.u. mol.$^{-1}$
            $\Delta H = 29{,}065 + 5{,}839 = 34{,}904$ B.t.u. mol.$^{-1}$
At 1600°F., $H = 1856.2$ to 1853.7 B.t.u. lb.$^{-1}$ (Steam Tables).
At 500°F., $H = 1279.1$ B.t.u. lb.$^{-1}$
            $\Delta H = 577.1$ to 574.6 B.t.u. lb.$^{-1}$
               $= 10{,}388$ to $10{,}343$ B.t.u. mol.$^{-1}$
Thus,   Total $\Delta H = 45{,}269.5 \pm 22.5$ B.t.u. mol.$^{-1}$

The Energy Equation.—The principle of conservation of energy, that energy cannot be created or destroyed but merely undergoes transformation, leads directly to the energy equation which accounts for all the energy involved in a process. If it is assumed that there is no change in energy content of the apparatus during a process occurring therein, the energy entering with the medium or by other means must be equal to the energy

[1] See also the chapter on The Combustion Process.

leaving the apparatus. Subscripts 1 and 2 being used to indicate energy entering or leaving with the medium,

$$C_1 + PE_1 + KE_1 + E_1 + P_1V_1 + W_{in} + Q_{in} + EE_{in} = \\ C_2 + PE_2 + KE_2 + \cdots \quad (1)$$

This is known as the steady-flow energy equation, which with proper modification may be applied to any thermodynamic process. Obviously, the dimensions for each term must be the same before an equality exists. Conversion factors for the various forms of energy are given in Table I (Appendix).

The principle of conservation of energy and the **First Law of Thermodynamics** are synonymous.

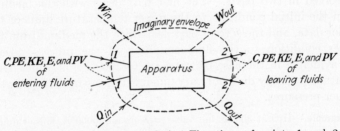

Fig. 9.—Diagram for process analysis. Elevations of points 1 and 2 not indicated.

Analysis of Processes.—The analysis of a thermodynamic process requires the evaluation of the various forms of energy and the application of the energy equation. An imaginary envelope (Fig. 9) thrown closely around the apparatus may be made to cut the flow lines to and from the apparatus at the desired points. At these points the various forms of energy that depend upon the condition, position, and motion of the medium should be determined. The work and heat-transfer values are determined as previously indicated, or one is the unknown in the energy equation into which the energy values are substituted.

Most processes do not involve all the energy terms indicated in the energy equation, given above, but are special cases as follows:

Nonflow processes eliminate KE, PE, and PV terms.

Nonreaction processes eliminate the C term.

the medium rejects heat to a *refrigerator*. It is usually necessary to introduce a pump or compressor mechanism to return the medium to its original condition.

The media used in the internal-combustion engine are not returned to their original condition. Consequently the internal-combustion-engine process should not be referred to as a "cycle."

Example.—The cycle indicated in Fig. 10 uses air as the medium. The low pressure is 1 atm. and the high pressure is 10 atm. The temperature of the air before compression is 600°R. One thousand B.t.u. are added per mol of air. Determine the net work and efficiency of the cycle.

During reversible adiabatic compression the entropy is constant [see Eq. (3) page 21].

At 600°R. (Table IV, Appendix) $S_1 = 1.00$ B.t.u. mol.$^{-1}$ °R.$^{-1}$

	1100	1200	1300
Assume T_2, °R.,			
$\int c_p \dfrac{dT}{T}$ (Table IV, Appendix)	5.30	5.95	6.54
$R \log_e 10$,	4.57	4.57	4.57
S_2,	0.73	1.38	1.97

By interpolation, T_2 is 1140°R.

During compression, since a steady-flow adiabatic process occurs,

$$W_{in} = E_2 + P_2V_2 - E_1 - P_1V_1 = 3171 + 2264 - 395 - 1192 = 3848$$
$$\text{B.t.u. mol.}^{-1}$$

Adding 1000 B.t.u. at constant pressure to H_2 results in $H_3 = 6435$ B.t.u., which indicates a temperature of $T_3 = 1275°R$. Solving the balance of the cycle results in $T_4 = 675°R$. after expansion, $W_{out} = 4323$ B.t.u., and $Q_{out} = 525$ B.t.u.

$$\text{Eff.} = \frac{\text{net work}}{Q_{in}} = \frac{4323 - 3848}{1000} = 0.475.$$

The same H values are used for both the work and heat-transfer determinations. Consequently, the equation

$$Q_{in} - Q_{out} = W_{out} - W_{in}$$

will check, even though an error is made in the compression or expansion computations.

Availability of Energy.—The first law of thermodynamics accounts for all the forms of energy involved in any process. The amount of mechanical work that can be obtained from a process may be evaluated either directly or from the energy

24   INTERNAL-COMBUSTION ENGINES

equation if all the other terms are known. However, no clue is given as to the conditions that will result in the maximum possible amount of work.

Carnot recognized, in 1824, the desirability of *adding* all of the *heat* to a medium undergoing a cyclic process *at the highest possible temperature* and of *rejecting* all the *heat* rejected from the medium *at the lowest possible temperature.* The Carnot cycle (Chap. IV) indicates the maximum availability of heat for given temperature limits and is the criterion for availability in heat-engine processes. Likewise, the higher the temperature at which energy is liberated by chemical reaction in the ideal internal-combustion-engine process and the lower the temperature to which the medium is expanded in the engine, the higher will be the availability of the energy released.

Unavailable Energy and Entropy.—The heat that cannot be transformed into work in the Carnot cycle is rejected and is called *unavailable energy.* Since this heat rejection occurs at constant refrigerator temperature T_2,

$$\text{Unavailable energy} = T_2 \Delta S \tag{1}$$

Thus, entropy is a measure of the unavailability of energy. An increase in entropy indicates an increase in unavailable energy, and vice versa.

The entropy change of media is determined by dividing the nonflow energy equation by T.

$$\text{Thus,} \qquad \Delta S = \frac{Q_{in} - Q_{out}}{T} = \frac{E_2 - E_1}{T} + \frac{W_{out} - W_{in}}{T}, \tag{2}$$

$$\text{and, if gaseous,} \qquad dS = \frac{c_v\, dT}{T} + \frac{P\, dV}{T} \tag{3}$$

$$= \frac{c_v\, dT}{T} + \frac{R\, dV}{V}, \tag{4}$$

$$\text{and} \quad S_{T,V} = \int c_v \frac{dT}{T} + R \log_e V + S_{0_{T,V}}. \tag{5}$$

Substituting, for V, its equal $RT \div P$,

$$S_{T,V} = \int (c_v + R) \frac{dT}{T} - R \log_e P + R \log_e R + S_{0_{T,V}} \tag{6}$$

$$\text{or} \quad S_{T,P} = \int c_p \frac{dT}{T} - R \log_e P + S_{0_{T,P}}; \tag{7}$$

from which it is obvious that

$$S_{0_{T,P}} = S_{0_{T,V}} + R \log_e R. \tag{8}$$

The terms involving specific heat in Eqs. (5) and (7) are functions of the temperature. Values of $\int c_p \frac{dT}{T}$ are given in Table IV (Appendix) as entropy values at 1 atmosphere. The entropy at any other pressure is determined by subtracting $R \log_e P$ from the tabulated value. The integration constant is neglected for it disappears in determining differences of entropy for a given medium.

Example.—Two mols of CO_2 at 1 atm. and 1000°R. undergo a process that changes the conditions to 10 atm. and 2000°R. Determine the entropy change.

From Table IV (Appendix), $S_{1000} = 6.50$ and $S_{2000} = 14.95$ B.t.u. °R.$^{-1}$, both values being for a pressure of 1 atm. and for 1 mol.
$R \log_e P = 1.986 \log_e 10 = 4.57.$
S_{2000} and 10 atm. $= 14.95 - 4.57 = 10.38.$
$\Delta S = 2(10.38 - 6.50) = 7.76$ B.t.u. °R.$^{-1}$

The entropy change for a chemical reaction involves the difference in entropy of two or more different media. The S_0 values do not disappear and must either be evaluated or eliminated by the use of chemical equilibrium relationships, or be determined from absolute values of entropy.

Using the equilibrium relationship[1] for gaseous constituents,

$$\Delta S = \frac{Q_p}{T} + R \log_e K_p, \tag{9}$$

the entropy change for a chemical reaction can be determined from the heat of reaction Q_p and the equilibrium constant K_p, at the temperature and pressure of the reaction.

Example.—Determine the entropy change for the reaction

$$(2H_2O)_l = (2H_2)_g + (O_2)_g$$

occurring at a temperature of 77°F. and a pressure of 1 atm. Subscripts l and g mean "liquid" and "gaseous," respectively.

---

[1] G. N. Lewis and M. Randall, "Thermodynamics," pp. 169, and 294, McGraw-Hill Book Company, Inc., New York, 1923. See Chap. III regarding K_p also.

For this reaction, $\log_{10} K_p = -80.22$.*

Also, from Table V (Appendix),

$$Q_p = 2(122{,}963 - 18 \times 1050.4) = 208{,}122 \text{ B.t.u.}$$

Then,

$$\Delta S = \frac{Q_p}{T} + R \log_e K_p = \frac{208{,}112}{536.6} - 1.986 \times 2.3026 \times 80.22$$

$$= 21.1 \text{ B.t.u. } °R.^{-1}$$

The entropy change for 2 mols of H_2O from saturated liquid to saturated vapor at 77°F. is

$$2 \times 18 \times 1.9569 = 70.4 \text{ B.t.u. } °R.^{-1}$$

Total $\Delta S = 21.1 + 70.4 = 91.5$ B.t.u. $°R.^{-1}$

The entropy change for all constituents in the gaseous state may be checked by the values of *absolute* entropy (Table IV) for the media involved in the reaction.

Thus, $\Delta S = 2S_{H_2} + S_{O_2} - 2S_{H_2O}$
$= 2 \times 31.23 + 49.03 - 2 \times 45.17 = 21.15$ B.t.u. $°R.^{-1}$

Reversibility and Irreversibility.—A process is thermodynamically reversible if it can be reversed and can return the medium and all other substances involved to their original condition existing before the process occurred. A medium with internal energy E_1 expands adiabatically against a mechanism and does work W which leaves the medium with internal energy E_2. Reversing the process should require the same work W done by the mechanism on the medium, which will be returned to the initial conditions with internal energy E_1 if the process is reversible.

A fast-moving piston-cylinder mechanism designed to cause pressure differences in the expanding medium results in an irreversible process (Fig. 11). Moving the piston very slowly, with the assumption that there is no heat transfer, produces the same pressure in all parts of the cylinder and clearance space; the result is represented by the reversible path AB. Moving the piston very fast causes a pressure difference between the clearance space and cylinder. The pressure in the clearance space is represented by the path AC, and that on the piston by the path AD. Obviously, the work done on the fast-moving piston

* B. Lewis and G. von Elbe, Heat Capacities and Dissociation Equilibria of Gases, A.C.S. Jour., **57**, 613 (1935). See also Table VI (Appendix).

(area under AD) is less than that done on the slow-moving piston (area under AB).

Thus, $\quad W_{out}$ (reversible) $> W_{out}$ (irreversible) $\qquad(1)$
and $\qquad E_2$ (reversible) $< E_2$ (irreversible). $\qquad(2)$

The pressure difference produces a flow of the medium from the clearance space into the cylinder. Some of the internal energy that was transformed into kinetic energy would have been transformed into work in the reversible process. The kinetic energy is eventually *transformed by friction* into internal

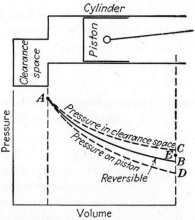

Fig. 11.—Reversible and irreversible expansion processes.

energy, and the pressures equalize at E (Fig. 11). This process is ordinarily termed an "adiabatic process with friction" and results in an increase in entropy.

Reversible adiabatic compression from E to the original volume would require more work than is obtained from any adiabatic expansion process shown. Also, such compression would return the medium to a point above A and would necessitate heat transfer to restore the medium to the original conditions.

Example.—Assume that the pressure on the piston in the irreversible process (Fig. 11) is represented by the relation $PV^{1.5} = $ const. The conditions at the beginning of expansion are 0.1 mol of air, 10 atm. pressure, and 2000°R. The volume increases to five times the initial volume. Determine the increase in entropy when the medium has attained equilibrium at E.

The pressure at the end of the expansion path $PV^{1.5}$ = const. is

$$P_2 = P_1\left(\frac{V_1}{V_2}\right)^{1.5} = 10\left(\frac{1}{5}\right)^{1.5} = 0.894 \text{ atm.}$$

$$V_1 = \frac{M_1 R T_1}{P_1} = \frac{0.1 \times 1545 \times 2000}{10 \times 14.7 \times 144} = 14.6 \text{ ft.}^3$$

$$\text{Work on piston} = \int_{V_1}^{V_2} P \, dV = \int_{V_1}^{V_2} P_1 V_1^{1.5} \frac{dV}{V^{1.5}} = \frac{P_2 V_2 - P_1 V_1}{-0.5}$$

$$= \frac{14.7 \times 144(0.894 \times 5 \times 14.6 - 10 \times 14.6)}{-0.5 \times 778} = 439.4 \text{ B.t.u.}$$

E_{2000}, (Table III, Appendix) = 808.7 B.t.u.

After equilibrium has been attained [Eq. (2), page 21],

$$E_2 = E_1 - W_{out} = 808.7 - 439.4 = 369.3 \text{ B.t.u.}$$

Interpolating in Table III (Appendix), $T_E = 1237°R$.
At 2000°R. and 10 atm. (Table IV, Appendix),

$$S_A = 0.988 - 0.1986 \log_e 10 = 0.531 \text{ B.t.u. °R.}^{-1}$$

At 1237°R. and 0.894 atm.,

$$S_E = 0.617 - 0.1986 \log_e 0.894 = 0.639 \text{ B.t.u. °R.}^{-1}$$

Increase in entropy = 0.639 − 0.531 = 0.108 B.t.u. °R.$^{-1}$

Heat transfer occurs because of temperature difference between media. It being assumed that two media remain at constant temperature during heat transfer, the hotter medium experiences the smaller entropy change (Fig. 12). This results in a net increase in entropy and unavailable energy. Thus, heat transfer is irreversible and results in a decrease in availability. The process approaches reversibility as the temperature difference between the two media approaches zero as a limit.

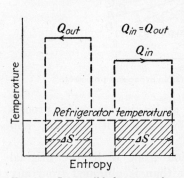

FIG. 12.—Irreversible heat-transfer process.

Example.—A medium at 1500°R. transfers 3000 B.t.u. of heat to a medium at 1000°R., both media remaining at constant temperature during the process. Determine the increase in entropy and decrease in available energy. The refrigerator temperature is 500°R. (see Fig. 12).

$$\Delta S = \frac{Q}{T} = \frac{3000}{1500} = 2 \text{ B.t.u. °R.}^{-1}, \text{ decrease in entropy for the medium}$$

from which the heat flows.

$$\Delta S = \frac{Q}{T} = \frac{3000}{1000} = 3 \text{ B.t.u. }°\text{R.}^{-1}\text{, increase in entropy for the medium}$$

to which the heat flows.
Net increase in entropy $= 3 - 2 = 1$ B.t.u. $°\text{R.}^{-1}$
Net increase in unavailable energy $= 500 \times 1 = 500$ B.t.u.
This is also the net decrease in available energy.

The two irreversible processes of *heat transfer* and *friction* are always present in practice and result in an increase in entropy and a decrease in availability. Reversible processes are assumed to occur only in limiting ideal cases, such as the Carnot cycle which consists of four reversible processes and is a reversible cycle.

Second Law of Thermodynamics.—This implies that all processes are irreversible and that energy naturally degrades from a high level of availability to the lowest possible level.

All actual processes result in net increases of unavailable energy and entropy.

Solution of Problems.—In the solution of problems, there is a great tendency to glance through the problem, pick out the data, look for a formula in which to put the values, and "grind out" an answer. Often the relation chosen does not apply, and many times the values substituted make the relation or equations inconsistent as to dimensions. Taking the nonflow energy equation

$$\Delta E + \Delta W + \Delta Q = 0,$$

it is obviously incorrect to substitute as follows:

$$\text{B.t.u.} + \text{ft.-lb.} + \text{B.t.u.} = 0.$$

Too often numerical values are substituted in an equation without substituting the dimensions of the values and noting if the equation is consistent as to dimensions or whether an inequality exists in this respect.

The following method is suggested for analyzing and solving problems:
1. Read the problem carefully.
2. List data with dimensions in a logical manner.
3. Also list questions or results asked for.
4. Draw a sketch, if possible, illustrating the problem.

5. Put data on the sketch.

6. Analyze the problem, and note the kind of process.

7. Note the principle that applies to this type of process, and indicate solution of problem.

8. If the analysis is correct and sufficient data are available, there should be no difficulty in making the solution.

9. Check the computations very carefully for possible errors, and compare with results from similar problems for possible discrepancies.

EXERCISES

1. Determine the weight of hydrogen (H_2) and also of helium (He) which would be required for an airship having a gas capacity of 10,000,000 ft.^{3} The gas pressure is 14.7 lb. in.$^{-2}$ abs., and the temperature is 60°F.

2. Assuming H_2 and He are mixed in equal proportions by weight, in Prob. 1, determine the partial pressure and weight of each gas used.

3. Three lb. of an air-water vapor mixture occupy a total volume of 35 ft.^{3} The partial pressure of the water vapor is 15 lb. in.$^{-2}$ abs. The mixture temperature is 300°F. Determine the amount of air and the total pressure of the mixture.

4. The vapor pressure of ethyl alcohol at 70°F. is 0.91 lb. in.$^{-2}$ abs. The volume of saturated vapor per pound of alcohol at this temperature is 135 ft.^{3} Determine the pounds of air required to have a mixture containing 1 lb. of saturated alcohol vapor at 70°F. Total pressure is 14.7 lb. in.$^{-2}$ abs.

5. One pound of air expands reversibly and at a constant temperature of 2000°R. from a pressure of 10 to 1 atm. Determine the entropy change and the work done.

6. The analysis of air is 21 per cent O_2, 78 per cent N_2, and 1 per cent A, by volume. Derive the specific heat equation for air, using this analysis as the basis.

7. Check the internal energy and PV values for air at 1000 and 3000°R. in Table III (Appendix).

8. The exhaust gases leave an engine at a temperature of 3000°R. and are cooled by heat transfer to the atmospheric temperature of 60°F., while remaining at constant pressure. The gases consist of 10 per cent CO_2, 4 per cent CO, and 13 per cent H_2O, by volume, and the remainder is assumed to be N_2. Determine the heat transferred out per mol of the gases.

9. A small internal-combustion engine drives a compressor for a refrigerator. Set up the energy equation for the engine and compressor as a unit and also for the engine, compressor, and refrigerator as a unit.

Note.—The envelope for analysis is drawn closely around the units mentioned.

10. Compute the example on page 23 for 4000, 8000 and 12,000 B.t.u. of heat added, and plot the thermal efficiency vs. heat added.

11. Check the S values for air at 1000 and 3000°R. in Table IV (Appendix).

12. Determine the entropy change for the reaction

$$2CO_2 = 2CO + O_2,$$

at a constant-temperature of 3000°R., and check with S values from Tables IV and VII (Appendix). Q_p at 3000°R. is 240,405 B.t.u.

13. Combine the reaction involving $2H_2O$ with the foregoing reaction involving $2CO_2$, and determine the entropy change at 3000°R. for the water-gas reaction

$$CO + H_2O = CO_2 + H_2.$$

14. During an irreversible compression process 100 B.t.u. of heat are transferred out of the medium which is 0.1 mol of air. The pressure on the piston is described by the equation $PV^{1.5}$ = const. The initial pressure is 1 atm., and the temperature is 600°R. The volume is decreased to $\frac{1}{10}$ the original volume. Determine the final condition of the air after equilibrium occurs.

15. One mol of water at 1 atm. pressure and 3000°R. transfers heat at constant pressure to 1 mol of carbon dioxide at 1 atm. pressure and 1000°R. until the temperatures of both media are the same. Determine the net change in entropy.

CHAPTER III

THE COMBUSTION PROCESS

Combustion.—Chemical reactions occur between various substances and result in the liberation or absorption of chemical energy. The rather rapid reaction between a fuel and oxygen which results in the liberation of chemical energy is called the "combustion" process. This process is used to liberate the energy supplied the internal-combustion engine.

The oxygen supply for the combustion process is usually obtained from the atmosphere which introduces nitrogen and other gases into the process. The nitrogen and other gases are usually assumed to be inert substances which do not change the reaction between the fuel and oxygen. However, under high-temperature conditions nitrogen may dissociate into atomic nitrogen and combine with atomic oxygen to form nitric oxide.

Air.—The volumetric analysis of dry air indicates small quantities of seven gases other than oxygen and nitrogen. Two of the more important ones from a standpoint of quantity, argon and carbon dioxide, are included in the following table of the constituents of air:

TABLE 1.—CONSTITUENTS OF AIR

Constituents	Symbol	Mol. wt.	Vol. analysis*	Relative wt.	Approx. vol. analysis	Relative vol.	Constituents
Oxygen.........	O_2	32	0.2099	6.717	0.21	1	Oxygen
Nitrogen.......	N_2	28	0.7803	21.848			Nitrogen
Argon..........	A	40	0.0094	0.376	0.79	3.76	and
Carbon dioxide..	CO_2	44	0.0003	0.013			other
Other gases.....	....	..	0.0001	......			gases
Total (air)....	....	..	1.0000	28.95	1.00	4.76	

* "International Critical Tables," Vol. I, 393, McGraw-Hill Book Company, Inc., New York, 1926.

The approximate volume analysis, $0.21O_2$ and 0.79 (N_2 and other gases), will be used in dealing with air in reaction equations, and the symbol N_2 will indicate nitrogen and other gases as shown in the foregoing tabulation.

Reaction Equations.—The principal fuels used in the combustion process are combinations of carbon and hydrogen which are called "hydrocarbons." There are also the alcohols which consist of combinations of carbon, hydrogen, and oxygen. Carbon is a monatomic substance, whereas hydrogen, oxygen, and nitrogen are normally diatomic gases. The weight of an atom of a diatomic substance is one-half the weight of a molecule; also, the weight of any molecule is the sum of the weights of the various atoms in the molecule, as indicated in the following table.

TABLE 2.—ATOMIC AND MOLECULAR WEIGHTS

Element	Mol. symbol	Mol. wt.*	Atomic symbol	Atomic wt.
Carbon	C	12	C	12
Hydrogen	H_2	2	H	1
Oxygen	O_2	32	O	16
Nitrogen	N_2	28	N	14

* These are the approximate molecular weights which are satisfactory for all practical purposes.

The elementary reaction equations for carbon and hydrogen uniting with the chemically correct amount of oxygen are as follows:

$$C + O_2 = CO_2. \tag{1}$$
$$2H_2 + O_2 = 2H_2O. \tag{2}$$

Thus, 1 mol of carbon unites with 1 mol of oxygen and forms 1 mol of carbon dioxide, while 2 mols of hydrogen unite with 1 mol of oxygen and form 2 mols of water. The substances indicated on both sides of reaction equations may or may not be at the same pressure and temperature. However, if all substances in a reaction are at the same pressure and temperature, and in the gaseous region in each case, the mols may be read as volumes. Thus, Eq. (2) may be read

$$2 \text{ ft.}^3 \text{ } H_2 + 1 \text{ ft.}^3 \text{ } O_2 = 2 \text{ ft.}^3 \text{ } H_2O.$$

The dimensions of a reaction equation may be changed from mols to pounds by multiplying the mols of each constituent by its molecular weight. Thus, Eq. (2) becomes

$$4 \text{ lb. } H_2 + 32 \text{ lb. } O_2 = 36 \text{ lb. } H_2O.$$

It should be noted that the sum of the weights of the mixture constituents always equals the sum of the weights of the products constituents in any reaction equation, whereas the mols and volumes of mixture and products constituents are usually different. However, the same number of atoms of each element such as C, H, O, and N, either alone or combined with others, will be found on both sides of a correct reaction equation.

Since air contains 3.76 mols of N_2 and other gases for each mol of O_2, there will be 3.76 mols of N_2 introduced into the reaction equation for each mol of O_2 required when the reaction occurs with air. The combustion of C and H_2 with the chemically correct amount of air results in the following equations:

$$C + O_2 + 3.76N_2 = CO_2 + 3.76N_2, \tag{3}$$

and
$$2H_2 + O_2 + 3.76N_2 = 2H_2O + 3.76N_2. \tag{4}$$

The foregoing interpretation regarding weight, volume, and mols may also be applied to these reactions.

Reactions may occur with more oxygen or air than is required for complete combustion. Denoting the excess oxygen by e, the reaction for a hydrocarbon becomes

$$C_nH_m + \left(n + \frac{m}{4} + e\right)O_2 + 3.76\left(n + \frac{m}{4} + e\right)N_2 = nCO_2 + \frac{m}{2}H_2O + eO_2 + 3.76\left(n + \frac{m}{4} + e\right)N_2. \tag{5}$$

The equation for a hydrocarbon reaction with the chemically correct amount of air is obtained by omitting the e terms.

Reactions with less oxygen or air than is required for complete combustion result in CO, H_2, and CH_4 in addition to the products of complete combustion. The products for such reactions can be obtained by applying the principles of chemical equilibrium.

Chemical Energy and Heating Value.—The chemical energy and heating value of a fuel are determined from the reaction of the fuel and oxygen in the presence of excess oxygen in a constant-volume or constant-pressure calorimeter.[1] The products of combustion are cooled to the initial temperature of the mixture.[2] Applying the energy equation to the constant-volume calorimeter process, Fig. 13, results in

$$C + E_m = E_p + Q_{out} \qquad (1)$$
or
$$C = Q_v + (E_p - E_m)_T \qquad (2)$$

where subscripts m and p indicate fuel-oxygen mixture and products, respectively.

Q_v is the heating value of the fuel for the constant-volume calorimeter process. It is an absolute value but varies with temperature as indicated by $(E_p - E_m)_T$.

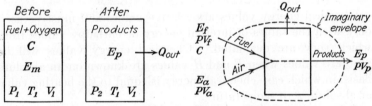

Fig. 13.—Constant-volume calorimeter process.   Fig. 14.—Constant-pressure calorimeter process.

The constant-pressure calorimeter usually has steady flow of mixture into the apparatus and steady flow of products out of it, sufficient heat being transferred out to reduce the temperature of the products to that of the mixture before reaction. Applying the energy equation to this process, Fig. 14, results in

$$C + E_m + PV_m = E_p + PV_p + Q_{out} \qquad (3)$$
or
$$C = Q_p + (H_p - H_m)_T. \qquad (4)$$

Q_p is the heating value of the fuel for the constant-pressure calorimeter process. It is an absolute value but varies with temperature as indicated by $(H_p - H_m)_T$. Thus, heating values vary with the process and the temperature of the process.

---

[1] For a comprehensive analysis of the bomb-calorimeter process, see *Nat. Bur. Stand., Jour. Res.*, **10**, 525 (1933).

[2] The cooling bath surrounding the constant-volume bomb rises a few degrees in temperature, for which a correction is made.

Example.—The heating value reported per mol of CO is 121,721 B.t.u. (Table V, Appendix). This was obtained at a constant pressure of 1 atm. and a temperature of 77°F. for the reaction

$$(CO)_g + \tfrac{1}{2}(O_2)_g = (CO_2)_g$$

where subscript g indicates the gaseous state. Determine the chemical energy per mol of CO.

The energy equation for this process is

$$C_{CO} + (H_{CO} + \tfrac{1}{2}H_{O_2})_{536.6} = (H_{CO_2})_{536.6} + Q_p.$$

Substituting values from Table III (Appendix) results in

$$C_{CO} + 1146 + 574 = 1180 + 121{,}721$$
$$C_{CO} = 121{,}181 \text{ B.t.u. mol}^{-1}.$$

Chemical energy values are relative values and must be used only with the tables of internal energy or enthalpy values from which they are evaluated.[1] Chemical energy values will be absolute values when E and H values are known on an absolute basis, in which case chemical energy is equal to the heating value at absolute zero temperature.

Fuels containing H_2 react with O_2 and form H_2O as one of the products. The humidity of the air or oxygen, pressure, temperature, and nature of the process determine the amount of H_2O condensed during the process. Under certain conditions, the amount of water condensed may be more than that formed. Calorimetric processes for the determination of heating value are usually conducted under conditions that result in the conden-

---

[1] Absolute values of energy should be substituted in Eqs. (1) to (4) for E_m and E_p. Such values are not available, and it is inferred that constants E_m^0 and E_p^0 are included which if added to the tabulated values would result in absolute values.

Thus for Eq. (1), $C + (E_m + E_m^0) = (E_p + E_p^0) + Q_{out}$,
or $\qquad (C + E_m^0 - E_p^0) = Q_v + (E_p - E_m)_T.$

Values of chemical energy as determined in the preceding example and given in Table V (Appendix), are actually values of $(C + E_m^0 - E_p^0)$. Obviously, substituting these *apparent* chemical energy values in an energy equation and making use of the same internal energy tables will eliminate the constants E_m^0 and E_p^0. Consequently, in the text, the terms in the energy equations for reactions are treated as absolute values.

THE COMBUSTION PROCESS

sation of all or nearly all the water formed. Results are reported on the basis that all the water formed has been condensed.[1]

Example.—The heating value reported per mol of H_2 is 122,963 B.t.u. This was obtained at a constant pressure of 1 atm. and a temperature of 77°F. for the reaction

$$(H_2)_g + \tfrac{1}{2}(O_2)_g = (H_2O)_l$$

when subscripts g and l represent gaseous and liquid states, respectively. Determine the chemical energy per mol of H_2.

The energy equation for this process is

$$C_{H_2} + (H_{H_2} + \tfrac{1}{2}H_{O_2})_{536.6} = [(H_{H_2O})_l]_{536.6} + Q_p.$$

Substituting values from Table III (Appendix),

$$C_{H_2} + 1145 + 574 = 1167 - 18.016 \times 1050.4^* + 122{,}963.$$
$$C_{H_2} = 103{,}486 \text{ B.t.u. mol}^{-1}.$$

Values of chemical energy and heating value for various fuels are given in Table V (Appendix).

Heating value represents heat transfer out of the calorimeter process operated under specified conditions. It is not an energy term in the energy equation for a combustion process at conditions different from those of the calorimeter process. Chemical energy, however, is a function of the substance. A mol of fuel

---

[1] This is commonly termed the "higher heating value." A purely fictitious lower heating value is obtained by assuming the H_2O is not liquid but is saturated vapor. This condition could not exist unless the calorimeter process was operated at a temperature corresponding to or higher than the dew point for the water vapor in the products of the reaction.

The difference between the higher and lower heating values for H_2 is $18.016 \times 1050.4 = 18{,}924$ B.t.u. mol^{-1} of H_2, where 1050.4 is the latent heat for 1 lb. of water at 77°F.

Then, Lower heating value $= 122{,}963 - 18{,}924 = 104{,}039$ B.t.u. mol^{-1}.

The principal argument for the use of lower heating value is that the internal-combustion engine exhausts the products of combustion at temperatures well above the dew-point temperature for the water vapor and, therefore, the engine is charged with the lower heating value. However, the mere fact that the engine fails to return the products to the original mixture temperature is an indication of inefficiency and the engine should be charged with the higher heating value in efficiency determinations.

* The latent heat of water is subtracted from the enthalpy value from Table III (Appendix), at the same temperature, to make the enthalpy value for liquid water on the same basis as the values in Table III.

at any pressure and temperature has associated with it a definite amount of chemical energy, for a given reaction, which can be used as an energy term in the energy equation for the given reaction process. Since heating value and chemical energy are related through the energy equation for the calorimeter process [Eq. (1)], heating value may be introduced by substituting for $C + (E_m)_{T_1}$ in the energy equation its equal $Q_v + (E_p)_{T_1}$ (see page 35).

Complete-combustion Temperatures.—The maximum temperature for the combustion process will be attained with a constant-volume, adiabatic, and complete-combustion process, on the assumption of no supply of energy other than that of the reaction. The energy equation for this process is

$$C + (E_m)_1 = (E_p)_2. \qquad (1)$$

All the chemical energy of the reaction and the internal energy of the mixture appear as internal energy of the products of combustion, which indicates the final and maximum temperature of the process. If work is done on the media or some other form of energy added during the combustion process, it will increase the energy of the products by a like amount.

Example.—Determine the temperature of the products of combustion of carbon monoxide with oxygen if the process occurs at constant volume with no heat loss. Initial mixture temperature is 60°F.

The reaction equation is

$$CO + \tfrac{1}{2}O_2 = CO_2.$$

The energy equation for this process is

$$C_{CO} + (E_{CO} + \tfrac{1}{2}E_{O_2})_{520} = (E_{CO_2})_T.$$

Substituting values from Tables III and V (Appendix)

$$121{,}181 + 0 + 0 = (E_{CO_2})_T.$$
$$(E_{CO_2})_T = 121{,}181 \text{ B.t.u.}$$

This is well above the internal energy of CO_2 at 5400°R.; and since some CO_2 will dissociate at this temperature, any estimate of final temperature by extrapolation of the internal energy values would be worthless.

The use of air instead of oxygen, in the foregoing example, appreciably reduces the final temperature of the products of combustion, since the energy is divided between the nitrogen and the products formed by the reaction.

Example.—If the required air instead of oxygen is used in the previous example, the reaction equation becomes

$$CO + \tfrac{1}{2}O_2 + 1.88N_2 = CO_2 + 1.88N_2.$$

The energy equation for this reaction is

$$C_{CO} + (E_{CO} + 2.38E_{air})_{520} = (E_{CO_2} + 1.88E_{N_2})_T.$$

Substituting values from Tables III and V (Appendix) results in.

$$(E_{CO_2} + 1.88E_{N_2})_T = 121{,}181 + 0 + 0 = 121{,}181 \text{ B.t.u.}$$

The left side of this equation is solved for various values of T, as follows:

T, °R.	E_{CO_2}	$1.88E_{N_2}$	Sum
5200	53,963	54,447	108,410
5300	55,265	55,738	111,003
5400	56,569	57,034	113,603

Extrapolating for 121,181 B.t.u. total energy results in a final temperature of about 5700°R. An appreciable amount of CO_2 would dissociate at this temperature and, therefore, it could not be attained.

The constant-pressure combustion process has work out during the reaction and results in a lower final temperature. The energy equation for this process with no heat loss is

$$C + (E_m)_1 = (E_p)_2 + W_{out} \qquad (2)$$
$$= (E_p)_2 + P[(V_p)_2 - (V_m)_1]$$
$$\text{or} \qquad C + (H_m)_1 = (H_p)_2. \qquad (3)$$

Example.—Determine the temperature of the products of combustion of CO with the required air if the process occurs at constant pressure with no heat loss. $T_1 = 60°F$.

The reaction equation for this process is

$$CO + \tfrac{1}{2}O_2 + 1.88N_2 = CO_2 + 1.88N_2.$$

The energy equation becomes

$$C_{CO} + (H_{CO} + 2.38H_{air})_{520} = (H_{CO_2} + 1.88H_{N_2})_T.$$

Substituting values from Tables III and V (Appendix) results in

$$(H_{CO_2} + 1.88H_{N_2})_T = 121{,}181 + 1033 + 2.38 \times 1033$$
$$= 124{,}673 \text{ B.t.u.}$$

The left side of this equation is solved for several T values as follows:

T, °R.	H_{CO_2}	$1.88 H_{N_2}$	Sum
4700	56,817	65,574	122,391
4800	58,308	67,223	125,531

Interpolating for 124,673 B.t.u. results in a final temperature of 4773°R. Appreciable dissociation of CO_2 would occur at this temperature and, therefore, it could not be attained.

Other combustion processes with heat transfer and dilution with other gases are solved in a similar manner. Complete combustion cannot occur unless heat loss or some other means of energy dissipation reduces the energy content of the products to a condition of about 3000°R., below which dissociation is usually negligible.

The chemical energy term may be eliminated from the foregoing equations in the following manner in the constant-volume process:

$$C + (E_m)_1 = (E_p)_2. \qquad \text{[Eq. (1), p. 38.]}$$

The constant-volume calorimeter equation is

$$C + (E_m)_{536.6} = (E_p)_{536.6} + (Q_v)_{536.6} \qquad (4)$$

Eliminating C between Eqs. (1) and (4) results in

$$(Q_v)_{536.6} + (E_p - E_m)_{536.6} + (E_m)_1 = (E_p)_2. \qquad (5)$$

Obviously, if the internal energy tables were based upon zero values at 536.6°R., the standard temperature for reporting heating values, the term $(E_p - E_m)_{536.6}$ would be eliminated and Q_v and C would be identical.

Writing Eq. (4) for subscript 1 condition reduces Eq. (5) to

$$(Q_v)_1 = (E_2 - E_1)_p. \qquad (6)$$

This shows that, if the fictitious process of burning the fuel at condition 1 and cooling the products to this condition to obtain (Q_v) is followed and then this heat is used to raise the products from condition 1, the correct final energy for the products will result. However, the actual process which starts with fuel and oxygen or air at T_1 and proceeds to products at T_2 is lost sight of in this method of analysis.

Chemical Equilibrium.—The maximum temperatures attained in combustion processes are appreciably lower than those based upon complete combustion, because of heat loss and dissociation of the products of combustion. Appreciable dissociation occurs

THE COMBUSTION PROCESS

at high temperatures and is accompanied by an absorption of internal energy which is transformed into chemical energy. Thus, the dissociated products will have chemical energy associated with them, and the net chemical energy liberated appears as internal energy or enthalpy of the products and as heat loss.

Hydrocarbon fuels burn to CO_2 and H_2O which appear in the products along with N_2 when air is used as the oxidizing agent. Dissociation results in the formation of appreciable quantities of CO, H_2, and O_2, according to the reactions

$$2CO_2 = 2CO + O_2, \qquad (1)$$
$$2H_2O = 2H_2 + O_2. \qquad (2)$$

Other constituents such as N, O, H, OH, NO, and C may be formed by further dissociation and combination of the various constituents.

Both combination of the various constituents into products of combustion and dissociation of the products are assumed to occur simultaneously in various parts of the mixture at chemical-equilibrium conditions, but the net result of the combination processes is exactly equivalent to the dissociation processes if the temperature remains constant. Writing the CO reaction equation for combustion or combination and then dissociating to chemical equilibrium conditions result in

$$2CO + O_2 = 2CO_2 = 2xCO + 2(1-x)CO_2 + xO_2 \qquad (3)$$

when x represents the part of the CO_2 that has been dissociated into CO and O_2. The H_2 reaction would be

$$2H_2 + O_2 = 2H_2O = 2yH_2 + 2(1-y)H_2O + yO_2. \qquad (4)$$

Similarly, for a hydrocarbon,

$$C_nH_m + \left(n + \frac{m}{4}\right)O_2 = nCO_2 + \frac{m}{2}H_2O = nxCO +$$
$$n(1-x)CO_2 + \frac{m}{2}yH_2 + \frac{m}{2}(1-y)H_2O + \left(\frac{n}{2}x + \frac{m}{4}y\right)O_2. \qquad (5)$$

The relation between the various constituents at chemical equilibrium is indicated by an equilibrium constant K_p. For any reaction equation

$$aA + bB = cC + dD \qquad (6)$$

where lower-case and capital letters indicate mols and constituents, respectively,

$$K_p = \frac{[C]^c[D]^d}{[A]^a[B]^b} \tag{7}$$

in which the brackets indicate the partial pressures, in atmospheres, of the various constituents. Inasmuch as all hydrocarbon fuels react with oxygen to form CO_2, H_2O, and then the dissociated constituents of these products of complete combustion, it is desirable to deal with the reactions as proceeding from the products to the dissociated constituents. Thus, for the CO_2 reaction,

$$2CO_2 = 2CO + O_2. \tag{8}$$

From Eq. (7)
$$K_{p_{2CO_2}} = \frac{[CO]^2[O_2]}{[CO_2]^2}. \tag{9}$$

The partial pressure of any constituent depends on the mols of that constituent, the total mols M_p, and the total pressure P, if all constituents are in the gaseous state. Thus, from Eq. (3),

$$[CO]^2 = \left(\frac{2x}{M_p}P\right)^2 = \left(\frac{2x}{2+x}P\right)^2. \tag{10}$$

In like manner the other partial pressures may be determined and all substituted in Eq. (9), the result being

$$K_{p_{2CO_2}} = \frac{x^3}{(1-x)^2(2+x)}P. \tag{11}$$

It should be noted that reversing the reaction inverts the equilibrium constant K_p, and dividing the reaction equation by 2 results in the square root of the original constant. Writing the CO reaction in this manner results in

$$CO + 0.5O_2 = CO_2 \tag{12}$$
$$= x_aCO_2 + (1-x_a)CO + 0.5(1-x_a)O_2 \tag{13}$$

in which x_a is defined as the part of CO_2 formed.

Now,
$$K_p = \frac{[CO_2]}{[CO][O_2]^{1/2}} = \frac{x_a}{1-x_a}\sqrt{\frac{3-x_a}{(1-x_a)P}}. \tag{14}$$

Substituting $(1 - x_a)$ in Eq. (11) for x, its equal by definition, extracting the square root, and inverting result in

$$K_p = \sqrt{\frac{(x_a)^2(3 - x_a)}{(1 - x_a)^3 P}} = \frac{x_a}{1 - x_a}\sqrt{\frac{3 - x_a}{(1 - x_a)P}}$$

which is the same as Eq. (14) and the expression used in the previous edition of this text.

The foregoing differences in reaction direction and mols of fuel are applied to the logarithms of the equilibrium constants in the usual manner and must be observed in using various equilibrium data in the literature.

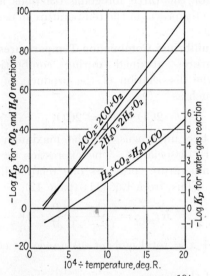

FIG. 15.—Log K_p plotted against $\dfrac{10^4}{T}$.

The reaction for the dissociation of H_2O

$$2H_2O = 2H_2 + O_2 \tag{15}$$

is similar to that for CO_2, Eq. (8). Consequently, K_p for this reaction will be similar to that for the CO_2 reaction, Eq. (11).

Thus,
$$K_{p_{2H_2O}} = \frac{y^3}{(1 - y)^2(2 + y)}P \tag{16}$$

where y represents the part of the H_2O dissociated.

Subtracting Eq. (15) from Eq. (8) and dividing by 2 result in

$$CO_2 + H_2 = CO + H_2O, \tag{17}$$

which is known as the "water-gas reaction." Applying these changes to the equilibrium constants results in

$$K_{p_{wg}} = \sqrt{\frac{K_{p_{2CO_2}}}{K_{p_{2H_2O}}}} = \frac{x(1-y)}{y(1-x)} \qquad (18)$$

since the O_2 term disappears on combining Eqs. (8) and (15).

Expressions for K_p and values for log K_p for various reactions at various temperatures are given in Tables VI and VII (Appendix). Values for the three foregoing reactions are plotted in Fig. 15 with the reciprocal of the temperature in degrees Rankine as the abscissa.

Chemical-equilibrium Combustion Temperatures.—The maximum temperatures attained during combustion processes depend upon the dissociation of the products of the reaction. For reactions such as

$$2CO + O_2 = 2CO_2$$

there will be two unknowns, T, the maximum temperature, and x, the part dissociated. The expression for $K_{p_{2CO_2}}$, which depends on temperature, provides one relationship with these two variables. Thus, from Eq. (11) (page 42),

$$K_{p_{2CO_2}} = \frac{x^3}{(1-x)^2} \frac{P}{M} \qquad (1)$$

where $M = 2 + x =$ total mols of substances at equilibrium for this reaction.

Regardless of the process,

$$\frac{P_1 V_1}{M_1 T_1} = \frac{PV}{MT} \qquad (2)$$

where subscript 1 represents the conditions at the beginning of the process. Eliminating P/M between Eqs. (1) and (2) results in

$$\frac{K_{p_{2CO_2}}}{T} = \frac{x^3}{(1-x)^2} \frac{P_1}{M_1 T_1} \frac{V_1}{V}. \qquad (3)$$

The nonflow energy equation applied to the process of combustion of CO, with chemical equilibrium [Eq. (3), page 41], is

$$2C_{CO} + (2E_{CO} + E_{O_2})_{T_1} + W_{in} + Q_{in} = 2xC_{CO} + [2xE_{CO} + 2(1-x)E_{CO_2} + xE_{O_2}]_T + W_{out} + Q_{out}, \qquad (4)$$

THE COMBUSTION PROCESS

which provides another relationship involving the two variables T and x. The simultaneous solution of Eqs. (3) and (4) provides a solution for the two variables.

Example.—Determine the temperature of combustion of carbon monoxide with air if the process occurs at constant volume with no heat transfer. Initial mixture temperature and pressure are 60°F. and 1 atm.

For this constant-volume process, Eq. (3) may be written

$$\log \frac{x^3}{(1-x)^2} = \log K_{p_{2CO_2}} - \log T + \log \frac{6.76 \times 520}{1}$$

where
$M_1 = 2$ mols CO $+ 1$ mol $O_2 + 3.76$ mols of N_2,
$P_1 = 1$ atm.,
$T_1 = 520°R$.

For every assumed value of T, a value of K_p for the dissociation of CO_2 can be obtained from Table VI (Appendix), and the right side of this equa-

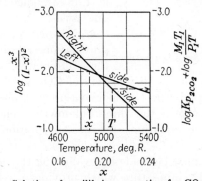

Fig. 16.—Solution of equilibrium equation for CO reaction.

tion can be evaluated and plotted (Fig. 16). Values of the left side are determined for various values of x. These are also plotted in Fig. 16, the same vertical ordinate being used since the left side equals the right side. Corresponding values of T and x can be read from this diagram and plotted as in Fig. 17, and labeled "equilibrium equation."

The energy equation for this process becomes

$$2(1-x)C_{CO} + (2E_{CO} + E_{O_2} + 3.76E_{N_2})_{520} = [2xE_{CO} + 2(1-x)E_{CO_2} + xE_{O_2} + 3.76E_{N_2}]_T.$$

Substituting values of C_{CO} and E_{520} in the left side and values of E at various assumed temperatures in the right side of this equation provides values of T and x that satisfy the energy equation. Thus, at 5000°R.,

$$2(1-x)121{,}181 + (0) = 55{,}814x + 2(1-x)51{,}365 + 29{,}616x \\ + 3.76 \times 27{,}589$$

or $\qquad x = 0.159.$

Plotting values of T and x obtained in the foregoing manner results in the curve labeled "energy equation" in Fig. 17. The intersection with the "equilibrium equation" indicates a solution at $T = 4900°R$. and $x = 0.183$.

The N_2 energy term in the foregoing solution is eliminated when the combustion occurs with O_2 alone. The partial-pressure expression for K_p is the same, since, with air, the partial pressure of the N_2 appears in both the numerator and denominator and cancels itself. However, the value of M_1 in the equilibrium equation is affected.

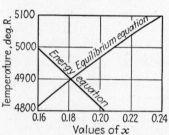

Fig. 17.—Solution for CO reaction.

Combustion processes dealing with hydrocarbons, as indicated in Eq. (5) (page 41), have three unknowns, T, x, and y. The energy equation for the process provides one relationship between these unknowns. The equilibrium expressions for the dissociation of CO_2 and H_2O, applied to the equilibrium mixture for the given process, provide two other relationships between the three unknowns. The simultaneous solution of these equations provides a solution for the combustion process.

The solution may be made by evaluating the energy equation for an assumed temperature which provides a relation between x and y. The combination of the two equilibrium relationships for the dissociation of CO_2 and H_2O, applied to this reaction, results in the equilibrium relationship for the water-gas reaction, since the O_2 term is the same in both cases. Substituting the values of $K_{p_{wg}}$ for the assumed temperature provides another relationship between x and y, which combined with that of the energy equation at the same assumed temperature results in the evaluation of x and y at this temperature. The assumed value of T and the determined values of x and y must satisfy either one of the equilibrium relationships applied to this reaction, which provides a check on the assumed temperature.

Example.—Determine the temperature of constant-volume, adiabatic combustion for C_8H_{18} and the required air. Initial temperature is $1100°R$. Initial pressure is 125 lb. in.$^{-2}$ abs.

THE COMBUSTION PROCESS

The reaction equation is

$$C_8H_{18} + 12.5O_2 + 47N_2 = 8xCO + 8(1-x)CO_2 + 9yH_2 + 9(1-y)H_2O + (4x + 4.5y)O_2 + 47N_2.$$

The energy equation for this reaction is

$$C_{C_8H_{18}} + (E_{\text{mixture}})_{1100} = (E_{\text{products}})_T + 8xC_{CO} + 9yC_{H_2}.$$

From Table V (Appendix),

$C_{C_8H_{18}} = 2{,}201{,}618;\quad 8xC_{CO} = 969{,}448x;\quad$ and $\quad 9yC_{H_2} = 931{,}374y$ B.t.u.

From Table III (Appendix), at 1100°R.,

$$E_{C_8H_{18}} + 59.5\, E_{air} = 32{,}241 + 175{,}942 = 208{,}183 \text{ B.t.u.}$$

From Table III (Appendix), at 5000°R.:

Products	Numerical terms	x terms	y terms
$8xCO$		$223{,}256x$	
$8(1-x)CO_2$	$410{,}920$	$-410{,}920x$	
$9yH_2$			$232{,}371y$
$9(1-y)H_2O$	$358{,}965$		$-358{,}965y$
$(4x + 4.5y)O_2$		$118{,}464x$	$133{,}272y$
$47N_2$	$1{,}296{,}683$		
Sum	$2{,}066{,}568$	$-69{,}200x$	$6{,}678y$

Substituting the various values in the energy equation produces

$$2{,}201{,}618 + 208{,}183 = 2{,}066{,}568 - 69{,}200x + 6{,}678y + 969{,}448x + 931{,}374y$$

or $\quad y = 0.3659 - 0.9597x \quad$ (A)

At 5000°R., for the water-gas reaction (Table VI, Appendix),

$$K_{pwg} = 6.547 = \frac{x(1-y)}{y(1-x)}$$

or $$y = \frac{x}{6.547 - 5.547x} \quad (B)$$

Eliminating y between Eqs. (A) and (B) and solving the quadratic equation result in,

$$x = 0.313$$

and, from Eq. (A), $\quad y = 0.065.$

These values have been plotted in Fig. 18.

The equilibrium relationship for the dissociation of CO_2 is applied to this combustion process as follows:

$$K_{p_{2CO_2}} = \frac{[CO]^2[O_2]}{[CO_2]^2} = \frac{\left(\dfrac{8x}{M_p}P\right)^2\left(\dfrac{4x+4.5y}{M_p}\right)P}{\left[\dfrac{8(1-x)}{M_p}P\right]^2} = \frac{x^2}{(1-x)^2}(4x+4.5y)\frac{P}{M_p}.$$

(C)

Substituting the determined values for x and y and values for P_1T/M_mT_1, the equivalent of P/M_p, results in

$$K_{p_{2CO_2}} = \frac{(0.313)^2}{(1-0.313)^2}(1.252+0.293)\frac{125\times 5000}{14.7\times 60.5\times 1100} = 0.203.$$

From Table VI (Appendix), $K_{p_{2CO_2}}$ at 5000°R. = 0.0200, which does not check the foregoing computed value. The correct and computed values for K_p are plotted in Fig. 18. Assuming other values of T and plotting the results indicate a solution at $T = 5220°R.$, $x = 0.216$, and $y = 0.040$.

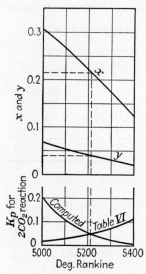

FIG. 18.—Solution for C_8H_{18}-air reaction.

Combustion Charts.—The foregoing combustion problems neglected the dissociation of O_2, N_2, and H_2 into their respective atoms and the formation of OH and NO. The inclusion of these constituents requires more relationships, which complicates the analysis. The solution of problems dealing with C_8H_{18} or fuels having the same (or nearly the same) carbon-hydrogen ratio when reacting with 85, 100, and 110 per cent required air has been simplified by the construction of diagrams[1] (Charts D, E, and F, Appendix), from which the thermodynamic properties of the equilibrium mixture can be obtained. The method followed[2] in computing the points for these diagrams indicates the method of solution for other combustion processes.

[1] R. L. Hersey, J. E. Eberhardt, and H. C. Hottel, "Thermodynamic Properties of the Working Fluid in Internal-combustion Engines," *S.A.E. Jour.*, **39**, 409 (1936).

[2] *Ibid.*, p. 423.

The four kinds of atoms present (C, H, O, N) are arranged as CO_2, H_2O, O_2, N_2, CO, H_2, OH, NO, H, and O in the equilibrium mixture. These 10 unknowns require 10 relationships for their solution. If it is assumed that sufficient fuel and air are used so that the total mols equal the total pressure at equilibrium, the mols of each constituent will represent its partial pressure at equilibrium. If A represents the total mols of air supplied per mol of carbon (6.322, 7.438, and 8.182 for 85, 100, and 110 per cent of required air, respectively), $0.79A$ = total mols of N_2 supplied per mol of C. Also, $9/8$ = total mols of H_2 per mol of C. Thus, the following three relationships exist between C and N_2, H_2 and N_2, and O_2 and N_2, representing both the mols and partial pressure of each constituent at equilibrium by the bracketed term:

$$\frac{C}{N_2} = \frac{1}{0.79A} = \frac{[CO_2] + [CO]}{[N_2] + 0.5[NO]}. \tag{1}$$

$$\frac{H_2}{N_2} = \frac{9/8}{0.79A} = \frac{[H_2O] + [H_2] + 0.5\{[OH] + [H]\}}{[N_2] + 0.5[NO]}. \tag{2}$$

$$\frac{O_2}{N_2} = \frac{1}{3.76} =$$
$$\frac{[CO_2] + [O_2] + 0.5\{[H_2O] + [CO] + [OH] + [NO] + [O]\}}{[N_2] + 0.5[NO]}. \tag{3}$$

The fourth relationship is obtained from Dalton's partial-pressure law. Thus,

$$[CO_2] + [H_2O] + [O_2] + [N_2] + [CO] + [H_2] + [OH] + [NO] + [H] + [O] = P. \tag{4}$$

The remaining six relationships are obtained from the equilibrium relationships (Table VII, Appendix).

Thus,
$$[CO_2] = \frac{1}{K_{10}} \frac{[CO][H_2O]}{[H_2]}. \tag{5}$$

$$[O_2] = K_7 \left\{ \frac{[H_2O]}{[H_2]} \right\}^2. \tag{6}$$

$$[H] = \sqrt{K_1[H_2]} \tag{7}$$

$$[O] = \sqrt{K_2[O_2]} \tag{8}$$

$$[\text{OH}] = \sqrt{K_8 \frac{[\text{H}_2\text{O}]^2}{[\text{H}_2]}} \tag{9}$$

$$[\text{NO}] = \sqrt{K_4[\text{O}_2][\text{N}_2]} = \sqrt{K_{11}[\text{N}_2] \frac{[\text{H}_2\text{O}]}{[\text{H}_2]}}. \tag{10}$$

Substituting Eq. (5) in Eq. (1) and solving for [CO] produce

$$[\text{CO}] = \frac{K_{10}[\text{H}_2]\{[\text{N}_2] + 0.5[\text{NO}]\}}{0.79A\{K_{10}[\text{H}_2] + [\text{H}_2\text{O}]\}}, \tag{11}$$

Substituting Eqs. (7), (9), and (10) into Eq. (2) and solving for [H$_2$O],

$$[\text{H}_2\text{O}] = \frac{\frac{9/8}{0.79A}[\text{N}_2] - [\text{H}_2] - 0.5\sqrt{K_1[\text{H}_2]}}{1 + 0.5\sqrt{\frac{K_8}{[\text{H}_2]}} - \frac{9/8 \sqrt{K_{11}[\text{N}_2]}}{1.58A[\text{H}_2]}}. \tag{12}$$

Since NO appears in small quantities even at high temperatures, the amount of N$_2$ in the products will be practically the same as in the mixture before the reaction. Hence, the N$_2$ can be estimated from the total pressure and will vary from $0.70P$ to $0.75P$ for the range of temperatures and air-fuel ratios considered.

Then, at the desired temperature, which fixes the various K values, a value for [H$_2$] is assumed. The higher the air-fuel ratio, the lower will be this value. Substituting the various values in Eq. (12) determines the [H$_2$O]. This is followed by determinations of [NO] from Eq. (10) and [CO] from Eq. (11). Then, values for [CO$_2$], [OH], [O$_2$], [H] and [O] can be obtained from Eqs. (5), (9), (6), (7), and (8), respectively. The assumed value of [H$_2$] is checked by substituting the foregoing results in Eq. (3), and the computations repeated with a different value of [H$_2$] until a check is indicated.

The total pressure of the equilibrium mixture is determined from Eq. (4). Then, the total mols, pressure, and temperature being known, values for V, E, C, H, and S can be determined for the equilibrium mixture.

Example.—Determine values of P, V, E, C, and S at 5000°R. for the reaction between C_8H_{18} and the chemically correct amount of air. The total pressure of the equilibrium mixture at this temperature is to be about 30 atm.

Nitrogen will be assumed to be 21.5 mols.

The computations are made as follows, the constants being obtained from Table VI (Appendix), and being given on page 52:

[H$_2$]	$0.5\sqrt{K_1[H_2]}$	Numerator, Eq. (12)	$0.5\sqrt{\dfrac{K_8}{[H_2]}}$	$\dfrac{0.00089}{[H_2]}$	$1 + 0.5\sqrt{\dfrac{K_8}{[H_2]}}$	Denominator, Eq. (12)	[H$_2$O], Eq. (12)
0.14	0.014	3.962	0.0558	0.0064	1.0558	1.049	3.777
0.16	0.015	3.941	0.0522	0.0056	1.0522	1.047	3.764
0.18	0.016	3.920	0.0492	0.0049	1.0492	1.044	3.755
[H$_2$]	$\dfrac{[H_2O]}{[H_2]}$	[NO], Eq. (10)	[N$_2$] + 0.5[NO]	$K_{10}[H_2]$	Numerator, Eq. (11)	$K_{10}[H_2]$ + [H$_2$O]	Denominator, Eq. (11)
0.14	26.979	0.250	21.625	0.917	19.846	4.694	27.582
0.16	23.525	0.218	21.609	1.048	22.646	4.812	28.275
0.18	20.861	0.193	21.597	1.178	25.441	4.933	28.986
[H$_2$]	[CO], Eq. (11)	$\dfrac{[CO][H_2O]}{[H_2]}$	[CO$_2$], Eq. (5)	$K_8[H_2O]$	$K_8\dfrac{[H_2O]^2}{[H_2]}$	[OH], Eq. (9)	$K_1[H_2]$
0.14	0.720	19.425	2.967	0.00657	0.1773	0.421	0.00084
0.16	0.801	18.844	2.878	0.00655	0.1541	0.393	0.00096
0.18	0.878	18.316	2.798	0.00653	0.1362	0.369	0.00108
[H$_2$]	[H], Eq. (7)	$\left\{\dfrac{[H_2O]}{[H_2]}\right\}^2$	[O], Eq. (6)	$K_2[O_2]$	[O], Eq. (8)	Numerator, Eq. (3)	N$_2$ ÷ O$_2$, Eq. (3)
0.14	0.029	727.9	0.342	0.000944	0.031	5.909	3.66
0.16	0.031	553.4	0.260	0.000718	0.027	5.739	3.77
0.18	0.033	435.2	0.205	0.000566	0.024	5.613	3.85

$T = 5000°R.$

[N_2] = 21.5 mols. $K_7 = 0.00047.$ $\dfrac{\%[N_2]}{0.79A} = 4.116.$

$0.79A = 5.876.$ $K_8 = 0.00174.$

$K_1 = 0.00603.$ $K_{10} = 6.547.$ $\sqrt{K_{11}[N_2]} = 0.00927.$

$K_2 = 0.00276.$ $K_{11} = 0.000004.$ $\dfrac{\%\sqrt{K_{11}[N_2]}}{1.58A} = 0.00089.$

Interpolating the results on page 51 for a N_2/O_2 ratio of 3.76, the ratio for these constituents in air, results in the values listed as follows:

Medium	Quantity	Internal energy		Chemical energy	
		Per mol	Per quantity	Per mol	Per quantity
H_2O	3.765	39,885	150,167		
H_2	0.158	25,819	4,079	103,486	16,351
CO_2	2.886	51,365	148,239		
CO	0.794	27,907	22,158	121,181	96,218
O_2	0.267	29,616	7,907		
H	0.031	13,346	414	144,565	4,482
O	0.027	13,346	360	105,833	2,857
OH	0.396	26,319	10,422	62,444	24,728
NO	0.221	28,570	6,314	38,746	8,563
N_2	21.500	27,589	593,164		
Sum	30.045 mols		943,224 B.t.u.		153,199 B.t.u.

Substituting the mols of the various constituents in Eq. (4) results in $P_4 = 30.05$ atm., which checks the original [N_2] assumption.

Then, $V = \dfrac{MRT}{P} = \dfrac{30.05 \times 1545 \times 5000}{30.05 \times 14.7 \times 144} = 3649$ ft.^{3}

Values of E and C from Tables III, V, and VII (Appendix) are used for evaluating E and C for the mixture of gases.

The entropy of each constituent is equal to

$$S = \int c_p \dfrac{dT}{T} - R \log_e P + S_0.$$

Table IV (Appendix) lists the entropy values per mol at 1 atm. The term $MR \log_e P_M$ for each constituent must be subtracted from the values determined from Table IV (Appendix). The entropy is evaluated as follows:

THE COMBUSTION PROCESS

Medium	Mols M and pressure P	S/mol, 1 atm.	S, 1 atm.	$1.986 M$	$\log_e M = \log_e P$	$MR \log_e P$
H_2O	3.765	22.42	84.411	7.477	1.3257	9.912
H_2	0.158	27.35	4.321	0.314	-1.8452	-0.579
CO_2	2.886	27.98	80.750	5.732	1.0599	6.075
CO	0.794	38.42	30.505	1.577	-0.2307	-0.364
O_2	0.267	18.49	4.937	0.530	-1.3205	-0.700
H	0.031	27.09	0.840	0.062	-3.4738	-0.215
O	0.027	24.78	0.669	0.054	-3.6119	-0.195
OH	0.396	26.12	10.344	0.786	-0.9264	-0.728
NO	0.221	21.03	4.648	0.439	-1.5096	-0.663
N_2	21.500	17.54	377.110	42.699	3.0681	131.005
Sum	30.045		598.535			143.548

$S = S_{1\ atm.} - \Sigma MR \log_e P = 598.535 - 143.548 = 454.987$ B.t.u. °R.$^{-1}$
Total $N_2 = N_2 + 0.5 NO = 21.5 + 0.11 = 21.61$ mols.
Total $O_2 = N_2 \div 3.76 = 21.61 \div 3.76 = 5.75$ mols.
Wt. of air $= 27.36 \times 28.95 = 792.1$ lb.

Dividing all the results by 792.1 lb. provides values as follows for a mixture containing originally 1 lb. of air:

$T = 5000$°R. $E = 1191$ B.t.u.
$P = 30.05$ atm. $C = 193$ B.t.u.
$V = 4.61$ ft.^{3} $E + C = 1384$ B.t.u.
$S = 0.574$ B.t.u. °R.$^{-1}$

These values check those obtained from Chart E (Appendix) at the same pressure and temperature.

The foregoing method may be used to develop combustion charts for reactions between other fuels and air.

EXERCISES

1. Write the reaction equation for complete combustion of ethyl alcohol in air.

2. The heating value of vaporized ethyl alcohol at 77°F. is reported as 606,204 B.t.u. per mol when determined in a constant-pressure calorimeter which condenses all the water vapor formed. Determine the chemical energy associated with the alcohol for this reaction.

3. Determine the net chemical energy and heating value associated with the reaction

$$H_2 + O_2 = 2OH$$

at 80°F. (see Tables V and VII, Appendix).

INTERNAL-COMBUSTION ENGINES

4. Determine the constant-pressure and constant-volume heating value of ethyl alcohol at 100 and at 140°F.

5. Determine the constant-pressure heat of reaction for the water-gas reaction if it occurs at a constant temperature of 1000°R.

6. Determine the combustion temperature for the constant-pressure reaction of C_2H_5OH and 110 per cent required air with 30 per cent of the chemical energy as heat transferred out of the reaction chamber during the process. Initial temperature is 1000°R. Complete combustion.

7. Determine the temperature of combustion of H_2 with the required air if the process occurs at constant volume with no heat transfer. Initial conditions are 60°F. and 1 atm. Consider chemical equilibrium, but neglect any dissociation of H_2, O_2, and N_2.

8. Determine the pressure at the end of the combustion process in Prob. 7.

9. Check the values on Chart E (Appendix) at $T = 5000$°R. and $P = 5$ atm.

10. The products of the reaction of 1 mol of C_8H_{18} and the required air expand adiabatically and reversibly from initial conditions of $T = 5000$°R. and 500 lb. in.$^{-2}$ to 100 lb. in.$^{-2}$ Determine the work of the process.

11. Assume that heat transfer occurs during the process in Prob. 10, which results in the entropy being decreased to 0.52 B.t.u. °R.$^{-1}$ for the amount upon which Chart E (Appendix) is based. Also assume that the path followed during the process can be described by the relation PV^n const. Determine the heat transfer and work of the process.

12. The products of the reaction of 1 mol of C_8H_{18} and the required air expand adiabatically from initial conditions of $T = 4600$°R. and $P = 400$ lb. in.$^{-2}$ through the nozzle of a rocket engine to an atmospheric pressure of 10 lb. in.$^{-2}$ Friction increases the entropy during the process to 0.58 B.t.u. °R.$^{-1}$ for the amount upon which Chart E (Appendix) is based. Determine the kinetic energy of the gases leaving the rocket. Also determine the propelling force if 1 mol of fuel is used per hour.

CHAPTER IV

AIR-STANDARD CYCLE ANALYSIS

The accurate analysis of the internal-combustion-engine process is a complex problem which led to simplifying assumptions and resulted in the air-standard cycle analysis. This hypothetical analysis implies that the entire medium is air and hence no chemical reaction occurs during the cycle. Energy is supplied and rejected to the air which may be used in successive cycles. The specific heat of the air is assumed to be constant. Also, losses due to heat transfer through the cylinder walls are assumed to be zero in the ideal analysis.

The foregoing assumptions result in an analysis that is far from correct for the actual internal-combustion-engine process but is of considerable value for indicating the hypothetical limit of performance when infinitely lean mixtures are being used. This analysis is also a simple means for showing that an increase in compression ratio increases the thermal efficiency of the cycle.

In every cycle, all the energy terms that depend upon the state of a medium are the same at the beginning and end of the cycle. This results in an energy equation for the cycle with work and heat-transfer terms only.

Thus, $$W_{out} - W_{in} = Q_{in} - Q_{out}. \qquad (1)$$

Hence, the net work of a cycle is equivalent to the difference of the heats added to and rejected from the medium during the cycle. Only the more important cycles are analyzed in the text, but the same method may be applied to any proposed cycle.

Carnot Cycle.—The Carnot cycle is the most efficient cycle for any given temperature limits. It consists of an adiabatic compression process from A to B (Fig. 19), a heat-addition process from B to C, an adiabatic expansion process from C to D, and a heat-rejection process from D to A, which returns the medium to the original conditions.

All the processes are assumed to be reversible which results in a rectangular cycle on the temperature-entropy diagram (Fig. 20). Since the area under any reversible path on this diagram represents the heat transferred,

$$Q_{in} = \int_{S_B}^{S_C} T_1 \, dS = T_1(S_C - S_B). \tag{2}$$

The heat rejected along the path D to A is

$$Q_{out} = -\int_{S_D}^{S_A} T_2 \, dS = T_2(S_D - S_A). \tag{3}$$

Since the entropy change is the same in both cases,

$$\text{Net work} = Q_{in} - Q_{out} = (T_1 - T_2)(S_2 - S_1), \tag{4}$$

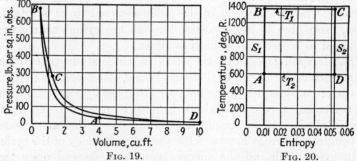

FIGS. 19 and 20.—P-V and T-S diagrams of the Carnot cycle.

which is equivalent to the area of the cycle $ABCD$ on the T-S diagram.

Also, $$\text{Thermal eff.} = \frac{\text{net work}}{Q_{in}} = \frac{T_1 - T_2}{T_1}. \tag{5}$$

It should be observed that T_1 and T_2 are the *highest* and *lowest* temperatures, respectively, attained during the cycle, and that *all* the heat is added at the *highest* temperature. Also, *all* the heat not transformed into work is rejected at the *lowest* temperature. No heat-engine cycle can attain a thermal efficiency higher than that of the Carnot cycle when operating between the same temperature limits, for no work area larger than the rectangular area $ABCD$ (Fig. 20) can be drawn between the two limiting temperatures for the given amount of heat added.

Example.—A new heat-engine cycle is reported to transform 75 per cent of the heat supplied into mechanical work. The medium used operates between temperatures of 1040 and 140°F. Analyze the performance of the new engine.

The Carnot cycle with the specified temperature limits would have a thermal efficiency of

$$\frac{T_1 - T_2}{T_1} = \frac{1500 - 600}{1500} = 0.60 = 60 \text{ per cent.}$$

The efficiency claimed for the new engine is incorrect since it is higher than the efficiency of a Carnot cycle for the same temperature limits. The efficiency might be lower than that of the Carnot cycle and still be impossible because of inherent losses necessary for actual engine operation.

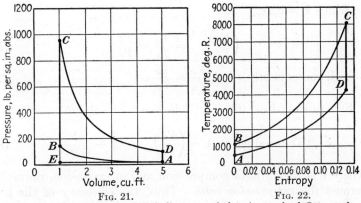

Figs. 21 and 22.—P-V and T-S diagrams of the air-standard Otto cycle.

Otto Cycle.—Air is compressed adiabatically from A to B (Fig. 21) in the air-standard Otto cycle. Heat is added to the air during the constant-volume heating process from B to C. Adiabatic expansion occurs from C to D, and the air is then cooled from D to A, which returns the air to the initial condition.

The heat added along the path B to C is

$$Q_{in} = Mc_v(T_C - T_B) \tag{1}$$

where M and c_v are the quantity and specific heat, respectively, of the medium.

The heat rejected along the path D to A is

$$Q_{out} = Mc_v(T_D - T_A) \tag{2}$$

Then, for the cycle,

$$\text{Net work} = Q_{in} - Q_{out} = Mc_v[(T_C - T_B) - (T_D - T_A)] \quad (3)$$

and

$$\text{Thermal eff.} = \frac{\text{Net work}}{Q_{in}} = 1 - \frac{T_D - T_A}{T_C - T_B} \quad (4)$$

$$= 1 - \frac{T_A}{T_B}\left(\frac{\frac{T_D}{T_A} - 1}{\frac{T_C}{T_B} - 1}\right). \quad (5)$$

Since both adiabatic processes of the cycle have the same volume ratio and have paths described by the equation

$$PV^k = \text{const.},$$

in which $k = c_p/c_v$,

$$\left(\frac{V_B}{V_A}\right)^{k-1} = \frac{T_A}{T_B} = \frac{T_D}{T_C} \quad (6)$$

and, consequently,

$$\frac{T_D}{T_A} = \frac{T_C}{T_B}. \quad (7)$$

Substituting Eqs. (6) and (7) in Eq. (5) results in

$$\text{Thermal eff.} = 1 - \left(\frac{V_B}{V_A}\right)^{k-1} = 1 - \frac{1}{(r)^{k-1}}, \quad (8)$$

where r is the ratio of compression and of expansion, normally termed the *compression ratio*. Thus, the efficiency of the air-standard Otto cycle is dependent only upon the compression ratio (see Fig. 41).

Example.—Determine the pressures at the points A, B, C, and D of the Otto cycle (Figs. 21 and 22), the net work, and thermal efficiency of the cycle for a compression ratio of 5:1. The conditions at the beginning of compression are 14.7 lb. in.$^{-2}$ abs. and 78°F. The heat supplied is 2,145,000 B.t.u. per 60.75 mols of air. The specific heat of air at constant volume is assumed to be 5 B.t.u. mol^{-1} °R.$^{-1}$

$$P_B = P_A\left(\frac{V_A}{V_B}\right)^k = 14.7 \times (5)^{1.4} = 139.9 \text{ lb. in.}^{-2} \text{ abs.}$$

$$T_B = T_A\left(\frac{V_A}{V_B}\right)^{k-1} = 538 \times (5)^{0.4} = 1024°\text{R.}$$

$$T_C = \frac{Q_{in}}{Mc_v} + T_B = \frac{2,145,000}{60.75 \times 5} + 1024 = 8086°\text{R.}$$

Then, $P_C = P_B \dfrac{T_C}{T_B} = 139.9 \times \dfrac{8086}{1024} = 1105$ lb. in.$^{-2}$ abs.

$T_D = T_A \dfrac{T_C}{T_B} = 538 \times \dfrac{8086}{1024} = 4248°$R.

Finally, $P_D = P_C \dfrac{P_A}{P_B} = 1105 \times \dfrac{14.7}{139.9} = 116.1$ lb. in.$^{-2}$ abs.

These computations indicate the attainment of pressures and temperatures that are absurdly high when a heat input is being used that is nearly equivalent to the energy liberated in a comparable actual engine process.

The heat rejected from D to A is

$Q_{out} = Mc_v(T_D - T_A) = 60.75 \times 5 \times (4248 - 538) = 1{,}126{,}900$ B.t.u.
Net work $= Q_{in} - Q_{out} = 2{,}145{,}000 - 1{,}126{,}900 = 1{,}018{,}100$ B.t.u.

$$\text{Thermal eff.} = \dfrac{\text{Net work}}{Q_{in}} = \dfrac{1{,}018{,}100}{2{,}145{,}000} = 0.475.$$

The same result can be obtained from Eq. (8).

Thus, Thermal eff. $= 1 - \dfrac{1}{r^{k-1}} = 1 - \dfrac{1}{(5)^{0.4}} = 0.475.$

Efficiencies obtained by the foregoing analysis are much too high. The maximum possible efficiency for the ideal Otto-engine process (Chap. V) with a compression ratio of 5:1 is 0.327 instead of 0.475. Equation (8) will result in an efficiency of 0.327 for this compression ratio if k is equal to 1.246. The use of this exponent makes Eq. (8) an empirical relationship and does not indicate that this exponent should be used in evaluating the compression and expansion curves. Exponents less than 1.4 indicate heat transfer on both the compression and expansion processes; consequently, efficiencies determined by using the energy equation will not check those obtained from Eq. (8) if the empirical exponent is used.

Diesel Cycle; *Constant-pressure Combustion.*—Air is compressed adiabatically from A to B (Fig. 23) in the air-standard Diesel cycle. Heat is added at constant pressure from B to C. Adiabatic expansion occurs from C to D, and heat is rejected from D to A, which returns the air to the initial condition.

The heat added along the constant-pressure path from B to C is

$$Q_{in} = Mc_p(T_C - T_B). \tag{1}$$

The heat rejected along the constant-volume path from D to A is

$$Q_{out} = Mc_v(T_D - T_A). \tag{2}$$

$$\text{Net work} = Q_{in} - Q_{out} = M[c_p(T_C - T_B) - c_v(T_D - T_A)]. \tag{3}$$

$$\text{Thermal eff.} = \frac{\text{Net work}}{Q_{in}} = 1 - \frac{1}{k}\left(\frac{T_D - T_A}{T_C - T_B}\right). \tag{4}$$

The efficiency of the air-standard Diesel cycle depends on the compression ratio, which determines T_B for a given T_A, and on the heat input, which determines T_C and T_D. Small heat inputs reduce the value of T_C and T_D. This increases the effective

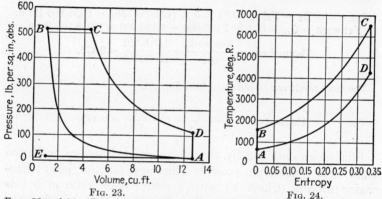

Fig. 23.   Fig. 24.

Figs. 23 and 24.—P-V and T-S diagrams of the air-standard constant-pressure Diesel cycle.

expansion ratio and the efficiency. The heat added at the end of the constant-pressure process, B to C, is added at the highest temperature. However, the possibility of transforming this heat into work is the smallest since the effective expansion ratio is the lowest at this point.

Example.—Determine the compression ratio for an air-standard constant-pressure Diesel cycle to have a compression pressure of 500 lb. in.$^{-2}$ gage. Also, compute the temperatures at the four points of the cycle (Figs. 23 and 24), the volume ratio at C, the pressure at D, and the efficiency of the cycle. $P_A = 14.7$ lb. in.$^{-2}$ abs. and $T_A = 78°$F.

$$r = \frac{V_A}{V_B} = \left(\frac{P_B}{P_A}\right)^{\frac{1}{k}} = \left(\frac{514.7}{14.7}\right)^{\frac{1}{1.4}} = 12.68.$$

$$T_B = T_A\left(\frac{V_A}{V_B}\right)^{k-1} = 538 \times (12.68)^{0.4} = 1486°\text{R}.$$

AIR-STANDARD CYCLE ANALYSIS

Assuming the same heat input, mols of air, and specific heats, as in the Otto-cycle analysis, and solving Eq. (1), results in

$$T_C = \frac{Q_{in}}{Mc_p} + T_B = \frac{2{,}145{,}000}{60.75 \times 7} + 1486 = 6530°R.$$

$$\text{Vol. ratio at } C = \frac{V_C}{V_B} = \frac{T_C}{T_B} = \frac{6530}{1486} = 4.39.$$

$$\text{Expansion ratio} = \frac{V_D}{V_C} = \frac{V_D}{V_B} \times \frac{V_B}{V_C} = 12.68 \div 4.39 = 2.89.$$

Then, $$P_D = P_C \left(\frac{V_C}{V_D}\right)^k = \frac{514.7}{(2.89)^{1.4}} = 116.5 \text{ lb. in.}^{-2} \text{ abs.}$$

Also, $$T_D = T_C \left(\frac{V_C}{V_D}\right)^{k-1} = \frac{6530}{(2.89)^{0.4}} = 4271°R.$$

Solving for the efficiency of the cycle, Eq. (4), results in

$$\text{Thermal eff.} = 1 - \frac{1}{k}\left(\frac{T_D - T_A}{T_C - T_B}\right) = 1 - \frac{1}{1.4}\left(\frac{4271 - 538}{6530 - 1486}\right) = 0.471.$$

The foregoing examples show that the air-standard efficiencies of the Otto cycle with a 5:1 compression ratio and the constant-pressure Diesel cycle with a 12.68 compression ratio are practically the same. The air-standard efficiency of the Otto cycle with a 12.68:1 compression ratio would be 0.638 (Fig. 42), or 36 per cent higher than the constant-pressure Diesel cycle for the same compression ratio, the Otto cycle being inherently more efficient than the Diesel cycle for a given compression ratio. However, fuel characteristics and maximum-pressure restrictions limit the compression ratios that may be used with the Otto cycle in practice.

Diesel Cycle; *Limited-pressure Combustion.*[1]—Air is compressed adiabatically in this cycle from A to B (Fig. 25). Part of the heat input is added at constant volume from B to C, and the balance at constant pressure from C to D. Adiabatic expansion occurs from D to E, and heat is rejected at constant volume from E to A, which returns the air to its initial condition.

The heat added along the constant-volume path from B to C is

$$Q_{in} = Mc_v(T_C - T_B). \quad (1)$$

---

[1] This analysis applies also to fuel-injection spark-ignition cycles of lower compression ratios, such as the Hesselman engine cycle.

The heat added along the constant-pressure path from C to D is

$$Q_{in} = Mc_p(T_D - T_C). \qquad (2)$$
$$\text{Total } Q_{in} = M[c_v(T_C - T_B) + c_p(T_D - T_C)]. \qquad (3)$$

The heat rejected along the constant-volume path from E

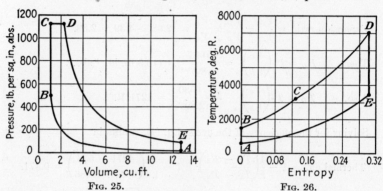

FIGS. 25 and 26.—P-V and T-S diagrams of the air-standard limited-pressure Diesel cycle.

to A is

$$Q_{out} = Mc_v(T_E - T_A). \qquad (4)$$

$$\text{Thermal eff.} = \frac{\text{Net work}}{Q_{in}} = \frac{Q_{in} - Q_{out}}{Q_{in}}$$

$$= 1 - \frac{T_E - T_A}{(T_C - T_B) + k(T_D - T_C)}. \qquad (5)$$

Example.—An air-standard limited-pressure Diesel cycle has a compression ratio of 12.68:1. One-fourth of the heat is added at constant volume and the remainder at constant pressure. $P_A = 14.7$ lb. in.$^{-2}$ abs. and $T_A = 78°F$. Determine the air-standard thermal efficiency.

From the constant-pressure Diesel example for the same conditions (page 60), $T_B = 1486°R$. and $P_B = 514.7$ lb. in.$^{-2}$ abs. Then, assuming the same total heat input, mols of air, and specific heats as in the previous examples, results in

$$T_C = \frac{0.25 Q_{in}}{Mc_v} + 1486 = \frac{0.25 \times 2{,}145{,}000}{60.75 \times 5} + 1486 = 3251°R.$$

$$P_C = P_B \frac{T_C}{T_B} = 514.7 \times \frac{3251}{1486} = 1126 \text{ lb. in.}^{-2} \text{ abs.}$$

$$T_D = \frac{0.75 Q_{in}}{Mc_p} + T_C = \frac{0.75 \times 2{,}145{,}000}{60.75 \times 7} + 3251 = 7034°R.$$

AIR-STANDARD CYCLE ANALYSIS

The volume ratio at D is

$$\frac{V_D}{V_C} = \frac{T_D}{T_C} = \frac{7034}{3251} = 2.16.$$

The adiabatic expansion ratio is

$$\frac{V_E}{V_D} = \frac{V_E}{V_C} \times \frac{V_C}{V_D} = \frac{12.68}{2.16} = 5.87.$$

Then, $\quad T_E = T_D \left(\dfrac{V_D}{V_E}\right)^{k-1} = 7034 \left(\dfrac{1}{5.87}\right)^{0.4} = 3466°\mathrm{R}.$

Substituting in Eq. (5),

$$\text{Thermal eff.} = 1 - \frac{3466 - 538}{(3251 - 1486) + 1.4(7034 - 3251)} = 0.585.$$

Comparison of Air-standard Cycles.—The foregoing cycles may be compared at the same compression ratio and same heat

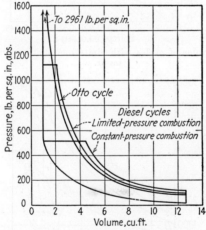

FIG. 27.—P-V diagrams of cycles with the same compression ratio and heat input.

input by plotting them on P-V and T-S diagrams (Figs. 27 and 28). These illustrations show that adding the heat at constant volume results in the highest maximum temperatures and pressures for the Otto cycle. Adding the heat at constant pressure results in the lowest maximum temperatures and pressures for the constant-pressure Diesel cycle, while the values for the limited-pressure Diesel cycle lie between.

On the T-S diagram (Fig. 28), the area under the lower constant-volume line, between the entropy limits for any given cycle, represents the heat rejected. For this case of the same compression ratio and the same heat input, the least heat is rejected by the Otto cycle and the most by the constant-pressure Diesel cycle. Consequently, the Otto cycle has the largest work area and highest efficiency, as indicated previously

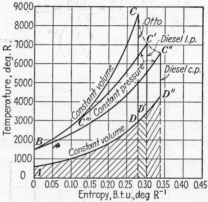

FIG. 28.—T-S diagrams of cycles with the same compression ratio and heat input.

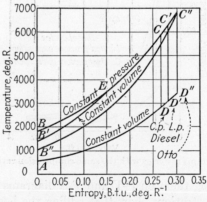

FIG. 29.—T-S diagrams of cycles with the same maximum pressure and heat input.

The foregoing shows that heat should be added to the working medium when the piston is in the position *permitting the maximum possible expansion*, for with such expansion the maximum amount of internal energy can be converted into work.

A comparison of the three air-standard cycles with the same maximum pressure and heat input (Fig. 29) shows the constant-pressure Diesel cycle to be the most efficient and the Otto cycle to be the least efficient. However, a much lower compression ratio must be used with the Otto than with the Diesel cycle.

A comparison of the three cycles with the same maximum pressure and temperature can be made by shifting the adiabatic expansion lines CD and $C'D'$ to coincide with $C''D''$ (Fig. 29). These cycles then have the same heat rejection under the lower constant-volume line; but the constant-pressure Diesel cycle has the largest work area and highest efficiency, and the Otto cycle the lowest of these values. Different compression ratios must be used when the adiabatic expansion ratio is the same in each case.

The Carnot cycle for the same temperature limits and heat input as in any of the cycles would have a much larger work area and higher efficiency. However, the range of pressure and volume limits is very large compared with that of any of the other three cycles.

Mean Effective Pressure.—The mean effective pressure is obtained by dividing the net work of the engine process by the piston displacement. For the air-standard Otto-cycle analysis,

$$\text{M.e.p.} = \frac{\text{Net work}}{\text{Displacement}} = \frac{Q_{in} \times \text{eff.}}{V\frac{r-1}{r}} = \frac{Q_{in}}{V}\left(\frac{1 - \frac{1}{r^{k-1}}}{\frac{r-1}{r}}\right), \quad (1)$$

where Q_{in} = the energy supplied,
V = the sum of the clearance and displacement volume,
r = the compression ratio.

Assuming the heat input is proportional to the displacement, which is the only reasonable assumption, makes Q_{in} proportional to $V(r-1)/r$, and makes the mean effective pressure proportional to the efficiency for the air-standard Otto cycle.[1]

Effect of Increasing Compression Ratio.—It has been shown that increasing the compression ratio increases the thermal efficiency of the cycle. An increase in compression ratio increases the displacement and heat input for a given total volume (Fig. 30).

For the lower compression ratio,

$$Q_{in} = Mc_v(T_3 - T_2). \quad (1)$$

[1] See also Mean Effective Pressure (Chap. V).

Assuming the same expansion curve, for the higher compression ratio, results in

$$Q'_{in} = Mc_v(T_{3''} - T_{2'}). \qquad (2)$$

The increase in work area, 22′3″32, is equal to

$$\int_3^{3''} P\,dV - \int_2^{2'} P\,dV = \frac{MR[(T_{3''} - T_3) - (T_{2'} - T_2)]}{k - 1}. \qquad (3)$$

Rearranging the terms in the parentheses and substituting values from Eqs. (1) and (2), result in

$$\Delta \text{ area} = \frac{\frac{R}{c_v}(Q'_{in} - Q_{in})}{k - 1} = Q'_{in} - Q_{in}. \qquad (4)$$

This indicates that an increase in compression ratio, with the same compression and expansion curves, results in a work area to the left of the original clearance line equal to the increase in heat supplied. Since the increase in heat supplied is directly proportional to the increase in displacement, it is obvious from a comparison of the increases in areas A and B for two equal increases in displacement that an increase in compression ratio lowers the expansion curve as indicated[1] (Fig. 30). Thus, an increase in compression ratio lowers point 4 to 4′, which decreases the percentage of heat rejected and increases the percentage of work of the air-standard Otto cycle. The same method may be used to show the effect of an increase in compression ratio for the other cycles.

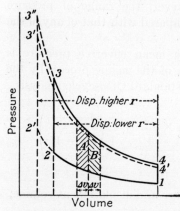

FIG. 30.—Effect of compression ratio.

EXERCISES

1. A Carnot cycle rejects heat at a temperature of 500°R. Determine and plot the thermal efficiencies for heat-input temperatures ranging from 500 to 5000°R.

2. A Carnot cycle and an Otto cycle are to operate with the same maximum volume. At the beginning of the compression stroke the cylinder is

---

[1] See also Effect of Compression Ratio on Efficiency (Chap. V).

filled with 1 mol of air at 14.7 lb. in.$^{-2}$ abs. and 78°F. The maximum pressure is to be 1000 lb. in.$^{-2}$ abs. in both cases. Determine the work and efficiency of each cycle per mol of air, using the same entropy change for both cycles. The compression ratio of the Otto cycle is 8:1. c_v is 5.0 B.t.u. mol^{-1} (°F)$^{-1}$.

3. Plot P-V diagrams of the two cycles in Prob. 2, first on logarithmic cross-section paper and then transferring them to rectangular cross-section paper.

4. Plot a T-S diagram for Prob. 2, and show how less heat input for the Carnot cycle changes the work and efficiency of the cycle. Determine and plot the relation between Carnot-cycle efficiency and heat input while satisfying the conditions of maximum volume and pressure imposed in Prob. 2.

5. Compute and plot the relationship between the air-standard Otto-cycle thermal efficiency and compression ratio for compression ratios up to 10:1.

6. Compute and plot the relationship between the air-standard thermal efficiency and compression ratio for one-third-, two-thirds-, and full-load heat additions for the constant-pressure Diesel cycle for compression ratios of 10, 13, 16, and 20:1. Assume full-load heat-addition amounts to 2,145,000 B.t.u. per 60.75 mols of air. Conditions at the beginning of compression are 14.7 lb. in.$^{-2}$ abs. and 78°F.

7. Compute and plot the relationship between the part of heat added at constant volume and thermal efficiency for the limited-pressure Diesel cycle with a 12.68:1 compression ratio. Conditions at the beginning of compression are 14.7 lb. in.$^{-2}$ abs. and 78°F. Note that data for one-fourth and zero heat additions at constant volume have been computed and are given in this chapter.

8. Compute and plot the relationships between thermal efficiency, part of heat added at constant volume, and compression ratio for the limited-pressure Diesel cycle for a maximum pressure of 1000 lb. in.$^{-2}$ abs. Conditions at the beginning of compression are 14.7 lb. in.$^{-2}$ abs. and 78°F.

9. On one diagram, plot all the efficiencies vs. compression ratios for the preceding problems.

10. Prove with both P-V and T-S diagrams that an increase in compression ratio increases the air-standard thermal efficiency for the Otto, the constant-pressure Diesel, and the limited-pressure Diesel cycles.

11. Compute the data and plot the P-V diagrams for the various cycles represented in Fig. 29.

CHAPTER V

INTERNAL-COMBUSTION-ENGINE-PROCESS ANALYSIS

Ideal Analysis.—The air-standard cycle analysis is based upon simplifying assumptions that are far from true. The actual internal-combustion-engine process deals with both fuel and air, the fuel being usually in the liquid or gaseous state. A chemical reaction occurs in the engine and results in products of combustion which are mainly CO_2, H_2O, and N_2. Variable specific heat and chemical equilibrium prevent the attainment of the high temperatures and pressures indicated by the air-standard analysis. The products are exhausted from the engine in a condition considerably different from that of the air and fuel entering the engine. The temperature and pressure of the products, after these have been discharged from the engine, soon decrease to that of the surrounding air. However, since the products are not transformed back to air and fuel in the engine or auxiliary apparatus, the internal-combustion-engine process cannot be termed a "cycle."[1]

In all theoretical engine-process analysis, it is customary to assume no heat transfer to or from the media and no fluid friction losses. Also, the valves are assumed to open and close on dead center, that is, at either one or the other end of the stroke of the piston in a reciprocating engine. Such analysis results in the maximum efficiency attainable and serves as a basis for comparison with actual engine efficiencies, which can approach but never reach the ideal standard established by this analysis.

IDEAL OTTO-ENGINE PROCESS

Constant-volume Combustion

The Actual Mixture.—The clearance space permits the retainment of some of the products of combustion at the end of the exhaust stroke. The amount of clearance gases depends upon

[1] The mechanism goes through a cycle, but the media do not.

the pressure and temperature of these gases and the volume of the clearance space. The clearance gases will have a high temperature and will mix with the air-fuel mixture inducted, the final temperature being the temperature of the mixture at the beginning of the compression process. The amount and temperature of the clearance gases have an appreciable effect on the temperature of the mixture at the end of the suction stroke, and, since suction temperature and the entire engine process fix the temperature of the clearance gases, it is impossible to compute the suction temperature without making an assumption of clearance-gas conditions.

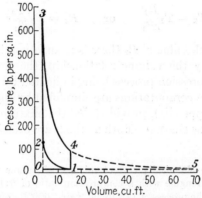

Fig. 31.—Pressure-volume diagram of Otto-engine process.

The best procedure appears to be that of assuming both the suction temperature and the amount of clearance gases. The clearance-gas conditions can be determined by analyzing the various parts of the engine process. Finally, both the amount of clearance gases and the suction temperature will be determined, and these should check the assumed values. Otherwise, the computations should be repeated with corrected assumptions.

The Compression Process.—The air-fuel mixture inducted and the clearance gases remaining in the cylinder are compressed adiabatically from 1 to 2 (Fig. 31) in the Otto-engine process. In the ideal case the compression will be adiabatic and reversible. Hence, the entropy of the total mixture remains constant during the process, and

$$MS_1 = (M_m S_m + M_c S_c)_1 = (M_m S_m + M_c S_c)_2 = MS_2 \quad (1)$$

where M = mols of total mixture,
S = entropy per mol,
m and c = subscripts indicating the air-fuel mixture and clearance gases, respectively.

The volume at the end of compression is determined by the compression ratio r of the engine. Thus, from Fig. 31,

$$V_2 = \frac{V_1}{r}. \quad (2)$$

Also,
$$V_2 = \frac{MRT_2}{P_2} = \frac{MRT_1}{rP_1} \quad (3)$$

or,
$$T_2 = T_1 \frac{P_2}{rP_1} \quad \text{or} \quad P_2 = P_1 \frac{rT_2}{T_1}. \quad (4)$$

Thus, for each value of P_2 there is a corresponding value of T_2 that will satisfy the volume relationship, the correct value for T_2 for the compression process being indicated by the solution of Eq. (1). These computations are simplified by the use of Charts A, B, and C (Appendix), provided that the mixture corresponds to that of one of the charts.[1] Both methods are used in the following example.

Example.—The chemically correct mixture of octane vapor and air is compressed in an engine with a compression ratio of 5:1. Conditions at the beginning of the process of 14.7 lb. in.$^{-2}$ abs., 600°R., and $M_c/M_m = 0.05$ are assumed. (Assume f instead of M_c/M_m if only the charts are used.) Determine the conditions at the end of compression.

From Table IV (Appendix), using 1 mol of correct mixture, the initial entropy is

$$M_m S_m = 1 \times 1.09 = 1.09.$$
$$M_c S_c = 0.05 \times 1.07 = 0.05.$$
$$MS_1 = \overline{1.14} \text{ B.t.u. mol}^{-1} \text{ °R.}^{-1}$$

Assume T_2, °R.	900	1000	1100
$P_2 = P_1 \frac{rT_2}{T_1}$, atm.	7.50	8.33	9.17
$M_m S_m$ at T_2 and 1 atm.	4.26	5.11	5.90
$M_c S_c$ at T_2 and 1 atm.	0.21	0.25	0.28
$-MR \log_e P_2 = -1.05 \times 1.986 \log_e P_2$.	-4.20	-4.42	-4.62
Sum = MS_2, B.t.u. mol^{-1} °R.$^{-1}$	0.27	0.94	1.56

[1] Not much error will be introduced if the mixture does vary somewhat from that of the charts.

Plotting the values of MS_2 against P_2 and T_2 results in $P_2 = 8.60$ atm., or 126.4 lb. in.$^{-2}$ abs., and $T_2 = 1030°$R. at $MS_2 = 1.14$ B.t.u. mol^{-1} °R.$^{-1}$
The clearance-gas weight fraction f of the total mixture is

$$f = \frac{M_c\left(\dfrac{M_m}{M_p}\right)}{M_m + M_c\left(\dfrac{M_m}{M_p}\right)} = \frac{0.05 \times \dfrac{60.5}{64}}{1 + 0.05 \times \dfrac{60.5}{64}} = 0.0452$$

where $(M_m/M_p) = 60.5/64$, which is the ratio of mols of correct air-fuel mixture to mols of products for complete reaction.

Considering a total equivalent mixture[1] of 0.0351 mol for which Chart B (Appendix) was constructed, the total charge is

$$M_m + M_c = 0.0351\left[(1-f) + f\frac{64.0}{60.5}\right]$$
$$= 0.0335 + 0.0017 = 0.0352 \text{ mol.}$$

Also, $\quad V_2 = \dfrac{MRT_2}{P_2} = \dfrac{0.0352 \times 1545 \times 1030}{126.4 \times 144} = 3.1$ ft.^{3}

If Chart B (Appendix) is used, point 1 is located at 600°R. and 14.7 lb. in.$^{-2}$ abs., which indicates a volume of 15.5 ft.^{3} Following a line of constant entropy from point 1 to a volume of 3.1 ft.^{3} (15.5/5) results in a pressure of about 125 lb. in.$^{-2}$ abs. and a temperature of about 1030°R., which check the foregoing computations.

The work done during adiabatic compression is equivalent to the difference of the internal energies at the end and beginning of the process.

Thus, $\quad$ Work of compression $= M(E_2 - E_1)$. $\quad$ (5)
Also,
$$M(E_2 - E_1) = (M_m E_m + M_c E_c)_2 - (M_m E_m + M_c E_c)_1. \quad (6)$$

Example.—Determine the work of compression for the preceding example.

---

[1] The equivalent mixture is the air-fuel mixture from which the products are formed. The total equivalent mixture is the sum of the equivalent mixture due to clearance gases and the inducted air-fuel mixture.

The reaction equation for C_8H_{18} and the correct amount of air indicates 59.5 mols of air per mol of fuel. Charts A to F (Appendix) were developed for 1 lb. of air and definite quantities of fuel. Thus, the equivalent mixture for Charts B and E (Appendix) is

$$\frac{M_a + M_f}{M_a \times \text{mol. wt.}} = \frac{59.5 + 1}{59.5 \times 28.95} = 0.0351 \text{ mol.}$$

From Table III (Appendix), $0.0335(E_m)_{1030} = 100.5$ B.t.u.
$0.0017(E_c)_{1030} = 4.9$ B.t.u.
$ME_2 = 105.4$ B.t.u.

Also, $0.0335(E_m)_{600} = 14.8$ B.t.u.
$0.0017(E_c)_{600} = 0.7$ B.t.u.
$ME_1 = 15.5$ B.t.u.
Work of compression $= 105.4 - 15.5 = 89.9$ B.t.u.

From Chart B (Appendix), for the conditions indicated by the chart solution in the previous example,

$$\text{Work} = 106 - 16 = 90 \text{ B.t.u.},$$

which checks the foregoing computed values.

The Combustion Process.—Chemical reaction is assumed to occur at constant volume at the end of compression in the ideal Otto-engine process. At the beginning of combustion the mixture contains chemical energy of the fuel and internal energy of the air, fuel, and clearance gases. At the end of the process, some of the dissociated products (H_2, CO, etc.) contain chemical energy and all the products contain internal energy. On the assumption that no heat transfer occurs during the process, the energy equation becomes

$$M_2(E_2 + C_2) = M_3(E_3 + C_3) \tag{7}$$
$$\text{or} \quad (M_f C_f + M_m E_m)_2 = (M_f C_f + M_p E_p)_3 \tag{8}$$

where subscripts f, m, and p indicate the fuel, mixture, and products constituents, respectively.

The left side of Eq. (8) is known, and the temperature at point 3 can be evaluated for the given process by application of the principles of chemical equilibrium (Chap. III). However, this lengthy computation may be eliminated by the use of one of the Charts D, E, and F in the Appendix.

Example.—The conditions at the end of compression (example, page 70) are $T_2 = 1030°$R., $P_2 = 125$ lb. in.$^{-2}$ abs., $V_2 = 3.1$ ft.^{3}, $M_m = 0.0335$ mol, and $M_c = 0.0017$ mol. Determine the conditions at the end of a constant-volume combustion process with no heat loss.

The foregoing data were obtained with the theoretically correct octane-air mixture. From Chart E (Appendix),

Chemical energy, $C = (1 - f) \times 1278 = (1 - 0.0452)1278 = 1220$ B.t.u.

From Chart B (Appendix), or the example on page 71,

$$\text{Internal energy } E = 106 \text{ B.t.u.}$$
$$C + E = 1220 + 106 = 1326 \text{ B.t.u.}$$

On Chart E (Appendix), at $V = 3.1$ ft.^{3} and $E + C = 1326$ B.t.u., T_3 and P_3 are found to be 4920°R. and 650 lb. in.$^{-2}$ abs., respectively.

The Expansion Process.—The gases do work on the piston during the expansion process which results in a decrease in internal energy and temperature. The decrease in temperature permits the recombination of some of the dissociated products, which occurs throughout the entire process. Actually, combination and dissociation of various molecules are taking place at the same time, the net result indicating the change that has taken place.

The expansion process is assumed to be adiabatic and reversible in the ideal case. Consequently, the process is one of constant entropy and may be solved analytically.[1] However, it is much simplified by the use of the combustion charts.

The work done during the adiabatic expansion process is equivalent to the difference between the internal and chemical energies at the beginning and end of the process.

Thus, $\quad \text{Work} = (ME + M_f C_f)_3 - (ME + M_f C_f)_4 \quad (9)$

where subscript f refers to the fuel constituents due to dissociation. These values are read directly from the charts, as well as values for the conditions of pressure, volume, and temperature.

Example.—The conditions at the beginning of expansion in an Otto-engine process with a 5:1 compression ratio are $P_3 = 650$ lb. in.$^{-2}$ abs., $V_3 = 3.1$ ft.^{3}, $T_3 = 4920°$R., and $(ME + M_f C_f)_3 = 1326$ B.t.u. (example, page 72). Determine the work done during the expansion process.

At the end of the expansion process, $V_4 = 5 \times 3.1 = 15.5$ ft.^{3}, and the entropy will be the same as that indicated by the conditions at the beginning of expansion. This indicates (Chart E, Appendix) values of $T_4 = 3640°$R., $P_4 = 98$ lb. in.$^{-2}$ abs., and $(ME + M_f C_f)_4 = 810$ B.t.u.

Thus, from Eq. (9), Work = $1326 - 810 = 516$ B.t.u.

The Clearance Gases.—The exhaust valve opens at the end of the expansion process, part of the gases escaping and part

[1] G. A. Goodenough and J. B. Baker, *Univ. Ill., Eng. Expt. Sta., Bull.* 160 (1927). A different arrangement of the same analysis is found in L. C. Lichty, "Thermodynamics," McGraw-Hill Book Company, Inc., New York, 1936. In neither reference is the dissociation so completely treated as in the construction of the combustion charts.

74 INTERNAL-COMBUSTION ENGINES

remaining in the cylinder. The part remaining in the cylinder expands adiabatically and reversibly in the ideal case to atmospheric pressure while the other part escapes. The condition of the part remaining in the cylinder is determined by continuing the expansion process (Fig. 31) to point 5. This is a constant-entropy process, and the use of the combustion charts results in the evaluation of the conditions at point 5.

At this condition, $$P_5 V_5 = M_5 R T_5 \tag{10}$$

where V_5 = the imaginary volume that would exist if all the gases in the cylinder at 3 expanded to 5.

At the end of the exhaust process the piston will have pushed a displacement volume of gases out of the cylinder, leaving M_c mols in the clearance space.

For these gases, $$P_c V_c = M_c R T_c \tag{11}$$

where $T_c = T_5$,
and $P_c = P_5$ in the ideal case.

Combining Eqs. (10) and (11) results in

$$\frac{M_5}{M_c} = \frac{M_p}{M_c} = \frac{T_c}{T_5}\frac{P_5}{P_c}\frac{V_5}{V_c} = \frac{V_5}{V_c}. \tag{12}$$

Also, $$M_p = M_m\left(\frac{M_p}{M_m}\right) + M_c, \tag{13}$$

where (M_p/M_m) = the ratio of mols of products formed to mols of air-fuel mixture inducted.

Thus, $$\frac{M_p}{M_c} = \frac{M_m}{M_c}\left(\frac{M_p}{M_m}\right) + 1, \tag{14}$$

from which the ratio M_m/M_c can be determined.

Example.—The conditions at the end of the expansion process in an engine with a compression ratio of 5:1 are $P_4 = 98$ lb. in.$^{-2}$ abs., $V_4 = 15.5$ ft.^{3}, and $T_4 = 3640°R$. Determine the ratio of fresh charge M_m to clearance gases M_c. The fresh charge is octane with the required air.

Extrapolating the expansion curve isentropically (Chart E, Appendix) from the conditions given to 14.7 lb. in.$^{-2}$ abs. results in a value of $V_5 = 68$ ft.^{3}, and $T_5 = 2510°R$.

Then, from Eq. (12), $\dfrac{M_p}{M_c} = \dfrac{V_5}{V_c} = r\dfrac{V_5}{V_4} = 5 \times \dfrac{68}{15.5} = 21.9 \left(=\dfrac{1}{f}\right).$

For the correct mixture of octane and air,

$$\left(\frac{M_p}{M_m}\right) = \frac{64.0}{60.5} = 1.058.$$

Hence, from Eq. (14), $\dfrac{M_m}{M_c} = \dfrac{\left(\dfrac{M_p}{M_c} - 1\right)}{\left(\dfrac{M_p}{M_m}\right)} = \dfrac{21.9 - 1}{1.058} = 19.8.$

The ratio M_c/M_m amounts to 0.0505 which nearly checks the assumed value of 0.05 (example, page 70).

Induction of the Charge.—The air and fuel (or only air), at atmospheric pressure and temperature outside the engine, flow into the intake system (Fig. 32) of the engine during the suction stroke 0 to 1 (Fig. 31). This flow into the intake system may be assumed to be steady in the case of the multicylinder engine, and with a negligible velocity of approach where the media enter the imaginary analysis envelope. Even in the case of intermittent flow as in the single-cylinder engine, the work of the atmosphere on the quantity of medium entering the intake system is equal to the flow-work term PV, which is used for steady flow. Thus, the energy entering with the air-fuel mixture exclusive of the chemical energy will be

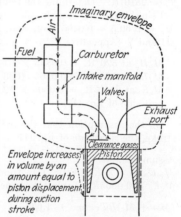

FIG. 32.—Diagram for analysis of suction process.

$$M_m(E + PV)_{atm} = M_m H_{atm} = (M_a H_a + M_f H_f)_{atm}. \quad (15)$$

The clearance gases left in the cylinder are at a high temperature T_c and have internal energy $M_c E_c$. During the suction stroke the inducted air-fuel mixture and the clearance gases do work on the piston that amounts to

$$\text{Work} = P_1(V_1 - V_0) = P_1 V_1 \frac{r-1}{r} = (M_m + M_c) R T_1 \frac{r-1}{r}.$$

$$(16)$$

At the end of the suction stroke the mixture will be at a temperature of T_1 and have internal energy of $(M_m E_m + M_c E_c)_1$.

Thus, the energy equation for the suction process is

$$(M_a H_a + M_f H_f)_{atm} + (M_c E_c)_5 = (M_m E_m + M_c E_c)_1 + (M_m + M_c) R T_1 \frac{r-1}{r}. \quad (17)$$

The solution of this equation determines the value of T_1 which should check the value assumed at the beginning of the analysis.

Example.—Assume that air and liquid octane at 14.7 lb. in.$^{-2}$ abs. and 60°F. are inducted into the intake system of an engine with a compression ratio of 5:1. It was found (example, page 74) that $M_c/M_m = 0.0505$ and $T_c = 2510°R$. Determine the temperature at the end of the suction stroke.

Using 1 mol of correct mixture for M_m results in $M_c = 0.0505$ mol, $M_f = 1/60.5 = 0.0165$ mol, and $M_a = 0.9835$ mol. Then, from Tables III and V (Appendix),

$M_a H_a$ at 520°R. = $0.9835 \times 1033 = 1016$ B.t.u.
$M_f H_f$ at 520°R. = $0.0165 (1033 - 17{,}800^*) = -277$ B.t.u.
$M_c E_c$ at 2510°R. = $0.0505 \times 12{,}746 = 644$ B.t.u.
Left side of Eq. (17) = $1016 - 277 + 644 = 1383$ B.t.u.

On the assumption that the fuel is vaporized at the end of the suction stroke, the right side of Eq. (17) is evaluated as follows:

Assume T_1, °R.	560	580	600
$M_m E_m = 1 \times E_m$ (Table III, Appendix),	220	332	443
$M_c E_c = 0.0505 E_c$ (Table III, Appendix),	11	16	22
$(M_m + M_c) R T_1 \frac{r-1}{r} = 1.67 T_1$,	935	969	1002
Right side of Eq. (17),	1166	1317	1467

Interpolation for 1383 B.t.u. results in $T_1 = 589°R$. compared with the assumed value of 600°R. Recomputation of all the examples with the new T_1 and M_c/M_m would result in a very close check but would not appreciably affect the work and efficiency values for the process.

A customary approximation of the relation for the suction process results from the assumption that the charge to be inducted is already in the cylinder and heat transfer occurs at constant pressure.

Thus, $\qquad Q = M_m \Delta H_m = M_c \Delta H_c. \qquad (18)$

However, with the volume of the charge to be inducted being equal to the piston displacement (100 per cent volumetric

* Approximately the latent heat of 1 mol of octane at 60°F.

efficiency), the flow work on the inducted charge is equal to the work on the piston. This assumption eliminates all except internal energy terms from Eq. (17) and results in

$$M_m \, \Delta E_m = M_c \, \Delta E_c. \tag{19}$$

This indicates that the heat transfer is equivalent to that which would occur at constant volume. With f representing the weight fraction of the total mixture that is clearance gas, Eq. (19) becomes

$$(1 - f) \, \Delta E_m = f \, \Delta E_c \tag{20}$$

or $\quad (1 - f)E_m + fE_5 = (1 - f)E_{m_1} + fE_{c_1} = E_1. \tag{21}$

This relation may be used with the various charts, for given weights of gases, and will be approximately correct for cases of or near 100 per cent volumetric efficiency.

Example.—From the example on page 74, $M_p/M_c = 21.9$ which is the ratio of mols as well as weights since the products and clearance gases are the same. From the example on page 70, $M_m = 0.0335$, from which $M_f = 0.0335 \div 60.5 = 0.000554$ mol. Determine the value for E_1.

E_5 at 2510°R. and 14.7 lb. in.$^{-2}$ abs. (Chart E, Appendix) = 475 B.t.u.
E_m at 520°R. and 14.7 lb. in.$^{-2}$ abs. (Chart B, Appendix) = 0 B.t.u.

Also, $\quad f = \dfrac{M_c}{M_p} = \dfrac{1}{21.9} = 0.0457.$

Then, Eq. (21), corrected for the internal latent heat of the fuel, becomes

$$E_1 = (1 - f)E_m + fE_5 = 0.9543 \times 0 - 0.000554 \times 16{,}700 + 0.0457 \times 475 = 13 \text{ B.t.u.}$$

This indicates a temperature of about 585°R. (Chart B, Appendix), which is 15° below the original assumption.

The Release Process.—The release process occurs with the piston assumed to be stationary at the end of the expansion stroke. The gases escaping from the cylinder undergo a *free expansion* or irreversible process, whereas the gases remaining in the cylinder are assumed to expand adiabatically and reversibly to atmospheric pressure before the piston begins the exhaust stroke.

Any small part of the gases, ΔM, occupying a volume of ΔV (Fig. 33) has energy $E + C$ at temperature T, immediately before escaping from the cylinder. The remainder of the gases in the cylinder expand and do work amounting to $\int P\,dV$. After escaping into the exhaust pipe, the element ΔM will have energy $E_5 + C_5$ and will do flow work $P_5 V_5$ in pushing other gases along

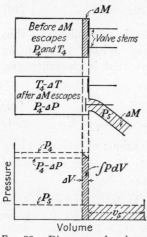

FIG. 33.—Diagram of release process.

the exhaust pipe. In equating the energies for the small element ΔM, it is assumed that all the kinetic energy acquired by the element in escaping from the cylinder is dissipated by friction into internal energy of the same element and, also, that no heat transfer occurs. Thus, the energy equation for the process for any element is

$$C + E + \int P\,dV = C_5 + E_5 + P_5 V_5. \qquad (22)$$

As the element ΔM approaches zero, the work area $\int P\,dV$ (Fig. 33) approaches a rectangle of height P and width V which is the volume of the element. At the limit, Eq. (22) for any element becomes

$$C + E + PV = C_5 + E_5 + P_5 V_5. \qquad (23)$$
$$\text{or} \qquad C + H = C_5 + H_5. \qquad (24)$$

Since the temperatures dealt with in this process are equal to or lower than T_4, $C = C_5$ (Charts D, E, and F). Thus Eq. (24) indicates that, for any infinitely small element, the enthalpy and temperature in the exhaust pipe will be the same as the enthalpy and temperature the instant before escaping from the cylinder. Thus, in Fig. 34, the element escaping at pressure P in the cylinder will have its condition changed from A to B, all other elements having conditions indicated by the P_5 line from the "first" to the "last" element. In using the combustion charts for this process, it should be noted that T, $H + C$, and $E + C$ lines are parallel only at temperatures low enough to make dissociation negligible.

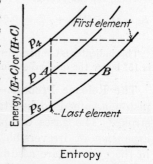

FIG. 34.—Energy-entropy diagram of release process.

Example.—The conditions of the gases in an engine cylinder at the end of the expansion process are $P_4 = 98$ lb. in.$^{-2}$ abs., $V_4 = 15.5$ ft.^{3}, and $T_4 = 3640°$R. (example, page 73). Determine the temperature-mass relationship for the gases escaping from as well as those remaining in the cylinder during the release process.

Pr. in cyl., lb. in.$^{-2}$ abs.,	98	75	50	25	14.7
T in cyl., °R, (Chart E, Appendix),	3640	3470	3220	2810	2510
$M_{cyl} = \dfrac{PV}{RT} = \dfrac{144 \times P \times 15.5}{1545 \times T}$ mols,	0.0389	0.0313	0.0225	0.0129	0.0085
M_{exh},	0	0.0076	0.0164	0.0260	0.0304
Percentage in exhaust pipe,	0	19.6	42.2	66.8	78.1

These results (Fig. 35) show the theoretical variation of temperature of the exhaust products. The first element to be released is at 3640°R. and

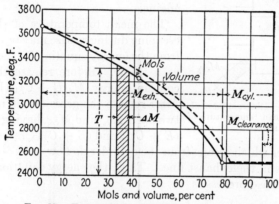

Fig. 35.—Temperature variation of exhaust gases.

the last at 2510°R., which is the temperature of the gases to be expelled from the cylinder during the exhaust stroke as well as of the gases to remain in the clearance space.

The foregoing analysis shows the variation of exhaust-gas temperature with regard to mass of gas released. The variation of exhaust-gas temperature with position in the exhaust pipe depends upon the volume occupied by each small element of gases released. Thus, for any element ΔM, having volume ΔV in the exhaust line at pressure P_{exh} and temperature T,

$$P_{exh} \Delta V = \Delta MRT, \tag{25}$$

or

$$\Delta V = \frac{R}{P_{exh}} T \Delta M, \tag{26}$$

from which

$$\int_0^V dV = \frac{R}{P_{exh}} \int_0^M T \, dM. \tag{27}$$

The term $\int T\,dM$ is the area beneath the mols curve (Fig. 35) down to $T = 0°R.$, from zero mol to the value desired. This integration resulted in the curve labeled "volume." Since one release and exhaust follows another, a series of such graphs would represent the variation in temperature as the exhaust gases flow past a given point in the exhaust pipe. Heat transfer between adjacent elements of exhaust gases as well as through the exhaust pipe, dissipation of kinetic energy of any given element into internal energy of adjacent elements, and overlapping of exhaust processes would tend to reduce the temperature variation encountered in actual practice with a multicylinder engine.

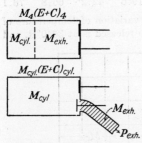

Fig. 36.—Diagram for analysis for mean temperature of gases released.

The mean temperature of the gases escaping during release may be obtained from the energy equation for the process as a whole. Before release the gases contain energy $M_4(E + C)_4$ (Fig. 36). After release the gases are divided into two parts M_{cyl} and M_{exh}, having energies $(E + C)_{cyl}$ and $(E + C)_{exh}$, respectively. The gases which escape from the cylinder do flow work $(PV)_{exh}$ on the gases in the exhaust line. Thus, the energy equation for the entire release process is

$$M_4(E + C)_4 = M_{cyl}(E + C)_{cyl} + M_{exh}(E + C + PV)_{exh}, \quad (28)$$
$$\text{or} \quad M_{exh}(H + C)_{exh} = M_4(E + C)_4 - M_{cyl}(E + C)_{cyl}. \quad (29)$$

The volume of the cylinder is determined from the amount and condition of the medium in the cylinder before release. Reversible adiabatic expansion fixes the condition of the gases remaining in the cylinder after release, and determines the mols of gas left in the cylinder as well as those released.

Example.—The $E + C$ energy in the gases at the end of the expansion process is 810 B.t.u. (example, page 73), and the volume is 15.5 ft.[3] At the end of the release process, the proportion of the gases left in the cylinder will be the ratio of the volume at the end of the expansion process and the volume, V_5, indicated if the expansion process had continued to exhaust pressure. Determine the mean temperature of the gases released.

From Chart E (Appendix), V_5 is 68 ft.[3], and the part left in the cylinder will be

$$\frac{15.5}{68} = 0.228 = \frac{M_{cyl}}{M_5}.$$

From Chart E (Appendix), $M_{cyl}(E+C)_{cyl} = 0.228 \times 475 = 108$ B.t.u.

Substituting in Eq. (29) gives

$$M_{exh}(H+C)_{exh} = 810 - 108 = 702 \text{ B.t.u.}$$

$$(H+C)_{exh} = \frac{702}{1 - 0.228} = 909 \text{ B.t.u.}$$

From Chart E (Appendix), at $H+C = 909$ B.t.u., $T_{exh} = 3210°$R., which is the mean temperature of the gases after escaping from the cylinder during the release process.

The Exhaust Process.—The piston pushes a displacement volume of gases out of the cylinder during the exhaust stroke, which, in the ideal case, occurs after the pressure in the cylinder has decreased to exhaust pressure. This process merely displaces the exhaust gases along the exhaust pipe, theoretically without changing the condition of the gas.

The mean condition of all the exhaust gases, including the part expelled during release is determined by applying the energy equation to the medium before release and after exhaust. Before release the gases contain energy amounting to $M_4(E+C)_4$. After exhaust the energy in the clearance gases will be $M_c(E+C)_c$, and in the exhausted gases will be $(M_4 - M_c)(E+C)_{exh}$. The exhaust gases will do work on gases previously exhausted in pushing them along the exhaust line. This amounts to

$$(M_4 - M_c)(PV)_{exh}.$$

The piston will have done work $P_{exh} \times$ displacement, in displacing the gases from the cylinder during the exhaust stroke. The energy equation for the process is

$$M_4(E+C)_4 + P_{exh} \text{ disp.} = M_c(E+C)_c + (M_4 - M_c)(E+C)_{exh} + (M_4 - M_c)(PV)_{exh}. \quad (30)$$

Combining E_{exh} and $(PV)_{exh}$ into H_{exh} and solving for $(H+C)_{exh}$ result in

$$(H+C)_{exh} = \frac{M_4(E+C)_4 + P_{exh} \text{ disp.} - M_c(E+C)_c}{M_4 - M_c}. \quad (31)$$

The mols of clearance gases decrease as the compression ratio increases, and Eq. (31) approaches the relationship usually set up for the ideal steam-engine process with zero clearance volume. Also, H_{exh} is equivalent to $\Sigma(H \Delta M)$, for the exhaust gases released, added to the H for the portion of gases displaced by the piston during the exhaust stroke.

Example.—Determine T_{exh} and H_{exh} for the conditions (example, page 79) $T_4 = 3640°$R., $V_4 = 15.5$ ft.³, and $r = 5:1$.

From the example on page 80, there was $0.228 M_4$ mol of gases remaining in the cylinder after release.

Then, $$M_c = \frac{0.228 M_4}{5} = 0.0456 M_4.$$

From Chart E (Appendix), $M_4(E + C)_4 = 810$ B.t.u.

$$P_{exh} \text{ disp.} = 14.7 \times 144 \times 15.5 \times \frac{5-1}{5} \div 778 = 34 \text{ B.t.u.}$$

From Chart E, $M_c(E + C)_c = 0.0456 \times 475 = 22$ B.t.u.

Then, from Eq. (31), $$(H + C)_{exh} = \frac{810 + 34 - 22}{1 - 0.0456} = 861 \text{ B.t.u.}$$

From Chart E, $T_{exh} = 3070°$R. and the entropy of the exhaust products has increased from 0.533 B.t.u. °R.⁻¹ at 3640°R. to 0.607 B.t.u. °R.⁻¹ at 3070°R. The latter values represent the condition of the exhaust gases if thermal equilibrium would be attained without heat transfer into or out of the gases during the process.

Summary of the Otto-engine Process.—The work done on the engine piston during the suction process is equal to the work done on the exhaust gases by the piston during the exhaust stroke in the ideal case with equal suction and exhaust pressures. Thus, the work values for only the compression, combustion, and expansion processes are required for evaluating the work of the ideal unsupercharged engine process.

The efficiency of the engine process is the ratio of the net work obtained to the energy supplied. If chemical energy were known as an absolute value, it should be used as the energy supplied the internal-combustion-engine process. However, at present, chemical energy values are based upon relative internal energy values. Consequently, it is desirable to use some absolute value, such as the heating value of the fuel at constant pressure, and some standard temperature, such as 77°F., for the energy supplied.[1]

[1] Goodenough and Baker, *op. cit.*, showed that the sum of the work of the engine process and the heat rejected in cooling the exhaust gases to

Example.—From the examples on pages 71, and 73, the work values for compression, combustion, and expansion are 90, 0, and 516 B.t.u., respectively, per 0.0335 mol of inducted charge. Determine the ideal thermal efficiency.

From Table V (Appendix), the heating value of liquid octane at standard conditions is

$$\frac{2{,}368{,}089 - 17{,}730}{60.5} \times 0.0335 = 1301 \text{ B.t.u.}$$

for the fuel in the inducted charge.

$$\text{Ideal Otto-engine eff.} = \frac{516 - 90}{1301} = 0.327.$$

Ideal efficiencies obtained by the foregoing method will be the theoretical maximum values which in practice may be approached but never attained.

IDEAL DIESEL-ENGINE PROCESS
Constant-pressure Combustion

The Diesel engine compresses air and clearance gases and then injects liquid fuel into the highly compressed gases. Combustion is initiated by the high temperature of the compressed gases, and the fuel is assumed to burn at constant pressure (Fig. 37) in the ideal process.

The compression process is solved in the same manner as in the previous analysis (page 69). The internal energy in the compressed gases at the end of the compression process is $(M_aE_a + M_cE_c)_2$, where subscripts a and c represent air and clearance gases, respectively.

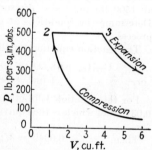

Fig. 37.—Pressure-volume diagram of Diesel-engine constant-pressure combustion process.

The internal and chemical energy of the liquid fuel injected are M_fE_f and M_fC_f at the fuel-supply conditions. The work done in injecting the fuel is approximately PV_f, where P is the injection pressure and V_f is the volume of fuel injected.

---

suction temperature was equal to the constant-pressure heat of combustion and that this was the heat supplied. However, the sum of the work and heat rejected could be the heat supplied only if the internal-combustion-engine process were a cycle, which it is not (see also page 68).

After constant-pressure combustion the products will have internal energy $M_p E_p$ and chemical energy $M_f C_f$, where subscript f now indicates the fuel constituents in the products of reaction at chemical equilibrium. During the combustion process the gases expand and do work on the engine piston amounting to $P_2(V_3 - V_2)$ (Fig. 37). Thus, the energy equation for this combustion process becomes

$$(M_a E_a + M_c E_c)_2 + M_f(E + C)_f + PV_f = (M_p E_p + M_f C_f)_3 + P_2(V_3 - V_2). \quad (32)$$

The left side of Eq. (32) is evaluated from the given conditions. P_2 is known from the compression process. An assumption of V_3 fixes the term $P_2(V_3 - V_2)$ and permits evaluation of the term $(M_p E_p + M_f C_f)_3$. The condition at the end of the combustion process is determined on one of the combustion charts by the pressure and energy term.

Example.—The temperature and pressure of the air and clearance gases at the end of the compression process in a Diesel engine are 1400°R. and 500 lb. in.$^{-2}$ abs., respectively. The ratio of air to clearance gases is $M_a/M_c = 74:1$. The required amount of liquid fuel ($C_{12}H_{26}$) is injected at 1500 lb. in.$^{-2}$ abs. pressure, the fuel-supply temperature being 60°F. Determine the conditions at the end of the constant-pressure combustion process.

The reaction equation for this process is

$$C_{12}H_{26} + 18.5O_2 + 69.56N_2 = 12CO_2 + 13H_2O + 69.56N_2$$

or   1 fuel + 88.06 air = 94.56 products.

For this reaction 1 lb. of air and the required amount of fuel amount to 0.0349 mol which is the equivalent mixture.

Then, $0.0349 = M_m + M_c \dfrac{89.06}{94.56}$

$$= M_m \left(1 + \dfrac{88.06 \div 74}{89.06} \times \dfrac{89.06}{94.56}\right)$$

$M_m = 0.0345$ mol of correct mixture.

$M_f = 0.0345 \div 89.06 = 0.000387$ mol of fuel.

$M_c = 0.0345 \times \dfrac{88.06}{89.06} \div 74 = 0.000461$ mol of clearance gases.

From Table III (Appendix),

$(M_a E_a)_{1400} = 0.0341 \times 4587 = 156.4$ B.t.u.
$(M_c E_c)_{1400} = 0.000461 \times 5107* = \underline{2.4}$ B.t.u.
   Total energy above 520°R. = 158.8 B.t.u.

* The value for the products for the C_8H_{18} reaction is used. It differs only slightly from the correct value.

INTERNAL-COMBUSTION-ENGINE-PROCESS ANALYSIS 85

This is appreciably lower than 192 B.t.u. indicated on Chart B (Appendix), which includes the energy of the vaporized octane.

At 520°R. (Tables III and V, Appendix),

$$M_f E_f = M_f(E_{gas} - E_{fg}) = 0.000387(0 - 21{,}930^*) = -8.5 \text{ B.t.u.}$$

The chemical energy of the fuel when reacting with O_2 is (Table V, Appendix),

$$M_f C_f = 0.000387 \times 3{,}258{,}667 = 1261 \text{ B.t.u.}$$

The work of injecting the fuel amounts to

$$PV_f = \frac{1500 \times 144 \times 0.000387 \times 3\dagger}{778} = 0.3 \text{ B.t.u.}$$

The left side of Eq. (32) amounts to

$$(M_a E_a + M_c E_c)_2 + M_f(E + C)_f + PV_f = 158.8 - 8.5 + 1261 + 0.3 = 1410.8 \text{ B.t.u.}$$

which is also the value for the right side of Eq. (32).

$$V_2 = \frac{M_2 R T_2}{P_2} = \frac{(0.0341 + 0.0005)1545 \times 1400}{500 \times 144} = 1.039 \text{ ft.}^3$$

Assume V_3, ft.^{3},	2.0	3.0	4.0
$P_2(V_3 - V_2)$, B.t.u.,	89	182	274
$(M_p E_p + M_f C_f)_3$ from Eq. (32),	1322	1229	1137
T_3 at 500 lb. in.$^{-2}$ abs., Chart E (Appendix),	4900	4700	4520
V_3 at 500 lb. in.$^{-2}$ abs., Chart E (Appendix),	4.00	3.80	3.65

The assumed and determined volumes are the same at 3.68 ft.^{3}, with values of $T_3 = 4590°\text{R.}$, and $P_2(V_3 - V_2) = 245$ B.t.u.

Also,
$$V_3 = \frac{M_3 R T_3}{P_3} = \frac{\left[(M_a + M_f)\dfrac{M_p}{M_m} + M_c \dfrac{M_p}{M_{p_c}}\right] R T_3}{P_3} = 3.68$$

$$= \frac{\left(0.0345 \dfrac{M_p}{89.06} + 0.0005 \dfrac{M_p}{94.56}\right) 1545 \times 4590}{500 \times 144} = 3.68$$

$$M_p = 95.14 \text{ mols at } T_3 \text{ and } P_3.$$

This is a fairly good check on the computations since M_p should be less than 1 per cent greater than M_{p_c},‡ the mols of products for complete combustion.

---

* Approximate value of internal latent heat per mol of fuel.
† Approximate volume per mol of liquid fuel.
‡ See Lichty, *op. cit.*, p. 276, Chart F.

The other parts of the Diesel-engine process are solved in the same manner as indicated in the analysis of the Otto-engine process.

IDEAL DIESEL-ENGINE PROCESS

Limited-pressure Combustion

The combustion in this process is assumed to occur first at constant volume until the limiting pressure is attained and then at constant pressure to the end of the combustion process. The first part of the process occurs with appreciable excess air, and the temperature attained will probably not be high enough to result in appreciable dissociation in the ideal case. However, the temperature at the end of the constant-pressure combustion process even with appreciable excess air will result in dissociation, and a solution for the final condition may be made by interpolating between or extrapolating beyond the combustion charts for excess air.

The solution for the condition at the end of the constant-volume part of the combustion process may be omitted.

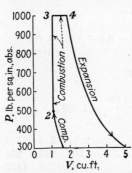

FIG. 38.—Pressure-volume diagram of Diesel-engine limited-pressure combustion process.

Example.—Assume a 20 per cent excess of air for the mixture of $C_{12}H_{26}$ and air in a limited-pressure Diesel-engine process in which part of the fuel is burned at constant volume and part at a constant pressure of 1000 lb. in.$^{-2}$ abs. The conditions at the end of compression are (Fig. 38) $P_2 = 500$ lb. in.$^{-2}$ abs., $T_2 = 1400°$R., and $V_2 = 1.039$ ft.^{3} Also, $M_a = 0.0341$, $M_f = 0.000387 \div 1.20 = 0.000322$ mol, and $M_c = 0.000461$ mol.* The internal energy at the beginning of the combustion process is 158.8 B.t.u. (see example, page 84). Determine the conditions at the end of combustion.

The chemical energy is $1261 \div 1.20 = 1051$ B.t.u.

$$M_f E_f = -8.9 \div 1.20 = -7.4 \text{ B.t.u.},$$
and $$PV_f = 0.3 \div 1.2 = 0.25 \text{ B.t.u.}$$

The left side of Eq. (32) amounts to

$$(M_a E_a + M_c E_c)_2 + M_f(E + C)_f + PV_f = 158.8 - 7.4 + 1051 + 0.3 = 1202.7 \text{ B.t.u.}$$

* The effect of the excess air on the clearance gases is neglected.

Assume V_4, ft.³ (Fig. 38),   1.5   2.0   2.5
$P_3(V_4 - V_3)$, B.t.u.,   85   178   271
$(M_p E_p + M_f C_f)_4 = 1203 - P_3(V_4 - V_3)$,   1118   1025   932
T_4, °R. (Chart E, Appendix),   4500   4270   4030
V_4, ft.³ (Chart E, Appendix),   1.83   1.73   1.63
T_4, °R. (Chart D, Appendix),   4600   4350   4070
V_4, ft.³ (Chart D, Appendix),   1.82   1.72   1.63
From Chart E data, $V_4 = 1.79$ ft.³, and $T_4 = 4370$
From Chart D data, $V_4 = 1.78$ ft.³, and $T_4 = 4460$

Extrapolating these data to 20 per cent excess air results in $V_4 = 1.77$ ft.³, and $T_4 = 4550$°R.

Also,
$$V_4 = \frac{M_4 R T_4}{P_4} = \frac{\left[(M_a + M_f)\frac{M_p}{106.67} + M_c \frac{M_p}{112.17}\right] R T_4}{P_4} = 1.77 \text{ ft.}^3$$

$$= \frac{\left[(0.0341 + 0.000322)\frac{M_p}{106.67} + 0.000461 \frac{M_p}{112.17}\right] 1545 \times 4550}{1000 \times 144} = 1.77$$

$M_p = 110.9$ mols mol⁻¹ of fuel.

This should be slightly higher than 112.17 mols which are the mols of products of complete combustion per mol of fuel for a mixture with 20 per cent excess air. This indicates a total error of 1 to 2 per cent in the values for T_4 and V_4 obtained in this manner.

The other parts of the limited-pressure Diesel-engine process may be solved in the same manner as indicated in the analysis of the Otto-engine process.

GENERAL ANALYSIS

Energy Distribution.—The analysis of the entire engine process, either ideal or actual, may be made by placing an imaginary envelope (Fig. 39) around the engine and setting up the energy equation for the energy entering and leaving the envelope. This results in

$M_a H_a + M_f(H + C)_f = W_{out} + Q_{out} + M_{exh}(H + C)_{exh}$ (33)

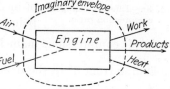

FIG. 39.—Flow diagram of engine process.

which is the steady-flow energy equation for the process. In the ideal case, heat transfer is zero and the work is computed as in foregoing ideal analyses. In the actual case, friction merely changes the distribution of energy between the three terms on the right side of Eq. (33), friction being a process and not an energy term.

Example.—One mol of correct air-fuel mixture, with air and liquid fuel at 60°F., flows into an engine; determine the work of the ideal engine process from data obtained in the examples on pages 71 and 73.

From page 76, $M_a H_a = 1016$ B.t.u.
$M_f H_f = -277$ B.t.u.
Also, $M_f C_f = 2,201,618 \div 60.5 = 36,390$ B.t.u.
From page 82, $T_{exh} = 3070°$R.

and from Table III (Appendix),

$$M_{exh} H_{exh} = 23,040 \times \frac{64}{60.5} = 24,373 \text{ B.t.u.}$$

Neglecting C_{exh} results in

Work $= 1016 - 277 + 36,390 - 24,373 = 12,754$ B.t.u.
For 0.0335 mol, Work $= 0.0335 \times 12,754 = 427$ B.t.u.

compared with 426 B.t.u. from the example on page 83.

Effect of Compression Ratio on Efficiency.—The work and efficiency of any of the ideal engine processes depend upon the difference between the total energy of the media in the cylinder at the beginning of the compression process and the energy of the media at the end of the expansion process. This difference represents the maximum work attainable with the process. An increase in compression ratio for the Otto-engine process (Fig. 40), without changing the nature or quantity of media, increases the internal energy at point 2 to that at 2' by the increase in work of compression, namely, $\int_{2'}^{2} P\, dV$. The work done in expanding from 3' to the original clearance volume is $\int_{3'}^{3''} P\, dV$. Applying the energy equation to the process 2-2'-3'-3'', results in

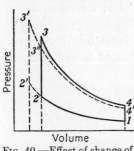

FIG. 40.—Effect of change of compression ratio.

$$(E + C)_2 + \int_{2'}^{2} P\, dV = (E + C)_{3''} + \int_{3'}^{3''} P\, dV. \quad (1)$$

For the process 2-3, $(E + C)_2 = (E + C)_3.$ (2)

But, $\int_{3'}^{3''} P\, dV > \int_{2'}^{2} P\, dV;$ (3)

then, $(E + C)_{3''} < (E + C)_2$ and $< (E + C)_3.$ (4)

This indicates that point 3″ is at a lower temperature and consequently at a lower pressure than point 3. Since both points are on reversible adiabatic paths, it is obvious that 4′ is at a lower temperature and pressure than 4 and consequently

$$(E + C)_{4'} < (E + C)_4. \qquad (5)$$

Thus, increasing the compression ratio decreases $(E + C)_4$ and increases the work and efficiency of the process.

The foregoing is based upon the same mixture of charge and clearance gases for each compression ratio. An increase in compression ratio actually decreases the ratio of quantity of clearance gases to inducted charge, and even for the same total volume the relative pressures for the two diagrams would not indicate relative temperatures. Decreasing the compression ratio toward 1:1 actually decreases the inducted charge toward zero as a limit and causes P_4 to approach P_1 as a limit. Increasing the compression ratio toward infinity decreases the clearance gases toward zero as a limit, and owing to infinite expansion causes P_4 again to approach P_1 as a limit.

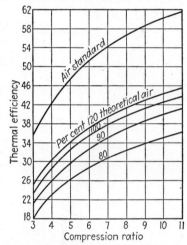

FIG. 41.—Ideal thermal efficiencies —Otto-engine process. (Based on heating value for liquid octane.)

This is due to the effect of increased compression ratio on the total energy at point 4. At low compression ratios the increase in charge quantity with increase in compression ratio more than offsets the lowering effect of increased compression ratio on P_4, whereas at high compression ratios the reverse is true. The actual diagram which takes into account the effect of compression ratio on ratio of clearance gases to inducted charge will show that P_4 (Figs. 40 and 43) passes through a maximum as the compression ratio is varied.[1] However, the energy in the media at point 4 per unit quantity of inducted charge decreases and the work of the process increases with an increase in compression ratio as indicated in the previous analysis.

[1] This was shown in the computations of Goodenough and Baker, *op. cit.*

The same method of analysis may be applied to any of the other types of engine processes to show that an increase in compression ratio increases the efficiency. In all cases, it is the liberation of chemical energy at a condition which permits the largest possible expansion of the medium thereafter that results in the highest efficiency.

The rate of increase in efficiency with an increase in compression ratio decreases as the compression ratio is increased (Figs. 41 and 42).

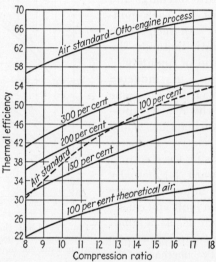

Fig. 42.—Ideal thermal efficiencies—Diesel-engine process—constant-pressure combustion. (Based on heating value for liquid dodecane.)

Effect of Air-fuel Ratio and Dilution.—The dilution of a correct air-fuel mixture with air, fuel, or clearance gases has an appreciable effect on the ideal engine process. The addition of air or clearance gases to a given air-fuel mixture decreases the temperature rise during the combustion process. However, the temperature and pressure rise per unit of energy supplied are increased since the specific heats are lower for lower temperatures. This results in larger work areas on the P-V diagram per unit of energy supplied, and consequently higher efficiencies are obtained. As dilution with air increases, the hypothetical air-standard efficiency is approached (Figs. 41 and 42).

The addition of fuel to a correct air-fuel mixture results in unliberated chemical energy which is charged against the process in the energy supplied. This excess fuel does not increase the work in proportion to the increase in fuel, and the ideal efficiency decreases as the mixture is made richer.

Increasing the compression ratio decreases the clearance-gas dilution effect on the inducted charge which tends to decrease the efficiency. This effect is more than offset by the effect of increased compression ratio on efficiency.

Mean Effective Pressure.—The amount of charge inducted and the work done in an engine process with a given compression ratio depend principally upon piston displacement. At very

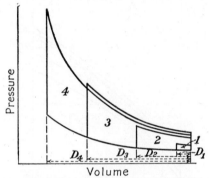

Fig. 43.—Effect of compression ratio on ideal work area.

low compression ratios only a small amount of charge would be inducted compared with the clearance gases, and a very small work area and low mean effective pressure (m.e.p.) would result (Fig. 43). Higher compression ratios increase the work area and m.e.p.

The m.e.p. is determined by dividing the net work by the displacement.

Thus,
$$\text{M.e.p.} = \frac{\text{Net work}}{\text{Disp.}}. \tag{1}$$

Example.—Determine the m.e.p. for the example, page 83. The net work per 0.0335 mol of inducted charge is 426 B.t.u. The clearance gases amount to 5.05 per cent of the inducted charge. Conditions at the end of the suction stroke are 590°R., and $P = 14.7$ lb. in.$^{-2}$ abs.

$$\text{Disp.} = \frac{r-1}{r}V_1 = \frac{5-1}{5}\left(\frac{1.0505 \times 0.0335 \times 1545 \times 590}{14.7 \times 144}\right) = 12.12 \text{ ft.}^3$$

$$\text{M.e.p.} = \frac{\text{Net work}}{\text{Disp.}} = \frac{426 \times 778}{12.12 \times 144} = 190 \text{ lb. in.}^{-2}$$

From the examples on pages 70 and 83.

$$\text{M.e.p.} = \frac{426 \times 778}{(15.5 - 3.1)144} = 186 \text{ lb. in.}^{-2},$$

which is about 2 per cent lower than the previous result.

The effect of compression ratio on m.e.p. for both the Otto- and Diesel-engine processes, with various air-fuel ratios, is shown in Figs. 44 and 45.

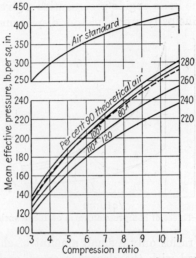

FIG. 44.—Ideal mean effective pressures—Otto-engine process.

Since work depends upon energy supplied and thermal efficiency,

$$\text{M.e.p.} = \frac{\text{Work}}{\text{Disp.}} = \frac{Q_p \times \text{disp.} \times \text{vol. eff.} \times \text{thermal eff.}}{\text{Disp.}} \quad (2)$$

or   M.e.p. = const. × vol. eff. × thermal eff.   (3)

where Q_p = the energy supplied per cubic foot of mixture, or air in the case of liquid fuel, at atmospheric conditions. Thus, m.e.p. for a given mixture should increase in direct proportion with thermal efficiency if volumetric efficiency remains constant.

Volumetric Efficiency.—The volumetric efficiency is defined as the ratio of the volume of the charge inducted, at the atmospheric conditions surrounding the engine, to the piston displacement. Only the volume of the air inducted is considered when using liquid fuel.

In the ideal analysis it is assumed that no fluid friction occurs during the suction stroke and that consequently the mixture is inducted at atmospheric pressure. It is also assumed that no

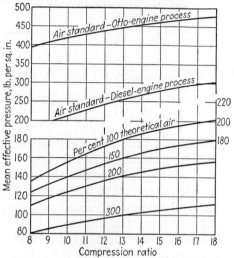

FIG. 45.—Ideal mean effective pressures—Diesel-engine process—constant-pressure combustion.

heat transfer occurs between the engine parts and the mixture or clearance gases. Thus, only the mixing of the inducted charge with the clearance gases can have an effect on volumetric efficiency.

It may be assumed for 100 per cent volumetric efficiency that the inducted air or air-gas charge, at atmospheric conditions, is in the cylinder and occupying the displacement volume without having mixed with the clearance gases (Fig. 46). The heat that will be transferred from the clearance gases is the heat transferred to the inducted charge.

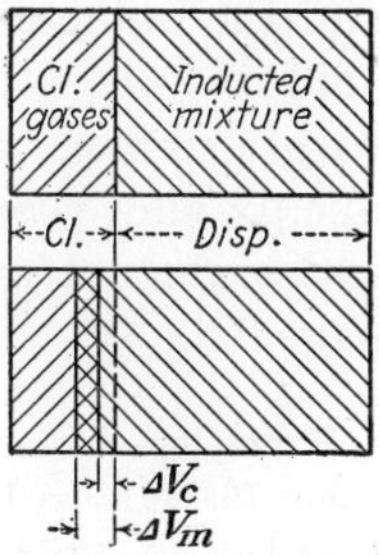

Assuming a constant-volume or constant-pressure process for the heat transfer (see page 76), results in

FIG. 46.—Effect of mixing inducted mixture and clearance gases.

$$Q = M_c c_c \, \Delta T_c = M_m c_m \, \Delta T_m \tag{1}$$

where subscripts c and m indicate clearance gases and inducted mixture, respectively.

The changes in temperature will result in a contraction of volume (or pressure) of the clearance gases and an increase in volume (or pressure) of the inducted mixture. These changes in volume for the constant-pressure process are

$$\Delta V_c = \frac{M_c R \, \Delta T_c}{P}, \quad \text{and} \quad \Delta V_m = \frac{M_m R \, \Delta T_m}{P}. \quad (2)$$

Eliminating ΔT between Eqs. (1) and (2), results in

$$\Delta V_c = \frac{M_c R}{P} \frac{Q}{M_c c_c} = \frac{KQ}{c_c}; \quad \text{also,} \quad \Delta V_m = \frac{KQ}{c_m}. \quad (3)$$

For any given air-fuel ratio and compression ratio, the heat transferred Q will be fixed and ΔV will depend on the specific heat in each case. Since c_c is greater than c_m for the correct vaporized octane-air mixture, ΔV_c is less than ΔV_m, and all the air and a proportional amount of vapor could not remain in the cylinder without increasing the pressure. This indicates ideal volumetric efficiencies lower than 100 per cent for vaporized mixtures. The effect of liquid fuel in the cylinder is to require more heat for a given temperature rise of the inducted charge, resulting in an increase of volumetric efficiency.

An increase in compression ratio reduces the amount of the clearance gases and the heat transferred to the inducted charge. It also reduces the mean temperature of the clearance gases and the inducted charge and thus lowers the mean specific heat values. The net result is a small change in volumetric efficiency with a change in compression ratio. This indicates that ideal m.e.p. and thermal-efficiency values at various compression ratios are not exactly proportional to each other [see Eq. (3), page 92].

Example.—Determine the ideal volumetric efficiency for the example on page 76. $M_m = 1$ mol; $M_c = 0.0505$ mol; $T_a = 520°$R.; $T_1 = 590°$R.; $P = 14.7$ lb. in.$^{-2}$ abs., $T_c = 2510°$R., and $r = 5:1$. The fuel is liquid octane at atmospheric conditions.

$$\text{Disp.} = \frac{r-1}{r} V_1 = \frac{r-1}{r} \times \frac{M_1 R T_1}{P_1}$$

$$= \frac{5-1}{5}\left(\frac{1.0505 \times 1545 \times 590}{14.7 \times 144}\right) = 361.9 \text{ ft.}^3$$

INTERNAL-COMBUSTION-ENGINE-PROCESS ANALYSIS

The volume of inducted air at atmospheric conditions is

$$V_a = \frac{M_a R T_a}{P_a} = \frac{59.5}{60.5}\left(\frac{1545 \times 520}{14.7 \times 144}\right) = 373.3 \text{ ft.}^3$$

$$\text{Vol. eff.} = \frac{373.3}{361.9} = 1.032 \text{ or } 103.2 \text{ per cent.}$$

This indicates the clearance gases contract in volume an amount equal to the difference between the displacement and V_a, in addition to the amounts equal to the expansion of the air upon being heated and the increase in volume of the fuel vaporizing and rising to suction temperature.

Thus, $\qquad V_a - \text{disp.} = 11.4 \text{ ft.}^3$

$$\Delta V_{air} = \frac{M_a R \, \Delta T_a}{P_a} = \frac{59.5}{60.5}\left[\frac{1545(590 - 520)}{14.7 \times 144}\right] = 50.2 \text{ ft.}^3$$

$$V_{fuel} = \frac{M_f R T_1}{P_1} = \frac{1}{60.5}\left(\frac{1545 \times 590}{14.7 \times 144}\right) = 7.1 \text{ ft.}^3$$

Neglecting the liquid-fuel volume, results in

$$\Delta V_c = 11.4 + 50.2 + 7.1 = 68.7 \text{ ft.}^3$$

Also, $\quad \Delta V_c = \dfrac{M_c R \, \Delta T_c}{P} = \dfrac{0.0505 \times 1545(2510 - 590)}{14.7 \times 144} = 70.7 \text{ ft.}^3$

which indicates T_1 should be about 593°R.

EXERCISES[1]

1. Using liquid C_8H_{18} and air at 60°F. and 14.7 lb. in.$^{-2}$ abs. pressure, determine the thermal efficiency, m.e.p., and volumetric efficiency for the ideal Otto-engine process for compression ratios from 5:1 to 10:1, with air-fuel mixtures having 85, 100, and 110 per cent of theoretical air requirement.

2. Plot the results obtained in Prob. 1, using both compression ratio and air-fuel ratio as the abscissae.

3. Using liquid $C_{12}H_{26}$ and air at 60°F. and 14.7 lb. in.$^{-2}$ abs. pressure, determine the thermal efficiency and m.e.p. for the ideal Diesel-engine process with constant-pressure combustion for compression ratios from 8:1 to 18:1. Use air-fuel ratios having 100, 150, 200, and 300 per cent of the theoretical air requirement.

4. Plot the results obtained in Prob. 3, using both compression ratio and air-fuel ratio as the abscissae.

5. Using liquid $C_{12}H_{26}$ and air at 60°F. and 14.7 lb. in.$^{-2}$ abs. pressure, determine the part burned at constant volume, the thermal efficiency, and the m.e.p. for the Diesel-engine process with limited combustion pressure of

[1] It is suggested that each exercise be broken into various parts with one compression ratio, air-fuel ratio, etc., and that each student be given one of these parts.

1000 lb. in.$^{-2}$ abs. in all cases. Use air-fuel ratios having 100, 150, 200, and 300 per cent of the theoretical air requirement. The compression ratios should cover the range requiring all to none of the fuel being burned at constant volume to attain the limiting combustion pressure.

6. Plot the results obtained in Prob. 5, using both compression ratio and air-fuel ratio as the abscissae.

7. Solve Prob. 1 with the engine supercharged or throttled so that the pressure at the end of the suction stroke is 29.4, 22, 10, 7, 5, 4, etc., lb. in.$^{-2}$ abs., the lowest suction pressure being that which makes the net work equal to zero. Use only the correct air-fuel ratio, and 5:1 and 10:1 compression ratios. See Fig. 47.

8. Plot the results obtained in Prob. 7, using suction pressure as the abscissa.

CHAPTER VI

DEVIATIONS FROM IDEAL PROCESSES

Ideal to Actual Process.—Any attempt to use an ideal process in an actual engine results in a performance appreciably less than that indicated by the ideal analysis. The lowered performance is caused by losses inherent in the transition from ideal to actual process and by losses due to engineering compromise necessary to attain the desired performance. Any attempt to analyze the losses on a basis of the hypothetical air-standard cycle results in a large amount of unaccounted-for loss. However, all the losses can be accounted for when the ideal analysis in Chap. V is used. This analysis includes a consideration of the actual mixture constituents, variable specific heat, and chemical equilibrium, all of which represent the discrepancy when the air-standard analysis is used.

The air-fuel ratio supplied the actual engine will vary from cycle to cycle and between the various cylinders of a multi-cylinder engine. The inducted mixture receives heat from the various hot surfaces with which it comes in contact. The pressure in the cylinder at the end of the suction stroke is usually less than atmospheric in the case of a normally aspirated engine, and particularly so at high speeds when large pressure drops are required to flow the charge into the cylinder in the short time available. Many engines are operated at part throttle which changes the exhaust-gas dilution and the pressure and temperature of the mixture in the cylinder. The intake valve closes appreciably after the end of the suction stroke (during the first part of the compression stroke) in order to obtain the desirable charging effect at high speeds.

Heat is usually transmitted from the cylinder walls to the mixture during the first part and is rejected to the walls during the latter part of the compression stroke. Any leakage of charge past the rings or valves is a total loss during either the compression or the expansion strokes.

The combustion process requires an appreciable amount of time during which the piston moves, heat is lost to the walls, and combustion will not be complete owing to chemical equilibrium, to poor mixing, and perhaps to the air-fuel ratio.. All these reduce the maximum pressure and temperature, and the efficiency of the actual process.

Heat loss and leakage occur during the expansion stroke. Combustion of some of the unburned and recombination of some of the dissociated constituents occur also during this stroke.

The exhaust valve opens considerably before the end of the expansion stroke. This reduces the pressure in the cylinder and the work of the exhaust stroke. The mean pressure during the exhaust stroke is usually above atmospheric. This increases the exhaust work and decreases the engine output.

The ratio of the actual indicated thermal efficiency of an engine to the efficiency of the ideal process, upon which the design of the engine is based, is a measure of the perfection of design and performance of the engine. This ratio is called the engine or *relative efficiency*.

Actual Mixture Conditions.—In the Otto-engine process, the air and fuel are brought into contact with each other in a carburetor or mixing chamber outside the engine. The liquid fuel must be vaporized or very finely atomized, and the particles of fuel must be so scattered throughout the mixture that sufficient particles of oxygen are close to each particle of fuel. It is practically impossible to accomplish this in the short time available. Consequently, it would be necessary to provide excess air to obtain complete combustion. However, it is common practice to use mixtures with excess fuel in liquid-fuel engines to obtain maximum power output.

The Diesel engine has the least time to accomplish thorough mixing of fuel and air, which ordinarily results in the use of appreciable excess air.

In general, a deficiency of fuel raises the thermal efficiency due to the effect of variable specific heat and lowers the power output due to less energy input. A deficiency of air lowers the thermal efficiency but increases the power output to a point beyond which any further deficiency of air reduces the power. The effect of various mixture ratios upon thermal efficiency and power output of the ideal process is dealt with in Chap. V.

DEVIATIONS FROM IDEAL PROCESSES

Effect of Throttling.—Throttling an engine increases the resistance to flow of the incoming charge and lowers the suction pressure. In the ideal case, the suction pressure is assumed to be atmospheric at wide-open throttle unless supercharged, and at some constant pressure below atmospheric for a partly closed throttle condition. The intake valve opens at the end of the exhaust stroke F (Fig. 47), and some of the clearance gases escape into the intake manifold. The pressure may be assumed to equalize at F' before the start of the suction stroke. The suction stroke reduces the pressure to the suction pressure as indicated by the dashed line $F'H$.

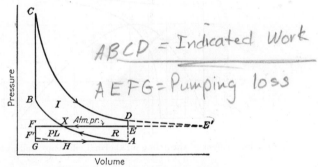

Fig. 47.—Ideal P-V diagram for throttled engine

The work done in exhausting the gases from the engine cylinder is equivalent to the area beneath the path EF, while the work done in inducting the charge is equivalent to the area beneath the path $F'HA$. The algebraic summation of these areas is termed the *pumping work*. The dashed path may be neglected, which makes the pumping work equivalent to the rectangular area $AEFG$.

The work area $ABCD$ represents the net work of the compression and expansion strokes and, when compared with the energy input, indicates any change in efficiency due to increased dilution and temperature changes caused by throttling. An ideal analysis, neglecting pump work, shows that this efficiency decreases slightly with throttling (Fig. 48). However, the ideal efficiency decreases rapidly with throttling when the pump work is included.

During compression from A to X the atmosphere does work on the piston equivalent to the area under EX, and the piston does work on the gases in the cylinder equivalent to the area under AX. Thus, the work area R is reclaimed from the atmosphere and the *pumping loss* amounts to the negative loop $XFGAX$, or the area PL. A decrease in suction pressure increases the height of the area PL but moves the intersection X to the left, which decreases the mean length of the area. Atmospheric suction pressure indicates zero pumping loss, and zero suction pressure also indicates zero pumping loss. Consequently, the pumping loss passes through a maximum (Fig. 48) as the throttle is closed.

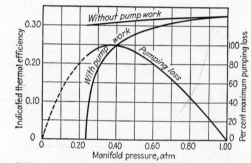

Fig. 48.—Effect of throttling on thermal efficiency and pumping loss.

Combustion Time.—The time of combustion has been assumed to be zero for the constant-volume process in the engine. In the constant-pressure process, combustion is assumed to occur at a rate necessary to maintain constant pressure during the process. Actually, every combustion process requires an appreciable amount of time which depends upon various factors and conditions. The process starts from a small nucleus between the electrodes of the spark plug and spreads progressively throughout the combustion chamber of a spark-ignition engine. Compression ignition causes combustion to start usually at several points from which the flame spreads in the usual manner.

It has been recognized for some time that flame is propagated by two distinctly separate processes:[1]

[1] C. Z. Rosecrans, An Investigation of the Mechanism of Explosive Reaction, *Univ. Ill., Eng. Expt. Sta., Bull.* 157 (1926).
F. W. Stevens, The Gaseous Explosive Reaction—The Effect of Inert Gases, *N.A.C.A. Rept.* 280 (1927).
W. A. Bone and D. T. A. Townend, "Flame and Combustion in Gases,"

DEVIATIONS FROM IDEAL PROCESSES

1. Propagation from molecule to molecule at *reaction velocity* relative to the unburned.

2. Propagation by mass movement due to expansion of the burned portion of the charge, effect of piston movement, and residual velocity from intake stroke, all of which produce a *gas velocity*.

Various factors such as composition, turbulence, and temperature of the charge influence the reaction velocity; while

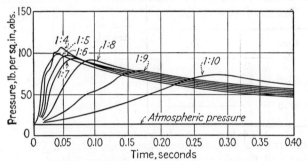

FIG. 49.—Clerk's experiments on London gas.

energy liberation, composition of the burned charge, heat loss, and combustion-chamber shape influence the gas velocity.

The classical experiments of Dugald Clerk,[1] in which he used a combustion chamber 7 in. in diameter and 8¼ in. long to which was attached a piston indicator, showed the effect of fuel and mixture ratio on combustion time. The results of his experiments (Fig. 49) on London gas proved that mixture ratio was an important factor influencing combustion time. He also found that hydrogen-air mixtures reached the maximum pressure much more rapidly than the artificial gas-air mixtures.

Clerk's method for determining the best mixture was based on the ability of the mixture to produce pressure and to resist cooling for 0.2 sec. He showed that hydrogen was a distinctly inferior fuel for internal-combustion engines since it necessitated too much cylinder volume for a given power output.

Longmans, Green & Company, New York, 1927.

B. Lewis and G. von Elbe, "Combustion, Flame, and Explosion of Gases," Cambridge University Press, London, 1938.

[1] D. Clerk, "The Gas, Petrol and Oil Engine," John Wiley & Sons, Inc., New York, 1909.

Combustion time has been studied by making use of the phenomenon that flame ionizes the gas when it arrives at a given position. Spark plugs are located at various places in the combustion-chamber wall and have a potential difference impressed on them which is insufficient to cause an electrical discharge until the flame reaches the spark-plug gap. MacKenzie and Honaman[1] obtained the following results on a single-cylinder 5- by 7-in. Liberty engine.

Distance between plugs		Time of flame travel, sec.	Mean flame velocity, ft. sec.$^{-1}$
Cm.	In.		
2.8	1.10	0.00434	21.13
9.5	3.74	0.00872	35.66
11.8	4.65	0.00939	41.27

These results indicate a straight-line relation between flame velocity and flame travel and also a flame velocity of approxi-

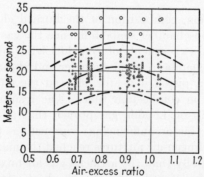

FIG. 50.—Combustion velocities of air-benzol mixtures. Compression ratio of 5 to 1; speed 1600 r.p.m.; larger circles indicate intake velocity. (*Schnauffer.*)

mately 15 ft. sec.$^{-1}$ very shortly after ignition occurs. However, these flame velocities appear to be low, other tests on 5-in. cylinders with one plug in the wall giving a time of pressure rise of approximately 0.006 sec. corresponding to a flame velocity of 70 ft. sec.$^{-1}$

[1] D. MacKenzie and R. K. Honaman, The Velocity of Flame Propagation in Engine Cylinders, *S.A.E. Trans.*, **15** (I), 299 (1920).

Schnauffer[1] used the same method in a series of experiments on a Siemens and Halske airplane-engine cylinder in which the flame distance was 5.1 in. The mean flame velocity was determined for various air-fuel ratios, benzol being used as the fuel. The outlining curves (Fig. 50) indicate a possible variation in combustion time of ±25 per cent. Thus, not only is considerable time required for combustion, but there will also be considerable variation in time required. The peak of the middle curve indicates a mean flame velocity of about 70 ft. sec.$^{-1}$

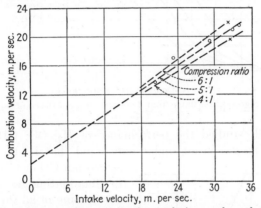

Fig. 51.—Relation between combustion velocity and turbulence. The average velocity of the entering charge at the intake valve is the unit of reference for expressing the degree of turbulence. (*Schnauffer.*)

An increase in engine speed increases the mean flame velocity (Fig. 51) as well as the average velocity of the entering charge at the intake valve. However, the mean flame velocity appears to increase less rapidly than the average intake velocity. Schnauffer has also made measurements of flame propagation with 24 ionization plugs activating as many neon lamps and found flame velocities of 870 to 980 ft. sec.$^{-1}$ with heavy detonation.

Marvin and Best have studied flame movement by means of a stroboscopic device[2] viewing 31 small quartz windows distributed over the head of a spark-ignition engine. Most of the results show low flame velocities at the beginning and end

---

[1] K. Schnauffer, Engine-cylinder Flame Propagation Studied by New Methods, *S.A.E. Trans.*, **29**, 17 (1934).

[2] C. F. Marvin, Jr., and R. D. Best, Flame Movement and Pressure Development in an Engine Cylinder, *N.A.C.A. Rept.* 399 (1931).

of the combustion process and appreciable differences between fuels (Fig. 52).[1]

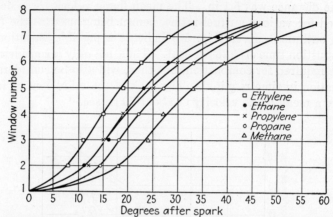

FIG. 52.—Progress of the flame front down the center of the combustion chamber. (*Marvin*.)

Burstall[2] studied the performance of H_2, CO, CH_4, and an illuminating gas in an engine and found that H_2 burned much faster than CO or CH_4 and that the illuminating gas which contained some of all three constituents had a combustion time

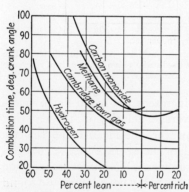

FIG. 53.—Effect of fuel and mixture strength on combustion time. (*Plotted from data by Burstall*.)

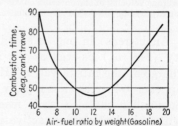

FIG. 54.—Effect of air-fuel ratio on combustion time. (*Rabezzana and Kalmar*.)

more than that of H_2 but less than that of other constituents (Fig. 53). The time of combustion tends to approach a minimum near the correct mixture.

[1] C. F. Marvin, Jr., Observations of Flame in an Engine, *S.A.E. Trans.*, **29**, 391 (1934).

[2] A. F. Burstall, Experiments on the Behavior of Various Fuels in a High-speed Internal-combustion Engine, *I. A. E. Proc.*, **22**, 358 (1927–1928); see also **19**, 670 (1924–1925), **21**, 628 (1926–1927).

DEVIATIONS FROM IDEAL PROCESSES

Rabezzana and Kalmar[1] used the ionization method and determined the effect of air-fuel ratio (Fig. 54) and engine speed (Fig. 55) as well as of various other factors upon combustion time.

Withrow, Lovell, and Boyd[2] used a sampling device and analyzed the gases at various points in a combustion chamber. They determined the time required for the flame to reach various positions in the combustion chamber with various engine speeds (Fig. 56). Later, Withrow and Boyd[3] inserted a quartz

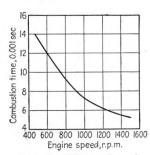

Fig. 55.—Effect of engine speed on combustion time. (*Rabezzana and Kalmar.*)

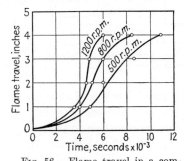

Fig. 56.—Flame travel in a combustion chamber. (*Withrow, Lovell, and Boyd.*)

window in the cylinder head and photographed the flame movement from the spark plug to a position at the opposite end of the combustion chamber. A similar method was used by Bouchard, Taylor, and Taylor.[4]

Withrow and Rassweiler[5] covered the entire combustion chamber with a quartz window and photographed the flame at intervals of 2.4 crankshaft degrees.

All investigations show that combustion requires an appreciable amount of time and that various factors affect the process. All investigators do not agree upon the effect of some of the

---

[1] H. Rabezzana and S. Kalmar, Factors Controlling Engine Combustion, *Auto. Ind.*, **72**, 324, 354, 394 (1935).

[2] L. Withrow, W. G. Lovell, and T. A. Boyd, Following Combustion in the Gasoline Engine by Chemical Means, *Ind. Eng. Chem.*, **22**, 945 (1930).

[3] L. Withrow and T. A. Boyd, *Auto. Ind.*, **65**, 4 (1931).

[4] C. L. Bouchard, C. F. Taylor, and E. S. Taylor, Variables Affecting Flame Speed in the Otto-cycle Engine, *S.A.E. Trans.*, **32**, 514 (1937).

[5] L. Withrow and G. M. Rassweiler, Slow Motion Shows Knocking and Non-knocking Explosions, *S.A.E. Trans.*, **31**, 297 (1936).

variables. This lack of agreement may be due partly to difference in apparatus and procedure, but it should be noted that it is practically impossible to change one variable in the normally operated internal-combustion engine without changing several others such as clearance-gas dilution, heat transfer, etc.

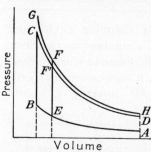

Fig. 57.—Effect of piston position on availability of energy liberation.

Effect of Combustion Time on Efficiency.—Ideal efficiencies are based upon instantaneous combustion at top-dead-center piston positions. If all the charge burned instantaneously before dead center at E (Fig. 57), the pressure would rise to F, compression would continue to G, and expansion to H which is above D of the ideal cycle. This can be proved by applying the energy equation to the two processes. For the constant-volume process, the total energy at E is equal to that at F.

Thus, $\quad (E + C)_E = (E + C)_F.$ (1)

For the process $EBCF'$;

$$(E + C)_E + {}_BW_E = {}_CW_{F'} + (E + C)_{F'}. \quad (2)$$

Since ${}_CW_{F'} > {}_BW_E$, $(E + C)_{F'} < (E + C)_F$ and points F and H lie on the higher adiabatic expansion path.

If combustion occurs instantaneously after dead center at E, the pressure will rise again to F and expansion occur to H with the same net result as before. Thus, the position of the piston, either before or after dead center, at which energy is released determines the availability of the energy released at that position. Obviously, maximum availability is possible for energy liberated at top dead center and zero availability for energy liberated at bottom dead center (Fig. 58).

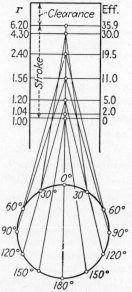

Fig. 58.—Expansion ratios and efficiencies for various piston positions.

Marvin[1] examined a number of combustion processes and established a curve of mass burned against combustion time (Fig. 59) that was considered typical of normal combustion.

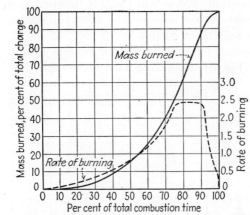

Fig. 59.—Mass burned and rate of burning of charge. (*Marvin.*)

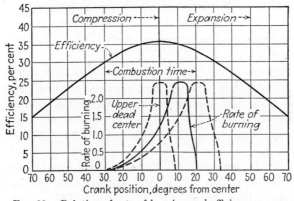

Fig. 60.—Relation of rate of burning and efficiency curve.

The rate of burning is indicated by the slope of the mass-burned curve. This rate curve was transferred to an efficiency vs. crank-angle diagram (Fig. 60) on the assumption of a definite spark advance and combustion time.[2] Integrating small areas under the rate curve, multiplying by the corresponding effi-

[1] C. F. Marvin, Jr., Combustion Time in the Engine Cylinder and Its Effect on Engine Performance, *N.A.C.A. Rept.* 276 (1927).

[2] A spark advance of 30 deg. and combustion times of 38, 50 and 63 deg. are shown in Fig. 60.

ciencies, and dividing the sum of the products thus obtained by the total area under the rate curve result in the ideal efficiency for the assumed combustion process. Later and earlier timing of the combustion process will result in different ideal efficiencies which, if plotted, will indicate the optimum timing or ignition advance for maximum output.

The foregoing has been done[1] for several combustion times, the results (Fig. 61) indicating a 5 per cent decrease in efficiency

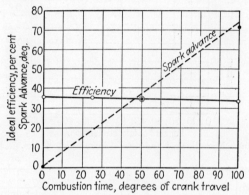

Fig. 61.—Effect of combustion time on efficiency when using optimum spark advance.

with combustion time increasing from 0 to 100 deg. of crank travel. The usual combustion process requires appreciably less time than 100 deg. of crank travel, so that the actual loss will probably be from 2 to 3 per cent. Obviously, any variation from optimum timing or increase in combustion time will increase this loss.

Timing of Ignition of Charge.—Optimum-spark-advance points, plotted in Fig. 61, nearly coincide with the dashed line based upon Upton's rule:

The optimum spark advance is such that the half pressure rise occurs at the dead center and that this stage of the pressure rise occurs practically at 75 per cent of the explosion time after ignition.

Minter and Finn[2] found optimum spark advance varied directly with clearance volume for any given engine speed.

[1] Marvin used air-standard efficiencies, whereas the author used those determined in Chap. V.

[2] C. C. Minter and W. J. Finn, Spark Advance—Compression Ratio, *Auto. Ind.*, **65**, 94 (1931).

Thus, Opt. sp. adv. $= k\,cl = k\dfrac{\text{disp.}}{r-1} = \dfrac{K}{r-1}$

for a given displacement. The constant K varies with the engine speed, being of the order of 100. However, there is no well-defined simple law for optimum spark advance since volumetric efficiency, fuel, speed, load, temperatures, and mixture ratio also influence the combustion process.

The appearance of the indicator card changes appreciably with variation in ignition timing. Too early timing increases the maximum pressure (Fig. 62), may cause detonation, and raises the expansion curve above that for the optimum timing. Too late timing lowers the maximum pressure and also raises the expansion line above that for the optimum timing, which is an indication of lower efficiency.

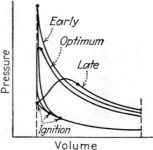

Fig. 62.—Effect of ignition timing on pressure-volume diagram.

Effect of Distribution.—It is practically impossible to obtain the same mixture ratio in all cylinders of a multicylinder engine, particularly when liquid fuel is being used. Optimum spark timing for the engine is too early for the fastest burning mixture and too late for the slowest burning mixture. Variation from optimum spark timing and variation in air-fuel ratio will reduce the ideal efficiency 3 to 5 per cent in the usual case.

Unequal mixture distribution may cause some cylinders to detonate and make necessary the retarding of ignition timing with still further loss of efficiency.

Heat Loss.—The high temperatures attained in the combustion chamber and cylinder of the internal-combustion engine make heat loss an inherent factor. During the early part of the compression process, heat may be transferred from the cylinder walls to the cooler incoming mixture; but during most of the process, heat is transferred from the gases to the cylinder walls. Heat transfer out during the compression process lowers the compression curve, AB to AB' (Fig. 63), which lowers the expansion curve and reduces the net work of the process. Lowering the compression curve reduces the work of compression but reduces the work of expansion by a larger amount. If the heat loss occurred only from B to B', the net work would be smaller since the work of compression is not reduced. Thus, the loss of availability depends upon the piston position when the heat transfer occurs.

Heat loss during the combustion process lowers the entire expansion curve and reduces the work of expansion. The loss of availability is again dependent upon the piston position when the heat transfer occurs.

Heat loss along the expansion curve reduces the work and efficiency of the process, having less effect as the end of the process

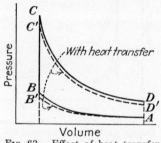

Fig. 63.—Effect of heat transfer on pressure-volume diagram.

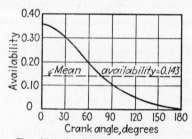

Fig. 64.—Availability of energy during a stroke. (Compression ratio, 6.2:1; connecting rod to crank, 3.5:1. See Fig. 58.)

is approached. Heat loss at the end of the expansion process or during the exhaust stroke has no direct effect on the work and efficiency. Thus, the availability of the energy lost by heat transfer during the expansion process would be about one-half the efficiency of the ideal engine process. Assuming a constant rate of heat loss, the mean availability for this energy can be determined by plotting the availability-crank-angle diagram (Fig. 64) and determining the mean height of the diagram. The mean availability amounts to about 40 per cent of the efficiency of the ideal engine process, which is the availability with the piston at top center.

The total effect of heat loss on availability is determined from an analysis of the heat loss from each part of the engine process. Thus, for a 6.2:1 compression ratio, with assumed heat losses the available loss is determined as follows:

DEVIATIONS FROM IDEAL PROCESSES

Period	Heat loss, %	Availability	Available loss, %
Compression	0.5	0.143	0.07
Combustion	7	0.359	2.51
Expansion	9	0.143	1.29
Exhaust	17	0.0	0.0
Total	33.5	.....	3.87

For 100 B.t.u. of energy input, of which 35.9 are available in the ideal case, a heat loss of 33.5 B.t.u. causes a loss of about 3.9 B.t.u. of work or a reduction of about 11 per cent in the available energy.

Valve Timing and Pumping Losses.—In all theoretical analyses, it is usually assumed that opening and closing of

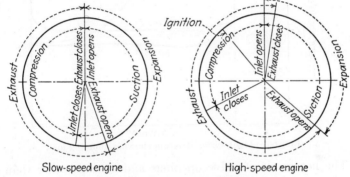

Fig. 65.—Valve-timing diagrams.

intake and exhaust valves occur on dead center (ends of the stroke), the valves remaining open for 180 deg. of crank travel. In actual engines the exhaust valve closes and the intake valve opens approximately on upper dead center, but the opening of the exhaust valve and the closing of the intake valve vary considerably from the bottom-dead-center position, depending principally on the desired speed.

The valve-timing diagrams (Fig. 65) show the intake valve opening at top dead center in both cases and before the exhaust valve closes in the high-speed engine, although this is reversed in some engines. Usually, the higher the speed, the later the intake valve closes. Late intake-valve closing permits the high intake velocity to continue to charge the cylinder even though the piston is moving against the incoming charge. Obviously, there will be one speed at which maximum charge weight and compression pressure will be attained. At lower speeds the piston will push some of the charge out of the cylinder before the valve closes, and at higher speeds the valve closes while charge is still flowing into the cylinder.

The exhaust valve is opened before dead center in all cases, and the higher the speed the earlier the opening. Early opening reduces the pressure in the cylinder to nearly atmospheric pressure before the exhaust stroke begins, which reduces the work of expansion slightly and the work of exhaust appreciably.

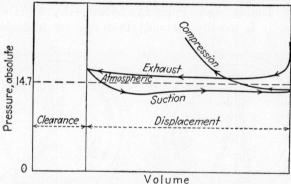

Fig. 66.—Light-spring diagram showing pumping loss.

The net result of valve openings and closings other than at dead centers is that the indicator card is "rounded" at the exhaust corner, the exhaust pressure is lowered, and more charge is inducted into the cylinder at one speed than at either lower or higher speeds. The rounding of the exhaust corner reduces the work of the indicator card about 1 to 2 per cent, which represents about ½ B.t.u. of available energy per 100 B.t.u. energy input.

Pumping Losses.—The work of getting the fresh charge into the cylinder and the exhaust products out is termed the *pumping work*. A difference in pressure is required to cause the gases to flow either into or out of the cylinder. The design of the carburetor, intake manifold, and intake passages affect the flow of the charge to the cylinder. Small areas mean high velocities. Sharp bends and rough passageways mean high fluid friction losses. The intake timing and port opening determine the velocities past the valve. All these factors reduce the suction pressure below atmospheric pressure.

Early opening of the exhaust valve reduces the pressure in the cylinder to nearly atmospheric pressure at the beginning of the exhaust stroke. Difference of pressure is required to produce flow and to overcome fluid friction losses through exhaust ports, exhaust manifold, and muffler.

The result of these losses is to produce on the indicator card a negative loop which is the *pumping loss* (Fig. 66). Different engines and different speeds produce considerably different diagrams, inertia effects of the gases often causing the exhaust and sometimes the suction line to reach or cross the atmospheric line. The exhaust line is often wavy owing to the various effects. In all cases, pumping losses increase with an increase in speed.

Although valve timings are adapted to the speeds desired, the suction pressure is always below atmospheric pressure in the normally aspirated engine and the exhaust pressure above atmospheric pressure. It is much more desirable in most cases, however, to keep the suction pressure, rather than the exhaust pressure, nearer the atmospheric pressure.

Example.—Determine the effect of a suction pressure of 1 lb. below atmospheric pressure on the charge weight and m.e.p.

$$\frac{P_s}{P_{atm}} = \frac{13.7}{14.7} = 0.932.$$

Thus, the charge weight would be decreased 6.8 per cent when the suction pressure is reduced from 14.7 to 13.7 lb.

Example.—Determine the effect of an exhaust pressure of 1 lb. above atmospheric pressure on the m.e.p. Assume an m.e.p. of 100 lb. in.$^{-2}$

One lb. back pressure above atmospheric reduces the m.e.p. 1 lb. Thus, 1 lb. back pressure decreases the m.e.p. 1 per cent.[1]

These examples indicate that in general for maximum power output it is much more important to maintain high intake-manifold pressures than low exhaust-manifold pressures.

Two-stroke-cycle Losses.—The piston of the common two-stroke-cycle engine (Fig. 67) uncovers the exhaust port before uncovering the intake port near the end of the expansion stroke.

---

[1] The effect of back pressure is more than this, as will be shown later in dealing with its effect on the clearance gases.

The slightly compressed fresh charge flows from the crankcase into the cylinder where it displaces some of the exhaust products and mixes somewhat with the products remaining in the cylinder. Some of the fresh charge may escape through the exhaust port.

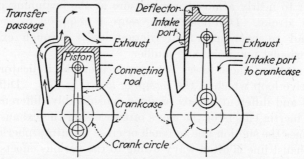

Fig. 67.—Three-port two-cycle engine.

Dilution of the fresh charge with clearance products tends to increase the efficiency of the process, this increase being offset by the tendency of the clearance products to increase the temperature of mixture. Dilution also increases the time required for com-

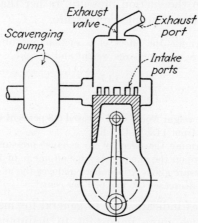

Fig. 68.—Two-stroke-cycle engine with scavenging blower.

bustion, which decreases the efficiency very appreciably in this type of engine particularly under part-throttle operation.

The piston covers the intake port before the exhaust port on the compression stroke, which may result in loss of mixture through the exhaust port. Any loss of mixture is a total loss of the available portion of its energy.

DEVIATIONS FROM IDEAL PROCESSES

The intake port to the crankcase is uncovered near the end of the compression stroke, which makes the pumping loss high and the volumetric efficiency low for this type of engine. Rotary or poppet intake valves reduce the pumping loss.

Fuel loss through the exhaust port is eliminated with the compression-ignition engine which may use an auxiliary scavenging pump (Fig. 68) for introducing air and displacing exhaust products from the cylinder. Dilution of the air with exhaust products may be practically eliminated if the scavenging system is efficient and the scavenging pump has a capacity larger than the piston displacement. Obviously, this increased capacity increases the pumping losses, and the auxiliary pump increases the total friction losses.

Incomplete Combustion.—The products of combustion of a chemically correct mixture of fuel and air are assumed to be CO_2, H_2O, and N_2. An analysis of the products from an engine process with such a mixture indicate that combustion is not complete, for CO, H_2, CH_4, and O_2 will be found. The following volume analysis contains these constituents although the analysis indicates a slight excess of air.

Constituent	Per Cent
CO_2	13.2
CO	0.5
O_2	0.6
CH_4	0.06
H_2	0.20
N_2	85.44
Aldehyde	Traces
Total	100.00

Assuming the fuel to be C_8H_{18} and considering the H_2 in the condensed H_2O which does not appear in the analysis, it is found that incomplete combustion amounts to about 2 per cent of the heating value of the fuel. This indicates a loss of ½ to ¾ B.t.u. of available energy per 100 B.t.u. input. Mixtures with excess air tend to reduce this loss to zero. Rich mixtures result in considerable unburned fuel due to O_2 deficiency, but this is included in the ideal analysis and efficiency. The loss due to that part which could but does not burn because of poor mixing, etc., will probably be about the same as the foregoing.

116 INTERNAL-COMBUSTION ENGINES

Summation of Losses.—The various deviations from the ideal process, which result in lowering the efficiency of the engine from the ideal to the actual indicated thermal efficiency, are as follows:

Factor	Approx. Loss, Available B.t.u. for 100 B.t.u. Input
1. Combustion time and variation	1
2. Distribution	1
3. Heat losses	4
4. Pumping losses	0.5
5. Incomplete combustion	0.5
Total loss	7

The ideal efficiency for a 6.2:1 compression ratio and an octane-air mixture with 90 per cent theoretical air is 32.9 per cent (Fig. 41). Subtracting the total loss due to the various possible deviations indicates an efficiency of 25.9 per cent. This represents a fuel consumption of about 0.47 lb. per indicated hp.-hr. which can be attained in practice. The relative efficiency is 25.9 ÷ 32.9 or about 77 per cent, which represents the degree of perfection by which the ideal cycle is approached.

The air-standard efficiency for a 6.2:1 compression ratio is 51.8 per cent (Fig. 41). The use of this efficiency introduces a discrepancy of 51.8 − 32.9 or 18.9 B.t.u. of available energy per 100 B.t.u. input, for which no accounting is possible. Thus, although the air-standard analysis is useful in making an approach to this subject, it is of no value when an accurate analysis of possible performance is desired.

EXERCISES

1. Determine the effect of charge dilution on the adiabatic compression process for a 6:1 compression ratio. Use octane-air mixtures with 0, 25, 50, and 75 per cent excess air for the diluent. Conditions at the beginning of compression are 1 atm. and 600°R.

2. Determine the effect of charge dilution on the constant-volume adiabatic-combustion pressure rise. Use octane-air mixtures with 0, 25, 50, and 75 per cent excess air, assuming complete combustion with all but 0 excess air. Compute and plot pressure rise per B.t.u. of energy supplied.

3. Assume the path AX (Fig. 47) is described by the equation $PV^{1.3} =$ const. Compute and plot the variation of pumping work and pumping loss with manifold pressure. Use 6:1 compression ratio.

4. Plot the flame-velocity values from MacKenzie and Honaman's experiments; compute the time of flame travel for 1, 2, 3, and 4 in., and plot on the same diagram. Then determine and plot the percentage of time required to traverse one-quarter, one-half, and three-quarters of the distance traveled. Take total distance as 5 in.

5. An engine is running at 5000 r.p.m., and 45 deg. of crank travel are required for combustion. Determine the time of combustion.

6. With a 3-in. bore and 4-in. stroke, and a 7.5-in. connecting rod, how far would the piston of the engine in Prob. 5 move during combustion (*a*) if ignition occurred on dead center; (*b*) if ignition occurred 30 deg. before dead center?

7. Determine the mean efficiencies for a combustion time of 50 deg. with ignition at 10, 20, 30, and 40 deg. before dead center. Plot these efficiencies against point of ignition. Use rate of burning curve given in Fig. 59 and a compression ratio of 7:1. Determine optimum ignition timing.

8. An engine has late ignition at dead center with compression pressure and temperature of 9 atm. and 1100°R., respectively. Combustion equilibrium is reached at a volume 2.5 times the clearance volume. Assume a pressure for combustion equilibrium, and sketch in a probable path (Fig. 62). Check the possibility of the assumed pressure and the assumed path. Assume 0 per cent excess air.

9. Plot the availability vs. crank-angle relations for a 7:1 compression ratio. Extrapolate the efficiency curve (Fig. 41) to zero efficiency. Determine the ratio of the mean to the maximum availability. Use a connecting-rod-to-crank ratio the same as in Fig. 58.

10. On the assumption that half of the heat loss to the walls could be suppressed, approximately how much would the brake thermal efficiency of an engine with a 7:1 compression ratio be increased?

11. With an exhaust opening at 45 deg. before dead center: (*a*) What would be the position of the piston? (*b*) What is the expansion ratio from this point? Use data given in Prob. 9.

12. Draw a valve-timing diagram for a two-stroke-cycle engine having inlet- and exhaust-port heights of ⅝ and ⅞ in., respectively. Use data in Prob. 6.

13. Sketch a valve-timing diagram for a blower-scavenged two-stroke-cycle engine having poppet exhaust valves. Discuss differences between this diagram and the one in Prob. 12.

CHAPTER VII

LIQUID AND GASEOUS FUELS

Fuels.—An internal-combustion engine may be operated on fuel in any of the three states—gaseous, liquid, or solid. The gaseous fuels present the least difficulty of the three from the standpoint of mixing with air and distributing to the various cylinders in a multicylinder engine. The liquid fuels must be vaporized, or atomized and at least partly vaporized during the process of mixing with air. The distribution of air-fuel mixtures with liquid particles in suspension presents considerable difficulty in the multicylinder carbureted engine and also in each cylinder of a fuel-injection engine. Diesel intended to use pulverized solid fuel in his engine, but there has been very little development of the use of fuel in this form.

Engines operated on gaseous fuels must be near the source of supply or distribution system. Gaseous fuel for automotive equipment necessitates the use of large containers for low-pressure gas, or tanks for high-pressure gas, and very much restricts the field of operation. Consequently, liquid fuels obtained from petroleum are used to the largest extent, primarily because of the large energy quantities per unit volume and the ease of handling, storing, and transporting.

Hydrocarbons.—Petroleum is a mixture of many different hydrocarbons, with some sulphur and other impurities. These hydrocarbons are grouped into three different classifications:

1. Paraffins (C_nH_{2n+2}).
2. Naphthenes (C_nH_{2n}).
3. Aromatics (C_nH_{2n-6}).

The *paraffin* series of hydrocarbons begins with CH_4, the next higher one having one more C atom and the corresponding number of H atoms, etc. The *normal* paraffin hydrocarbons have *straight-chain* structures with one bond between each atom. Thus, for three of the normal series,

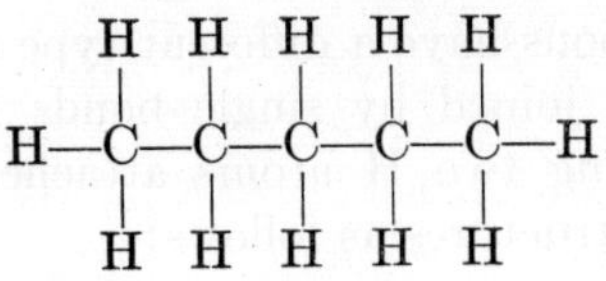

LIQUID AND GASEOUS FUELS

Pentane

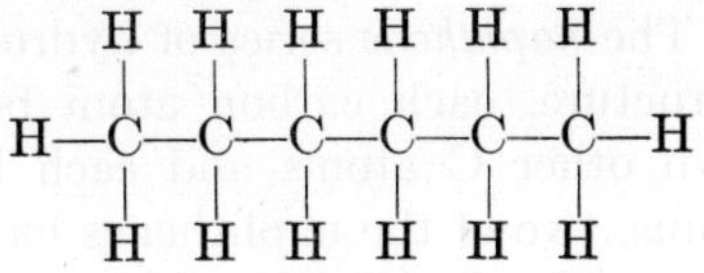

Hexane

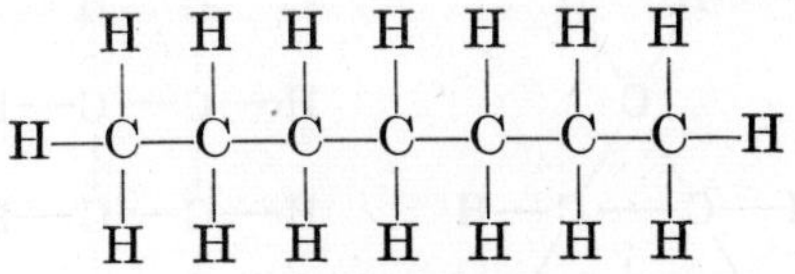

Heptane

the carbon molecules are connected to each other by a chainlike structure.

Any of the more complicated paraffin hydrocarbons may have a number of *isomers*, which have the same number of C and H atoms but have a different structure. Thus, three of the heptane isomers have structures as follows:

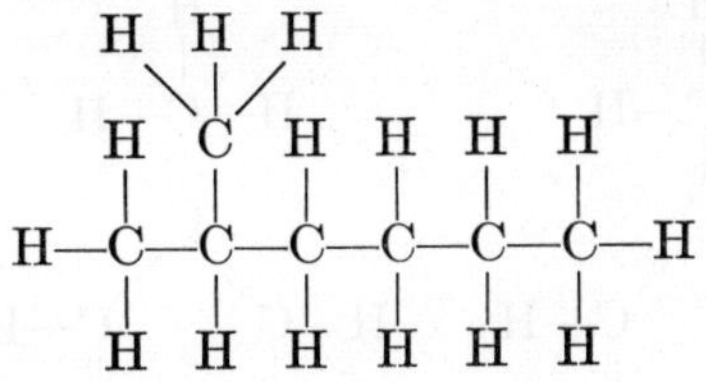

Methyl hexane (2)

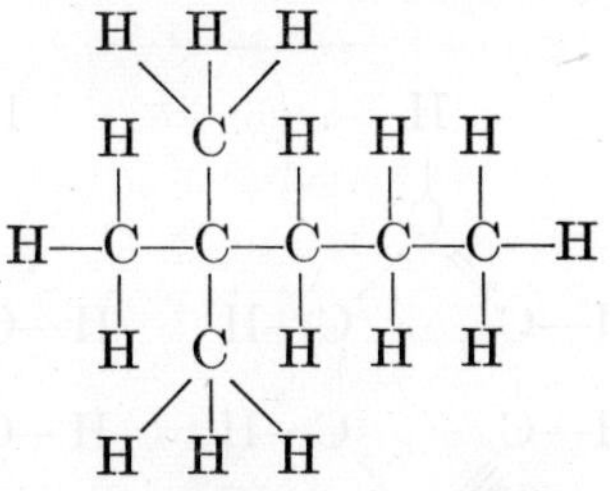

Dimethyl pentane (2) (2)

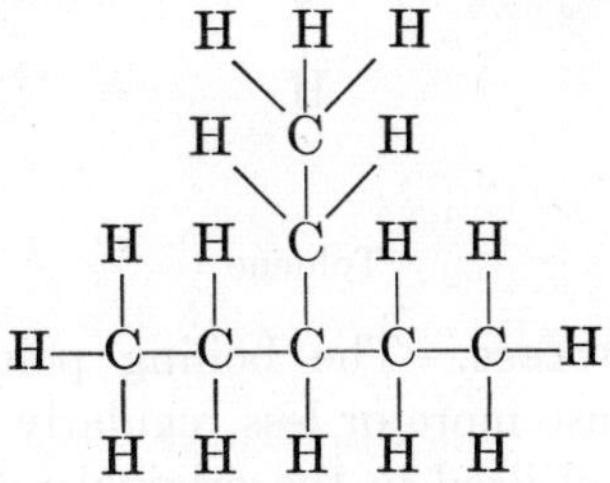

Ethyl pentane (3)

The numbers indicate the positions of the carbon atoms to which the methyl or ethyl groups are attached. The difference in structure of the molecules results in different physical properties and reaction characteristics although the same molecular weight and C/H ratio are retained.

The *naphthene* series of hydrocarbons have a different type of structure, each carbon atom being joined by single bonds to two other C atoms and each having two H atoms attached. Thus, two of the naphthenes have structures as follows:

Cyclopropane   Cyclobutane

The *aromatic* series of hydrocarbons have a *ring* type of structure for most or all of the carbon atoms, to which are attached H or groups of C and H atoms. Thus, three hydrocarbons with the typical benzene ring have structures as follows:

Benzene   Toluene   Xylene (para)

The Refining Process.—The boiling points of the various hydrocarbons increase more or less regularly with the molecular weight. This was utilized in the original refining process, that of *fractional distillation*. In this process of refining the lightest fractions are driven off at comparatively low temperatures. The first vapor given off is called "petroleum ether" and is only a fractional percentage of the whole. This ether is driven off below 100°F. Next come the gasolines, naphthas, kerosene, gas and fuel oils, and lubricating oils in order. After distillation

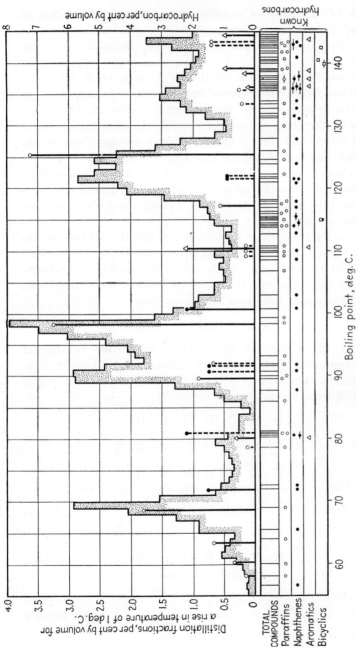

FIG. 69.—Distillation of a fraction of Mid-Continent petroleum and hydrocarbons isolated or suspected. (*R. T. Leslie and J. D. White, Nat'l. Bur. of Stand., Jour. of Res.*, **15**, 211, 1935.)

Lines in space entitled "Total Compounds" show distribution of boiling points. Different symbols indicate the various series of hydrocarbons. Distillation graph has scale at left. Solid vertical lines, with scale at the right, indicate hydrocarbon isolated. In some cases the boiling points do not check those previously published and indicated in the space entitled "Total Compounds." Dashed lines indicate suspected hydrocarbons.

there is left a residue of paraffin wax or asphalt, depending on the base of the oil.

The distillation of a fraction of a Mid-Continent crude oil, boiling from 55 to 145°C. (Fig. 69) indicates the complexity of the composition of even a fraction of a given crude oil. This illustration shows the volume distilled for each degree centigrade rise in temperature, the boiling points of all the hydrocarbons in the given temperature range, the distribution in the various series, and those which have been isolated from or suspected in the given petroleum fraction.

The close grouping of the boiling points at the higher temperatures indicates the greater number of hydrocarbon structures possible with the molecules having the greater number of carbon atoms.

The demand for more of the lighter fractions than can be obtained from the straight-run fractional-distillation process resulted in the *cracking process* which breaks down the larger molecules and produces the more desirable lighter molecules. A typical example, indicated by the reaction

$$C_{14}H_{30} \rightarrow C_7H_{16} + C_7H_{14}$$

shows the dissociation of the heavy molecule into two lighter ones. Assuming that the heavy molecule is a straight-chain hydrocarbon, the breaking of the chainlike structure into several parts with some recombination results in

$$H-\underset{\underset{H}{|}}{\overset{\overset{H}{|}}{C}}-\underset{\underset{H}{|}}{\overset{\overset{H}{|}}{C}}-\underset{\underset{H}{|}}{\overset{\overset{H}{|}}{C}}-\underset{\underset{H}{|}}{\overset{\overset{H}{|}}{C}}-\underset{\underset{H}{|}}{\overset{\overset{H}{|}}{C}}-\underset{\underset{H}{|}}{\overset{\overset{H}{|}}{C}}-\underset{\underset{H}{|}}{\overset{\overset{H}{|}}{C}}-H$$

Heptane (C_7H_{16})

$$H-\underset{\underset{H}{|}}{\overset{\overset{H}{|}}{C}}-\underset{\underset{H}{|}}{\overset{\overset{H}{|}}{C}}-\overset{}{C}=\overset{}{C}-\underset{\underset{H}{|}}{\overset{\overset{H}{|}}{C}}-\underset{\underset{H}{|}}{\overset{\overset{H}{|}}{C}}-\underset{\underset{H}{|}}{\overset{\overset{H}{|}}{C}}-H$$

Heptylene (3) (C_7H_{14})

Heptylene is an *olefin* (C_nH_{2n}), which is a series of unsaturated aliphatic[1] hydrocarbons. These are similar to the paraffin

---

[1] Compounds having an open chain structure.

series except for the double bond between two C atoms and the absence of two H atoms. The double bond of the olefins may appear between any two carbon atoms, the position being designated by a number indicating the smaller number of carbon atoms at one side of the double bond.

Some *diolefin* hydrocarbons (C_nH_{2n-2}) may be formed as indicated by the following reaction:

$$C_{21}H_{44} \rightarrow 2C_7H_{16} + C_7H_{12}$$

The diolefins have the following type of structures,

$$H-\underset{H}{\overset{H}{C}}=\underset{H}{\overset{}{C}}-\underset{H}{\overset{H}{C}}-\underset{H}{\overset{}{C}}=\underset{H}{\overset{}{C}}-\underset{H}{\overset{H}{C}}=\underset{H}{\overset{}{C}}-\underset{H}{\overset{H}{C}}-H$$

Heptadiene (1) (5) C_7H_{12}

the numbers indicating the location of the double bonds, which are arranged in various ways. This series of hydrocarbons is more unsaturated than the olefins. The unsaturated condition is indicated by the absence of four H atoms.

The amount of any given product obtained by the cracking process depends upon the conditions imposed and the equilibrium constant for the given reaction at the imposed conditions, in a manner similar to dissociation of the products of combustion.

The process of *hydrogenation* of petroleum is primarily one of adding hydrogen to unsaturated hydrocarbons. It usually consists of the cracking of the heavier hydrocarbons in the presence of hydrogen. The products obtained are more saturated than are those obtained from the cracking process alone. The hydrogenation process is also used to produce liquid fuel from coal and other similar materials.

The cracking process produces some olefins of low molecular weight such as ethylene (C_2H_4), propylene (C_3H_6), and butylene (C_4H_8), which are normally in the gaseous state. The process of *polymerization* is now being used extensively to combine two or more molecules of the same substance to form a heavier molecule.

Thus, $2C_4H_8 \rightarrow (C_4H_8)_2$, etc.

Other reactions may occur, depending on the process, resulting in paraffins, naphthenes, and aromatics.

124   INTERNAL-COMBUSTION ENGINES

Gasoline.—Gasoline varies considerably in characteristics depending upon the nature of the crude oil and the process of preparation. In general, gasoline is prepared from crude oil by one or more of the processes already mentioned. It is also produced by compressing and condensing the more volatile petroleum vapors that are found in natural gas. This is called *natural gasoline* and must be blended with a heavier product before it can be marketed.

TABLE 3.—HEAT OF VAPORIZATION OF GASOLINE*

Temperature	Aviation gasoline	Motor gasoline
	"Average volatility"†	
	176 to 230°F.	230 to 284°F.
°F.	B.t.u. lb.$^{-1}$	B.t.u. lb.$^{-1}$
0	161	157
50	155	151
100	148	145
150	142	138
200	136	132
250	130	126
300	123	120

* R. S. Jessup, *Nat. Bur. Standards, Jour. Research*, **15**, 227 (1935).
† Average volatility is defined as the average of the three temperatures at which 10, 50, and 90 per cent of the fuel is evaporated in the A.S.T.M. (American Society for Testing Materials) distillation test. Evaporated = distilled + loss.

Three grades of gasoline are marketed in this country, namely, premium-price, regular-price, and third-grade gasoline. The specific gravity of the majority of these gasolines varies from about 55 to 65° A.P.I.* The latent heat of vaporization depends upon the temperature and "average volatility" of the gasoline

* The specific gravity of a petroleum product is the ratio of the weight of a given volume of the product at 60°F. to the weight of an equal volume of distilled water at the same temperature, both weights corrected for air buoyancy. The relation between the American Petroleum Institute scale and specific gravity is,

$$°\text{A.P.I.} = \frac{141.5}{\text{sp. gr. } 60°/60°\text{F.}} - 131.5.$$

(Table 3). The heating value of average gasoline is about 20,300 B.t.u. lb.$^{-1}$*

The properties that determine the value of a gasoline as a motor fuel are:

1. Volatility:
   a. Starting.
   b. Warming up and acceleration.
   c. Vapor lock.
   d. Crankcase dilution.
2. Detonation.
3. Gumming.
4. Sulphur.

A gasoline should have sufficient constituents with low-temperature boiling points to provide easy starting under low-temperature conditions but not to cause vapor lock under high-temperature conditions.

The combustion characteristics of the fuel should permit operation in the given engine without resulting in detonation (see Chap. VIII).

A gasoline should be noncorrosive and have a low sulphur content. Also, there should be little tendency to form gum while in storage.

Table 4 gives the principal details of gasoline specifications[1] as set up by the U. S. Army, the U. S. Navy, and the Federal Specifications Board.

Volatility of Gasoline.—The volatility of a gasoline is determined by its A.S.T.M. distillation curve (Fig. 70) which indicates the temperatures at which the various amounts of a given sample are distilled under specified test conditions.[2] However, gasoline will be completely evaporated in the presence of air at a temperature much lower than the "end point" of the A.S.T.M.

---

* For other thermal properties of petroleum products, see *Nat. Bur. Standards, Misc. Pub.* 97 (1929) and *Jour. Research,* **7,** 1133 (1931).

[1] See also W. H. Hubner, G. Egloff, and G. B. Murphy, Aircraft-fuel Specifications, *Nat. Petroleum News,* **29,** 57 (1937).

Information regarding test methods for petroleum products and their significance is found in the following publications of the A.S.T.M.:

"A.S.T.M. Standards on Petroleum Products and Lubricants."
"The Significance of Tests of Petroleum Products."

[2] *A.S.T.M. Standard,* No. 86.

TABLE 4.—GASOLINE SPECIFICATIONS

Service	U. S. Army			U. S. Navy		Fed. Spec. Board	
				High octane base	Low octane base		U. S. motor gasoline
Specification	2-93-A	2-95-A	2-92-A	M 302-a	M 222-a	VV-M-571-a	VV-G-101-a
Effective date	4-15-38	6-19-35	6-20-35	2-1-36	9-15-34	4-16-36	4-16-36
Grade designation				High octane base	Low octane base	Motor fuel V	
Octane no.	65	92	100	87,[a] 87, 100	73, 87	65-75	
Test engine	C.F.R.[b]	C.F.R.[b]	C.F.R.[b]	C.F.R.[c]	C.F.R.[c]	C.F.R.[c]	
Max. PbEt$_4$, cm^3 gal.$^{-1}$	0	6	3	0, 0.5, 3	0, 3.27	3	
Corrosion and residue	[d]	[d]	[d]	[d]	[d]	[e]	
Gum		[f]	[f]	[f]	[f]		
Sulphur, max, %	0.10	0.10	0.10	0.10	0.10	0.10	0.10
Distillation, not less than							
10% evap. at °F	167	167	167	167	167	149[g]	167
50% evap. at °F	230	212	212	212	212	257[h]	248
90% evap. at °F	320	275	275	275	275	356[i]	392
10 + 50% points °F	307	307	307	307	307		
Residue max, %	2	2	2	2	2	2	2
Reid vapor pr., lb. in.$^{-2}$	7	7	7	7	7	10[j]	12
Freezing point, °F	−76[k]	−76[k]	−76[k]	−76	−76		
Water tolerance							

[a] This clear unleaded grade is used for postoverhaul testing of high-output aviation engines and for operation prior to placing engines in storage.
[b] Special C.F.R. (Cooperative Fuel Research) engine cylinder; hot-plug method, U. S. Army specification 2-94 (7-12-35).
[c] A.S.T.M. test method D 357-38T standard.
[d] One hundred cubic centimeters of the fuel evaporated to dryness in a weighed polished copper dish shall cause no gray or black corrosion, and the increase in the weight of the dish shall not exceed 5 mg.
[e] A.S.T.M. test method D130-30.
[f] The fuel shall be subjected to an accelerated aging test at 212°F. with oxygen at approximately 100 lb. pr. and in the presence of iron for 5 hr. On evaporation of 100 cm.^{3} of a mixture of oxidized sample and gum solvent, the amount of residue shall not exceed 6 mg. (see original specification for further details).
[g], [h], and [i]. These temperatures are raised 14, 17, and 19°F., respectively, for an altitude of 8000 ft. above sea level.
[j] The Reid vapor-pressure limit is lowered 1.5 lb. for each 5000 ft. increase in altitude above sea level.
[k] Eighty cubic centimeters of the finished fuel shall be shaken with 20 cm.^{3} of distilled water at room temperature. On settling after shaking, the volume of the aqueous layer shall not have increased or decreased by more than 2 cm.^{3}

distillation curve, since the vapor pressure is less than atmospheric pressure.

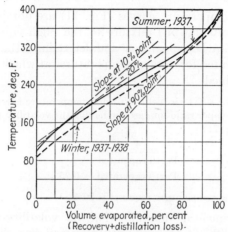

FIG. 70.—Average A.S.T.M. distillation curves of "regular-price" gasoline sold in the United States. (*C.F.R. Motor-gasoline survey; Bur. of Mines, Rept. of Investigation* 3408.)

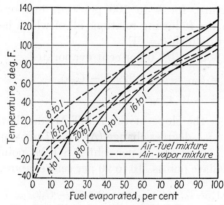

FIG. 71.—Volatility-temperature curves for a gasoline.

Bridgeman[1] defines volatility in terms of

... the temperature at which a given air-vapor mixture is formed under equilibrium conditions at a pressure of one atmosphere, when a given percentage is evaporated. According to this definition, one gaso-

[1] O. C. Bridgeman, Equilibrium Volatility of Motor Fuels from the Standpoint of Their Use in Internal-combustion Engines, *Nat. Bur. Standards, Research Paper* 694 (1934).

line is more volatile than another for any given percentage evaporated if it forms the given air-vapor mixture at a lower temperature.

Volatility-temperature curves (Fig. 71) are obtained by supplying a definite air-fuel mixture to the Sligh equilibrium-air-distillation (E.A.D.) apparatus[1] and determining the percentages of fuel evaporated for various temperatures at which the apparatus is maintained.

Example.—For a 4:1 air-fuel ratio, what temperature is required to produce a 16:1 air-vapor ratio for gasoline having the volatility characteristics indicated in Fig. 71?

$$\frac{4}{\text{Part evaporated}} = 16.$$

Part evaporated $= 0.25$.

From Fig. 71 the temperature required is 20°F.

The experimental determination of volatility-temperature curves is a difficult task which Bridgeman has eliminated by correlating the A.S.T.M. distillation curves with the volatility curves. He has shown[2] that the volatility curve for a 16:1 air-vapor mixture, for 0 to 100 per cent evaporated, can be obtained from the following relations all of which have the same general form, in which $t = °F$.

For bubble points BP, 0 per cent evaporated,

$$1.5 t_{BP_{16}} = t_{10\% ASTM} - (229.8 - C_{BP}), \tag{1}$$

and $\quad C_{BP} = -13.42\sqrt{S} + 70, \tag{2}$

where $\quad S =$ the slope of the A.S.T.M. distillation curve in degrees change (ΔT) per 1 per cent evaporated at the 10 per cent A.S.T.M. point.

For intermediate points (10 to 90 per cent evaporated),

$$1.5 t_{EAD_{16}} = t_{ASTM} - (229.8 - C_{10\ to\ 90\%}) \tag{3}$$

where $t_{EAD_{16}}$ and $t_{ASTM} =$ temperatures, °F., for the same percentage evaporated,

$$C_{10\ to\ 90\%} = \sqrt{(828 S)} \log \left[\frac{(100 - P)}{50}\right] + 70, \tag{4}$$

and $P =$ the percentage evaporated.

[1] Bridgeman, *op. cit.*
[2] *Ibid.*

For dew points DP, 100 per cent evaporated,

$$1.5 t_{DP_{16}} = t_{90\%ASTM} - (229.8 - C_{DP}) \qquad (5)$$

where C_{DP} = the right side of Eq. (2), also.

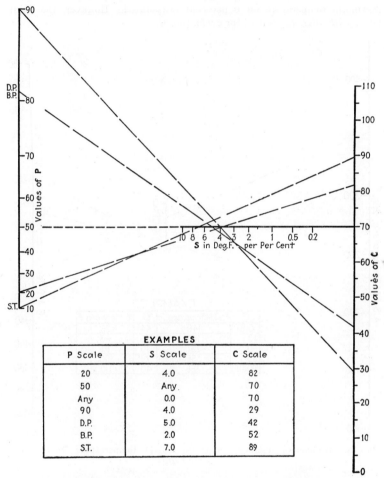

Fig. 72.—Bridgeman alignment chart for evaluating C.

Equations (2) and (4) were used as the basis for the alignment chart (Fig. 72) for the evaluation of the parameter C, while Eqs. (1), (3), and (5) may be solved with the alignment chart for E.A.D. (equilibrium-air-distillation) temperatures (Fig. 73). In making use of the charts, it should be noted:

130 INTERNAL-COMBUSTION ENGINES

1. The E.A.D. temperatures obtained are the minimum temperatures required to obtain a 16:1 air-vapor mixture with a given percentage of the fuel evaporated.

2. The temperature and slope at the 10 per cent point of the A.S.T.M. distillation curve are used to determine the bubble point, which is the maximum temperature for 0 per cent evaporated. However, the point labeled *BP* (Fig. 72) is used for evaluating *C*.

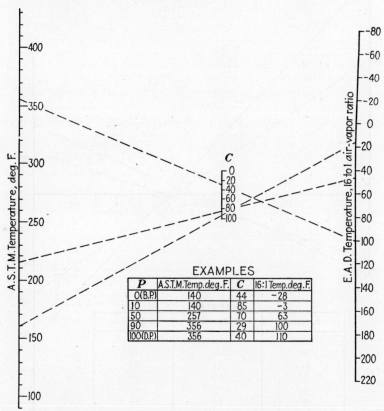

Fig. 73.—Bridgeman alignment chart for E.A.D. temperatures (16 to 1 air-vapor mixtures).

3. The temperature and slope at any point on the A.S.T.M. distillation curve between 10 and 90 per cent evaporated are used to determine the E.A.D. temperature for the percentage evaporated at the given point.

4. The temperature and slope at the 90 per cent point of A.S.T.M. distillation curve are used to determine the dew point, which is the minimum temperature for 100 per cent evaporated. However, the point labeled *DP* (Fig. 72) is used for evaluating *C*.

Example.—Determine the bubble point, the 10 per cent E.A.D. temperature, and the dew point for a 16:1 air-vapor mixture; use the distillation curve labeled "Summer, 1937" in Fig. 70.

The A.S.T.M. temperature and the slope at the 10 per cent point are 140°F. and 3.3°F. per 1 per cent evaporated, respectively. A line from point BP (Fig. 72) through $S = 3.3$ indicates a value of $C = 47$. A line from 140°F. (Fig. 73) through $C = 47$ indicates -25°F. as the bubble point.

A line from the 10 per cent point (Fig. 72) through $S = 3.3$ indicates a value of $C = 83$. A line from 140°F. (Fig. 73) through $C = 83$ indicates -4°F. as the E.A.D. temperature to form a 16:1 air-vapor mixture with only 10 per cent of the fuel evaporated. Other points from 10 to 90 per cent evaporated are determined in the same manner.

The A.S.T.M. temperature and the slope at the 90 per cent point are 350°F. and 4.3°F. per 1 per cent evaporated, respectively. A line from point DP (Fig. 72) through $S = 4.3$ indicates a value of $C = 43$. A line from 350°F. through $C = 43$ (Fig. 73) indicates 106°F. as the dew point.

Plotting these as well as other values against the percentage evaporated results in the 16:1 air-vapor volatility curve for this fuel.

Air-vapor Ratios Other than 16:1.—The charts (Figs. 72 and 73) were developed for a 16:1 air-vapor ratio. However, Bridgeman[1] has shown that there is a definite temperature difference between the 16:1 air-vapor volatility curve and other air-vapor curves. Other volatility curves for any gasoline for which the 16:1 curve is known may be determined by adding the temperature correction in the following table:

Air-vapor ratio.	5	8	9	10	11	12	13	14	15	16	17	18	19	20	30	40	50
Temp. correction, °F	41	24	20	16	13	10	7	4	2	0	-2	-4	-6	-7	-21	-32	-39

Example.—The temperature corrections for 8:1 and 20:1 air-vapor ratios are 24 and -7°F., respectively. These are the differences between the 16:1 and the other air-vapor curves mentioned (Fig. 71).

Starting Mixtures and Temperatures.—A combustible air-vapor mixture is required to start an engine under any conditions. It has been found[2] that an air-vapor mixture of 13.3:1 under equilibrium conditions is desirable for satisfactory starting. The carburetor will supply at least a 1:1 air-fuel mixture with the choke closed. If 7.5 per cent of the fuel leaving the carbu-

---

[1] *Ibid.*

[2] C. S. Cragoe and J. O. Eisinger, Fuel Requirements for Engine Starting, *S.A.E. Trans.*, **22**, 1 (1927)

retor is evaporated, the result will be a 13.3:1 air-vapor mixture. Bridgeman has found that the E.A.D. temperatures for 13.3:1 air-vapor mixture with 7.5 per cent evaporated corresponds very closely to the temperatures for 16:1 mixture with 10 per cent evaporated. Thus, the minimum satisfactory starting temperature can be obtained from the two nomograms (Figs. 72 and 73) by making use of the necessary A.S.T.M. distillation data.

Example.—Determine the minimum satisfactory starting temperature for the fuel labeled "Winter, 1937–1938" in Fig. 70. Assume the choke provides a 1:1 air fuel ratio.

The tangent at the 10 per cent point indicates a slope of 3.1°F. per 1 per cent evaporated. A line from the 10 per cent point (also labeled *ST*, Fig. 72) through a slope value of 3.1 indicates a value of 82 for *C*. A line from the 10 per cent point temperature (120°F.) through *C* = 82 (Fig. 73) indicates a minimum satisfactory starting temperature of −19°F.

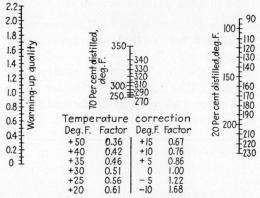

Fig. 74.—Nomogram for warming-up period. (*Rendel.*)

The condition and capacity of the starting battery as well as the engine friction, which is determined by the viscosity characteristics of the engine lubricant, have an appreciable effect on the ease of starting at low temperatures.

Warming Up and Acceleration.—The warming-up period is defined as the time required, after starting and running at a definite speed on the level road, before a given speed and load can be maintained with the choke in the nonchoking position. Rendel[1] has found that the warming-up period is indicated by

[1] T. B. Rendel, Warming-up Quality as a Measure of Fuel Volatility, Discussion, *S.A.E. Trans.*, **30**, 365 (1935).

the 20 and 70 per cent points on the A.S.T.M. distillation curve. The nomogram (Fig. 74) includes a table of correction factors for variation in atmospheric temperature.

Example.—Determine the relative warming-up periods for summer and winter gasoline as indicated by the data on Fig. 70. Assume summer temperature is 50°F. and winter temperature is 0°F.

From Fig. 70, the following data are obtained:

	20%, °F.	70%, °F.
Winter gasoline	148	277
Summer gasoline	172	290

From Fig. 74, drawing lines through the given temperatures indicates values of 0.76 and 1.12 for the winter and summer gasoline, respectively. Correcting for atmospheric temperature results in 0.76 and 0.40, respectively, indicating about 47 per cent less warming-up period for the summer fuel and conditions.

Eisinger and Barnard[1] have shown that the ability to accelerate can be related to the 20 and 90 per cent evaporated points of the

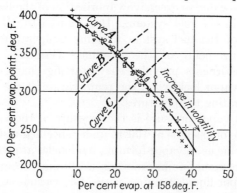

FIG. 75.—Diagram of equivalent volatilities. (*Eisinger and Barnard.*)

For curve A the abscissa represents a series of gasolines with a 90 per cent point of 350°F., while the ordinates represent a series of gasolines with a 20 per cent point of 158°F. Curves B and C indicate the variation in both points for constant volatility.

A.S.T.M. distillation curve. Two series of fuels were used, one with the 20 and the other with the 90 per cent point fixed. Curve A (Fig. 75) is based upon equal accelerating ability (from a speed of 10 m.p.h.) and shows the change in amount evaporated

[1] J. O. Eisinger and D. P. Barnard, A Forgotten Property of Gasoline, *S.A.E. Trans.*, **30**, 293 (1935).

at 158°F. which is equivalent to a given change in the 90 per cent point.

Curves B and C (Fig. 75) represent constant volatility and equal accelerating characteristics, curve B indicating higher intake-manifold temperature than curve C. Thus, for any given intake-manifold temperature there will be various combinations of percentages evaporated at 158°F. and the temperature at 90 per cent evaporated that will result in equal volatility and accelerating characteristics. The practical limits for equal volatility curves, such as B and C, are determined by starting requirements at the one end and by dilution restrictions at the other.

Example.—Curve B (Fig. 75) indicates that a gasoline with 10 per cent evaporated at 158°F. and 90 per cent evaporated at 300°F. is equivalent in volatility, as measured by an acceleration test, to a gasoline with 28 per cent evaporated at 158°F. and 90 per cent evaporated at 380°F. The first gasoline would be "hard-starting" compared with the second one.

Vapor Lock.[1]—Vapor lock is the "locking" or interrupting of the fuel flow due to excessive vaporization of the light fractions of gasoline in the fuel system. It is caused by the heating of some part of the fuel system to some temperature above the initial boiling point of the gasoline. Gasoline with highly volatile constituents provides easy starting characteristics but may cause vapor lock when the temperature of the fuel system rises to operating temperatures.

The Reid vapor pressure is the pressure of the vapor from a given sample of gasoline at 100°F. under specified test conditions. It gives no clue as to vapor-forming tendencies at other temperatures. More complete information is obtained from data showing the vapor-liquid volume ratio for various temperatures (Fig. 76).

The temperature at which a given fuel system will vapor-lock will depend upon the vapor-forming characteristic of the gasoline and the ability of the system to get rid of the vapor without seriously affecting the fuel flow. This can be determined by

---

[1] References:

O. C. Bridgeman, H. S. White and F. B. Gary, *S.A.E. Trans.*, **28**, 157 (1933).

N. MacCoull and E. M. Barber, *S.A.E. Trans.*, **30**, 237 (1935).

E. M. Barber and B. A. Kulason, *S.A.E. Trans.*, **31**, 351 (1936).

operating a car under specified test conditions, such as 40 m.p.h. and definite grade conditions, with fuels of known vapor-liquid volume-ratio characteristics. When vapor lock occurs, the temperatures of the fuel system and the vapor-liquid volume-ratio characteristics of the fuel indicate the vapor-venting capacity of the fuel system.

Example.—Determine the vapor-venting capacity for fuel systems that will vapor-lock with the fuels in Fig. 76, when the fuel system is at 143°F.

Reference to Fig. 76, indicates that fuels 1 and 3 have a V/L ratio of 7, fuel 5 a ratio of 13, and fuel 2 a ratio of 27. These values would

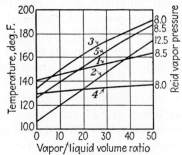
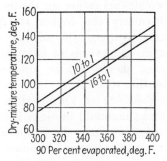

Fig. 76.—Vapor-forming characteristics of various gasolines.
1 and 2, blends of natural and refinery gasoline; 3, butane-refinery gasoline; 4, butane-natural gasoline; 5, typical commercial gasoline.

Fig. 77.—Dry-mixture temperature for given 90 per cent point (slope of 4.3°F. per 1 per cent evaporated used).

represent the vapor-venting capacities of the various fuel systems that would vapor-lock under the given conditions.

Fuels 1 and 3 should not cause vapor lock at 143°F. in cars with vapor-venting capacities of 10, whereas fuels 2, 4, and 5 would cause vapor lock. At 170°F., fuel 4 should be the worst offender, and fuel 3 should give the least trouble.

The loss of vapor from the fuel system increases the fuel consumption. Engines with fuel systems at high temperatures, having large vapor-venting capacities and using fuels with highly volatile constituents, should have considerable fuel loss if operated slightly below vapor-locking conditions. Under actual conditions, the loss is not large except in unusual cases.

Obviously, the desirable condition is a low-temperature fuel system and a fuel with low vapor-liquid volume characteristics. Otherwise, the vapor-venting capacity of the fuel system, which ranges from below 10 to above 50, should be large.

Crankcase Dilution.—The dilution of the lubricating oil with the heavier ends of gasoline is more or less inevitable since, usually, the mixture entering the cylinder does not have the heavier ends completely evaporated. This condition is exaggerated during the starting and warming-up periods but is reduced to a minimum with proper heat application to the manifold and normal operating engine temperature.

The dew point indicates the lowest mixture temperature required for complete evaporation of the fuel in a given air-fuel ratio. Various dew points have been determined (Fig. 77) for two air-fuel ratios, for the fuel labeled "Summer, 1937" (Fig. 70). Obviously the higher the 90 per cent point the higher the temperature of the mixture should be for complete evaporation or given wetness of mixture. For this reason the 90 per cent point should not be abnormally high, or excessive crankcase dilution will result.

Volatility at Various Pressures.—In automotive engines the intake manifold pressure is usually below 14.7 lb. in.$^{-2}$ abs. In aircraft engines the manifold pressure may be either below or above the standard barometric pressure. The relation between the air-vapor ratio M'_R, formed at pressure p for a given percentage of fuel evaporated at a given temperature, and the air-vapor ratio M_R which will be formed at the same temperature and percentage evaporated under a pressure of 760 mm. of Hg is

$$M'_R = M_R\left(1.04\frac{p}{760} - 0.04\right). \qquad (6)[1]$$

Example.—Assume a 16:1 air-vapor mixture is formed at 86°F. with 50 per cent of the fuel supplied evaporated and a total pressure of 1 atm. Determine the air-vapor ratio that will correspond to a 16:1 air-vapor mixture at a total pressure of ½ atm. Then determine the temperature at which this mixture will be formed compared with the 86°F. for the assumed conditions.

Substituting in Eq. (6) results in

$$M_R = \frac{16}{[1.04 \times \frac{1}{2} - 0.04]} = 33.3.$$

Reference to the correction table, page 131, indicates that an air-vapor mixture of 33.3 will be formed at a temperature about 25°F. lower than that required for a 16:1 mixture. Thus, a 16:1 mixture at ½ atm. total

[1] *Bridgeman, op. cit.*

LIQUID AND GASEOUS FUELS

pressure and with 50 per cent evaporated will be formed at 86 − 25 or 61°F.

Heat Required for Mixture.—The volatility curves indicate that higher mixture temperatures are required for more complete evaporation of the fuel. The heat that must be supplied to produce a given mixture can be divided into three parts:

1. Heat to raise the temperature of the air.
2. Heat to raise the temperature of fuel and vapor.
3. Latent heat to evaporate the given percentage of gasoline.

Thus, $Q = M_a c_{p_a}(T_2 - T_1) + M_f c_{p_f}(T_2 - T_1)* + M_f PL$ (7)

where subscripts a, f, and p = air, fuel, and const. pr., resp.,
P = the percentage evaporated,
L = the latent heat of the fuel.

Example.—How much heat must be supplied per pound of fuel to make possible 50 per cent evaporation in the intake manifold? Assume air temperature to be 60°F. Use correct air-fuel ratio. Assume distillation characteristics of "Summer, 1937" gasoline (Fig. 70).

The temperature at the 50 per cent point is 247°F. Making use of the alignment charts (Figs. 72 and 73) indicates that the E.A.D. mixture temperature is 36°F. for a 30.2 air-vapor ratio. Then

$Q = 15.1 \times 0.24(36 - 60) + 1 \times 0.50(36 - 60) + 1 \times 0.50 \times 150$
  $= -24$ B.t.u. lb.$^{-1}$ of gasoline.

Example.—Determine the amount of heat to be supplied for 100 per cent evaporation. Use the same data as in the previous example.

The temperature for 100 per cent evaporation is the dew point. From the example, page 131 the dew point is 106°F. The minimum heat requirement is

$Q = 15.1 \times 0.24(106 - 60) + 1 \times 0.50(106 - 60) + 1.0 \times 150$
  $= 340$ B.t.u. lb.$^{-1}$ of gasoline.

Actually higher temperatures and more heat application are required to produce the desired results since equilibrium conditions cannot be attained in the short time the mixture is in the intake manifold.

The heat may be supplied to the entering air, but it is customary to supply the heat to the intake manifold so that the change in temperature of the carburetor is reduced to a minimum.

* This is assuming the specific heat of liquid and vapor to be the same, since the fuel is vaporized from T_1 to T_2, and the fuel-vapor mixture which is changing to more vapor and less liquid must all be heated through the rise in temperature.

Effective Volatility.—E.A.D. temperatures would be obtained if sufficient time were allowed for the evaporation process to reach equilibrium conditions. The time between discharge of gasoline from the carburetor jet and the ignition of the charge is very small. Consequently E.A.D. temperatures will never be attained in practice but form an excellent guide as to the volatility characteristics of gasolines. In practice, it will be necessary to use temperatures higher than the E.A.D. temperatures to secure the desired result. This increase in temperature will vary with the conditions and the characteristics of the gasoline.

G. G. Brown[1,2] has shown that to get complete vaporization for average gasolines under actual engine conditions about 40 to 50°F. should be added to E.A.D. temperatures. It would seem that the increase in temperature should approach zero as the percentage evaporated decreases to zero. However, according to Brown, the effective volatility approaches the equilibrium volatility for anything below 40 per cent evaporated, varying considerably with mixture ratio.

Gum Content.—Some cracked gasolines may deposit a gummy material if kept in storage for a long period of time. The rate of gum formation is slow at first but usually accelerates rapidly with time. Such gasolines may deposit gum on the intake valves and in the intake manifold if subjected to moderate or high manifold temperatures if the gasoline is particularly unstable with regard to gum formation. Unusual cases have been reported in which the intake manifold was practically filled with a gummy deposit.

Various methods have been devised to determine gum content[3] and gum stability[4] of gasolines. However, the correlation between laboratory test results and gum depositions in engines has not been satisfactory. Obviously, the lower the gum content, the more desirable is the gasoline from this standpoint.

Gum inhibitors or stabilizers are added to gasoline to delay or prevent the formation of gum.

[1] G. G. Brown, The Relation of Motor-fuel Characteristics to Engine Performance; *Univ. Mich., Eng. Research Bull.* **7** (1927).

[2] G. G. Brown, The Volatility of Motor Fuels, *Univ. Mich., Eng. Research Bull.* 14 (1930).

[3] O. C. Bridgeman, and E. W. Aldrich, A Comparison of Methods for Determining Gum Contents of Gasolines, *S.A.E. Trans.*, **26**, 476 (1931).

[4] E. W. Aldrich, and N. P. Robie, The Gum Stability of Gasolines, *S.A.E. Trans.*, **27**, 198 (1932).

Sulphur.—High sulphur content is undesirable because of the formation of corrosive substances in the presence of water vapor. The Federal, U. S. Army, and U. S. Navy specifications for gasoline limit the sulphur content to a maximum of 0.10 per cent as determined by the A.S.T.M. method D90.

The sulphur content in the samples of commercial gasoline reported by the Bureau of Mines in the C.F.R.[1] motor-gasoline surveys from the winter of 1935–1936 to the winter of 1937–1938 varied as follows:

	Per Cent
Premium-price gasoline	0.006 to 0.25
Regular-price gasoline	0.007 to 0.36
Third-grade gasoline	0.005 to 0.27

Benzene.—Benzol is a coal-tar distillate that consists principally of benzene (C_6H_6) with small amounts of toluene and thiophene. It is particularly desirable as an internal-combustion-engine fuel on account of its high antiknock characteristic.

The specific gravity of benzene is about 0.88; the heating value of gaseous benzene is 18,160 B.t.u. (Table VIII, Appendix); the latent heat at atmospheric pressure is 186 B.t.u. lb.$^{-1}$, and the boiling point is 176°F. The correct air-fuel ratio is 13.25:1, and at this ratio the heat of combustion per cubic foot of air at 77°F. is 101 B.t.u., which is about 1 per cent lower than for octane.

Some of the properties of air-benzene mixtures have been computed as follows and are given in Table 5.

Example.—Determine the possible air-vapor ratio for a mixture temperature of 50°F., when using benzene as a fuel.

At 50°F., benzene vapor pr. = 0.43 lb. in.$^{-2}$ abs.
Partial pr. of the air = 14.27 lb. in.$^{-2}$ abs.
Sp. vol. of air is

$$V = \frac{1545 \times 510}{28.95 \times 144 \times 14.27} = 13.24 \text{ ft.}^3 \text{ lb.}^{-1}$$

This is also the volume of the saturated benzene vapor which has a specific volume of approximately

$$\frac{1545 \times 510}{144 \times 0.43 \times 78} = 163.4 \text{ ft.}^3 \text{ lb.}^{-1}$$

---

[1] Cooperative Fuel Research. This committee is composed of members of the automotive and petroleum industries and of the National Bureau of Standards.

140   INTERNAL-COMBUSTION ENGINES

The weight of air per pound of benzene vapor is

$$\frac{163.4}{13.24} = 12.3.$$

This represents the lowest air-fuel ratio that can exist with benzene vapor at a temperature of 50°F. and a total pressure of 1 atm.

From Table 5 it can be seen that the minimum mixture temperature for an air-fuel ratio of 13.25:1 is about 50°F.

The temperature drop of the air due to vaporization of the fuel is 186 ÷ (13.25 × 0.24) or 58.5°F. The minimum entering air

TABLE 5.—PROPERTIES OF BENZENE AND AIR-BENZENE MIXTURES

Temperature, °F.	Vapor pressure, lb. in.$^{-2}$	Air pressure, lb. in.$^{-2}$	Volume of air, ft.^{3} lb.$^{-1}$	Volume of vapor ft.^{3} lb.$^{-1}$	Air-vapor ratio, lb. of air lb.$^{-1}$ of C_6H_6
45	0.21	14.49	12.91	330.4	25.6
50	0.43	14.27	13.24	163.0	12.3
55	0.70	14.00	13.63	101.1	7.4
60	0.97	13.73	14.03	73.7	5.3
65	1.26	13.44	14.47	57.3	4.0
70	1.55	13.15	14.93	47.0	3.1

temperature should be 109°F., if all the latent heat of the benzene is supplied by the air.

Benzene has a freezing point of 42°F., which is a distinct disadvantage during the winter season. In mixture with gasoline the temperature at which benzene will separate from the gasoline and solidify is considerably lowered, depending on the percentage of benzene in the mixture. The National Bureau of Standards[1] reported that "a benzol-gasoline mixture containing 40 per cent benzol will not separate or solidify above −6°F."

Alcohol.—There are three alcohols that can be used as engine fuel: ethyl or grain alcohol (C_2H_6O); methyl or wood alcohol CH_4O), and butyl alcohol ($C_4H_{10}O$). A few of the properties of the alcohols are given in Table 6.

The low heating values of the alcohols, due to the partial oxidation as indicated by the O atom, is one of their chief disadvantages when they are compared with gasolines on a weight or

[1] Notes on Benzol-gasoline Mixtures for Automobile Fuels, Jan. 22, 1925.

volume basis. Higher compression ratios may be used with the alcohols, which tend to offset this disadvantage. However, as the octane number of gasoline approaches that of alcohol, (99 for ethyl alcohol), the difference in permissible compression ratio is reduced.

TABLE 6.—PROPERTIES OF VARIOUS ALCOHOLS

Name	Symbol	Sp. gr.	Boiling point, °F.	Latent heat, B.t.u.	Correct air-fuel ratio	Heating value, B.t.u. lb.$^{-1}$, liq. fuel
Ethyl	C_2H_6O	0.79	173	396	8.99	12,780
Methyl	CH_4O	0.79	151	503	6.46	9,790
Butyl	$C_4H_{10}O$	0.81	242	254	11.18	

The heating value per cubic foot of correct air-fuel mixture is practically the same for ethyl alcohol and gasoline (Table 13). Consequently, the power output should be the same in a given engine if the volumetric efficiency is not affected by the fuel.

The air-fuel ratios that can be obtained with alcohol at any pressure and temperature depend upon the vapor pressure of the alcohol. Values of vapor pressure and resultant air-fuel ratios for methyl and ethyl alcohols at atmospheric pressure are given in Table 7, the air-fuel ratios being computed in the same manner as for benzene.

By plotting the air-vapor ratios in Table 7, it will be found that mixture temperatures of 68 and 71°F. are required for methyl and ethyl alcohols, respectively, to produce the correct air-vapor ratios.

The latent heat of ethyl alcohol is 396 B.t.u. lb.$^{-1}$ If this heat must be provided by 8.99 lb. of air, the temperature to which the air should be heated for 100 per cent evaporation in the manifold is determined as follows:

$$8.99 \times 0.24(t_2 - 71) = 396 \text{ B.t.u.}$$
$$t_2 - 71 = 183° \text{ drop in temperature;}$$
$$t_2 = 254°F.$$

This is based upon the assumption that the fuel is at the temperature of the mixture. For other fuel temperatures the specific heat of the fuel must be considered.

For methyl alcohol with a latent heat of 503 B.t.u., the temperature of the air for 100 per cent evaporation should be 393°F.

TABLE 7.—PROPERTIES OF ALCOHOL AND AIR-ALCOHOL MIXTURES

Temperature, °F.	Methyl alcohol			Ethyl alcohol		
	Vapor pr., lb. in.$^{-2}$	Vapor, ft.^{3} lb.$^{-1}$	Ratio of air to sat. alcohol vapor*	Vapor pr., lb. in.$^{-2}$	Vapor, ft.^{3} lb.$^{-1}$	Ratio of air to sat. alcohol vapor*
30	0.56	...	....	0.19		
40	0.79	212	15.9	0.33		
50	1.08	158	11.4	0.45	264	19.9
60	1.43	122	8.7	0.64	189	13.8
70	1.92	93	6.0	0.91	135	9.4
80	2.53	72	4.4	1.26	100	6.7
90	3.38	55	3.0	1.70	75	4.8

* Pounds of air per pound of vapor.

Fuel Oil for Compression-ignition Engines.—The compression-ignition engine should be able to burn any of the fuels between gasoline and fairly heavy residual oils. Large slow-speed engines are usually more able to handle the variety of fuels than the small high-speed engines, although much depends on the injection system.

The gravity of fuel oil usually ranges from about 25 to 35° A.P.I., the lower value indicating the higher weight and heating value per gallon. The lower A.P.I. gravity fuels are generally of higher viscosity to which most fuel-injection systems are quite sensitive. A survey[1] of engine manufacturers' recommendations indicated an average viscosity, Saybolt universal[2] at 100°F., of 36 to 78 sec. for high-speed engines, 40 to 87 sec. for medium-speed engines (500 to 1000 r.p.m.), and 40 to 106 sec. for low-speed engines. Large slow-speed engines have larger dimensions for all the fuel-injection parts and are less sensitive to fuel viscosity.

The A.S.T.M. distillation curves usually lie in the range indicated in Fig. 78.

[1] W. H. Hubner and G. Egloff, *Univ. Oil Products Booklet* 209 (1937).

[2] The viscosity in Saybolt universal seconds is the time required for a given quantity of liquid to flow through an orifice in the Saybolt viscosimeter under specified conditions.

High-speed engines have little time for the combustion process and should have fuels with more of the low-boiling and less of the high-boiling constituents than the larger low-speed engines.

The flash point of fuel oils, the temperature at which the vapor will flash when exposed to flame, ranges from about 135 to 190°F. It is of little value except as an indication of fire hazard.

The ignition quality of a fuel oil, or the ability of the fuel to ignite in the presence of air in an engine under starting conditions and the delay between injection and ignition under operating conditions, influences the combustion characteristics in an engine. Engines vary considerably in sensitiveness to ignition quality which in terms of cetane number[1] amounts to about 30 to 60. High-speed engines usually require fuels with high ignition quality.

The pour point, the temperature below which the fuel will not flow under specified conditions, should be 10 to 15°F. below the minimum operating temperature expected.

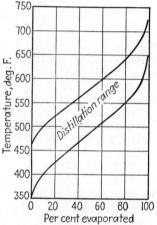

Fig. 78.—A.S.T.M. distillation range for Diesel fuels.

Low sulphur content, of less than 0.5 per cent, is usually desirable. The recommendations of engine manufacturers[2] range from 0 to 2 per cent, except that in slow-speed engines the upper limit is 4 per cent. Any condensation of water vapor in the exhaust products results in the formation of sulphuric acid. This possibility makes low sulphur content extremely desirable.

Carbon residue of fuel oils (Conradson method) is a good indication not so much of carbon formation in the engine as of the high boiling fractions that are difficult to determine. High carbon residue is permitted for slow-speed engines, recommendations[1] ranging from 0 to 10 per cent. Average values of 0.6, 1.0, and 1.7 per cent are recommended for high- , medium- , and low-speed engines, respectively.

[1] See Chap. VIII for cetane number and other measures of ignition quality.

[2] Hubner and Egloff, *op. cit.*

Water and sediment values should be low, less than 0.1 per cent. However, in the case of low-speed engines higher limits may be used.

The ash content of a fuel oil should be less than 0.1 per cent; it represents noncombustible material some of which is abrasive.

Natural Gas.—Natural gas is found in various localities in oil- and gas-bearing sand strata located at various depths below

TABLE 8.—CHARACTERISTIC PROPERTIES OF LIGHTER HYDROCARBONS*

Substance	Ethane	Propane	Iso-butane†	Normal butane†	Pentane	Propylene	Butylene
Formula	C_2H_6	C_3H_8	C_4H_{10}	C_4H_{10}	C_5H_{12}	C_3H_6	C_4H_8
Boiling point, °F	−127	−44	14	33	97	−53	20–34
Lb. gal.$^{-1}$ of liquid at 60°F	3.11	4.24	4.72	4.85	5.25	4.37	5.0–5.1
Heating value of the gas:							
B.t.u. lb.$^{-1}$	22,330	21,670	21,280	21,320	21,100	21,050	20,840
B.t.u. gal.$^{-1}$	69,500	91,900	100,400	103,400	110,800	92,000	105,200
B.t.u. ft.$^{-3}$ at 77°F. and 14.7 lb. in.$^{-2}$	1,731	2,488	3,254	3,282	3,891	2,301	3,086
Sp. gr. of gas	1.05	1.55	2.08	2.14	2.49	1.46	1.98
Vapor pr., lb. in.$^{-2}$ gage:							
−44°F	88	0	−9	−12	−14	3	−12
0	206	24	−4	7	−13	32	−6
33	343	54	7	0	−11	69	4
70	553	112	27	16	−6	135	21
100		196	55	37	4	218	43
130		271	93	64	11	323	74
150		346	128	87	21	420	116
Ft.^{3} of air ft.$^{-3}$ of gas	16.7	23.9	31.0	31.0	38.2	21.5	28.6

* *Nat. Bur. Standards, Letter Circ.* 503.
† See the section on liquid fuels.

the earth's surface. The gas is usually under considerable pressure and flows naturally from the well to the connecting pipe lines. Vacuum pumps are used to extract the gas when the pressure has decreased to or below atmospheric pressure.

Any entrained sand must be separated from the gas before being supplied to an engine located near a gas well. No other cleaning is required since the gas usually does not contain sulphuric or other acid-forming constituents.

The composition of natural gas varies considerably but usually contains a large amount of methane (CH_4), a smaller amount of ethane (C_2H_6), traces of other hydrocarbons, and small amounts of CO_2 and N_2. Average compositions from various fields in this country are listed in Table 12.

The natural gas used in the production of natural gasoline, and the petroleum refinery gases, contain appreciable quantities of propane and butane which are easily condensed at atmospheric temperatures. These fuels are supplied in liquid state under pressure, in containers, and evaporate into a superheated vapor or gas before being supplied to an engine. The atmospheric temperature must be appreciably higher than the boiling temperature at atmospheric pressure in order to promote rapid evaporation. Commercial propane and butane contain other hydrocarbons which are listed in Table 8.

Producer Gas.—This gas is made by flowing air or air and steam through a thick coal or coke bed which ranges in temperature from red-hot to low-temperature fuel. The oxygen in the air burns the carbon to CO_2 which is reduced to CO by contacting hot carbon above the combustion zone. The steam is dissociated which introduces H_2, while the freed O_2 combines as above with the carbon. Producer gas has a high percentage of N_2 (Table 12) since air is used, which results in a low heating value.

Illuminating Gas.—*Blue water gas* is made by blowing steam through a hot bed of coal or coke, CO and H_2 being the principal gases formed (Table 12). This process reduces the temperature of the fuel bed so that alternate runs with air are required to raise its temperature during which time the gases are wasted.

Carbureted water gas contains hydrocarbons (Table 12) formed by spraying oil into a carburetor filled with hot brick checkerwork through which the gases pass.

Coal and *oil gas* are formed by applying heat to coal and oil, typical analyses being given in Table 12.

Coke-oven Gas.—This gas is similar to the coal or bench gas but is obtained from a process designed to produce coke. The composition varies appreciably depending upon the type of coal used in the process (Table 12). The volatile portion of the coal is driven off by the application of heat and the heavier hydrocarbons are cracked, which results in a gas high in H_2 and CH_4.

Blast-furnace Gas.—This gas is a by-product of the steel plants. It consists principally of CO and N_2 (Table 12), since it is formed by blowing air through cupolas containing alternate layers of hot coke and pig iron. It is similar to producer gas and has a low heating value.

Gas Cleaning.—All manufactured gases must be cleaned to reduce the dust and solid impurities to a desired minimum. Tar and ammonia, also, are removed by washing or scrubbing the gases. Sulphur is usually removed by passing the gases through iron oxide beds.

Ignition Temperature.—The ignition temperature of an air-fuel mixture depends upon the method and apparatus used.

TABLE 9.—IGNITION TEMPERATURES OF AIR-FUEL MIXTURES

Fuel	Symbol	With lag of about 8 to 10 sec.,* °F.	Instantaneous,† °F.
Hydrogen	H_2	1080	1380
Carbon monoxide	CO	1190	1710
Methane	CH_4	1200 (1030)‡	1830
Ethane	C_2H_6	970 (970)‡	
Propane	C_3H_8	910‡	
Acetylene	C_2H_2	760	
Ethylene	C_2H_4	1010	1830
Benzene	C_6H_6	....	1940
Ether	$C_4H_{10}O$	....	1890

* H. B. Dixon and H. F. Coward, *Trans. Chem. Soc.*, **98**, 514 (1909).
† J. W. McDavid, *Trans. Chem. Soc.*, **111**, 1003 (1917).
‡ With oxygen only.

An appreciable increase in pressure usually decreases the ignition temperature. Hence, ignition temperatures determined at atmospheric conditions may not be reliable indicators even of relative ignition temperatures in the engine.

Low ignition temperatures usually require an appreciable time lag before the appearance of flame. High ignition temperatures are reported for instantaneous appearance of flame. The lower temperature for any gas (Table 9) may be considered as an upper safety limit, below which the gas will not ignite in appreciable time, while the higher temperature may be considered as an upper limit which may be approached but should not be reached by the unburned part of the charge in an engine before the flame reaches it, if detonation is to be avoided.

Compression ratios are usually low enough so that the instantaneous ignition temperature is not reached during compression. However, the burned portion of the charge compresses the unburned portion which approaches and may reach the instantaneous ignition temperature. Thus, a reaction may start in the unburned portion before the flame from the burning portion

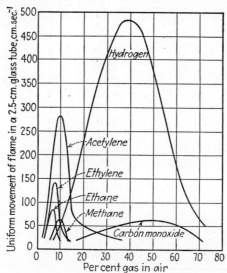

FIG. 79.—Effect of gas and mixture ratio on flame speed. (*W. R. Chapman, Trans. Chem. Soc.*, **119**, 1677, 1921.)

Note.—Maximum flame velocities for C_3H_8, C_4H_{10}, and C_5H_{12} are the same as for ethane but occur at 2.9, 3.7, and 4.7 per cent of gas in air, resp. (*W. Payman, Jour. Chem. Soc.*, **115**, 1446, 1919.)

reaches it. Rapid flame movement may sweep through the unburned portion before the reaction therein results in spontaneous inflammation.

Flame Speed.—Considerable differences exist between the flame speeds of the various air-fuel mixtures. For all fuels the flame speeds are low near the limits of inflammability and have a maximum near the correct mixture (Fig. 79).

Limits of Inflammability.—The range of air-fuel ratios that may be used in engines depends on the limits of inflammability of the fuel in air (Table 10). An increase in pressure increases the range of inflammable mixtures. However, mixture ratios near the limits of inflammability have very low rates of burning (Fig. 79) and are undesirable for engine use.

TABLE 10.—APPROXIMATE LIMITS OF INFLAMMABILITY OF GASES AND VAPORS IN AIR AT ATMOSPHERIC CONDITIONS*

Fuel	Symbol	Limits, % by vol.		
		Low	Correct mix.	High
Hydrogen	H_2	4.1	29.6	74
Carbon monoxide	CO	12.5	29.6	74
Methane	CH_4	5.3	9.5	14
Ethane	C_2H_6	3.2	5.7	12.5
Propane	C_3H_8	2.4	4.0	9.5
Butane	C_4H_{10}	1.9	3.1	8.5
Pentane	C_5H_{12}	1.4	2.6	8.0
Ethylene	C_2H_4	3.3	6.5	
Acetylene	C_2H_2	2.5	7.7	
Benzene	C_6H_6	1.4	2.7	8.0
Toluene	C_7H_8	1.3	2.3	7.0
Methyl alcohol	CH_3OH	6.0	12.3	
Ethyl alcohol	C_2H_5OH	3.5	5.7	19.0
Ethyl ether	$C_4H_{10}O$	1.7	3.4	
Gasoline	See	1.4	1.7	6.0
Water gas	the	9.0	31.8	55
Natural gas	reference	4.8	9.0	13.5
Illuminating gas	for	5.3	17.7	31
Blast-furnace gas	analyses	35–57	50	65

* H. F. Coward and G. W. Jones, *U. S. Dept. Com., Bull.* 279 (1928).

Compression-ratio Ranges.—Besides fuel characteristics, other factors such as size and speed of engine, combustion-chamber design, and cooling influence the determination of the permissible compression ratio. Consequently, the various fuels may be used in engines with compression ratios that vary through a given range (Table 11) depending upon combustion-chamber design and operating conditions.

Hydrogen* has a great tendency to detonate and also to backfire if the air-fuel mixture does not contain 25 per cent or more excess air, according to Burstall.[1] Carbon monoxide and

* Hydrogen has been used as an auxiliary fuel with fuel oil in a compression-ignition engine. See *N.A.C.A. Rept.* 535, 1935.

[1] A. F. Burstall, Experiments on the Behavior of Various Fuels in a High-speed Internal-combustion Engine, *I.A.E. Proc.*, **22**, 365 (1927–1928).

methane have much less tendency to detonate. In comparing the ranges of compression ratios it should be noted that illuminating gas has the highest percentage of H_2 and blast-furnace gas the lowest; that natural gas has the highest percentage of hydrocarbons, some of which may appreciably influence the tendency to detonate; and that both producer gas and blast-furnace gas are diluted with a large amount of N_2, which decreases the tendency to detonate.

TABLE 11.—PERMISSIBLE COMPRESSION RATIOS

Fuel	Range of r
Hydrogen	4 to 7*
Carbon monoxide	5 to 8
Methane	4 to 7
Propane (commercial)	6 to 7
Butane (commercial)	6.5 to 10
Illuminating gas	4 to 5
Natural gas	4 to 6
Producer gas	5 to 7
Blast-furnace gas	5 to 7

* Lean mixtures only (25 per cent excess air). See text, also.

Heating Value of Correct Mixture.—The heating value of a fuel depends upon the heating value of the various constituents of the fuel. Gases with constituents having low or no heating values, such as CO and N_2, respectively, will have low heating values. However, only small amounts of air are required to burn fuels with low heating values. Thus, the heating value per cubic foot of correct or desired air-fuel mixture, rather than the heating value of the fuel, is required to determine the possible performance of a fuel in an engine.

Example.—Blast-furnace gas contains $0.050H_2$, $0.258CO$, $0.005CH_4$, $0.094CO_2$, and $0.593N_2$ by volume analysis. Determine the heating value at constant pressure per cubic foot of fuel, cubic feet of air per cubic foot of fuel, and the heating value per cubic foot of correct mixture at 77°F.

From Table VIII (Appendix),

$$\begin{aligned}
&\text{H.V. of } 0.05 \text{ ft.}^3 \text{ of } H_2 = 0.05 \times 314 = 15.7 \text{ B.t.u.} \\
&\text{H.V. of } 0.258 \text{ ft.}^3 \text{ of } CO = 0.258 \times 311 = 80.2 \text{ B.t.u.} \\
&\text{H.V. of } 0.005 \text{ ft.}^3 \text{ of } CH_4 = 0.005 \times 978 = \underline{4.9 \text{ B.t.u.}} \\
&\hspace{6cm} 100.8 \text{ B.t.u.}
\end{aligned}$$

From the reaction equations for the various constituents, the air required is

$$(0.05 + 0.258)2.38 + 0.005 \times 9.52 = 0.78 \text{ ft.}^3 \text{ ft.}^{-3} \text{ of fuel.}$$

The heating value per cubic foot of correct mixture is

$$\frac{100.8}{1 + 0.78} = 57 \text{ B.t.u.}$$

The heating values and air requirements for the fuels in Table 12 were computed in the foregoing manner. The heating values for these fuels vary from 101 to 1189 B.t.u. ft.$^{-3}$ of fuel, whereas the heating values per cubic foot of correct mixture vary only from 57 to 95 B.t.u.

H_2 and CO are two constituents that result in less mols of products than mols of mixture (Table VIII, Appendix). This decrease causes lower pressure along the expansion curve in the engine than if there were no molal contraction. CH_4 and C_2H_4 have no change in mols during combustion, while the heavier hydrocarbons have an increase in mols of products compared with mols of mixture. Such an increase raises the pressure along the expansion curve and is a desirable characteristic. The product

TABLE 12.—DATA ON COMMON FUEL GASES

Kind of gas	Constituents of gas by vol., %								Q_p, B.t.u. ft.$^{-3}$ at 77°F.	Air, ft.^{3} ft.$^{-3}$ of gas	B.t.u. ft.$^{-3}$ of corr. mixture	Mp/Mm
	H_2	CO	CH_4	C_2H_4	C_2H_6	O_2	CO_2	N_2				
Natural gas:[1]												
California			68.9		19.3		11.3	0.5	1004	9.77	93	1.009
Ohio			79.7		17.0		0.1	3.2	1070	10.42	94	1.007
Oklahoma			83.1		10.5		0.7	5.7	992	9.66	93	1.005
Pennsylvania			68.4		30.4		0.1	1.1	1189	11.58	95	1.012
Texas			57.7		9.0		0.2	33.1	718	6.99	90	1.006
Producer gas:												
Anthracite coal	20.0	25.0				0.5	5.0	49.5	140	1.05	68	0.890
Bituminous coal	10.0	23.0	3.0	0.5		0.5	5.0	58.0	140	1.12	66	0.922
Coke	10.0	29.0				0.5	4.5	56.0	122	0.90	64	0.898
Illuminating gas:												
Blue water gas	50.0	43.3	0.5				3.0	3.2	297	2.27	91	0.857
Carbureted water gas	40.0	19.0	25.0	8.5		0.5	3.0	4.0	561	4.97	94	0.951
Coal or bench gas	46.0	6.0	40.0	5.0		0.5	0.5	2.0	632	5.74	94	0.961
Oil gas	32.0		48.0	16.5		0.5		3.0	826	7.66	95	0.981
By-product gas:												
Coke-oven gas	53.0	6.0	35.0	2.0			2.0	2.0	558	5.02	93	0.951
Coke-oven gas	50.0	6.0	36.0	4.0		0.5	1.5	2.0	590	5.31	94	0.956
Blast-furnace gas	5.0	25.8	0.5				9.4	59.3	101	0.78	57	0.914
Blast-furnace gas	5.2	26.8	1.6			0.2	8.2	58.0	115	0.90	61	0.916

[1] The analyses given are the averages of samples from 3 to 10 fields. There is considerable variation in analyses, some wells in Mexico having a very high percentage of CO_2 and very little fuel constituents.

of this ratio and the heating value per cubic foot of correct mixture results in an equivalent heating value by which the energy input to an engine with various fuels may be compared.

Volumetric Efficiency. *Gaseous Fuels.*—The volumetric efficiency of a gas engine is defined as the ratio of the volume of the charge inducted, at the atmospheric conditions surrounding the engine, to the piston displacement. Analyzing the engine process with any fuel, as indicated in Chap. V, results in data from which the volumetric efficiency can be obtained. It has been shown that clearance-gas temperature, atmospheric temperature, and compression ratio can have only a small influence on volumetric efficiency owing to variation in specific heats with variation in temperature ranges for the cooling of the clearance gases and the heating of the incoming charge. The gaseous fuel for which the ratio of mean specific heats of the clearance gases and the incoming charge is the smallest should have the highest volumetric efficiency, since the relative contraction of the clearance gases will be the largest and the relative expansion of the incoming charge will be the smallest for a given energy transfer between the two. Inspection of energy data in the Appendix indicates that the H_2-air mixture has an advantage over the CO-air mixture, but is about equivalent to the CH_4-air mixture, whereas the C_2H_6-air mixture is more desirable than the foregoing mixtures.

Effect of Specific Heats and Pressures.—The characteristic equation for the clearance gases is

$$P_c V_c = M_c R T_c \qquad (1)$$

where P_c = pr. of the clearance gases at the end of the exhaust stroke,

V_c = clearance vol.,

M_c = mols of clearance gases,

T_c = temperature of exhaust products in the clearance space at the end of the exhaust stroke.

The characteristic equation for the gases at the end of the suction stroke is

$$P_1 V_1 = (M_m + M_c) R T_1 \qquad (2)$$

where M_m = mols of fuel and air inducted.

Combining Eqs. (1) and (2) and noting that $V_1/V_c = r$, the compression ratio, results in

$$\frac{M_m}{M_c} = r\frac{T_c}{T_1}\frac{P_1}{P_c} - 1. \tag{3}$$

The heat-transfer process during the suction stroke is approximated by assuming constant-volume conditions. Thus, the heat given up by the clearance gases is equal to that absorbed by the fresh charge,

or
$$M_c(E_{T_c} - E_{T_1})_c = M_m(E_{T_1} - E_{T_a})_m \tag{4}$$

or
$$\frac{M_m}{M_c} = \frac{(E_{T_c} - E_{T_1})_c}{(E_{T_1} - E_{T_a})_m}. \tag{5}$$

The suction temperature may be determined by solving Eqs. (3) and (5).

Example.—Determine the suction temperature when using methane as a fuel in an engine with a compression ratio of 5:1. Temperature of products in clearance space = 1040°F. $T_a = 60°F.$, $P_1 = 14.2$ lb. in.$^{-2}$ abs., and $P_c = 15.2$ lb. in.$^{-2}$ abs.,

Mixture		Products	
CH_4	1.00	CO_2	1.00
O_2	2.00	H_2O	2.00
N_2	7.52	N_2	7.52
Total	10.52 mols		10.52 mols

M_m/M_c in Eqs. (3) and (5) is determined in the following manner:

Assumed values of T_1, °R.	580	600	620
$r\dfrac{T_c}{T_1}\dfrac{P_1}{P_c}$	12.1	11.7	11.3
M_m/M_c from Eq. (3)	11.1	10.7	10.3
E_{T_c} for clearance gases	5702	5702	5702
E_{T_1} for clearance gases	319	428	538
$(E_{T_c} - E_{T_1})_c$	5383	5274	5164
E_{T_1} for air-fuel mixture	306	410	515
E_{T_a} for air-fuel mixture	0	0	0
$(E_{T_1} - E_{T_a})_m$	306	410	515
M_m/M_c from Eq. (5)	17.6	12.9	10.0

Plotting the values of M_m/M_c at the assumed temperatures (Fig. 80) results in $T_1 = 616°R.$ and $M_m/M_c = 10.4$.

The effect of a 0.5 lb. increase in clearance-gas pressure and the same decrease in suction pressure is also shown in this illustration.

The volume of the charge inducted at atmospheric conditions is

$$V_m = \frac{M_m R T_a}{P_a}. \qquad (6)$$

The volume displaced is determined from the total mols of mixture in the cylinder at the conditions at the end of the suction stroke.

Thus
$$V_{disp} = \frac{(M_m + M_c)RT_1}{P_1}\left(\frac{r-1}{r}\right) \qquad (7)$$

and   Vol. eff. $= \dfrac{V_m}{V_{disp}} = \dfrac{M_m}{M_m + M_c}\left(\dfrac{r}{r-1}\right)\dfrac{P_1}{P_a}\dfrac{T_a}{T_1}.$ (8)

Example.—Determine the ideal volumetric efficiency and the effect of 0.5 lb. decrease in suction pressure and 0.5 lb. increase in back pressure for the data in the example on page 152.

From Fig. 80, $T_1 = 609°R.$ and $M_m/M_c = 11.3$ with ideal conditions of $P_1 = P_c = 14.7$ lb. in.$^{-2}$ Using $M_c = 1$ mol, results in

Vol. eff. $= \dfrac{11.3}{12.3} \times \dfrac{5}{4} \times \dfrac{520}{609} = 0.981$

which is the ideal volumetric efficiency.

In the same manner, it is found that a decrease in suction pressure to 14.2 lb. in.$^{-2}$ reduces the volumetric efficiency to about 0.94 (Fig. 81), and an increase in clearance-gas pressure to 15.2 lb. in.$^{-2}$ with the lowered suction pressure reduces the volumetric efficiency to about 0.93.

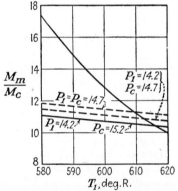

Fig. 80.—Suction temperature solution.

The relative effect of variations of suction pressure and clearance-gas pressure on volumetric efficiency (Fig. 81) shows the importance of maintaining high suction pressures rather than low clearance-gas pressures.

Effect of Heat Transfer.—The transfer of heat from the cylinder walls to the gases during the suction stroke adds the term $-M_m Q_{in}$ to the right side of Eq. (4). Thus Eq. (5) becomes

$$\frac{M_m}{M_c} = \frac{(E_{T_c} - E_{T_1})_c}{(E_{T_1} - E_{T_a} - Q_{in})_m}. \qquad (9)$$

When applied to the previous example the volumetric efficiency is reduced from 0.93 to about 0.86 for 200 B.t.u. of heat transfer per mol of incoming charge. Thus heat transfer from the cylinder walls to entering mixture during the suction stroke has an appreciable effect (Fig. 81) on the volumetric efficiency. Heat transfer usually amounts to 100 to 300 B.t.u. mol^{-1} of incoming charge, depending on conditions.

The effects of variations in atmospheric temperature T_a, and compression ratio r are negligible, and the effect of clearance-gas temperature is almost negligible (Fig. 81).

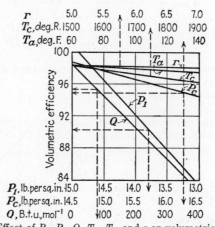

Fig. 81.—Effect of P_1, P_c, Q, T_a, T_c, and r on volumetric efficiency.

The fixed values for the various curves are $r = 5$, $T_c = 1500°\text{R}$., $T_a = 60°\text{F}$., $P_1 = 14.7$ lb. in.^{-2}, $P_c = 14.7$ lb. in.^{-2}, and $Q = 0$ B.t.u. mol^{-1}, excepting the value for the curve considered. For the foregoing conditions the volumetric efficiency is 0.982.

Effect of Liquid Fuels.—Liquid fuels are usually not completely evaporated when the air-fuel mixture enters the cylinder. This may result in lower mixture temperature than when gaseous fuels are being used. Also, heat transfer to the entering charge during the suction stroke will further the evaporation of the fuel, which minimizes the rise in mixture temperature. This tends to result in lower mixture temperatures and higher volumetric efficiencies, but fuel evaporation tends to reduce the quantity of mixture entering, particularly in the case of fuels with large fuel-vapor volumes compared with the air volume. Thus, each case requires separate analysis to determine the net effect of heat transfer, but the effect of the other factors is the same as for gaseous fuels.

Effect of Value Timing.—The intake valve closes when the piston has completed 5 to 15 per cent of the compression stroke.

LIQUID AND GASEOUS FUELS

The charge continues to flow into the cylinder while the valve is closing until the difference in pressure between the port and the cylinder is equalized and until the inertia effect of the flowing charge is reduced to zero by the tendency of the charge to flow out of the cylinder. Thus, the volumetric efficiency depends upon the pressure and temperature of the charge in the cylinder at intake-valve closure.

Example.—The pressure in the cylinder at intake-valve closure is 13.7 lb. in.$^{-2}$ abs. Clearance-gas pressure is 15.2 lb. in.$^{-2}$ abs. The mixture enters the cylinder at a temperature of 100°F., and 200 B.t.u. mol^{-1} of mixture are transferred to the charge before the intake valve closes at 12 per cent of the compression stroke. Estimate the volumetric efficiency.

The ideal volumetric efficiency has been shown to be about 103 per cent with octane as the fuel (page 94). The effect of the various factors is obtained from Fig. 81, if it is assumed that the net effect of heat transfer is the same as for gaseous fuels.

Thus, Vol. eff. = ideal vol. eff. $[P_1][P_c][T_a][Q]$[valve timing]

$$= 1.03 \times \frac{89.5}{98.2} \times \frac{97.2}{98.2} \times \frac{97.9}{98.2} \times \frac{90.6}{98.2} \times 0.88 = 0.75.$$

Comparison of Liquid Fuels.—The fuels that may be used in the carbureted spark-ignition engine have different properties which appreciably affect the mixture temperature and percentage evaporated. This results in different volumetric efficiencies and power outputs in an engine with a given compression ratio.

TABLE 13.—COMPARISON OF FUELS*

Fuel	Correct air-fuel ratio by wt.	Heat supplied, 100% evap., B.t.u.	% evap., 200 B.t.u. heat lb.$^{-1}$ fuel	Manifold mixture temperature, °F.		Relative vol. eff.		Heating value, B.t.u. ft.$^{-3}$, of mixture	
				100% evap.	200 B.t.u.	100% evap.	200 B.t.u.	100% evap.	200 B.t.u.
Gasoline†	15.11	340	79	106	81	100	100	95	99
Benzene	13.25	149	100	50	64	111	103	104	101
Ethyl alcohol	8.99	425	55	71	55	107	105	102	108
Methyl alcohol	6.46	519	47	68	44	108	107	105	118

* See Table VIII (Appendix).
† Volatility characteristics of Summer gasoline, Fig. 70. Other data those of octane.

The effect based upon complete evaporation and also upon a given heat input is given in Table 13.

The higher volumetric efficiencies are due to the lower mixture temperatures, the effect being estimated from Fig. 81. In a given engine the lower mixture temperatures would tend to cause more heat to be transferred to the mixture, which would reduce the differences indicated.

The fuels rate[1] in the following order, based on the heating value of the mixture for 100 per cent evaporated and used in engines of the same compression ratio: gasoline, 100; ethyl alcohol, 109; benzene, 107; methyl alcohol, 111. Considering 200 B.t.u. heat input, the fuels rate[1] in the following order: gasoline, 100; benzene, 102; ethyl alcohol, 109; methyl alcohol, 119.

In general, liquid fuels having the higher latent heats will have the lower mixture temperatures and the higher volumetric efficiencies for a given heat supply. Also, a low percentage of fuel evaporated in the manifold indicates a low mixture temperature at the end of the suction stroke.

All the foregoing ratings are based upon the assumption that the distribution characteristics are not impaired by any of the changes indicated. If insufficient heat is added to the manifold, the lack of evaporation will result in poor distribution and affect the power output.

Blends of Fuels.—Blends of fuels usually result in performance that is better than that of the poorer fuel but poorer than that of the better fuel in the blend. Thus, a 10 per cent blend of ethyl alcohol and gasoline should increase the power output less than 1 per cent compared with the results obtained with gasoline. Actually, very little difference is observed.[2,3]

An engine having poor distribution with a given fuel may have its performance improved by blending with the fuel a fraction of the light ends or very volatile portion of gasoline. Also, detonation characteristics may be improved to such an extent by the addition of fractions of certain fuels that operating conditions may be changed and power output increased.

[1] These ratings are approximate since only the effect of equilibrium manifold mixture temperature is considered.

[2] L. C. Lichty and E. J. Ziurys, Engine Performance with Gasoline and Alcohol, *Ind. Eng. Chem.*, **28**, 1094 (1936).

[3] L. C. Lichty and C. W. Phelps, Gasoline-alcohol Blends in Internal-combustion Engines, *Ind. Eng. Chem.*, **30**, 222 (1938).

LIQUID AND GASEOUS FUELS

The various hydrocarbon fuels are readily blended and do not separate. However, methyl alcohol is not miscible with gasoline, and ethyl alcohol is miscible only to a degree. Small quantities of water cause the ethyl alcohol and water to separate into their respective layers, which is one of the problems encountered in the use of this fuel. The water tolerance varies somewhat with the various gasolines, but more particularly with the percentage of gasoline in the mixture, and can be represented by the equation[1]

$$\log (\text{per cent water}) = f_1(P) + \frac{f_2(P) + Kf_3(P)}{410}\left[1 - \frac{738}{T}\right] \quad (1)$$

where the various $f(P)$ terms = functions of the percentage of gasoline in the mixture (Table 14),

K = a constant that depends upon the gasoline (varying from about 420 to 670, having an average value of about 550),

and $T = °R.$

TABLE 14.—VALUES OF P FUNCTIONS

P, % of gasoline in mixture	Value of $f_1(P)$	Value of $f_2(P)$	Value of $f_3(P)$
10	1.310	−123.0	0.639
20	1.246	−76.5	0.783
30	1.172	−43.8	0.892
40	1.086	−20.0	0.964
50	0.986	0.0	1.000
60	0.862	21.0	1.000
70	0.703	48.1	0.964
80	0.479	86.3	0.892
90	0.096	140.5	0.783
95	−0.287	175.1	0.716

Water tolerances for 5 and 10 per cent blends of ethyl alcohol and average gasoline show (Fig. 82) the tendency of the fuels to separate at winter temperatures.

[1] O. C. Bridgeman and Elizabeth W. Aldrich, Water Tolerances of Mixtures of Gasoline with Ethyl Alcohol, *Nat. Bur. Standards, Jour. Res.*, **20**, 1 (1938).

Exhaust-gas Analysis.—The products of complete combustion of a hydrocarbon are CO_2 and H_2O. N_2 is always present in the products since air is used for the oxygen supply. Actually, complete combustion never occurs even with correct mixtures of air and fuel, but it is approached with lean mixtures.

Rich mixtures result in CO, H_2, CH_4, and O_2 in addition to CO_2 and H_2O, whereas lean mixtures even considerably removed from the correct mixture have the same constituents.[1,2]

The maximum possible percentage of CO_2, as obtained by Orsat analysis, can be determined from the reaction equation for complete combustion.

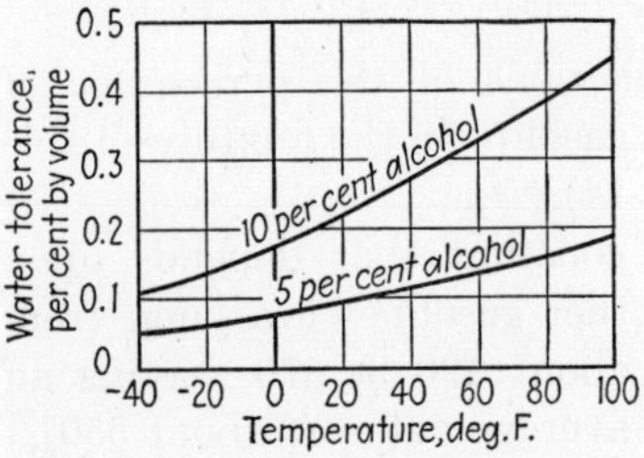

FIG. 82.—Water tolerances of two blends of average gasoline and ethyl alcohol.

Example.—The products of combustion for the correct mixture of 1 mol of octane and air are $8CO_2$, $9H_2O$, and $47N_2$. The Orsat apparatus eliminates the H_2O and gives the dry-gas analysis.[3]

Thus,

$$CO_2 = \frac{8}{8 + 47} \times 100 = 14.5 \text{ per cent.}$$

The theoretical analysis of products of combustion of lean mixtures can be determined in the same manner.

Comparatively recent analyses of the exhaust products of engines using several grades of aviation gasoline and Diesel fuels indicate the following simple approximate relationships[4] for H_2, CO, and CH_4:

$$H_2 = 0.51 CO. \tag{1}$$
$$CH_4 = 0.22 \text{ per cent.} \tag{2}$$

---

[1] S. H. Graf, G. W. Gleason, and W. H. Paul, Interpretation of Exhaust Gas Analysis, *Ore. State Agric. Coll., Eng. Expt. Sta., Bull.* 4 (1934).

[2] H. C. Gerrish and A. M. Tessmann, Relation of H_2 and CH_4 to CO in Exhaust Gases from Internal Combustion Engines, *N.A.C.A. Rept.* 476 (1933).

[3] L. C. Lichty, "Thermodynamics," p. 149, McGraw-Hill Book Company, Inc., New York, 1936.

[4] Gerrish and Tessmann, *op. cit.* Graf, Gleason, and Paul, *op. cit.*, considering all published data available, concluded that $H_2 = 0.40 CO$ up to 5 per cent CO; also that CH_4 is almost directly proportional to the CO. The authors call attention to marked deviations probably due to poor conditions and possible errors in sampling and analysis of gases. See this reference for a bibliography.

The average value of the H/C weight ratio for the foregoing fuels was 0.175.

The water vapor formed will be indicated by the mols of the exhaust-gas constituents. Thus,

$$H_2O = 5.955\frac{H}{C}(CO_2 + CO + CH_4) - H_2 - 2CH_4. \quad (3)$$

The consumed O_2 appears in the mols containing O and is also given by the N_2 and O_2 appearing in the products, which are equated as follows:

$$\text{Consumed oxygen} = CO_2 + \frac{1}{2}CO + \frac{1}{2}H_2O = \frac{20.9}{79.1}N_2 - O_2. \quad (4)$$

Eliminating H_2O between Eqs. (3) and (4) results in

$$N_2 = 3.785(O_2 + CO_2 - CH_4) + 1.892(CO - H_2) + 11.268\frac{H}{C}(CO_2 + CO + CH_4). \quad (5)$$

N_2 may also be obtained by difference, 100 mols being assumed as the total quantity.

$$N_2 = 100 - O_2 - CO_2 - CO - H_2 - CH_4. \quad (6)$$

Eliminating the N_2 between Eqs. (5) and (6), results in

$$O_2 + CO_2\left(1 + 2.355\frac{H}{C}\right) + CO\left(0.604 + 2.355\frac{H}{C}\right) - 0.186H_2$$
$$CH_4\left(0.582 - 2.355\frac{H}{C}\right) = 20.9, \quad (7)$$

which is the general theoretical relationship between the products of combustion of the hydrocarbons.

Inserting the values of H_2, CH_4 and H/C in Eq. (7) produces

$$CO = 22.733 - 1.086O_2 - 1.533CO_2. \quad (8)$$

Inserting this value of CO in Eq. (1) results in,

$$H_2 = 11.594 - 0.554O_2 - 0.782CO_2. \quad (9)$$

Inserting the values of H_2, CH_4, CO, and H/C in Eq. (3) yields

$$H_2O = 11.885 - 0.578O_2 + 0.226CO_2. \quad (10)$$

Also, Eq. (6) becomes

$$N_2 = 65.453 + 0.6395O_2 + 1.3148CO_2. \quad (11)$$

The air-fuel ratio by weight is

$$\frac{\text{Air}}{\text{Fuel}} = \frac{28.84(N_2/0.791)}{12(CO_2 + CO + CH_4) + 2.015(H_2 + H_2O + 2CH_4)}. \quad (12)$$

Inserting the values for N_2, CO, CH_4, H_2 and H_2O, produces

$$\frac{\text{Air}}{\text{Fuel}} = 1.523\left(\frac{102.350 + O_2 + 2.056 CO_2}{21.139 - O_2 - 0.491 CO_2}\right). \quad (13)$$

The air-fuel ratios determined from exhaust-gas analysis do not include the carbon formation that occurs with rich mixtures and, consequently, are higher than actual air-fuel ratios by about 3 per cent[1] for rich mixtures.

The ratio of the weight of H_2O to the weight of fuel from which the products were formed is

$$\frac{\text{Water}}{\text{Fuel}} = \frac{18.015 H_2O}{12(CO_2 + CO + CH_4) + 2.015(H_2 + H_2O + 2CH_4)}. \quad (14)$$

Substituting the relations for H_2O, CO, CH_4, and H_2 produces

$$\frac{\text{Water}}{\text{Fuel}} = 0.680\left(\frac{20.562 - O_2 + 0.391 CO_2}{21.139 - O_2 - 0.491 CO_2}\right). \quad (15)$$

All the foregoing relationships are taken either from the work of Gerrish and Tessman[2] or of Gerrish and Voss.[3] The CO_2 curve (solid line, Fig. 83) represents experimental results,[3] whereas the other solid lines were computed from the foregoing equations. Table 15 contains the computed results for various air-fuel ratios.

Example.—The CO_2 in the exhaust analysis is 13.02 per cent for a fuel having an H/C ratio of 0.175 and a 14:1 air-fuel ratio (Table 15). Compute the various other constituents.

From Eq. (13), $\quad 14 = 1.523\left(\dfrac{102.350 + O_2 + 2.056 \times 13.02}{21.139 - O_2 - 0.491 \times 13.02}\right)$

or $\quad O_2 = 0.62$ per cent.

[1] B. A. D'Alleva and W. G. Lovell, Relation of Exhaust-gas Composition to Air-fuel Ratio, *S.A.E. Trans.*, **31**, 90 (1936).

[2] Gerrish and Tessmann, *op. cit.*

[3] H. C. Gerrish and F. Voss, Interrelation of Exhaust-gas Constituents, *N.A.C.A. Rept.* 616 (1937).

From Eq. (8),

$CO = 22.733 - 1.086 \times 0.62 - 1.523 \times 13.02 = 2.10$ per cent.

From Eq. (1), $\quad H_2 = 0.51 \times 2.10 = 1.07$ per cent.
From Eq. (11),

$N_2 = 65.453 + 0.6395 \times 0.62 + 1.3148 \times 13.02 = 82.97$ per cent.

From Eq. (10),

$H_2O = 11.885 - 0.578 \times 0.62 + 0.226 \times 13.02 = 14.47$ per cent.

and from Eq. (15),

$$\frac{\text{Water}}{\text{Fuel}} = 0.680 \left(\frac{20.562 - 0.62 + 0.391 \times 13.02}{21.139 - 0.62 - 0.491 \times 13.02} \right) = 1.205$$

TABLE 15.—N.A.C.A. GAS-ANALYSIS TABLE
(For fuels having an H/C ratio of 0.175)

Air / Fuel	CO_2, %	O_2, %	CO, %	H_2, %	N_2, %	H_2O, %	H_2O / CO_2	Water / Fuel	Combustion eff., %
11	8.76	0.15	9.14	4.66	77.08	13.78	1.57	0.972	66.7
12	10.18	0.44	6.65	3.39	79.13	13.93	1.37	1.043	73.8
13	11.60	0.59	4.31	2.20	81.09	14.16	1.22	1.122	81.5
14	13.02	0.62	2.10	1.07	82.97	14.47	1.11	1.205	89.6
15	13.23	1.35	0.99	0.50	83.72	14.09	1.06	1.247	93.8
16	12.62	2.49	0.68	0.35	83.65	13.30	1.05	1.256	94.8
17	12.00	3.55	0.48	0.25	83.51	12.54	1.05	1.261	95.5
18	11.45	4.49	0.30	0.16	83.39	11.88	1.04	1.267	96.2
19	10.90	5.36	0.20	0.10	83.23	11.25	1.03	1.269	96.5
20	10.40	6.15	0.11	0.06	83.07	10.68	1.03	1.272	96.8
21	9.92	6.86	0.08	0.04	82.90	10.16	1.03	1.271	96.9
22	9.44	7.55	0.06	0.03	82.71	9.65	1.02	1.268	96.8
23	9.00	8.18	0.05	0.03	82.53	9.19	1.02	1.266	96.7
24	8.60	8.74	0.06	0.03	82.37	8.78	1.02	1.264	96.6

The curves of Graf, Gleason, and Paul (Fig. 83) were based upon the results of various experiments with various engines and fuels. The D'Alleva and Lovell[1] curves represent the results of their experiments on three multicylinder engines with one fuel. Their curves are plotted against the measured air-fuel ratio instead of the air-fuel ratio indicated by the products analysis.

[1] D'Alleva and Lovell, *op. cit.*

The usual effect of poor distribution between cylinders and poor mixing in any cylinder is to reduce the percentage of CO_2. This is obvious from an inspection of Fig. 83. At 14.4:1 air-fuel ratio the per cent of CO_2 is 13.6. If half the cylinders get a 12.4:1 air-fuel ratio, the per cent of CO_2 in the exhaust from these cylinders will be 10.8. The other half of the cylinders then would get an air-fuel ratio of 16.4, resulting in the CO_2 from those cylinders being 12.3 per cent. Obviously, the exhaust from all the cylinders will show about 11.5 per cent CO_2, instead

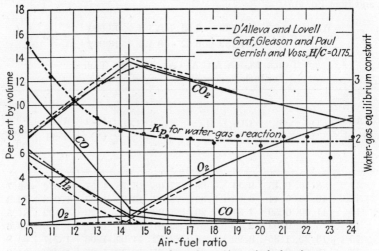

Fig. 83.—Exhaust products for various air-fuel ratios.

of the expected value, 13.6 per cent. If, however, the leanest and richest cylinders are on the same branch of the CO_2 curve (Fig. 83), the exhaust-gas analysis would not be affected appreciably by distribution.

Exhaust-gas analyses of products of combustion of mixtures of two fuels, such as alcohol-gasoline blends, lie between the analyses obtained with the individual fuels.[1]

Lovell and Boyd[2] found that the water-gas equilibrium constant for a number of gas analyses varied in most cases from 3 to 4, with an average value of 3.6. D'Alleva and Lovell[3]

---

[1] L. C. Lichty and C. W. Phelps, Carbon Monoxide in Engine Exhaust Using Alcohol Blends, *Ind. Eng. Chem.*, **29**, 495 (1937).

[2] W. G. Lovell and T. A. Boyd, Chemical Equilibrium in Gases Exhausted by Gasoline Engines, *Ind. Eng. Chem.*, **17**, 1216 (1925).

[3] D'Alleva and Lovell, *op. cit.*

LIQUID AND GASEOUS FUELS 163

found an average value for their experiments of 3.8, the values ranging mostly from 3 to 5. The values of this constant for the values in Table 15, plotted in Fig. 83, show an appreciable decrease with an increase in air-fuel ratio for rich mixtures and a practically constant value of about 2.0 for lean mixtures.

Combustion Efficiency.—The combustion efficiency is defined as the ratio of the energy liberated to that which could be liberated under ideal conditions. CO_2, CO, and H_2O indicate energy liberated, whereas H_2, CO, and CH_4 indicate unliberated energy. Using the heating value at constant pressure (Tables V and VIII) as the energy liberation, results in

Energy liberated = $(174.2CO_2 + 52.5CO + 123.0H_2O)1000$.
Energy unliberated = $(123.0H_2 + 121.7CO + 383.0CH_4)1000$.
Combustion efficiency =
$$\frac{174.2CO_2 + 52.5CO + 123.0H_2O}{174.2CO_2 + 174.2CO + 123.0(H_2 + H_2O) + 383.0CH_4}. \quad (16)$$

Substituting the equations for CO, H_2O, H_2 in terms of O_2 and CO_2, the value of 0.22 for CH_4, and simplifying, result in

$$\text{Comb. eff.} = 0.390\left(\frac{20.73 - O_2 + 0.949CO_2}{21.11 - O_2 - 0.491CO_2}\right). \quad (17)$$

Example.—Determine the combustion efficiency for the 20:1 air-fuel ratio. H/C ratio = 0.175.

From Table 15, $CO_2 = 10.40$ and $O_2 = 6.15$ per cent.

$$\text{Comb. eff.} = 0.390\left(\frac{20.73 - 6.15 + 0.949 \times 10.40}{21.11 - O_2 - 0.491 \times 10.40}\right) = 96.8 \text{ per cent.}$$

EXERCISES

1. Gasoline varies in specific gravity from 50 to 70° A.P.I. Determine the variation in specific gravity compared with water.

2. Determine and plot the 8:1, 16:1, and 20:1 volatility-temperature curves for the winter gasoline (Fig. 70).

3. Determine the minimum satisfactory starting temperature for the fuel labeled "Summer, 1937" in Fig. 70.

4. Determine and plot the relative warming-up period for variations in the 20 per cent point of the winter gasoline (Fig. 70), maintaining the same 70 per cent point. Also, vary the 70 per cent point while holding the 20 per cent point constant.

5. Estimate the amount that must be evaporated at 158°F. of a gasoline with a 90 per cent point temperature of 375°F. if it is to have equal accelerating characteristics with the winter gasoline (Fig. 70).

164   *INTERNAL-COMBUSTION ENGINES*

6. Determine the dew points for various 90 per cent points, using the winter gasoline (Fig. 70). Estimate and plot the probable change in slope as the 90 per cent point temperature is varied.

7. Assume a 12:1 air-fuel ratio mixture flows into the manifold of a supercharged aircraft engine using the winter gasoline (Fig. 70). Determine the mixture temperature at which a 12:1 air-vapor ratio can be obtained with a manifold pressure of 18 lb. in.$^{-2}$ abs.

8. Assume the mixture in Prob. 7 is compressed along the path $PV^{1.3}$ = const. Determine the amount of heat that must be added to the mixture. Initial air and fuel temperature = 40°F.

9. A gasoline engine operates with an intake-manifold pressure of 18.0 lb. in.$^{-2}$ abs. and an exhaust pressure of 14 lb. in.$^{-2}$ abs. The mixture temperature entering the cylinder is 150°F. During the suction stroke, 150 B.t.u. of heat transfer per mol of mixture occur. The intake valve closes after 10 per cent of the compression stroke occurs. Estimate the volumetric efficiency.

10. Assume a 13:1 air-benzene mixture is at 70°F. Obviously, from Table 5, the vapor will be superheated. Assume the pressure-volume relationship between the saturated and superheated vapor is PV = const. Determine the volume of the mixture per pound of air.

11. A gas has the following volumetric analysis: H_2 = 0.35, CO = 0.30, CH_4 = 0.20, C_2H_4 = 0.05, O_2 = 0.02, CO_2 = 0.03 and N_2 = 0.05. Determine the heating value ft.$^{-3}$ of fuel, ft.^{3} of air ft.$^{-3}$ of fuel, and the heating value per cubic foot of correct mixture at conditions of 1 atm. pressure and 77°F. temperature.

12. Solve Prob. 11 for air-fuel ratios representing 10 and 20 per cent excess air.

13. Determine the suction temperature when methane is used as a fuel in an engine with a compression ratio of 5:1. P_c = 14.7 lb. in.$^{-2}$, T_c = 1040°F. and T_a = 60°F. when P_1 = 14.7 lb. in.$^{-2}$ Use values for P_1 of 15.7, 17.7, and 19.7 lb. in.$^{-2}$ Assume supercharger increases T_a due to compression along the path $PV^{1.3}$ = const. Also, assume 150 B.t.u. of heat transfer from the cylinder walls to each mol of correct mixture.

14. Determine the volumetric efficiency and estimate the displacement required for a 100-hp. gas engine with a compression ratio of 5:1, running at 500 r.p.m. M_m/M_c = 15:1, atm. temperature = 520°R., suction temperature = 650°R., and suction pr. = 16.7 lb. in.$^{-2}$

15. Check the values in Table 13.

16. Determine the H/C ratio for octane. Then derive Eqs. (8), (9), (10), (11), (13), (15), and (17), pages 159, 160 and 163, for octane.

17. Determine the effect on the equations listed in Prob. 16 when the approximate molecular and atomic weights listed in Table 2 are used.

18. Check the values in one line in Table 15, at some air-fuel ratio other than 14:1.

CHAPTER VIII

DETONATION AND KNOCK TESTING

DETONATION

The Detonation Process.—Detonation in the internal-combustion engine was not given serious consideration until Kettering[1] in this country and Ricardo[2] in England recognized that this characteristic of the combustion process limited the compression ratio at which an engine could be operated. Early attempts to use kerosene instead of gasoline necessitated the lowering of the compression ratio with a corresponding loss of power.

A gasoline engine not knocking at wide-open throttle may be made to knock by raising the compression ratio. A small increase in compression ratio usually causes incipient knocking or detonation. Further increases in compression ratio increase the severity of the knocking until preignition, overheating, and loss of power result. Continued severe detonation usually damages the engine.

The combustion process starts from a small nucleus between the spark-plug points and spreads nearly radially from this point.[3] Combustion is thought by some[4] to take place in a comparatively narrow zone, which may be termed the "flame front," and by others[5] to continue after the flame front has passed through the charge. The high temperature of the products formed in the flame front results in an expansion

[1] C. F. Kettering, More Efficient Utilization of Fuel; *S.A.E. Jour.*, **4**, 263 (1919).

[2] H. R. Ricardo, Paraffin as Fuel, *Auto. Eng.*, **9**, 2 (1919).

[3] C. F. Marvin, Jr., and R. D. Best, Flame Movement and Pressure Development in the Engine Cylinder, *N.A.C.A. Rept.* 399 (1931).

C. F. Marvin, Jr., A. Wharton, and C. H. Roeder, Further Studies of Flame Movement and Pressure Development in the Engine Cylinder, *N.A.C.A. Rept.* 556 (1936).

[4] L. Withrow, W. G. Lovell, and T. A. Boyd, Following Combustion in the Engine by Chemical Means, *Ind. Eng. Chem.*, **22**, 945 (1930).

[5] A. M. Rothrock and R. C. Spencer, A Photographic Study of Combustion and Knock in a Spark-ignition Engine, *N.A.C.A. Rept.* 622, 1938.

which compresses the burned fraction back of the flame front, and the unburned fraction ahead of the flame front (see Chap. XII). This raises the temperature of the unburned fraction which may autoignite and cause a local pressure rise which is apparently the source of the knocking sound characteristic of detonation. This theory was originally advanced by Ricardo[1] in 1919.

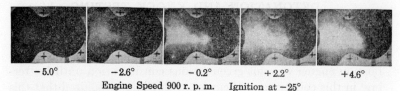

-5.0°   -2.6°   -0.2°   +2.2°   +4.6°
Engine Speed 900 r. p. m.   Ignition at -25°

FIG. 84.—Nonknocking combustion process. Only part of the combustion process is shown, the more or less orderly flame movement continuing to 16° after top center. (*Withrow and Rassweiler.*)

Withrow and Boyd[2] photographed the combustion process through a fused-quartz window in the cylinder head. When detonation occurred, there was a sudden inflammation of the last part of the charge to burn. In some cases the flame appeared at the same instant throughout the last part of the charge, whereas in others combustion started in the unburned fraction ahead of the

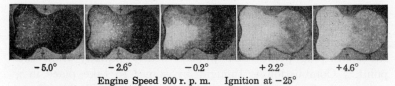

-5.0°   -2.6°   -0.2°   +2.2°   +4.6°
Engine Speed 900 r. p. m.   Ignition at -25°

FIG. 85.—Knocking combustion process. (*Withrow and Rassweiler.*)

flame front and burned back toward the original flame as well as on toward the cylinder wall.

Withrow and Rassweiler[3] photographed the combustion process through a fused-quartz cylinder head. The nonknocking process (Fig. 84) has a more or less orderly progress of the flame front across the combustion chamber. The knocking process (Fig. 85) shows practically the same flame progress for part of

[1] Ricardo, *op. cit.*
[2] L. Withrow and T. A. Boyd, Photographic Flame Studies in the Gasoline Engine, *Ind. Eng. Chem.* **23**, 539 (1931).
[3] L. Withrow and G. M. Rassweiler, Slow Motion Shows Knocking and Non-knocking Explosions, *S.A.E. Trans.*, **39**, 297 (1936).
Note.—See also Rothrock and Spencer, *op. cit.*

the way but terminates with the very sudden inflammation of the balance of the unburned fraction.

A comparatively high local pressure must occur as the result of the almost instantaneous combustion of the last fraction of the charge to burn when detonation occurs. The inequality of pressure in the combustion chamber is of very brief duration, for the expansion of the last fraction of the charge almost immediately equalizes the pressure. This creates a disturbance that increases the heat transfer and may result in loss of power in the case of severe detonation.

The factors that influence the process of detonation are:
1. Fuel characteristics:
    a. Molecular structure.
    b. Self-ignition temperature.
    c. Rate of burning.
2. Mixture conditions:
    a. Air-fuel ratio.
    b. Distribution.
    c. Mixture temperature.
    d. Charge density.
3. Compression ratio:
    a. Clearance-gas dilution.
    b. Valve timing.
4. Ignition timing.
5. Combustion chamber:
    a. Design.
    b. Spark-plug location.
    c. Material.
    d. Surface condition.

Fuel Characteristics. *Molecular Structure.*—Petroleum fuels usually consist of many hydrocarbons of different molecular structure or arrangement of atoms in the molecule (Chap. VII). Some hydrocarbons have a great tendency to produce knock in an internal-combustion engine whereas others are excellent antiknock fuels. An investigation[1] of the detonation characteristics of six normal paraffin hydrocarbons and some of their isomers indicates certain general relationships (Fig. 86) between molecular structure and knocking tendency in an engine.

[1] W. G. Lovell, J. M. Campbell, and T. A. Boyd, Detonation Characteristics of Some Paraffin Hydrocarbons, *Ind. Eng. Chem.*, **23**, 26 (1931).

1. Increasing the length of the carbon chain increases the knocking tendency.
2. Centralizing the carbon atoms decreases the knocking tendency.

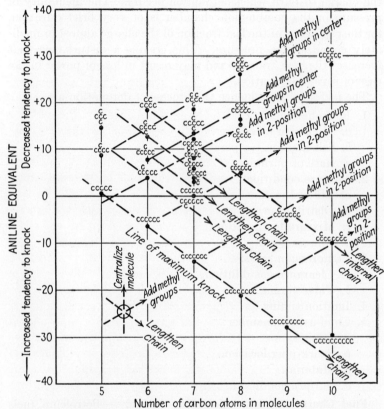

Fig. 86.—Molecular structure and knocking tendency for various paraffin hydrocarbons. (*Lovell, Campbell and Boyd.*)[1] (H atoms omitted in indicating structure.)

3. Adding methyl groups (CH_3) to the side of the carbon chain in the 2- or center position decreases the knocking tendency.

The unsaturated aliphatic hydrocarbons on investigation[2] showed less knocking tendency than did the corresponding

[1] Lovell, Campbell, and Boyd, *op. cit.*
[2] W. G. Lovell, J. M. Campbell, and T. A. Boyd, Detonation Characteristics of Some Aliphatic Olefin Hydrocarbons, *Ind. Eng. Chem.*, **23**, 555 (1931).

hydrocarbons, with "notable exceptions" in the case of ethylene, acetylene, and propylene. Thus, acetylene ($C\equiv C$) knocks much more readily than ethane($C-C$), the corresponding saturated hydrocarbon.

Investigations[1,2] of the naphthenes and aromatics showed the following relationships between the molecular structure and knocking tendency:

1. The naphthenes have distinctly greater knocking tendency than have the corresponding aromatics. Thus, cyclohexane (C_6H_{12}) has a much greater knocking tendency than has benzene (C_6H_6).

2. One double bond has little antiknock effect, whereas two and three double bonds generally result in appreciably less knocking tendency.

3. Lengthening the side chain increases the knocking tendency in both groups of fuels, whereas branching of the side chain decreases the knocking tendency.

In general, it appears that the more compact molecular structures of most hydrocarbons are associated with the lower knocking tendency.

Self-ignition Temperatures.—Self-ignition of a combustible mixture occurs when its temperature is sufficiently high to cause any reaction to take place whereby chemical energy will be liberated faster than it is dissipated by heat transfer or other methods. The excess energy liberated increases the internal energy and temperature of the mixture, which increases both the rate of preflame reaction and temperature rise until flame and rapid reaction occur. Thus an appreciable amount of time elapses between self-ignition and the appearance of flame, which depends on the mixture, apparatus, and conditions imposed.

The compression process in an engine increases the temperature of the combustible charge. The combustion process further compresses the unburned fraction to such a degree, in all spark-ignition engines operated at compression ratios near that causing incipient detonation, that the self-ignition temperature of the

[1] W. G. Lovell, J. M. Campbell, and T. A. Boyd, Knocking Characteristics of Naphthene Hydrocarbons, *Ind. Eng. Chem.*, **25**, 1107 (1933).

[2] W. G. Lovell, J. M. Campbell, F. K. Signaigo, and T. A. Boyd, Knocking Characteristics of Aromatic Hydrocarbons, *Ind. Eng. Chem.*, **26**, 475 (1934).

Note.—See also W. G. Lovell and J. M. Campbell, Molecular Structure of Hydrocarbons and Engine Knock, *Chem. Reviews*, **22**, 159 (1938).

last fraction to burn is always attained long before the flame from the ignition spark reaches it.

Experiments[1] on various fuels (Fig. 87) show that the compression pressures required to cause spontaneous combustion of the mixtures are much less than the maximum pressures attained at top-center position of the piston with spark ignition and incipient detonation. These data indicate for the given engine conditions that the flame front passed through about 45 to 60 per cent of the combustion-chamber volume and that

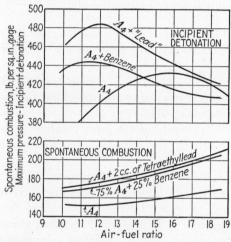

Fig. 87.—Spontaneous combustion and incipient detonation in a C.F.R. engine. (Speed 900 r.p.m., manifold mixture temperature 212°F., no clearance-gas dilution.)

about 20 to 30 per cent of the charge was burned[2] when the unburned fraction has reached the self-ignition temperature. Evidence of appreciable reaction when the mixture is compressed sufficiently to be on the verge of spontaneous combustion is shown in Fig. 88.

A blend of reference fuel A-4 and benzene, and A-4 plus 2 cm.³ of tetraethyllead, with the same octane number required practically the same compression pressure to cause spontaneous combustion (Fig. 87). The compression pressures for spontaneous combustion increased with an increase in air-fuel ratio.

[1] G. F. McDermott, *Thesis for Master of Engineering*, Yale University, 1936.

[a] See Chap. XIII.

The determination of incipient detonation by ear is rather difficult and is probably affected by changes in atmospheric conditions, which may account for the crossing of the A-4 and A-4 plus benzene curves. However, the maximum pressures attained with incipient detonation at all air-fuel ratios with spark ignition are much higher than the compression pressures required for spontaneous combustion.

Rate of Burning.—The time required for the flame to travel from the spark plug to the last fraction of the charge to burn

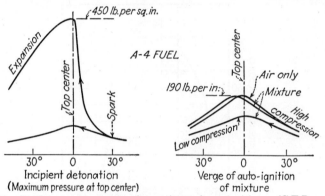

Fig. 88.—Indicator cards showing detonation and autoignition. (C.F.R. engine, 900 r.p.m., no clearance-gas dilution.)

Note the shift of the peak of the mixture curve to the left of top center as the compression is raised. This indicates the effect of preliminary reactions prior to sudden inflammation which is prevented by rapid expansion after top center.

depends upon the fuel, mixture ratio, temperature and pressure conditions, turbulence, and combustion-chamber design. A high rate of burning allows less time for the unburned fraction to lose energy by heat transfer to the combustion-chamber walls. This tends to promote detonation. A high rate of burning allows less time for the unburned fraction to attain the temperature required for spontaneous inflammation by the self-ignition process. This tends to prevent detonation. However, high rates of burning require small optimum spark advances which indicate the burning of the last fraction of the charge nearer top-center piston position in an engine than with low rates of burning. This also influences the tendency to detonate.

Methane- and ethylene-air mixtures have identical "instantaneous" ignition temperatures, but the ethylene-air mixture has a lower ignition temperature for a given time lag before inflammation (Table 9). Ethylene-air mixtures have the higher flame speeds (Fig. 79), but the methane-air mixtures can be used in engines with much higher compression ratios (Table 11). Both ethane- and propane-air mixtures have lower ignition temperatures than hydrogen-air mixtures. However, hydrogen-air mixtures, which have much higher flame speeds, detonate much more readily than the other mixtures. Thus, the effect of rate of burning on detonation is involved with the effect of the various other factors. Obviously, there will be no detonation if the

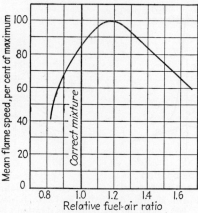

FIG. 89.—Effect of air-fuel ratio on mean flame speed. (*L. A. Peletier; N.A.C.A. Tech. Memo.* 853, 1938.)

flame progresses through the unburned fraction before the self-ignition process results in its sudden inflammation and vice versa.

Mixture Conditions. *Air-fuel Ratio.*—Both bomb and engine experiments have shown the highest rates (Fig. 89) of combustion for mixtures somewhat richer than the chemically correct mixture. Leaner or richer mixtures have lower rates of combustion. An increase in combustion time, even with optimum spark timing, results in more of the expansion stroke occurring before the combustion process is completed. Thus, more energy is removed from the unburned fraction, by both heat transfer and work processes, with lean and rich mixtures than with those nearly chemically correct. This energy reduction and its temperature effect may partly account for the lessened tendency to detonate when the air-fuel ratio is made rich or lean.

The suction pressure may be varied with air-fuel ratio to obtain the maximum allowable m.e.p. with incipient detonation in engines with high compression ratios. Tests on a supercharged C.F.R. engine do not show so much effect as do those on a single-cylinder airplane test engine (Fig. 90). The speed and heat load are much higher in the airplane engine, which may account for the difference in effects noted.

Distribution.—Unequal distribution of air and fuel in any cylinder or between various cylinders in a multicylinder engine results in some cylinders having greater tendency than others to detonate. Usually some retardation of ignition timing or enrichment of carburetor mixture is required to reduce the knocking to incipient detonation in those cylinders having the most tendency to knock. Obviously, such adjustments may reduce the power output appreciably and will increase the specific fuel consumption.

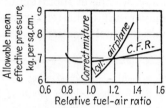

FIG. 90.—Effect of air-fuel ratio on allowable m.e.p. (*L. A. Peletier; N.A.C.A. Tech. Memo.* 853, 1938.)

1-cyl. airplane test engine; 2250 r.p.m., 7 to 1 compression ratio, 100°C. air. C.F.R. test engine, supercharged; 900 r.p.m., 5 to 1 compression ratio, 40°C. air.

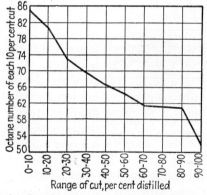

FIG. 91.—Antiknock value of ten equal cuts of a typical commercial base gasoline separated by simple distillation. (*Bartholomew, Chalk, and Brewster.*)

C.F.R. engine, standard manifold; mixture temp. 140°F., heated intake air only, spark advance, 18°, cylinder jacket temp., 212°F., and engine speed, 600 r.p.m.

The various fractions of a gasoline have different antiknock characteristics (Fig. 91).[1] The more volatile fractions are completely evaporated in the manifold and are usually uniformly

[1] E. Bartholomew, H. Chalk, and B. Brewster, Carburetion, Manifolding, and Fuel Antiknock Value, *S.A.E. Trans.*, **33**, 141 (1938).

distributed to the various cylinders. The less volatile fractions are in liquid form and are unevenly distributed. Consequently, the knocking tendency of the various cylinders may vary appreciably (Fig. 92).[1]

Mixture Temperature.—An increase in mixture temperature increases the evaporation of the fuel in the intake manifold and may result in better distribution of the fuel supplied or of the various fuel fractions. This would minimize the tendency of some cylinders to detonate more than others because of poor distribution.

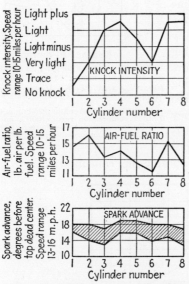

Fig. 92.—Variation of knock intensity, air-fuel ratio and spark advance for an automobile engine in the speed range of maximum knock. (*Bartholomew, Chalk, and Brewster.*)

An increase in mixture temperature increases the temperature of the unburned portion of the charge for any given position of the flame front. This increases the tendency of a larger unburned fraction to detonate. Combustion pressures decrease with an increase in initial mixture temperature. Also, the density of the charge is lowered as the temperature is increased. Consequently, the possible detonation pressures are lowered with an increase in mixture temperature, which tends to offset the detonation of a larger unburned fraction.

The effect of an increase in intake temperature on detonation may be offset by an increase in the richness of the mixture (Fig. 93) but this increases the specific fuel consumption.

Charge Density.—Throttling an engine reduces the tendency to detonate, whereas supercharging increases this tendency. In both cases the clearance-gas dilution is varied by varying charge density. The diluent affects the rate of burning, whereas the variation in charge density affects the heat transfer to the

---

[1] Bartholomew, Chalk, and Brewster, *op. cit.*

walls and the temperature conditions of the burned and unburned mixture.

Compression Ratio. *Clearance-gas Dilution.*—The tendency of a fuel to detonate is most susceptible to a change in compression ratio. Three factors are changed when the compression ratio is increased:

1. Compression pressure is increased.
2. Compression temperature is increased.
3. Dilution with exhaust products is decreased.

An increase in the compression pressure causes an increase in the maximum pressures attained during either normal combustion or combustion with detonation. The rise in temperature that accompanies increased compression tends to offset the foregoing factor. The third factor is, no doubt, the most important, as Ricardo[1] has shown that adding cooled clearance gases decreases the tendency to detonate, whereas scavenging the clearance space with fresh charge resulted in violent detonation.

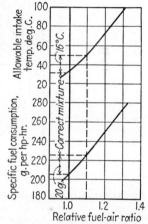

Fig. 93.—Effect of air-fuel ratio on allowable intake temperature for incipient detonation. (*L. A. Peletier, N.A.C.A. Tech. Memo.* 853, 1938.)

Single-cylinder air-cooled engine, 2250 r.p.m., 7:1 compression ratio, constant m.e.p. of 7.9 kg. cm.$^{-2}$

Valve Timing.—Late intake-valve closing lowers the compression ratio from the nominal to the actual compression ratio. The charge density at various speeds varies with valve timing. Also, overlapping the closing of the exhaust valve and the opening of the intake valve varies the percentage dilution with clearance gases. Thus, valve timing has an appreciable effect on detonation.

Allowable Compression Ratio.—The allowable compression ratio with incipient detonation increases rapidly with an increase in octane number of the fuel, particularly with the fuels of higher octane number (Fig. 94). This is probably due to the smaller decrease in clearance volume to obtain a given increase in compression ratio at the higher compression ratios along with the decrease in clearance-gas temperature.

[1] H. R. Ricardo; "The Internal Combustion Engine," Vol. 2, D. Van Nostrand Company, Inc., New York, 1923.

There is a considerable spread in compression ratio for a given octane-number fuel (4.1 to 6.6:1 compression ratio for 70 ON fuel). This indicates the differences in engines and operating conditions.

The effect of air-fuel ratio on allowable compression ratio for a given octane-number fuel is also shown in Fig. 94, curve a. Curve c is obtained from the curves a and b, as indicated.

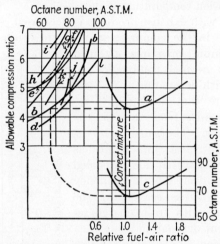

Fig. 94.—Relation between compression ratio for incipient knock, air-fuel ratio, and octane number.

    a and c Fuel-air ratio for use with b only.
    b C.F.R. engine, special camshaft, 900 r.p.m., mixture temp. 150°C., 66 octane fuel.
    d C.F.R. engine, research method, heptane and octane mixtures, 600 r.p.m.
    e C.F.R. engine, standard knock intensity for motor method, standard barometer.
    f Single-cylinder L-head 3¼- by 5-in. engine, 600 r.p.m.
    g Same as e, but with aluminum head.
    h Eight-cylinder automobile engine, 1500 r.p.m.
    i Same as g except at 2250 r.p.m.
    j Six-cylinder 4⅞- by 5½-in. engine, 1000 r.p.m.
    k Same as i except at 2000 r.p.m.
    l C.F.R. engine, 1000 r.p.m.
    [a, b, and c, from L. A. Peletier, *N.A.C.A. Tech. Memo.* 853, 1938; d, *Courtesy of Ethyl Gas. Corp.*; e, based on "C.F.R. exchange group data (A. E. Becker); f through k, from E. Bartholomew and C. D. Hawley, "The Relation of Fuel Octane Number to Engine Compression Ratio," *A.P.I. Proc.*, **15M**, III, 55(1934); l, from Kents' "Mech. Eng. Handbook," vol. 2, pp. 14–69.]

The critical compression ratio or that for incipient detonation in a single-cylinder engine under definite operating conditions indicates the effect that compression ratio has on the detonation characteristics of various fuels. Only a few of the hydrocarbons found in the original reference are included in Table 16.

Ignition Timing.—Any variation in ignition timing varies the amount of the charge burned before and after top-dead-center position of the piston. Early spark timing increases the work of compression, which increases the temperature of the charge and also causes the last part of the charge to be burned nearer dead center. Both factors increase the tendency to detonate. Hence, detonation can be eliminated by retarding the spark, which may decrease the power output appreciably. The relation between optimum spark advance and that required

TABLE 16.—CRITICAL COMPRESSION RATIOS FOR VARIOUS HYDROCARBONS*

Hydrocarbon	Critical compression ratio	Hydrocarbon	Critical compression ratio
Saturated paraffins:		Naphthenes:	
Methane	>15	Cyclopentene	7.9
Ethane	14	Cyclopentane	10.8
Propane	12	Cyclohexene	4.8
n-Butane	6.4	Cyclohexane	4.5
n-Pentane	3.8	Methylcyclohexane	4.6
n-Hexane	3.3	Ethylcyclohexane	3.8
n-Heptane	2.8	n-Butylcyclohexane	3.3
2-,2-,4-Trimethylpentane	7.7	n-Amylcyclohexane	3.1
Unsaturated paraffins:		Aromatics:	
Acetylene	4.6	Benzene	>15
Ethylene	8.5	Toluene	13.6
Propylene	8.4	o-Xylene	9.6
1-Pentene	5.8	m-Xylene	13.6
1-Hexene	4.6	p-Xylene	14.2
1-Heptene	3.7	n-Propylbenzene	10.1
1-Octene	3.4	n-Butylbenzene	7.7

* W. G. Lovell, J. M. Campbell, and T. A. Boyd, Knocking Characteristics of Hydrocarbons, *Ind. Eng. Chem.*, **26**, 1105 (1934).

for incipient detonation depends on the fuel. The spark advance for incipient detonation will be above or below the optimum spark advance at any engine speed if the octane number of the fuel is above or below, respectively, that required for optimum spark advance (Fig. 95).

The variation in spark timing from a given setting limits the power that may be obtained from an engine with spark advance set for incipient detonation. A spark dispersion of ±5 crankshaft degrees compared with ±¾ deg. necessitated considerably later spark advance for engine performance with incipient knock at a given inlet-air pressure (Fig. 96). Also, for a given spark advance, considerably greater inlet pressure and indicated power output can be obtained with the smaller spark dispersion.

Combustion Chamber. *Design.*—The combustion-chamber shape depends principally upon the valve arrangement. The various shapes of combustion chambers (Fig. 97) can be classi-

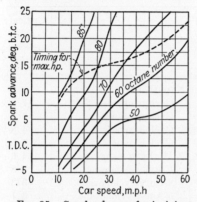

Fig. 95.—Spark advance for incipient detonation and maximum power. (*Hebl and Rendel.*)[1]

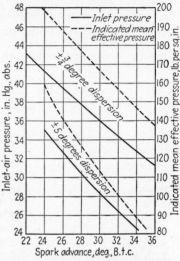

Fig. 96.—Effect of spark advance on inlet pressure and i.m.e.p. for incipient detonation. (*A. E. Biermann, N.A.C.A. Tech. Note* 651, 1938.)

fied as follows, some being adapted for either two or four valves per cylinder:

1. T-head:
2. L-head:
    a. Inlet valve beside exhaust valve.
    b. Inlet valve over exhaust valve or in cylinder head.
    c. The same as (a) with high-turbulence head.
3. Valve-in-head:
    a. Vertical valves.
    b. Valves at an angle.
4. Sleeve-valve head:
    a. Knight, or double sleeve.
    b. Burt-McCollum, or single sleeve.

The T-head design necessitates the use of the lowest compression ratios to prevent detonation with a given fuel, whereas

[1] L. E. Hebl and T. B. Rendel, Spark Timing and Its Relation to Road Octane Numbers and Engine Performance, *S.A.E. Trans.*, **34**, 210 (1939).

DETONATION AND KNOCK TESTING

the sleeve-valve design permits the use of the highest compression ratios. The poppet-valve engine has an unusually hot spot at the exhaust valve, which is eliminated in the sleeve-valve engine.

The nonturbulent L-head engine is normally of lower compression ratio than the valve-in-head engine. However, the introduction of the high-turbulence L-head, developed by Ricardo, made these two designs about equal from the standpoint of detonation characteristics. Since then, turbulence has been increased in the valve-in-head engine by changing the shape of the piston head. (See Chap. XIII.)

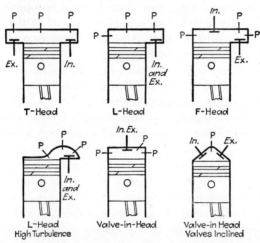

FIG. 97.—Combustion-chamber shapes. *P* indicates possible plug location.

Spark-plug Location.—The detonating characteristics of any one of the various combustion chambers previously described may be changed appreciably with a change in location of spark plug or plugs. In general, the spark plug should be located so as to reduce the length of flame travel to a minimum. This decreases the combustion time to a minimum so that the flame may pass through the unburned fraction before detonation occurs, if too high a compression ratio is not used.

The plug should also be located near the hottest spot in the combustion chamber which is the head of the exhaust valve. The reason for this location is that, during the combustion process, compressing the unburned part of the mixture next to the hot-exhaust valve will raise its temperature much higher than will compressing it against a cooler surface. Hence, it is much more desirable to start the combustion process near the exhaust valve rather than to end the process there.

When a high turbulence is produced in the mixture, the central spark-plug location may result in a very rapid combustion process and a "rough" engine. When this occurs, the spark-plug location should be changed so that the flame travel is lengthened and the time of combustion increased.

In the T-head engine (Fig. 97), plug location over the intake valve is extremely bad, central location is better, and a location somewhat nearer the exhaust valve gives the best results. In the valve-in-head design, two plugs located on opposite sides of the cylinder are better than one plug located in the side. The domed combustion chamber with inclined valves provides a good central location for the spark plug.

In most L-head and some valve-in-head combustion chambers the last part of the charge to burn is located in the small space between the cylinder head and the piston. The spark plug should be farther removed from this than from any other portion of the combustion chamber. This tail-end space with a large surface-volume relationship dissipates some of the energy of compression and also delays the burning of the last unburned fraction. This permits higher compression ratios and tends to offset the effect of turbulence on flame speed.

Material and Surface Condition.—Materials with high heat-transfer coefficients such as aluminum alloys are desirable for high-compression cylinder heads, since a cool combustion-chamber wall is essential for high compression without detonation. However, hot spots may develop because of poor circulation of the coolant or improper distribution of the metal.

Combustion chambers with highly polished surfaces tend to detonate more readily than do those coated with a light carbon deposit. This may be due to heat-absorption characteristics. Carbon deposition results in hotter surfaces and smaller clearance space, both of which increase the detonation tendency.

Detonation Suppressors.—The blending of a good antiknock fuel with a poor antiknock fuel results in a mixture with detonation characteristics usually between those of the two constituents. The effect of such fuel additions in terms of critical compression ratio is not directly proportional to the amount of high antiknock fuel added (Fig. 98). Thus, large percentages of the high antiknock fuel are usually required to raise the critical compression ratio much above that of the low antiknock fuel.

Midgley and Boyd[1] investigated the effect of numerous substances on the detonating tendencies of fuels in internal-combustion engines and found tetraethyllead, $Pb(C_2H_5)_4$, to be the most effective material for suppressing detonation. Aniline is an excellent antiknock compound, but more than thirty times

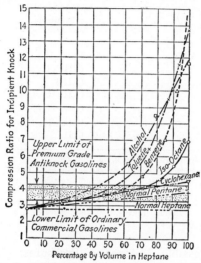

FIG. 98.—Effect of addition of various fuels to n-heptane on permissible compression ratio. (*J. M. Campbell, W. G. Lovell, and T. A. Boyd; S.A.E. Trans.* **25**, 126, 1930.)

as much material is required to produce the same effect as is obtained with tetraethyllead (Table 17).

Ethylene dibromide $C_2H_4Br_2$ and ethylene dichloride $C_2H_4Cl_2$ are mixed with tetraethyllead to prevent lead deposition in the combustion chamber. The lead combines with the bromine and chlorine during the combustion process and is expelled as a vapor.

Ethyl gasoline-air mixtures contain 1 molecule of tetraethyllead to about 100,000 molecules of air and gasoline. Tetra-

---

[1] T. Midgley, Jr., and T. A. Boyd, Detonation Characteristics of Blends of Aromatic and Paraffin Hydrocarbons, *Ind. Eng. Chem.*, **14**, 589 (1922).

T. Midgley, Jr., and T. A. Boyd, The Chemical Control of Gaseous Detonation with Particular Reference to the Internal-combustion Engine, *Ind. Eng. Chem.*, **14**, 894 (1922).

ethyllead has no apparent effect on the flame progress during the first part of the combustion process but prevents the sudden inflammation of the last part of the charge to burn. Consequently, it is thought to inhibit the preflame oxidation of the hydrocarbons in the unburned charge.

TABLE 17.—RELATIVE EFFECTS OF DETONATION-SUPPRESSING COMPOUNDS*

Compound	Chemical symbol	Weight for a given effect, gm.
Tetraethyllead	$Pb(C_2H_5)_4$	0.0295
Aniline	$C_6H_5NH_2$	1.0000
Ethyl iodide	C_2H_5I	1.55
Ethyl alcohol	C_2H_5OH	4.75
Xylene	$C_6H_4(CH_3)_2$	8.00
Toluene	$C_6H_5CH_3$	8.8
Benzene	C_6H_6	9.8

* "International Critical Tables," Vol. 2, pp. 162–163.

The foregoing values were determined by using kerosene as a fuel and apply only qualitatively to current gasolines. Thus, these data indicate that 1 cm.³ of tetraethyllead per gallon of gasoline is equivalent to about an 8 per cent blend, by volume, of ethyl alcohol with the gasoline. Actually, more than 10 per cent alcohol in current gasolines is required to equal 1 cm.³ of tetraethyllead. Kuring, *Can. Jour. Research*, **2**, 489 (1934). Brown and Singer, *Nat. Petroleum News*, **26**, 21 (1934). L. C. Lichty and C. W. Phelps, *Ind. Eng. Chem.*, **30**, 222 (1938).

Maximum Possible Detonation Pressures.—The maximum possible detonation pressure will be the pressure attained during constant-volume instantaneous combustion of the last part of the charge to burn. During the normal part of the combustion process the burned portion expands and compresses the unburned portion which increases its temperature and pressure. These conditions vary with the flame-front position and may be determined as indicated in Chap. XIII. Obviously, the pressure and temperature of the unburned fraction will be the highest at the end of the process (Fig. 99).

Constant-volume combustion of the last infinitesimal portion of the charge should theoretically result in a local pressure of about 2200 lb. in.$^{-2}$ for the octane-air mixture. The maximum pressure attained if all the charge burned instantaneously should theoretically be about 560 lb. in.$^{-2}$ for the given conditions. Local pressures due to detonation must lie somewhere between these two limits, depending on flame position when

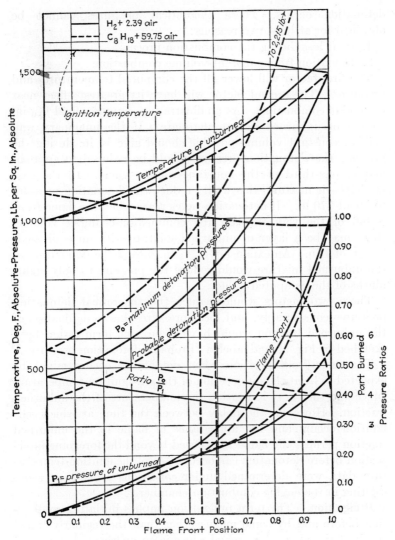

Fig. 99.—Maximum detonation pressures.
Initial pressure 100 lb. in.$^{-2}$, initial temperature 1000°R. No piston movement during the process.

detonation occurs. These theoretical pressures cannot be obtained for at least two reasons:

1. Heat loss during the combustion process.
2. Deviation from constant-volume combustion.

The first factor will lower all the computed temperatures and pressures. The second factor will have its greatest effect near the end of combustion. As an illustration, assume a final $\frac{1}{16}$-in. section of unburned gas to detonate. If the imaginary wall making constant volume possible should give $\frac{1}{8}$ in. during the process, the detonation pressure would be reduced to approximately one-third of the constant-volume value. In the case of C_8H_{18}-air mixture, this would reduce the detonation pressure to about 770 lb. If detonation takes place when the flame front is 0.6 of the way through the combustion-chamber volume, there may be a layer of unburned mixture with an equivalent length of 1 in. An expansion of $\frac{1}{8}$ in. during the detonation of the unburned mixture would reduce the pressure to only eight-ninths of the computed value.

These two factors would reduce the theoretical detonation pressures considerably, and local pressures due to detonation of the unburned fraction would have some such form as has been sketched in Fig. 99 and labeled "Probable detonation pressures."

The flame-front position when detonation occurs might be expected to be at the intersection of the self-ignition-temperature curve and the curve indicating the temperature of the unburned fraction. However, the delay between the time at which self-ignition occurs and the appearance of flame in the unburned fraction permits the flame to proceed beyond the foregoing intersection before detonation can occur. Also, the ignition-temperature data were determined under conditions other than those existing in the engine combustion chamber.

Preignition.—The term preignition implies initiating combustion before and by some means other than the regularly timed spark at the plug. The usual cause is an overheated spot such as an exhaust-valve head, the center of the piston, the center electrode of the spark plug, or carbon deposit.

The ideal engine process is indicated by diagram $ABCD$ (Fig. 100) in which combustion occurs from B to C at top dead center. Actually, ignition occurs at B' and combustion ends beyond C', resulting in less work than is obtained from the ideal process. Preignition occurring at B'' results in diagram $AB''C''D''$ which indicates less work than is obtained from the actual spark-ignition process.

The limiting case would be preignition at A with combustion from A to E. Compression would occur along the path E to F, the point F indicating the maximum possible pressure due to preignition. The expansion process would be from F to E, irreversible effects being neglected, and the net work would be zero.

The result of preignition is to increase the work of the compression stroke, decrease the net work of the process, subject the engine to excessive pressures, increase the heat loss from the engine, and decrease the efficiency. Obviously, too early ignition of the charge will have the same effect as preignition. In either case the engine will knock, owing either to high pressures occurring before the end of the compression stroke or to the detonation of the last part of the charge to burn. Retarding the spark will eliminate the knock if the engine is not preigniting.

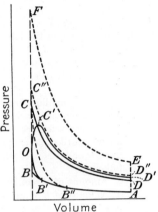

FIG. 100.—P-V diagram illustrating preignition.

KNOCK TESTING

Object of Knock Testing.—Optimum engine performance can be obtained only with fuels having knock characteristics that will permit optimum spark advance with little or no detonation. Consequently, it is desirable to rate fuels for knocking tendency in order to control and maintain this fuel characteristic so that satisfactory engine operation is possible.

Methods of Knock Rating Fuels.—The numerous hydrocarbons in various proportions that constitute gasoline have eliminated chemical analysis as a method of rating fuels. Combustion in a bomb does not simulate engine operating conditions and has not proved satisfactory in rating fuels. Consequently, most of the rating methods that have appeared promising were based upon the knock intensity as obtained under engine operating conditions. These are classified into two principal groups as follows:

Methods of Rating Fuels[1]

A. Knock intensity for each fuel under identical test conditions:
  1. Knock intensity by sound:
     a. By human ear.
     b. By microphone or other sound apparatus.
  2. Knock intensity by blow struck on diaphragm in combustion-chamber wall.
  3. Knock intensity from temperature of plug in combustion-chamber wall.
B. Rating at equal knock intensity:
  1. Varying one of the following test conditions:
     a. Compression ratio.
     b. Throttle opening.
     c. Spark advance.
     d. Speed.
     e. Cooling medium temperature.
  2. Varying fuel composition. Assume an unknown fuel X which knocks more than reference fuel C and less than reference fuel A, both of known antiknock characteristics. Also assume S is a knock suppressor and I is a knock inducer. Then the ratings may be made as follows:
     a. Amount of S required to make X match C.
     b. Amount of I required to make X match A.
     c. Mixture of C and A required to match X.

In the first group an engine is run under fixed test conditions and fuels are rated according to the degree of intensity of knock. However, the intensity of knock changes from time to time, owing to some change in engine or operating conditions when apparently all conditions have been maintained constant.

The variation of one of the test conditions to obtain a given knock intensity also presents the difficulty of knock-intensity evaluation.

Ricardo[2] originated the rating of fuels by the highest useful compression ratio (H.U.C.R.) that could be employed in a given engine under a given set of conditions. This is the compression ratio at which an increase in ratio will not result in an increase in power. Usually the power is not measured, but the compression ratio is increased until the detonation is frequent and moderate.

[1] S.A.E. *Jour.*, **25**, 80 (1929).
[2] H. R. Ricardo, The Influence of Various Fuels on the Performance of Internal-combustion Engines, *Auto. Eng.*, **11**, 92 (1921).

Campbell, Lovell, and Boyd[1] have used a method similar to that of Ricardo but based upon incipient knock. (See Fig. 98.)

Varying the fuel composition and matching the knock intensity of the unknown fuel eliminate much of the effect of changing conditions over which the operator has no control. Any such change in conditions is assumed to have a similar effect on all fuels, which is not always true.

Ricardo and Thornycroft[2] made use of toluene and heptane as high and low antiknock fuels, respectively. Campbell, Lovell, and Boyd[1] used aniline as a knock suppressor to be added to either the unknown fuel or the reference fuel, giving the unknown fuel a positive or negative "aniline equivalent" rating.

The method developed by the C.F.R. Committee and adopted by the A.S.T.M. is to rate a fuel by matching the knock intensity of the fuel with a mixture of two primary reference fuels. The two reference fuels are normal heptane, which has low antiknock characteristics, and iso-octane (2-, 2-, 4-trimethylpentane),[3] which has high antiknock characteristics. The matching is done in a C.F.R. engine equipped with certain apparatus and under specified test conditions.

Octane Number.—The octane number of a fuel is the percentage by volume of iso-octane in a mixture of iso-octane and normal heptane which produces the same knock intensity as the given fuel. Thus, a mixture of 70 per cent octane and 30 per cent heptane has an octane number of 70. Obviously, the higher the octane number, the higher the antiknock characteristics of the fuel. One hundred per cent octane, meaning 100 octane number, is the highest rating that can be made with these reference fuels, but the scale is arbitrarily extended beyond 100 by using mixtures of iso-octane and tetraethyllead or pure hydrocarbons with antiknock characteristics higher than octane, such as benzene (C_6H_6).

Secondary Reference Fuels.—The high cost of the primary reference fuels prohibits their use for routine testing. However, secondary reference fuels have been developed by the Standard

---

[1] Campbell, Lovell, and Boyd, *op. cit.*

[2] H. R. Ricardo and O. Thornycroft, Petrol Engines and Their Fuels, World Power Conference (London); **3**, 662 (1928).

[3] Originally suggested by G. Edgar, Detonation Specifications for Automotive Fuels, *S.A.E. Trans.*, **22**, 1, 55 (1927).

Oil Development Company and are used for practically all knock testing. The secondary reference fuels are calibrated (Fig. 101) against the primary reference fuels in the C.F.R. engine, so that the octane number for any mixture of secondary reference fuels is known.

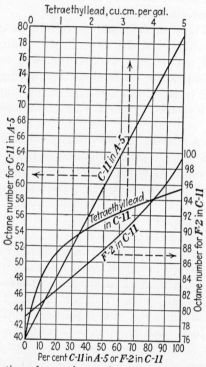

FIG. 101.—Calibration of secondary reference fuels. (Volume percentages.)

C.F.R. Knock-testing Engine.—The C.F.R. engine (Fig. 102) is a single-cylinder four-stroke-cycle valve-in-head engine, with a $3\frac{1}{4}$-in. bore and $4\frac{1}{2}$-in. stroke. It is of the variable-compression type having a mechanism for raising or lowering the cylinder which varies the compression ratio from 3:1 to more than 15:1. A unique valve mechanism maintains a constant valve clearance for any position of the cylinder.

The carburetor used is of the horizontal-draft type with a single air-bled jet. Three individual self-contained fuel-container and float-chamber units may be raised or lowered to provide the maximum-knock air-fuel ratio. One unit is used for the unknown fuel, which is to be rated, and the two other

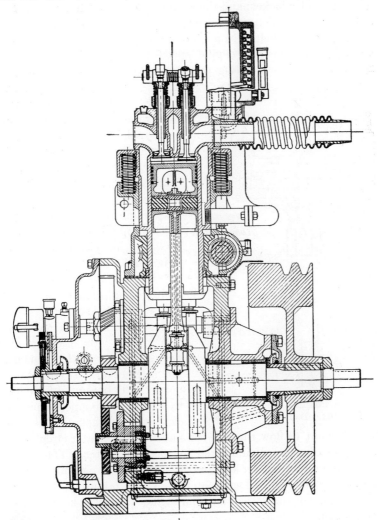

Fig. 102.—C.F.R. knock-testing engine. (*Waukesha Motor Co.*)

units for mixtures of the reference fuels, which will bracket the unknown fuel in knock intensity.

Magneto or battery ignition is used, and the spark advance is automatically decreased with an increase in compression ratio.

The evaporative cooling system is used to maintain constant cylinder-jacket temperature. Water is allowed to boil in the cylinder jacket; and the steam formed rises to the reflector condenser, is condensed, and returns to the system.

The engine is belted to a synchronous induction motor which maintains a constant engine speed for knock testing. This motor is used for starting purposes and then "floats" on the line. A direct-current generator to furnish current for the knock meter is belt-driven from the synchronous motor.

Bouncing Pin and Knock Meter.—The bouncing pin, originally suggested by Dr. H. C. Dickinson,[1] consists of a diaphragm, a steel rod insulated at the upper end, and two contacts in an electrical circuit (Fig. 103). The mechanism is screwed into the cylinder head, and the diaphragm is exposed to the combustion pressures. Normal combustion without knock does not have an appreciable effect on the bouncing pin. However, detonation flexes the diaphragm sufficiently to cause the bouncing pin to close the contacts. This permits current to flow through the knock-meter heater circuit, the duration of which is indicated by a knock meter. The knock meter is a milliammeter actuated by a thermocouple embedded in the heater element. The greater the knock intensity, the longer the circuit remains closed and the higher will be the knock-meter indication.

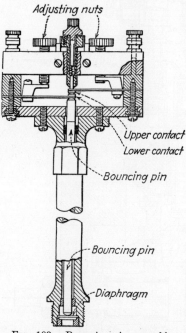

Fig. 103.—Bouncing-pin assembly.

Rating Procedure.—Fuels are rated under specified test conditions, as outlined in the A.S.T.M. test method D357, at a knock intensity equivalent to that obtained when using a given octane-number blend of secondary reference fuels with the

[1] *Nat. Bur. of Standards.*

engine at a compression ratio equivalent to a given ratio with an atmospheric pressure of 760 mm. of Hg. The knock-meter reading for the standard knock intensity is determined, and the compression ratio is varied until this knock-meter reading is attained with the unknown fuel to be rated. The unknown fuel is then rated at this knock intensity by comparing its knock intensity with that of two blends of either primary or secondary reference fuels, one of which knocks more and the other less than the unknown fuel.

Example.—The knock-meter reading is 55 for the standard knock intensity. Determine the octane number of fuel with knock-meter readings on the fuel and reference blends as follows:

				Average
80 per cent C-11 and 20 per cent A-5.....	42	40	41	= 41.0
Unknown fuel........................	55	54 55	54	= 54.5
75 per cent C-11 and 25 per cent A-5......	66	66	65	= 65.7

Interpolating, $80 - \dfrac{54.5 - 41.0}{65.7 - 41.0} \times 5 = 77.3$ per cent C-11 in A-5.

This represents about 70 O.N. (Fig. 101) which is the rating given the fuel.

Any variation in test conditions will vary the ratings of fuels that differ in knocking characteristics from the reference fuels. The effect of three of the principal variables, speed, spark advance, and mixture temperature, that affect knock intensity has been studied[1] in an effort to develop desirable sets of test conditions.

Knock-rating Accuracy.—Campbell and Lovell[2] found the mean value of 72 A.S.T.M. ratings of a fuel in six different laboratories was 70.7 O.N. The spread of ratings was 2.4 O.N., and the standard deviation was 0.49 O.N.

Brooks[3] examined the results of 1882 tests on 95 fuels and found the average probable error to be about 0.47 O.N.

Apparently, the average deviation and error are less than 1 O.N. for a number of ratings whereas an individual rating may vary by several octane numbers from the average of a series of ratings on the same fuel.

[1] *S.A.E. Trans.*, **33**, 63 (1938). See also H. W. Best, The Knock Rating of Motor Fuels, *Chem. Reviews*, **22**, 143 (1938).

[2] J. M. Campbell and W. G. Lovell, Application of Statistical Concepts to the Knock-rating Problem, *S.A.E. Trans.*, **33**, 421 (1938).

[3] D. B. Brooks, The Precision of Knock Rating, *S.A.E. Trans.*, **31**, 22 (1936).

192 INTERNAL-COMBUSTION ENGINES

Road Rating of Fuels.—The method adopted by the C.F.R. Committee for rating fuels on the road[1] depends upon the knock-intensity method. It consists of bracketing the unknown fuel between two blends of reference fuel each having a maximum

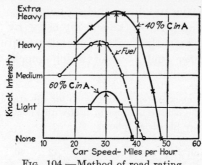

Fig. 104.—Method of road rating

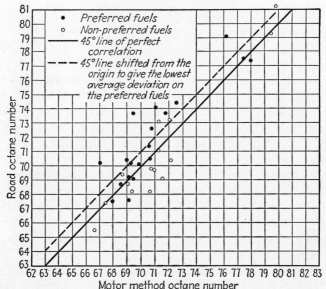

Fig. 105.—Relation between road octane number and motor method octane number.

knock intensity, irrespective of speed, one of which is heavier and one lighter than the maximum knock intensity of the unknown fuel.

[1] C. B. Veal, H. W. Best, J. M. Campbell, and W. M. Holaday, Antiknock Research Coordinates Laboratory and Road Tests, *S.A.E. Trans.*, **28**, 105 (1933).

Figure 104 illustrates the knock-intensity method of rating of fuels on the road. The maximum knock intensity for the fuel occurred at 28 m.p.h., while for the 60 per cent C and 40 per cent C in A it occurred at 30 and 33 m.p.h., respectively. It is obvious that the fuel should be rated nearer the 40 per cent C blend than the 60 per cent C blend, say about 48 per cent C. This is equivalent to about 59 O.N. (Fig. 101).

The differences in fuels, engines, engine conditions, and operating conditions result in large variations in ratings obtained for a given fuel in various cars on the road. The average of the ratings of a given fuel in various cars is called the "road rating" of the fuel.

Plotting the road rating against the laboratory rating for various types of fuel indicates the correlation between the two methods (Fig. 105).[1] Since each point is the average of ratings, some of which were two or more octane numbers higher or lower and one or more octane numbers to the right or left, it is obvious that the variation in road-rating conditions and difference in fuels make perfect correlation an impossibility.[2]

IGNITION QUALITY

The Compression-ignition Process.—The compression-ignition engine compresses air and clearance gases, which increases the energy and temperature of the gases. Fuel is injected into the hot compressed mixture of gases near the end of the compression stroke. The fuel leaves the injection nozzle at a high velocity and is broken up or atomized by friction with the compressed gaseous medium. This atomization promotes rapid vaporization around the various fuel particles, that tends to reduce the temperature of the compressed mixture. However, the temperature of the gases is sufficiently high to cause preliminary reactions between the vaporized fuel and the oxygen that liberate energy and raise the temperature in various spots in the combustion chamber until sudden inflammation at one or more spots causes rapid combustion to take place.

The lapse of time between the beginning of injection and sudden inflammation is termed "ignition delay" or "lag."

[1] C. F. R. Report, *S.A.E. Trans.*, **33**, 416 (1938).
[2] G. Edgar, Knock Testing, *S.A.E. Jour.*, **43**, 7 (1938).

Fuels with small ignition lags are desirable for the compression-ignition process but very undesirable for the spark-ignition process.

The rapid combustion that follows the appearance of flame results in a rapid pressure rise which should occur near top dead center to obtain the maximum work effect. Consequently, fuel injection should begin ahead of top dead center, the exact

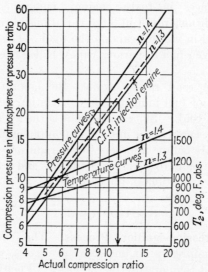

Fig. 106.—Compression pressure and temperature conditions. ($P_1 = 1$ atm. and $T_1 = 500°R.$)

position being determined by the ignition lag for the given conditions and the duration of injection.

Compression Ratio.—The ignition of the fuel injected into an engine cylinder depends upon the temperature attained by the compression of the air and clearance gases during the compression stroke. This temperature depends upon the compression ratio, the rate at which the charge is compressed, the surface-volume relationship of the combustion chamber, and the initial temperature of the mixture to be compressed.

The relation between temperatures, pressures, and volumes for the compression stroke is closely approximated by the relation

$$\frac{T_2}{T_1} = \left(\frac{P_2}{P_1}\right)^{\frac{n-1}{n}} = \left(\frac{V_1}{V_2}\right)^{n-1} = r^{n-1} \qquad (1)$$

DETONATION AND KNOCK TESTING

where subscripts 1 and 2 = the conditions at the beginning and end of compression, resp., and

r = the compression ratio.

Under starting conditions the suction temperature may be quite low, and consequently the compression ratio should be

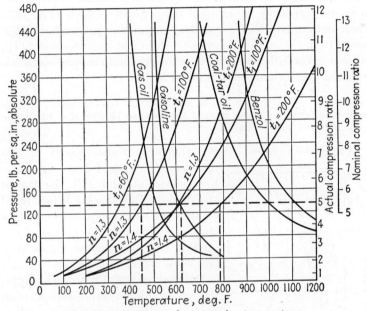

Fig. 107.—Ignition and compression temperatures.

high enough to ignite the charge under this condition. Assuming T_1 to be 40°F., Eq. (1) becomes

$$T_2 = 500r^{n-1}.$$

This relation has been plotted in Fig. 106 and shows that considerably higher compression ratios are required to reach ignition temperatures if the compression is slow and follows the path indicated by $n = 1.3$ instead of the adiabatic path indicated by $n = 1.4$.

Example.—Assume that a temperature of 1100°R. is required at the end of the compression stroke. Determine the minimum compression ratio required and that which would be required if the compression followed the path $n = 1.3$.

From Fig. 106 a compression ratio of 7.2 would be required for adiabatic compression from an initial temperature of 40°F. However, a compression ratio of 14.0:1 is required for the path $n = 1.3$.

Effect of Compression Pressure on Self-ignition Temperatures.—Self-ignition temperatures are usually determined at some standard condition such as 1 atm. of pressure. An increase in pressure reduces the ignition temperature of fuels (Fig. 107). The conditions at the end of the compression process are determined from the *actual* compression ratio and heat loss during the process. These factors determine the minimum compression ratio required for the ignition of a given fuel.

Example.—With reference to Fig. 107, with $n = 1.3$, the minimum actual compression ratio for benzol is much greater than 12:1 for $t_1 = 100°F$. and is about 10.7:1 for $t_1 = 200°F$.

Compression ratios higher than those determined in this manner are required if the engine is to operate at any but the lowest speed, since the time lag before rapid combustion would permit the piston to start the down stroke and stop the preliminary reactions or cause very late combustion (Fig. 88).

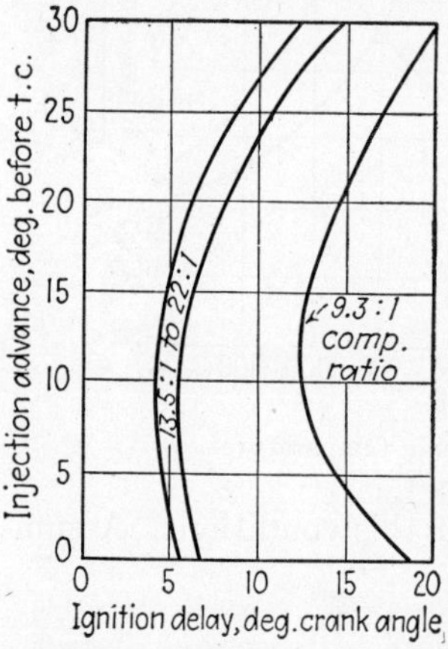

Detonation with Compression Ignition.—The ideal compression-ignition process would have rapid combustion of fuel without any time delay after the fuel leaves the injection nozzle. Actually, there is always some time delay which in some cases permits most or all of the fuel to be injected before inflammation occurs. The sudden inflammation and combustion of a large portion of the injected fuel result in detonation that is similar to that obtained in the spark-ignition engine.

Fig. 108.—Effect of compression ratio and injection advance on ignition delay. (*Fenney*).[1]

Increasing the compression ratio reduces the time lag. However, the time lag approaches a minimum very rapidly with an

---

[1] W. Fenney, Jr., Nozzle Characteristics and the Combustion Process in a Compression-ignition Engine, *Thesis for Master of Engineering*, Yale University, 1938.

increase in compression ratio for a given fuel and engine. Thus, increasing the compression ratio from 9.3:1 to 13.5:1 in the C.F.R. Diesel engine at 900 r.p.m. and using Shell high-reference fuel reduced the ignition lag 7 to 8 deg. of crank angle (Fig. 108). Increasing the compression ratio from 13.5:1 to 22:1 reduced

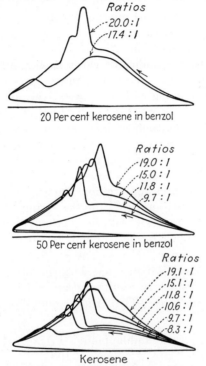

Fig. 109.—Effect of compression ratio on the combustion process. (*Pope and Murdock.*)[1]

the lag about 1 deg. of crank travel.[2] The reduced time lag with high compression ratios reduces or eliminates detonation since the fuel burns nearly as fast as injected and prevents the accumulation of a large quantity of fuel. Elimination of the knocking characteristic[3] is due not only to higher compression temperatures but also to the higher compression pressures.

[1] A. W. Pope, Jr., and J. A. Murdock, Compression-ignition Characteristics of Injection-engine Fuels, *S.A.E. Trans.*, **27**, 136 (1932).

[2] Fenney, *op. cit.*

[3] C. B. Dicksee, Some Problems Connected with High-speed Compression-ignition Engine Development, *Proc. I.A.E.*, **26**, 309 (1932).

Figure 107 indicates that an increase in pressure lowers the ignition temperature of a fuel. This characteristic makes the temperature difference between compression and ignition temperatures increase faster than compression temperature increases, and speeds up the initiation of combustion. Higher pressures reduce the tendency of the fuel to vaporize and form an appreciable quantity of explosive mixture before combustion gets well under way.

The effect of compression ratio on time lag and rate of pressure rise is shown by the indicator cards in Fig. 109, reported by Pope and Murdock.[1] With a high compression ratio there was very little time lag with a comparatively slow pressure rise, at least when kerosene was used as the fuel. As the compression ratio was lowered the time lag increased and the rate of pressure rise increased. However, with still lower compression ratios, excessive time lags occurred, accompanied by slow pressure rises during the expansion stroke.

Rating of Compression-ignition Fuels.—Various methods have been proposed[2] and used for the rating of fuels for compression-ignition engines. All the methods may be classified in one of the groups termed "bomb," "chemical," "physicochemical," or "engine" tests. The bomb and chemical tests have not been very satisfactory, but the other test methods provide fairly satisfactory results and are widely used.

Physicochemical Tests.—The "Diesel index"[3] is defined by the equation

$$\text{Diesel index} = \frac{\text{aniline point (°F.)} \times \text{A.P.I. gr.}}{100},$$

aniline point being the lowest temperature at which equal parts by volume of the test sample and freshly distilled, water-free aniline are completely miscible.

---

[1] Pope and Murdock, *op. cit.*

[2] T. B. Hetzel, The Development of Diesel Fuel Testing, *Penn. State Coll., Bull.* 45 (1936).

W. H. Hubner and G. B. Murphy, Effect of Crude Source on Diesel Fuel Quality, *Nat. Petroleum News*, **28**, 22, 29 (1936).

P. H. Schweitzer, Methods of Rating Diesel Fuels, *Chem. Reviews*, **22**, 107 (1938).

[3] A. E. Becker and H. G. M. Fischer, A Suggested Index of Diesel Fuel Performance, *S.A.E. Trans.*, **29**, 376 (1934).

A high Diesel index indicates a fuel of high ignition quality.

The "viscosity-gravity number"[1] (VG) for compression-ignition fuels is obtained from the specific gravity G at 60°F.

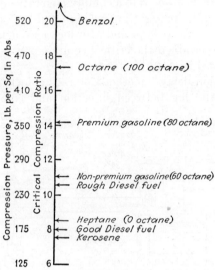

Fig. 110.—Critical compression ratios and pressures for various fuels. (*Pope and Murdock.*)

and the kinematic viscosity KV in millistokes at 100°F., which are related in the following equation:

$$G = 1.082 VG + (0.776 - 0.72 VG)[\log \log (KV - 4)] - 0.0887. \tag{1}$$

Viscosity-gravity number is an index of chemical composition; a high value indicates a high naphthenic content, and a low value indicates a high paraffinic content. Also, a low number indicates a high ignition quality of the fuel.

Other tests are termed the "boiling-point-gravity number," the "blending octane number," and the "characterization factor" (Hubner and Murphy, *op. cit.*).

[1] Developed by J. B. Hill and H. B. Coats, The Viscosity Gravity Constant of Petroleum Lubricating Oils, *Ind. Eng. Chem.*, **20**, 641 (1928).

Proposed for ignition quality by C. C. Moore, Jr. and G. R. Kaye, Viscosity-gravity Constant Held to Be Index of Diesel Fuel Ignition Quality, *Oil Gas Jour.*, **33**, 108 (1934).

Engine Tests.—The "critical-compression-ratio"[1] method makes use of an engine that is motored under definite conditions. Fuel is injected for 3 sec., and if audible combustion occurs the compression is lowered to a compression ratio at which audible combustion can be brought in or eliminated by a definite change in compression ratio. Pope and Murdock used a modified C.F.R. engine and determined the critical compression ratio for various common fuels (Fig. 110).

The "ignition-delay" method[2] of rating fuels depends upon determining the lag between injection and combustion. The

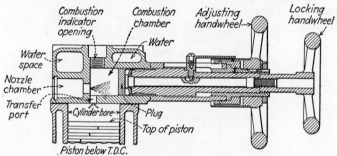

Fig. 111.—C.F.R. high-turbulence variable-compression-ratio combustion chamber.

beginning of the delay period is indicated by the lifting of the stem of the injection nozzle. Various methods are used for indicating the end of the delay period, such as a modified bouncing pin[2] a balanced pressure diaphragm[3] which closes when a given pressure is attained in the combustion chamber, and a magnetic pickup[4] operated from a diaphragm in the combustion-chamber wall.

The modified bouncing pin or "combustion indicator" is so designed that the pin is held in contact with the diaphragm (Fig. 103) for accelerations up to the maximum rate of pressure rise due to compression. Rapid combustion of an appreciable part of the fuel causes the bouncing pin to leave the diaphragm and close an electric circuit which is indicated by the flash of a neon lamp attached to the flywheel. In the proposed method,

[1] Pope and Murdock, *op. cit.*
[2] Proposed Method of Test for Ignition Quality of Diesel Fuels, *A.S.T.M. Proc.*, **38**, 392 (1938).
[3] Socony-Vacuum Oil Co. method.
[4] Schweitzer, *op. cit.*, and Hetzel, *op. cit.*

injection occurs at 13 deg. B.T.C., but the neon lamp is set 13 deg. ahead so the flash indicating injection occurs at T.D.C. The compression ratio is varied until both the injection and combustion flashes are coincident. Thus, fuels are rated with a fixed ignition lag, the unknown fuel being bracketed between two blends of reference fuels by means of the compression ratios required for coincident flashing of the two neon lamps with each fuel.

A C.F.R. engine is used, having a special cylinder and cylinder head with the Ricardo high-turbulence combustion chamber (Fig. 111). The compression ratio is varied by moving a plug in or out of the combustion chamber.

Compression-ignition Reference Fuels.—The primary reference fuels are cetane and α-methylnaphthalene, the percentage by volume of cetane in a mixture of the two fuels being the cetane number of the mixture. Secondary reference fuels are Shell high-reference fuel and a mixture of α- and β-methyl-naphthalene, a low-reference fuel. The calibration curve between per cent Shell reference fuel and cetane number is a straight line (Fig. 112).

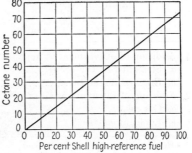

FIG. 112.—Calibration curve for mixtures of shell secondary reference fuel and a mixture of α- and β-methyl-naphthalene. [*T. B. Rendel, Report of Vol. Group. for C.-I. Fuel Research, S.A.E. Trans.*, **31**, 225 (1936).]

Example.—A compression ratio of 10.35:1 is required for the unknown fuel to cause coincident flashing of the neon lamps indicating injection and combustion. A mixture of secondary fuels with 65 per cent Shell high-reference fuel required a compression ratio of 10.60, whereas a 75 per cent mixture required a compression ratio of 10.25. Determine the cetane rating of the fuel.

Straight-line interpolation based on compression ratios indicates that the unknown fuel is equivalent to about 72 per cent Shell high-reference fuel. Reference to Fig. 112 indicates a cetane number of 53.

Ignition-quality "Dopes."[1]—Various materials have been suggested and used for improving the ignition quality of a fuel, such as

[1] W. H. Hubner and G. Egloff, Ample Future Supply of Diesel Fuel Obtainable from Various Sources, *Nat. Pet. News*, **28**, 4-22, 5-25 (1936).

1. Alkyl nitrates and nitrites.
2. Nitro and nitroso compounds.
3. Peroxides.
4. Oxidizing agents.

The effectiveness of ethyl nitrate and tetralin peroxide is shown in Table 18.

TABLE 18.—EFFECTIVENESS OF TWO FUEL "DOPES"

Addition agent	Cetane no. for additions of various percentages of dope						
	0.0	0.5	1.0	1.5	2.0	4.0	5.0
1. Addition of ethyl nitrate* to							
a. Straight-run Diesel fuel.........	55	..	..	..	74	93	
b. Cracked distillate...............	40	..	..	..	56	70	
2. Addition of tetralin peroxide† to							
a. Gas oil.......................	42	50	56	60			
b. Edeleann extract...............	14	..	..	..	..	..	33

* Hubner and Murphy, *op. cit.*
† U. S. Patent No. 2011297.

Comparison of Ratings by Various Methods.—A comparison[1] of the results obtained by several methods shows that a general

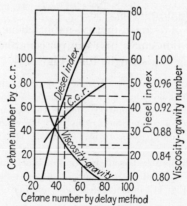

FIG. 113.—Comparison of several rating methods. (*Hubner and Murphy.*)

relationship exists between each method and cetane number by the ignition-delay method (Fig. 113) when using straight-run,

[1] Hubner and Murphy, *op. cit.*

cracked, and 50-50 blends of the two fuels. Fuels doped with ethyl nitrate and blends of cetane and α-methylnaphthalene showed appreciable variation.

EXERCISES

1. Compute and plot on logarithmic paper the pressures and temperatures at the end of compression for compression ratios from 5:1 to 15:1, for values of n ranging from 1.3 to 1.4 and for initial temperatures of 500, 550, and 600°R.

2. Assuming air is the medium compressed in Prob. 1, determine and plot the density of the air at the end of compression.

3. For a given value of T_1, n, and r, plot the temperature, pressure, and density of the air being compressed, at various crank angles. Connecting-rod-to-crank ratio is 4:1. See Table 30, Chap. XVII, for piston positions at various crank angles.

4. Compute and plot the clearance volume in percentages of displacement for various compression ratios.

5. Determine the maximum detonation pressures, assuming constant-volume combustion of the unburned fraction when the flame-front position is 0, 0.5, 0.8, and 1.0. Use the values of pressure and temperature of the unburned fraction as given in Fig. 99. Use Charts B and E in the Appendix. Note that Fig. 99 is determined from data other than those for Charts B and E and that slightly different results will be obtained.

6. Assume the conditions at A (Fig. 100) are 14.7 lb. in.$^{-2}$ and 600°R. Determine the conditions at E, and compare with the conditions at D in the illustrative example in Chap. V. Also, estimate the conditions at F, and compare with the conditions at C in the same example.

7. Assume B'' (Fig. 100) is located at the middle of the stroke of the piston, and choose a probable path like $B''C''$. Check the conditions at C'', and change the path until a check is obtained. Compression ratio is 5:1. Other conditions are to be the same as in Prob. 6.

CHAPTER IX

CARBURETION AND FUEL INJECTION

CARBURETION

Mixture Formation.—The formation of an air-fuel mixture consists of subdividing or atomizing the fuel and mixing the finely divided particles of fuel with air. Atomization is accomplished by spraying the liquid fuel through nozzles or jets into a stream of moving air. The suction stroke of an engine reduces the pressure in the cylinder and causes a pressure gradient from atmospheric pressure outside the carburetor to the pressures that exist in the carburetor, manifold, and cylinder. This pressure drop causes air to flow through the induction system and fuel to be sprayed from the fuel nozzles. The pressure drop at the exit of the fuel nozzle is accentuated by a restricted or venturi section which increases the air velocity and pressure drop at this point.

The fuel leaves the jet more or less in a stream which is torn apart into ligaments that break up and contract into various sizes of drops. During this process, vaporization from the surfaces of the drops and particles is taking place, which causes the disappearance of the finest particles and the reduction in size of the others. The ideal situation would be to have all the particles vaporized and uniformly distributed before the mixture enters the cylinders. Actually, some fuel particles enter the cylinders in liquid form and must be vaporized and mixed during the compression stroke.

The degree of atomization depends upon the relative velocity of air and fuel streams, the density of the fuel, and its surface tension. For a given fuel, the degree of atomization is thought to be a function of the square of the relative velocity.

Desirable Air-fuel Ratios.—The mixture ratio must be between the limits of inflammability. Practically, these limits are about 7:1 on the rich side and about 20:1 on the lean side for average gasoline in single-cylinder engines.[1] The limits for

[1] L. C. Lichty and E. J. Ziurys, Engine Performance with Gasoline and Alcohol, *Ind. Eng. Chem.*, **28**, 1094 (1936).

multicylinder engines lie within this range depending upon the mixture distribution between cylinders. There is no reason for operating with excessively rich mixtures, but very lean mixtures may be desirable in order to obtain low fuel-consumption rates (Fig. 41).

Maximum power should be obtained theoretically from mixtures slightly richer than the correct air-fuel ratio. (See Fig. 44.) Also, poor distribution and poor mixing of the air and fuel vapor necessitate the use of rich mixtures to obtain maximum power.[1]

The dilution of the charge by the exhaust products left in the clearance space reduces the probability of the fuel particles reacting with the oxygen particles. This is not a serious condition under wide-open throttle conditions, but under throttled conditions the probability of a combustion reaction decreases as the throttle is closed. Under the completely closed throttle condition, air will not flow into the engine since the expanded clearance gases fill both the clearance space and the displacement volume at a pressure equal to the intake pressure. Also, the quantity of mixture flowing into the engine at any given speed is proportional to the intake pressure when the suction temperature is constant. Thus, the line A (Fig. 114) represents the mixture entering the cylinders. The weight of clearance gases would be constant if its temperature and pressure were constant for various intake pressures. The total weight of gases in the cylinder at the end of the suction stroke is proportional to the intake pressure and is zero at zero pressure. Thus, line C represents the total gases for a constant suction temperature and constant weight of clearance gases. Actually, the weight of clearance gases will be greater at low pressures, owing to lower exhaust temperatures at low loads, so that line B, the actual total weight of gases, lies between lines A and C.

[1] A correct gasoline-air mixture has about 1.6 per cent gasoline vapor and 98.4 per cent air by volume. An increase of 10 per cent in fuel volume would decrease the air and oxygen content by 0.16 per cent, a negligible amount. This should result in complete consumption of the oxygen and about maximum power output. A decrease of 10 per cent of the fuel reduces the energy input by 10 per cent even though all the fuel is consumed. Thus, maximum power is always obtained with rich mixtures which consume all the oxygen if the correct air-vapor ratio by volume is appreciably greater than unity. The reverse would be true if the vapor or gaseous volume of the fuel is appreciably larger than the air volume for the correct mixture.

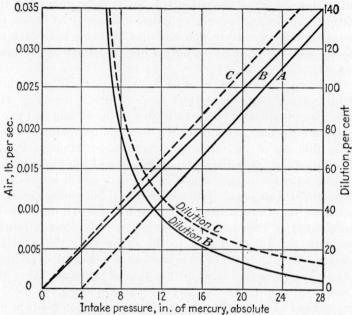

Fig. 114.—Effect of intake pressure on charge dilution. (*Tice.*)[1]

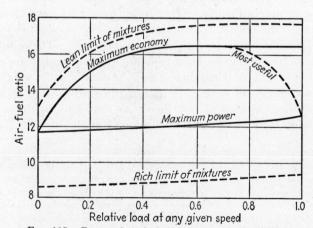

Fig. 115.—Range of air-fuel mixtures required. (*Tice.*)[1]

[1] P. S. Tice, Fuel-supply Requirements, *Automotive Industries*, **53**, 53 (1925).

CARBURETION AND FUEL INJECTION

The dilution is the ratio of the weight of clearance gases to the weight of mixture and is equal to $B - A$ divided by A. Two dilution curves are plotted, based upon lines B and C, and show considerable dilution at low intake pressures which is the condition under low loads.

Under idling or no-load conditions, where dilution is high, an extra rich mixture must be supplied so that there will be more fuel particles than required to react with the oxygen particles. This increases the probability that fuel particles will find all the oxygen particles. During part throttle conditions, an economy mixture is desired, whereas, at full load, maximum power is required. Zero relative load (Fig. 115) represents idling conditions and at this load the maximum economy and maximum power mixtures are the same (12:1), because of the effect of dilution. As the load increases and more charge is admitted, the effect of dilution diminishes and the maximum economy mixture rises to above 16:1 air-fuel ratio. The most useful mixture follows the maximum economy mixture to about three-fourths load and then is enriched until reaching maximum power at wide-open throttle, under which condition maximum power is usually desired.

Thus, the desired air-fuel ratios range, not from one limit of inflammability to the other, but from those mixtures resulting in maximum economy to those of maximum power. The primary object of carburetion is to produce under all operating conditions the desired air-fuel ratio. This ratio between the air and fuel depends on the flow characteristics of the two fluids.

Flow of Fluids.—The fundamental equation for fluid flow is

$$v = c'\sqrt{2gh} \qquad (1)$$

where v = velocity of flow, ft. sec.,$^{-1}$
c' = coefficient of velocity,
g = acceleration due to gravity, 32.2 ft. sec.,$^{-2}$
h = intake depression, head of fluid causing flow, ft.

The quantity of fluid flowing is given by the expression

$$M = c''dAv = dAc\sqrt{2gh} \qquad (2)$$

where M = weight of fluid flowing per sec., lb.,
d = density of the fluid flowing, lb. ft.,$^{-3}$
A = area of passageway, ft.^{2}
c and c'' = coefficients of discharge and jet contraction, respectively.

The head causing the flow is usually measured in inches of water. The relation between h_w in inches of water and h in feet of fluid flowing is

$$h = \frac{h_w}{12} \cdot \frac{d_w}{d}. \tag{3}$$

Substituting in Eq. (2), using a value of 62.4 for d_w, results in

$$M = 18.3 A c \sqrt{h_w d}. \tag{4}$$

Applying this equation to both air and fuel results in

$$\text{Air} = 18.3 A_a c_a \sqrt{h_w d_a}, \tag{5}$$
and
$$\text{Fuel} = 18.3 A_f c_f \sqrt{h_w d_f}. \tag{6}$$

Then,
$$\text{Air-fuel ratio} = \frac{A_a}{A_f} \cdot \frac{c_a}{c_f} \sqrt{\frac{d_a}{d_f}}. \tag{7}$$

Example.—Determine the intake depression at the carburetor throat for a 200-cu. in. four-stroke cycle engine running at 2700 r.p.m. having a volumetric efficiency of 100 per cent. Diameter of carburetor throat is 1.25 in. The barometer is 30 in. Hg and the temperature is 60°F. Venturi discharge coefficient = 0.83. The engine has four or more cylinders.

The volume displaced per second is

$$\frac{200 \times 2700 \times 1.0}{1728 \times 2 \times 60} = 2.60 \text{ ft.}^3$$

The mols and pounds of air under given conditions are

$$M = \frac{30 \times 0.491 \times 144 \times 2.60}{1545 \times 520} = 0.00686 \text{ mol} \equiv 0.199 \text{ lb.}^*$$

Substituting in Eq. (5) results in

$$0.199 = 18.3 \times \frac{0.7854 \times \overline{1.25}^2}{144} \times 0.83 \sqrt{h_w d_a},$$

from which
$$h_w d_a = 2.36.$$

d_a, at 30 in. Hg and 60°F., is equal to

$$M \times 28.95 = \frac{30 \times 0.491 \times 144 \times 1}{1545 \times 520} \times 28.95 = 0.076 \text{ lb. ft.}^{-3}$$

*This is one-fourth the rate for a single-cylinder engine of the same displacement.

CARBURETION AND FUEL INJECTION

Values of $h_w d_a$, are determined in the following manner:

h_{Hg}, in.	Abs. pr., in. Hg	h_w, in.	d_a	$h_w d_a$
0	30.0	0	0.0764	0.0
1	29.0	13.6	0.0739	1.01
2	28.0	27.2	0.0713	1.94
3	27.0	40.8	0.0688	2.81
4	26.0	54.4	0.0662	3.60

Plotting values of $h_w d_a$ against h_{Hg} gives an intake depression of 2.48 in. Hg or 33.7 in. of water, and a value for d_a of 0.0701 when $h_w d_a = 2.36$.

Coefficients of Discharge.—The coefficient of discharge for air flow through orifices varies from 0.60 to 0.98, depending on the kind of orifice and the relation between the diameters of the orifice and pipe in which it is located. A sharp-edged orifice, one-half the size of the pipe, will have a coefficient of about 0.60, whereas, if the orifice were well rounded, the coefficient would be about 0.98. The venturi tubes used in carburetors have coefficients of discharge of about 0.83. These coefficients remain fairly constant under all conditions except for velocities below 145 ft. sec.$^{-1}$ For lower velocities the coefficient decreases.

The coefficient of discharge for liquid fuels depends upon a number of factors:

1. Size of the orifice.
2. Shape of the fuel orifice (circular, annular, etc.).
3. Proportions of the orifice (ratio of length to diameter).
4. Entrance to orifice.
5. Viscosity of the fuel.
6. Temperature of the fuel.
7. Rate of flow.

For a small orifice there is considerable rubbing surface compared with the volume of the passageway. As the diameter of the orifice is increased, the proportion of rubbing surface becomes smaller and the coefficient of discharge increases.

The annular orifice presents more rubbing surface than a circular orifice and, consequently, has a lower coefficient of discharge.

The greater the ratio of length to diameter, the smaller the coefficient.

A well-rounded entrance to the fuel orifice will increase the coefficient of discharge.

Viscosity and temperature effects are closely related. The higher the temperature, the lower the viscosity, the greater the flow for a given head, and the higher the coefficient of discharge.

The coefficient of discharge increases with an increase in rate of flow, for laminar flow.

Coefficients of discharge, for typical fuel jets with circular orifices, computed from preceding formulas, range from below 0.70 to well above unity, depending upon the various factors involved.

Submerged square-edged jets, with the diameter of the jets below 0.09 in. and with a length-diameter ratio of about 0.33, have coefficients of discharge that vary from 0.63 to 0.76, depending upon the rate of flow, the diameter, and the viscosity of the liquid.[1]

Example.—Assume all the fuel in the example, page 208, is provided by one orifice having a coefficient of discharge of 0.75. Determine the diameter of fuel orifice to provide a 12:1 air-fuel ratio. Sp. gr. of the fuel = 0.74.

Substituting in Eq. (7) results in

$$12 = \frac{\overline{1.25}^2}{\overline{D_f}^2} \cdot \frac{0.83}{0.75} \sqrt{\frac{0.0701}{0.74 \times 62.4}}$$

$$D_f = \sqrt{0.00565}$$
$$= 0.075 \text{ in., diameter of fuel orifice.}$$

The Elementary Carburetor.—The simplest form of carburetor (Fig. 116) consists of an air passageway with a fuel jet located therein. A float chamber is usually required for maintaining the level of the fuel in the jet, and a throttle provided for controlling the amount of mixture. The level of the fuel is slightly below the top of the jet to prevent leakage when not operating.

When the air begins to flow past the jet, the fuel begins to rise because of reduced pressure at that point. The reduction in pressure is accentuated by restricting the air passageway with a venturi tube. However, the reduction in pressure must reach about 0.5 in. of fuel before fuel will flow from the jet. This is due not only to the level of the fuel being below the top of

[1] M. J. Zuchrow, Discharge Characteristics of Submerged Jets, *Purdue Univ., Eng. Expt. Sta., Bull.* 31 (1931).

the jet when not operating but also to the viscosity of the fuel. Thus, Eq. (6) becomes

$$\text{Fuel} = 18.3 \, A_f c_f \sqrt{(h_w - h'_w) d_f} \qquad (8)$$

where h'_w = head, in. of water required to cause the fuel to begin to flow out of the jet.

Then, Air-fuel ratio $= \dfrac{A_a}{A_f} \cdot \dfrac{c_a}{c_f} \sqrt{\dfrac{d_a}{d_f} \cdot \dfrac{h_w}{(h_w - h'_w)}}.$ \qquad (9)

For fixed air and fuel openings, and on the assumption of con-

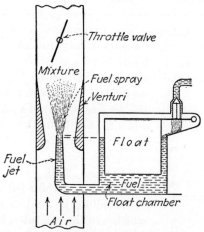

Fig. 116.—Elementary updraft carburetor.

stant coefficients of flow, the air-fuel ratio is proportional to

$$\sqrt{\dfrac{h_w \, d_a}{h_w - h'_w}}.$$

As the load increases, h_w increases and d_a decreases. If h'_w is neglected the air-fuel ratio will decrease with load, or the mixture becomes richer as more air is inducted. The effect of h'_w is appreciable at low loads, increasing the air-fuel ratio and necessitating a larger fuel orifice for a given air-fuel ratio. The effect of h'_w is negligible at high loads, but the larger orifice required at low loads will result in richer mixtures at high loads.

Solving Eq. (9) for the foregoing assumptions and an air density of 0.0760 lb. ft.$^{-3}$ results in the following tabulation:

h_w, in. of water	d_a, air density	$h_w\, d_a$	$\sqrt{\dfrac{h_w\, d_a}{h_w - 0.5}}$	Air-fuel ratios	
				15:1 at $h_w = 5$	15:1 at $h_w = 0.75$
0.5	0.0759	0.0380	∞	∞	∞
0.75	0.0758	0.0568	0.476	24.8	15.0
1.0	0.0758	0.0758	0.388	20.2	11.4
2	0.0756	0.1512	0.317	16.5	10.0
5	0.0751	0.3755	0.288	15.0	9.1
20	0.0723	1.4460	0.272	14.2	8.6
40	0.0686	2.7440	0.264	13.7	8.3

A fuel jet of sufficient size to produce a 15:1 air-fuel ratio when $h_w = 5$ in. of water results in mixtures that would be satisfactory at the heavier loads but unsatisfactory at the lighter loads. Actually, the engine would not be able to run under no-load or idling condition. Increasing the jet size to obtain a 15:1 air-fuel ratio when $h_w = 0.75$ in. of water, or richer for idling conditions, results in mixtures much too rich at or even near full-load conditions. This characteristic of the single-jet carburetor led to development of the idling jet and its air passageway, which are incorporated in an idling and low-load carburetor operating in parallel with the main carburetor.

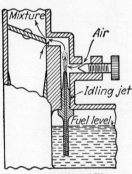

Fig. 117.—The idling or low-load carburetor.

Equation (9) also applies to a single-jet gas carburetor. However, d_a/d_f will remain practically constant and the air-fuel ratio will be proportional to

$$\sqrt{\dfrac{h_w}{h_w - h'_w}}$$

where h'_w = head required to cause the gas pressure regulator to permit gas to flow to the fuel jet.

The Idling or Low-load Carburetor.—The idling carburetor discharges mixture at the edge of the throttle plate in the closed position (Fig. 117). The manifold depression may amount to 200 in. of water with the throttle closed, but the depression at the idling jet will be appreciably less because of the flow of air from below the throttle plate into the discharge passageway and also to air entering at the idling adjustment. The desirable idling mixture is obtained by adjustment of the idling screw.

The manifold depression decreases as the throttle is opened until at wide-open throttle and low speed the depression may amount to 5 in. of water or less. Thus, the idling carburetor operates at maximum capacity when the throttle is closed and is inoperative when the throttle is wide open. Since the idling carburetor is an elementary single-jet carburetor, it supplies a rich mixture at closed throttle and, in general, a progressively leaner mixture as the throttle is opened.

A slight opening of the throttle may expose more of the idling discharge passageway to the manifold depression and tend to maintain the richness of the idling mixture until the air flowing through the venturi tube causes fuel to flow from the main jet. As the throttle is opened, the air-fuel ratio increases because of the decreasing richness characteristic of the idling carburetor and also since the percentage of charge supplied by the idling carburetor decreases as more mixture is delivered from the main carburetor.

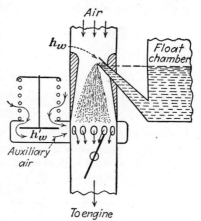

Fig. 118.—Auxiliary air valve and single-jet downdraft carburetor (idling jet omitted).

The Main Carburetor.—The main carburetor should supply a practically constant air-fuel ratio throughout its range of operation. The single-jet carburetor supplies a mixture that increases in richness as the air flow through the venturi is increased, as indicated heretofore. This has given rise to the various compensating devices for maintaining constant air-fuel ratios.

The Auxiliary Air Valve.—The flow of air through the auxiliary air valve (Fig. 118) from Eq. (5) is

$$\text{Air}_{aux} = 18.3 A_{aux} c_{aux} k_1 \sqrt{h_w d_a} \qquad (10)$$

on the assumption that $k_1^2 h_w = h_{aux}$, the head causing flow through the air valve.

The area of opening of the auxiliary air valve is equal to the circumference times the lift, and with a fixed valve-spring constant the lift is proportional to the head h_{aux}.

Thus, $\quad A_{aux} = \pi DL = \pi D k_1' h_{aux} = \pi D k_1' k_1^2 h_w = k_2 h_w \quad$ (11)

where $k_1' = $ const. relating h_{aux} and L.

The density also depends upon carburetor suction. For an assumed barometer of 30 in. of mercury or 408 in. of water,

$$d_a = \frac{d_{30}(408 - h_w)}{408}. \quad (12)$$

Substituting these values and inserting the equation for air flow through the venturi, with area A and coefficient c,

$$\text{Total air} = 18.3(Ac + k_1 k_2 c_{aux} h_w)\sqrt{\frac{h_w\, d_{30}(408 - h_w)}{408}}.$$

Collecting all the constants, on the assumption of constant coefficients of discharge, results in

$$\text{Total air} = K(A + k' h_w)\sqrt{h_w(408 - h_w)}. \quad (13)$$

Dividing by Eq. (6) results in

$$\text{Air-fuel ratio} = K'(A + k' h_w)\sqrt{(408 - h_w)}. \quad (14)$$

The factor $(A + k' h_w)$ will increase with h_w, whereas the factor $\sqrt{(408 - h_w)}$ will decrease. The air-fuel ratio will vary with the product of the two factors.

Example.—Take values of h_w of 0, 10, 20, 30, and 40, and values of 1 for A and 1 for $k' h_w$ when h_w is 40 in. Compute the effect of the two foregoing factors on air-fuel ratio.

$$k' = \tfrac{1}{40} = 0.025.$$

h_w	$k'h_w$	$A + k'h_w$	$408 - h_w$	$\sqrt{408 - h_w}$	Air-fuel ratio $\dfrac{}{K'}$
0	0.00	1.00	408	20.20	20.20*
10	0.25	1.25	398	19.95	24.94
20	0.50	1.50	388	19.70	29.55
30	0.75	1.75	378	19.44	34.02
40	1.00	2.00	368	19.18	38.36

* Actually, no air will flow with $h_w = 0$, and fuel will begin to flow at a head of about 0.5 in. of water.

CARBURETION AND FUEL INJECTION

Since the product of the two factors increases, the mixture becomes leaner the greater the flow. This is true, at least, for a relation of 1:1 between the primary air opening A and the factor $k'h_w$.

Example.—Determine the ratio of $k'h_w : A$ in the foregoing example that will result in an approximately constant air-fuel ratio.

$$\frac{(\sqrt{408 - h_w})_0}{(\sqrt{408 - h_w})_{40}} = \frac{20.20}{19.18} = 1.053.$$

The ratios of $(A + k'h_w)$ at $h_w = 40$ to A at $h_w = 0$ should be approximately 1.053.

Then, $\qquad k' = \dfrac{0.053}{40} = 0.001325.$

The product of the two variable factors is as follows:

h_w	$A + k'h_w$	$\sqrt{408 - h_w}$	Air-fuel ratio $\overline{K'}$
0	1.000	20.20	20.20
10	1.013	19.95	20.21
20	1.026	19.70	20.21
30	1.040	19.44	20.22
40	1.053	19.18	20.20

The foregoing examples indicate that auxiliary air valves with comparatively large openings result in leaner mixtures as the flow increases through a single-jet carburetor. Air valves of the correct size, with opening proportional to the suction head, should maintain practically constant air-fuel ratios.

Unrestricted Air-bled Jet.—The bleeding of air into a fuel jet prevents the suction head at the venturi from acting on the fuel orifice (Fig. 119). The head h, causing the flow of fuel into the unrestricted air-bled jet, remains constant as soon as the air passageway is emptied of fuel. Thus, the fuel flow remains constant regardless of air flow, and the unrestricted air-bled

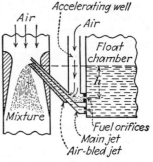

Fig. 119.—Unrestricted air-bled jet and single-jet down-draft carburetor (idling jet omitted).

jet provides leaner mixtures with the higher loads. A combination of the elementary jet and the unrestricted air-bled jet may be used for maintaining constant or nearly constant air-fuel ratios at various loads.

The flow of air through the venturi is given by Eq. (5):

$$\text{Air} = 18.3 A_{ven} c_{ven} \sqrt{h_w \, d_a}.$$

The flow of fuel will be the sum of flow from both jets, the flow from the air-bled jet under steady-flow conditions attaining a maximum when the accelerating well is empty. Then, h_w (Eq. 6) is equivalent to h, the head of fuel on the orifice, and

$$\text{Fuel} = 18.3 A_{f_1} c_{f_1} \sqrt{(h_w - h'_w) d_f} + 18.3 A_{f_2} c_{f_2} \sqrt{h \, d_f} \quad (15)$$

where subscript 1 = the primary jet,
subscript 2 = the fuel orifice of the air-bled jet,
h = the maximum head on the air-bled orifice, in. of water.

Then, on the assumption of constant orifice discharge coefficients,

$$\text{Air-fuel ratio} = \frac{k''_1 \sqrt{h_w \, d_a}}{k''_2 \sqrt{h_w - h'_w} + \sqrt{h}}. \quad (16)$$

Example.—Determine the values for k''_1 and k''_2, assuming a constant air-fuel ratio of 15:1. Also, show the variation in air-fuel ratio with variation in h_w. Assume $d_a = 0.075$ at $h_w = 0$, $h = 1$ in. of water, and $h'_w = 0.5$ in. of water.

For $h_w = 2$ in., Eq. (16) becomes

$$\frac{15}{1} = \frac{0.386 k''_1}{1.225 k''_2 + 1}.$$

For $h_w = 40$ in., Eq. (16) becomes

$$\frac{15}{1} = \frac{1.644 k''_1}{6.285 k''_2 + 1}.$$

Equating the fractions containing k''_1 and k''_2 results in $k''_1 = 184.2$ and $k''_2 = 3.053$. Solving for air-fuel ratios at various heads is accomplished as follows:

h_w	$h_w\,d_a$	$\sqrt{h_w\,d_a}$	$k_1''\sqrt{h_w\,d_a}$	$k_2''\sqrt{h_w - h_w'} + \sqrt{h}$	Air-fuel ratio
2	0.1492	0.386	71.1	4.74	15.0
10	0.7320	0.856	157.7	10.41	15.1
20	1.4260	1.194	219.9	14.48	15.2
30	2.0850	1.443	265.8	17.58	15.1
40	2.7040	1.644	302.8	20.19	15.0

Restricted Air-bled Jet.—A restricted air-bled jet has an orifice (A_b, Fig. 120) which restricts the flow of air into the fuel jet. The introduction of the bled air reduces the suction on the fuel from h_w at the nozzle orifice to h_w'' in the upper part of the nozzle. Thus, the head $h_w - h_w''$ causes fuel and air to flow through the orifice A_n.

The head causing fuel to flow through the orifice A_f is $h_w'' + h$, in which h is the variable difference between the fuel level in the float chamber and the level in the fuel nozzle under various operating conditions. The weight of fuel flowing per second is

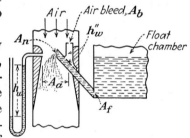

FIG. 120.—Restricted air-bled jet.

$$M_f = 18.3 A_f c_f \sqrt{(h_w'' + h)\,d_f}. \quad (17)$$

Since both air and fuel flow through A_n, it will be assumed that A_n' is the equivalent area occupied by fuel flow, and then $A_n - A_n'$ is the equivalent area for air flow. The quantity of fuel that flows through A_f, because of a suction head of $h_w'' + h$, also flows through the equivalent area A_n' because of the head $h_w - h_w''$. The equivalent velocity of this fuel through the equivalent area A_n' is

$$\text{Fuel velocity} = 18.3 c_{n_f} \sqrt{\frac{h_w - h_w''}{d_f}}. \quad (18)$$

Also, $\quad$ Area $A_n' = \dfrac{\text{volume}}{\text{velocity}} = \dfrac{M_f/d_f}{\text{velocity}}$

$$= \frac{18.3 A_f c_f \sqrt{(h_w'' + h)\,d_f}}{18.3 c_{n_f} \sqrt{\dfrac{h_w - h_w''}{d_f}} \cdot d_f} = k_f A_f \sqrt{\frac{h_w'' + h}{h_w - h_w''}} \quad (19)$$

where $k_f = c_f/c_{n_f}$.

218   INTERNAL-COMBUSTION ENGINES

The weight of air bled, flowing through area $A_n - A'_n$, is

$$M_{ab} = 18.3 c_{n_a} \left[A_n - k_f A_f \sqrt{\frac{h''_w + h}{h_w - h''_w}} \right] \sqrt{(h_w - h''_w) d_{a_{h_w}}}. \quad (20)$$

This is the same weight of air flowing through A_b, and is

$$M_{ab} = 18.3 c_b A_b \sqrt{h''_w d_{a_{h''_w}}}. \quad (21)$$

Equating Eqs. (20) and (21) results in

$$\sqrt{\frac{d_{a_{h''_w}}}{d_{a_{h_w}}}} = k_a \frac{A_n}{A_b} \sqrt{\frac{h_w - h''_w}{h''_w}} - k_b \frac{A_f}{A_b} \sqrt{\frac{h''_w + h}{h''_w}}, \quad (22)$$

where $k_a = c_{n_a}/c_b$,
and, $k_b = c_{n_a} k_f / c_b = k_a k_f$.

Values of h''_w for various values of h_w can be determined from this relationship, and the air-fuel ratio can then be found from the air flow through the venturi A_a and the fuel flow through the fuel orifice A_f.

Thus, Air-fuel ratio $= \dfrac{18.3 A_a c_a \sqrt{h_w d_a}}{18.3 A_f c_f \sqrt{(h''_w + h) d_f}}. \quad (23)$

Or assuming the coefficients of discharge to remain constant, results in

$$\text{Air-fuel ratio} = k \sqrt{\frac{h_w d_a}{h''_w + h}}. \quad (24)$$

Example.—The orifices in a restricted air-bled jet have the following relations: $A_n = 4 A_l$, and $A_f = A_b$. Also $h = 0.25$ in. of water. Determine the relative air-fuel ratios for various values of h_w. Assume $k_a = k_b = 1.0$. Also, air density = 0.0760 lb. ft.$^{-3}$ Then, Eq. (22) can be written as follows:

$$\sqrt{\frac{408 - h''_w}{408 - h_w}} = 4 \sqrt{\frac{h_w - h''_w}{h''_w}} - \sqrt{\frac{h''_w + 0.25}{h''_w}},$$

from which h''_w can be determined for each value of h_w. The results for various values of h_w and the computations for air-fuel ratios are given in the following tabulation, assuming a 13:1 air-fuel ratio for $h_w = 2$ in. H_2O:

h_w, in. H_2O	h''_w, in. H_2O	d_a, lb. ft.$^{-3}$	$h_w d_a$	$\sqrt{\dfrac{h_w d_a}{h''_w + h}}$	Air-fuel ratio
2	1.57	0.0756	0.1512	0.289	13.0
5	3.9	0.0751	0.3755	0.301	13.5
10	8.0	0.0742	0.7420	0.300	13.5
20	15.9	0.0723	1.4460	0.299	13.5
30	23.9	0.0705	2.1150	0.296	13.3
40	31.9	0.0686	2.7440	0.292	13.1

CARBURETION AND FUEL INJECTION

This shows that the air-fuel ratio remains fairly constant throughout the range of h_w chosen. However, if A_n is decreased in relation to A_b, the air-fuel ratio can be made to increase. Thus, with $A_n = A_b = A_f$, the air-fuel ratio at $h_w = 40$ will be about 14.8 whereas the air-fuel ratio at $h_w = 2$ is 13:1. It should be noted that this type of jet acts the same as the elementary jet until the suction causes air to be bled. Consequently, the air should be bled into the nozzle just under the fuel level maintained in the nozzle when not operating.

Variable Fuel Orifice.—It has been shown that a single-jet carburetor provides an air-fuel ratio that becomes richer with increased suction, the mixture ratio changing more rapidly at low suctions than during the major portion of the suction range. The air-fuel ratio at any suction depends upon the area of the

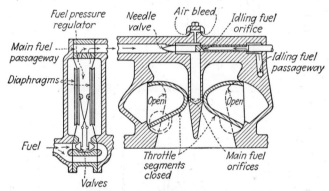

FIG. 121.—Variable fuel orifice in aircraft carburetor. (Holley C-G aircraft carburetor.)

fuel orifice. Varying the area of the fuel orifice with the suction at the throat of the venturi results in the desired air-fuel-ratio characteristic. A tapered metering pin inserted in the orifice of the fuel jet provides a means for varying the annular space through which the fuel is discharged.

Small annular orifices usually have erratic discharge coefficients, which is an undesirable characteristic. Also, the fuel volume is about $\frac{1}{9000}$ of the air volume, which makes the desired variations in orifice area relatively small and requires great accuracy for both the metering pin and fuel orifice.

A unique variable-fuel-orifice aircraft carburetor (Fig. 121) uses throttle segments that are rotated through part of a circle from closed to wide-open positions. The throttle segments are shaped to provide a variable rectangular venturi for various throttle positions. The needle valve is closed when the throttle is closed, and idling fuel flows through the idling-fuel orifice and then through the main fuel orifices along with the bled air.

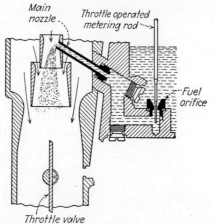

Fig. 122.—Carter high-load enriching device.

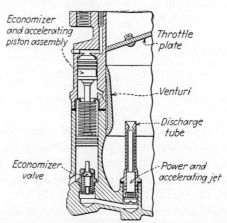

Fig. 123.—Zenith high-load enriching and accelerating device.

Highest suction on the main fuel orifices occurs with closed throttle; consequently, the needle valve is opened as the throttle is opened. A partial opening of the needle valve closes the idling-fuel orifices.

The float chamber is eliminated in this carburetor by the use of a diaphragm pressure regulator. The regulator chamber and the main fuel passageway are full of fuel during operation. This feature eliminates interruption of fuel flow arising from splashing when the aircraft is undergoing violent maneuvers, inverted flight, etc.

High-load Conditions.—High-load conditions occur as the throttle approaches the wide-open position and require maximum-power air-fuel ratios. The mixture may be enriched by enlarging the fuel orifice (Fig. 122) (withdrawing a metering

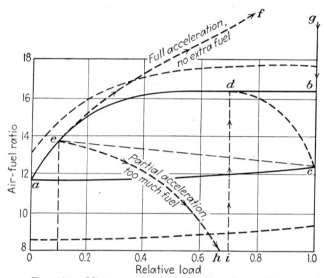

FIG. 124.—Mixture ratios during acceleration. (*Tice.*)

pin) or by opening a secondary fuel orifice as the throttle nears the wide-open position.

The enriching device (Fig. 123) may be operated by a spring and the manifold vacuum. As the throttle is closed from wide-open position, the manifold vacuum closes the high-load fuel orifice. The orifice is opened by means of a spring when the throttle is opened and the manifold vacuum decreases.

Acceleration.[1]—The pressure in the intake manifold is considerably reduced with partial throttle openings and the fuel is vaporized quite readily. At low speeds, with the throttle wide open, part of the fuel, such as the heavier fractions of gasoline, will condense or collect on the walls of the intake manifold and

[1] Tice, *op. cit.*

form a fuel film. During acceleration a change from the first to the second condition takes place, resulting in leaner air-fuel mixtures reaching the cylinders, owing to the loss of fuel to build up the fuel film. As the wall film is built up, the mixture returns to the desirable ratio. With sudden accelerations the mixture ratios would follow a path similar to *efgc* (Fig. 124), the mixtures becoming so lean that firing back in the carburetor (see page 228) or failure of the mixture to ignite would result.

Various devices are used to provide extra fuel for acceleration:

1. Damped air valve.
2. Accelerating well (Fig. 119).
3. Displacement pump:
    a. Throttle-operated.
    b. Suction-operated (Fig. 123).

The characteristics of these various devices are shown in Fig. 125. The air valve, even with very little damping, usually

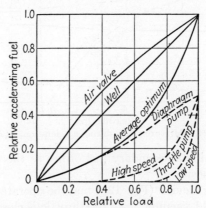

Fig. 125.—Characteristics of accelerating devices. (*Tice.*)

provides too much accelerating fuel. This is also true of the accelerating well, but to a less degree. The throttle-operated pump provides less than the average optimum amount. The suction-operated pump provides about the desirable amount for partial accelerations but must be augmented by some other device for accelerations to full-load conditions. A combination of the suction-operated pump and a throttle-operated pump, which comes into action as the throttle approaches wide-open

CARBURETION AND FUEL INJECTION 223

position, approaches fairly well the requirements indicated by the average optimum curve. This average optimum curve applies only to carburetors following the "most useful" curve in Fig. 115.

When an accelerating device provides too rich a mixture during partial acceleration, the mixture ratio follows a path similar to *ehid* indicated in Fig. 124. Such carburetor action is termed "gobbing" and results in soft and uneven engine operation.

Carburetor Choking.—Starting in cold weather is accomplished by "choking" the carburetor. The closing of the choke restricts the air supply as the engine is motored. This provides an over-supply of fuel, only the lighter fractions of which can evaporate at the low temperatures and form a combustible air-vapor mixture. Choking should be kept to the minimum necessary to provide starting mixtures, as the unvaporized heavy fractions of the fuel wash the lubricant from the cylinder walls and also run into the crankcase and dilute the oil supply. Automatic choking is accomplished by a thermostatic element (Fig. 126) which closes the choke. As soon as the engine begins to operate, the manifold vacuum causes the vacuum piston to tend to open the choke which is held in a nearly closed position by the tension of the spring thermostat. Exhaust gases heat the thermostat casing, which decreases the tension and causes the thermostat to permit the choke to open to wide-open position.

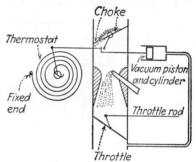

Fig. 126.—Sketch of automatic choke.

A fast idle device, usually incorporated with the automatic choking mechanism, keeps the throttle partly open and hastens the warming-up period.

Commercial Carburetors.[1]—Commercial carburetors have incorporated in their design a manual or automatic choking device, an idling carburetor, a single- or multiple-jet carburetor

---

[1] For a study of the application of the foregoing principles to commercial carburetors the reader should obtain descriptive material from the various carburetor manufacturers.

with some method of compensation, a device for providing extra fuel for acceleration, and a device for increasing the normal fuel flow at wide-open throttle. Actual carburetor characteristics (Fig. 127) obtained from bench flow tests show how closely the ideal has been approached in practice. Obviously, manufacturing tolerances would cause the results from a number of carburetors of the same model to have an appreciable spread. Consequently, the lines in Fig. 127 should be considered as lying in a band representing a maximum spread of about 1 air-fuel ratio.

Air Cleaner and Filter Restriction.—All air cleaners and filters introduce some resistance to air flow into the carburetor.

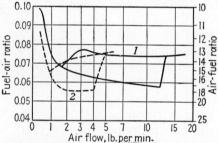

FIG. 127.—Actual carburetor characteristics. 1, Stromberg carburetor—Ford 85 hp., V-8. (*Private communication.*) 2, Zenith carburetor—Vauxhall 10 hp., 4 cyl. (*A. Taub, S.A.E. Trans.*, **42**, 229, 1938.)

The devices which depend upon change of direction of air flow to throw dirt particles out of the air stream have constant resistance characteristics, whereas those which accumulate the dirt removed from the air have changing resistance characteristics. In the latter type of device the resistance to air flow increases with the accumulation of dirt. This increase in resistance changes the carburetor characteristics since it decreases the quantity of air entering the carburetor for a given depression at the venturi, while the fuel flow is the same for a given depression at the venturi regardless of how the depression is attained. Thus, the effect of the added resistance is to make the mixture richer.

This effect can be offset by connecting the space over the gasoline in the float chamber, as well as all air bleeds, to the air passage between the cleaner and the choke. This causes the reduction in pressure in the air intake, due to the air filter, to be imposed, not only on air flow, but also on fuel flow, and consequently does not materially affect carburetor characteristics.

Tests on a Zenith carburetor with a restricted air cleaner (Fig. 128) show the effect on mixture ratio. With a pressure connection between the float chamber and air intake giving a balanced condition, the air-fuel ratio remains in the maximum power range for restrictions from 1.5 to 9 in. Hg. In the unbal-

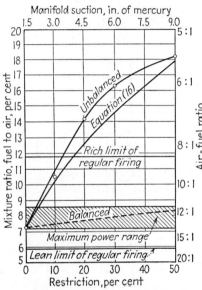

Fig. 128.—Effect of air-cleaner resistance on Zenith carburetor characteristics.

anced condition the air-fuel ratio decreases with the restriction and soon passes beyond the rich limit of regular firing. A solution for this condition, constant velocity through the venturi being assumed, was made with Eq. (16) and also plotted. It shows fairly good agreement with the experimental results, considering the assumptions made in arriving at the equation used.

Effect of Altitude on Carburetion.—The air pressure above sea level decreases with altitude according to the relation[1]

$$h = 62,900 \log_{10} \frac{14.7}{p} \qquad (25)$$

[1] H. C. Dickinson, W. S. James, and G. V. Anderson, Effect of Compression Ratio, Pressure, Temperature, and Humidity on Power, *N.A.C A. Rept.* 45 (1919).

where h = altitude, ft. above sea level,
   p = atmospheric pr., lb. in.$^{-2}$

The temperature of the air also decreases with altitude; and, while this is not a straight-line relation, the average decrease per 1000 ft. of altitude is approximately 3.4°F.

The decrease in both pressure and temperature affect the density of the air, the first decreasing the density and the second increasing it. However, the decrease is greater than the increase, and, according to Eq. (9), the air-fuel ratio would decrease with altitude.

Example.—Determine the air-fuel ratio at 20,000 ft. altitude if the carburetor is adjusted for a 15:1 ratio at sea level. Temperature of air at sea level is 60°F.

$$20{,}000 = 62{,}900 \log_{10} \frac{14.7}{p},$$
$$p = 7.06 \text{ lb. in.}^{-2},$$
$$t = 60 - (20 \times 3.4)$$
$$= -8°F. \text{ or } 452° \text{ abs.},$$
$$d_a = \frac{7.06}{14.7} \times \frac{520}{452} \times d_{\text{sea level}}$$
$$= 0.553 d_{\text{sea level}},$$
$$\text{Air-fuel ratio} = \sqrt{0.553} \times 15$$
$$= 11.16.$$

This is equivalent to about 34 per cent too much fuel.

The effect of altitude on mixture ratio may be compensated by admitting more air or less fuel to the induction system or by maintaining the air at constant density in the carburetor intake. Thus, there are four ways in which altitude compensation may be accomplished:

1. An altitude air port or valve.
2. A variable fuel orifice.
3. Pressure reduction in the float chamber.
4. Constant density of entering air.

The first two methods may be manually or automatically operated. In the latter case, sylphon bellows may be used to provide the necessary action with change in altitude.

The third method bleeds air into the space above the fuel in the float chamber and then from the float chamber to the carburetor barrel around the venturi.[1] This altitude control

---

[1] R. W. Young, Air-cooled Radial Aircraft Engine Performance Possibilities, *S.A.E. Trans.*, **31**, 234 (1936).

automatically regulates the air flow into or out of the float chamber, which controls the pressure or head causing the fuel to flow.

The constant-density air control[1] throttles the air flowing to the carburetor to maintain an atmospheric pressure equivalent to a given altitude. A temperature control maintains a desired air temperature by proper proportioning of the heated and atmospheric air flowing to the altitude throttling device. Throttling the air flowing to the carburetor increases the head, causing fuel to flow, by the difference between the atmospheric pressure and intake air pressure. This head difference decreases to zero as the given altitude is approached, thus maintaining a constant air-fuel ratio up to the given altitude.

The volume of the charge inducted into the engine cylinder at wide-open throttle is practically constant regardless of altitude. Thus, the velocity through the venturi should remain practically constant. With reference to Eqs. (2) and (4), it will be seen that

$$\text{Air velocity} \propto \sqrt{\frac{h_w}{d_a}}. \tag{26}$$

This equation indicates that h_w varies directly with d_a for a constant air velocity through the venturi. Thus, h_w decreases with an increase in altitude for wide-open throttle conditions. However, this factor should not appreciably affect the air-fuel ratio characteristics of a commercial carburetor.

Ice Formation in Aircraft-engine Carburetors.—The evaporation of fuel in an induction system of an engine necessitates the addition of heat to the fuel. This heat flows from anything the fuel contacts during its evaporation process. The result is a cooling effect on the air, water vapor, and fuel, and under certain conditions of pressure, temperature, humidity of the entering air, fuel characteristics, mixture ratio, and heat transfer from the engine and surrounding air to the carburetor, may result in the condensing and freezing of water into ice on the throttle plate. An appreciable accumulation of ice throttles the engine, and may be disastrous.

The process in the carburetor may be represented by the following steady-flow energy equation:

[1] G. E. Beardsley, Jr., An Automatic Power and Mixture Control for Aircraft Engines, *S.A.E. Trans.*, **30**, 301 (1935).

$$(H_{air} + H_{wv} + H_f)_{entering} + Q_{in} = (H_{air} + H_{wv} + H_{ice} + H_{fv} + H_f)_{leaving}$$

where f, fv, and wv = fuel, fuel vapor, and water vapor, respectively.

The entering conditions are known, and the ice formed may be determined from an assumed temperature below 32°F. which must be checked by the volatility relationships and percentage of fuel evaporated.

Tests[1] indicate that appreciable ice formation will occur when the temperature of the carburetor venturi is about 9°F. below the dew point of the atmosphere if this is below 32°F. Ice formation increases with rich mixtures and highly volatile fuels.

Heating of the intake air or the introduction of a liquid, such as alcohol, which will mix with the water vapor and lower its freezing point, will reduce or eliminate the formation of ice in carburetors.

Backfiring.—Under certain conditions of mixture ratio there will be backfiring in the intake manifold or exhaust manifold. During the starting of an engine under cold-weather conditions the usual manipulation of the choke varies the mixture from too lean to too rich. A very lean mixture will burn very slowly, and flame may still exist in the cylinder when the exhaust valve is closing and the intake valve is about to open. The fresh charge in the intake manifold is not so diluted as when inducted into the cylinder and mixed with the clearance gases and consequently will burn more rapidly than the charge in the cylinder. If the lean charge, upon being inducted, comes in contact with flame existing in the cylinder, there will be a flash of flame back through the intake manifold, burning the charge therein and causing the customary popping or backfiring in the carburetor.

Rich mixtures burn faster than lean ones, and under starting conditions the extra fuel that must be supplied to form a rich mixture is at least partly evaporated by the heat of combustion and extinguishes the flame before the next charge is inducted. Thus, under cold starting conditions, backfiring occurs in the carburetor only when the mixture is too lean.

Backfiring occurs in the exhaust system under two conditions of operation. The most common occurrence is the intermittent

[1] H. H. Allen, G. C. Rogers, and D. C. Brooks, Ice Formation in Aircraft Engine Carburetors, *S.A.E. Trans.*, **29,** 417 (1934).

CARBURETION AND FUEL INJECTION

but somewhat regular backfiring that occurs in descending a grade while the engine is being used as a brake. Under this condition, the throttle valve is closed and the idling jet is supplying the mixture. It can be shown that the manifold depression with closed throttle will not vary much with speed so that the quantity and quality of the mixture supplied are practically the same as under normal idling speeds. However, the car is turning the engine at considerably higher speeds, so that the quantity of charge entering the cylinders per suction stroke is much less than under normal idling conditions. The clearance space is filled with exhaust products at about atmospheric pressure regardless of speed. Consequently, as the engine speed is increased while descending a grade with the throttle closed, the amount of charge per stroke becomes smaller and the dilution greater until firing ceases. The succeeding unburned charges are pushed out into the exhaust pipe, the dilution with clearance gases is decreased, and after a few cycles with no firing a charge will be fired. It will be a much diluted and slow-burning charge, and at the opening of the exhaust valve the persisting flame will ignite the unburned mixture in the exhaust manifold and result in a mild or more serious explosion in the exhaust system. In a multicylinder engine the unburned charges passed into the exhaust system by some cylinders will be ignited by the flame from another cylinder. This type of backfiring can be eliminated by increasing the richness of the idling mixture.

The other condition which results in backfiring in the exhaust system is usually that of faulty fuel flow. Under normal operating conditions a faulty float in the carburetor causes an enriching of the mixture under part throttle operation which causes misfiring. Opening the throttle reduces the richness, and firing of the charge is resumed. In the meantime the unburned mixtures that have been pushed out of the cylinders into a cooler exhaust system have become combustible, probably because of condensation of some of the heavier ends of the gasoline. These are ignited by the flame from the exhaust of a cylinder that fires, and a rather violent explosion occurs which in some cases may destroy the muffler.

FUEL INJECTION

Atomization of Liquid Fuels.—A liquid fuel is atomized when it has been broken up into many small particles of various sizes.

In this condition a given quantity of fuel presents the maximum amount of surface and is most readily vaporized and burned.

There are two methods by which atomization is accomplished, the first being that of inducting fuel into a moving stream of air as is done in the carburetor. Castleman[1] has shown that the effect of the relative motion between the air and liquid streams is to form fine ligaments or threads of fuel, which collapse because of surface tension and form the various sizes of drops found in the fuel spray.

The second method is to inject the fuel under comparatively high pressure into the combustion chamber of air which has comparatively little movement. This is known as "solid injection" and was invented by McKechnie in 1910. This method relies also upon the relative motion between the air and the fuel to obtain atomization, and it appears that the ligament theory also satisfactorily explains the mechanism in this case.

A combination of the two methods is used in the case of air injection which was proposed by Diesel. In this case high-pressure air is used to inject the fuel, and relative motion was established between the high-pressure injection air and the fuel as well as between the fuel stream and the hot compressed air in the combustion chamber.

Spray Form.—The fuel issues from the jet in a liquid stream. The surface of the liquid stream comes in contact with the air, and the friction between the two results in the formation of ligaments or threads that break down into small particles and form an envelope surrounding the core of the spray. The core of the spray is made up of fuel threads or particles having the highest velocity which is diminished when the threads are torn into smaller particles and thrust into the outer conical envelope of slower moving particles.

Lee[2] investigated the dimensions of both the core and the outside diameter of the spray for various conditions of injection pressure and air density. The increase in diameter of the core (Fig. 129) indicates the breakup of the stream issuing from the jet into the heavier particles, while the reduction in core size

---

[1] R. A. Castleman, Jr., Mechanism of Atomization of Liquids, *Nat. Bur. Standards, Jour. Research*, **6**, 281 (1931).

[2] D. W. Lee, Experiments on the Distribution of Fuel in Fuel Sprays, *N.A.C.A. Rept.* 438 (1932).

near the end of the spray indicates a greater loss to the outer envelope of finer particles than occurs nearer the fuel orifice.

Schweitzer[1] investigated the effect of injection pressure and fuel viscosity upon the distance the jet travels before breakup

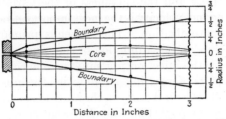

FIG. 129.—Typical fuel spray. Injection pr., 4000 lb. in.$^{-2}$. Air density, 1.1 lb. ft.$^{-3}$. (*Lee.*)

begins when injecting into the atmosphere. An increase in pressure or a decrease in viscosity, which increases the Reynolds number, decreased the distance before jet breakup begins.

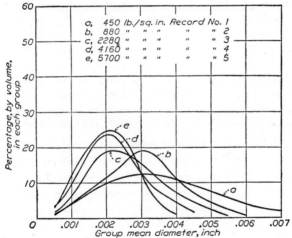

FIG. 130.—Effect of injection pressure on atomization. Orifice diameter, 0.020 in. (*Lee.*)

However, the Reynolds number was not the sole determining factor, for an increase in injection pressure with the same Reynolds number decreased the distance before breakup begins.

[1] P. H. Schweitzer, Mechanism of Disintegration of Liquid Jets, *Jour. App. Physics*, **8**, 513 (1937).

Obviously, high injection pressures are desirable for rapid atomization.

Degree of Atomization. *Injection Pressure*—The degree of atomization is indicated by the smallness of the size of the particles in the spray and the smallness of the variation in size of particles. The fuel velocity is the most important factor affecting the degree of atomization. This depends primarily upon the injection pressure, being a function of the square root of the difference between the injection pressure and the compression pressure. The results of experiments[1] in which the injection pressure was varied while the air density in the injection chamber remained constant at 0.94 lb. ft.$^{-3}$ (Fig. 130) show that increasing the injection pressure reduces the mean diameter of the particles as well as the variation in size. It is particularly noticeable that the percentage of particles with the large diameters decreases very markedly with an increase in injection pressure.

Lee[1] states:

... these atomization curves express both the degree of fineness and the uniformity of the atomization. The closer the curves are to the vertical axis the finer is the atomization, and the smaller the range of drop diameters included the more uniform is the atomization. Specific values can be obtained from the curves only at points for the "group mean diameters," 0.0005 in., 0.0010 in., etc. For example, at a group mean diameter of 0.0010 in. the curves should be read: So many per cent of the total number (or volume) of the drops larger than 0.00075 in. in diameter were found to be between 0.00075 and 0.00125 in. in diameter.

Orifice Diameter.—A small orifice diameter results in a large surface-volume relationship for the fuel stream. This should result in better atomization, and some evidence of this was found by Lee[1] (Fig. 131), the largest percentages of the smaller sizes of fuel particles being obtained with the smallest orifice. Other experiments by Lee on the effect of orifice length-diameter ratio show no appreciable differences in atomization but do show some effect on penetration.

Air Density.—The effect of air density in the combustion chamber upon the degree of atomization is somewhat uncertain. Some experiments indicate a smaller mean drop size as the air density is increased, but others contradict this result. Theoret-

[1] D. W. Lee, The Effect of Nozzle Design and Operating Conditions on the Atomization and Distribution of Fuel Sprays, *N.A.C.A. Rept.* 425 (1932).

ically, it would seem that the mean drop size should decrease with an increase in air density. However, the air density in the combustion chamber is fixed primarily by the compression ratio that will cause ignition of injected fuel, and consequently there will be only a comparatively small variation in air density in the actual engine. Until more significant data are available, it can be assumed that the effect of air density in the combustion chamber on degree of atomization is negligible.

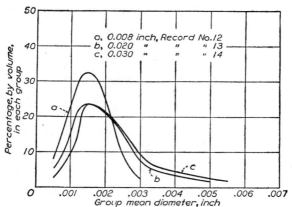

Fig. 131.—Effect of orifice diameter on fuel atomization. Mean effective pressure, 3913 lb. per sq. in. (*Lee.*)

Penetration.—The distance to which the tip of a fuel spray will penetrate the air in the combustion chamber in a given time depends primarily upon the jet velocity, combustion-chamber pressure, and orifice size. Fuel viscosity has a small effect on penetration.

An increase in injection pressure increases the spray-tip penetration (Fig. 132); the higher the pressure, the less the increase in penetration. An increase in combustion-chamber pressure decreases the penetration (Fig. 133). These results check the earlier results of Miller and Beardsley.[1]

An increase in orifice diameter increases the penetration of the spray tip (Fig. 134). The orifice should have a high coefficient of discharge and should be of such length that the fuel leaves the orifice as a stream flowing along the axis of the orifice if high

[1] H. E. Miller and E. G. Beardsley, Spray Penetration with a Simple Fuel-injection Nozzle, *N.A.C.A. Rept.* 222 (1926).

penetration is desired. An orifice length-diameter ratio[1] of from 4:1 up to 6:1 results in about the maximum penetration,

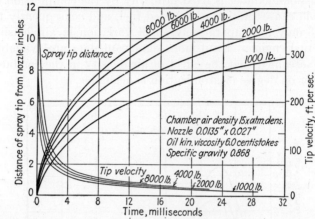

FIG. 132.—Effect of injection pressure on spray-tip penetration. (*Schweitzer*.)

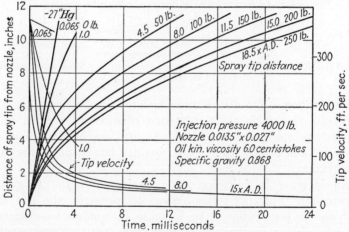

FIG. 133.—Effect of air density in combustion chamber on spray-tip penetration. (*Schweitzer*.)[2]

while the minimum penetration is reached with a ratio of from 1:1 up to 3:1.

[1] A. G. Gellales, The Effect of Orifice Length-diameter Ratio on Fuel Sprays for Compression-ignition Engines, *N.A.C.A. Rept.* 402 (1932).

[2] P. H. Schweitzer, Penetration of Oil Sprays, *Penn. State Coll., Eng. Expt. Sta., Bull.* 46 (1937).

Schweitzer[1] plotted the results from tests, using a given orifice and combustion-chamber pressure with variations in injection pressure, and found the results to lie on a single curve when plotting penetration s against the factor $t\sqrt{\Delta P}$, where t is the time in seconds and ΔP is pressure difference between injection pressure and chamber pressure in pounds per square inch.

Thus,
$$s = f(t\sqrt{\Delta P}). \tag{1}$$

This relation may be used to compute the penetration curve

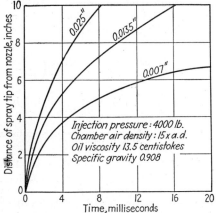

Fig. 134.—Effect of orifice diameter on spray-tip penetration. (*Schweitzer.*)

for any injection pressure when such a curve is known for one injection pressure.

Example.—A spray penetration of 8 in. in 15.7 millisec. is obtained with an injection pressure of 2000 lb. in.$^{-2}$ (Fig. 132). Determine the time required for the spray to penetrate this distance when an injection pressure of 6000 lb. in.$^{-2}$ abs. is used. Orifice and combustion-chamber air density remain constant. Combustion-chamber pressure is 200 lb. in.$^{-2}$ abs.

$$\frac{s_1}{s_2} = \frac{f(t_1\sqrt{\Delta P_1})}{f(t_2\sqrt{\Delta P_2})}$$

or
$$t_1\sqrt{\Delta P_1} = t_2\sqrt{\Delta P_2},$$
$$15.7\sqrt{2000 - 200} = t_2\sqrt{6000 - 200},$$

and
$$t_2 = 8.7 \text{ millisec.}$$

This checks fairly well with the data presented in Fig. 132.

[1] Schweitzer, *op. cit.*

Schweitzer[1] also plotted the results from various tests in which the orifice diameter d was varied and determined the relationship

$$\frac{s}{d} = f\left(\frac{t}{d}\right) \qquad (2)$$

from which the spray-tip penetration can be computed for any orifice diameter d if data are available for any similar orifice under the same conditions.

Example.—The penetration for an orifice of 0.0135 in. diameter and 0.027 in. long is 8 in. in 10 millisec. for 6000 lb. injection pressure (Fig. 132). Determine a similar point for the penetration curve of an orifice having a diameter of 0.007.

$$\frac{s_1}{d_1} = f\left(\frac{t_1}{d_1}\right) \quad \text{and} \quad \frac{s_2}{d_2} = f\left(\frac{t_2}{d_2}\right).$$

Then, $\quad s_2 = d_2 \dfrac{s_1}{d_1} = 0.007 \times \dfrac{8}{0.0135} = 4.15$ in.

and $\quad t_2 = d_2 \dfrac{t_1}{d_1} = 0.007 \times \dfrac{10}{0.0135} = 5.18$ millisec.

Thus, a penetration of 4.15 in. in 5.18 millisec. will be obtained with a similar 0.007-in. diameter orifice under the same conditions.

Schweitzer[1] also plotted the results of tests in which the combustion-chamber air density d_a (in atmospheres) was varied and determined the relationship

$$s(1 + d_a) = f(t\, d_a) \qquad (3)$$

from which the spray-tip penetration for any combustion-chamber pressure can be computed if known for one combustion-chamber pressure.

Example.—A spray-tip penetration of 4.50 in. in 4 millisec. is obtained when the combustion-chamber air density is fifteen times that of the normal atmosphere (Fig. 133). Determine a similar point for a relative density of 8.

Also, $\quad\begin{aligned}s_1(1 + d_{a1}) &= f(t_1\, d_{a1}).\\ s_2(1 + d_{a2}) &= f(t_2\, d_{a2}).\end{aligned}$

Then, $\quad s_2 = \dfrac{s_1(1 + d_{a1})}{(1 + d_{a2})} = \dfrac{4.50(1 + 15)}{(1 + 8)} = 8.0$ in.

and $\quad t_2 = \dfrac{t_1\, d_{a1}}{d_{a2}} = \dfrac{4 \times 15}{8} = 7.5$ millisec.

[1] Schweitzer, *op. cit.*

Thus, a penetration of 8.0 in. in 7.5 millisec. will be obtained with a relative density of 8 for the air in the combustion chamber and the other conditions remaining the same.

Dispersion.—The dispersion of the fuel spray depends upon the injection-nozzle design. The type of orifice that gives a large spray cone produces better dispersion of the fuel than does a small spray cone. This indicates a small length-diameter ratio for the orifice. However, an increase in spray dispersion decreases the penetration.

Various types of fuel orifices have been used to obtain better dispersion, such as annular, slit, impinging jet, and helical groove orifices. Such orifices have not caused an appreciable gain in the degree of atomization. Better dispersion but less penetration is produced with the centrifugal spray.[1]

The flow through the nozzle orifice may be either laminar or turbulent. It has been suggested that the dispersion characteristics of a spray can be improved by providing turbulent flow. The following ways of accomplishing turbulent flow have also been suggested: higher injection pressure, lower viscosity, larger orifice, sharp-edged orifice, divergently tapered orifice, and irregularities in approach and orifice.

The flow through long smooth tubes changes abruptly from laminar to turbulent flow at a Reynolds number of about 2000. For flow through short orifices, the dividing line is not so sharp and turbulent flow may occur with a Reynolds number less than 2000 if the nozzle design produces a disturbance in the fuel prior to entering the orifice.[2] However, high Reynolds numbers indicate conditions favorable to turbulent flow.

The Reynolds number is defined as

$$R = \frac{vd}{\nu} = \frac{vd\delta}{\mu g} \qquad (4)$$

where v = fluid velocity, in. sec.$^{-1}$,
 d = orifice diameter, in.,
 δ = fuel density, lb. in.$^{-3}$,
 μ = abs. visc., lb. sec. in.$^{-2}$,
 g = force of gravity, in. sec.$^{-2}$

---

[1] Lee, *op. cit.*, Rept. 425.
[2] Schweitzer, *op. cit.*

Example.—Determine the Reynolds number for flow through an orifice 0.006 in. in diameter with a pressure difference of 12,500 lb. in.$^{-2}$ Fuel density is 52 lb. ft.$^{-3}$ and $\mu = 25$ centipoises.

The velocity of flow through the orifice is given by the expression

$$v = c\sqrt{\frac{2g\,\Delta P}{\delta}}. \tag{5}$$

Assuming a coefficient of discharge of 0.94, results in

$$v = 0.94\sqrt{\frac{2 \times 32.2 \times 12 \times 12{,}500}{52 \div 1728}} = 1.68 \times 10^4 \text{ in. sec.}^{-1}$$

Then, $\quad R = \dfrac{vd\delta}{\mu g} = \dfrac{1.68 \times 10^4 \times 0.006 \times 52}{25 \times 1.45 \times 10^{-7} \times 32.2 \times 12 \times 1728} = 2160$

where the factor 1.45×10^{-7} lb. sec. in.$^{-2}$ = 1 centipoise.

Fuel-injection Systems.—A fuel-injection system is designed to inject a definite quantity of fuel at the desired time and at a definite rate into the combustion chamber of an engine.[1] The system must also atomize the fuel and distribute it to the various parts of the combustion chamber. Consequently, the performance of fuel-injection engines of good design depends largely upon the fuel-injection system.

Fuel is injected into the combustion chamber of an engine near the end of the compression process, at which time the pressure in the cylinder is usually between 300 and 500 lb. in.$^{-2}$ It is desirable to burn the fuel at or near top dead center, which makes the duration of injection very small. Also, good atomization depends upon high injection velocities which require high pressure differences. Thus, high pressures are required for fuel-injection systems.

There are two main classifications for fuel-injection systems,[2] namely, *solid* injection and *air* injection. Solid injection applies to systems injecting only liquid fuel, whereas air injection applies to systems injecting air along with the liquid fuel.

---

[1] A low-pressure injection system (Marvel) has been developed for injecting fuel into the intake port of each cylinder of an engine during the induction stroke. See J. F. Campbell, Fuel Injection as Applied to Aircraft Engines, *S.A.E. Trans.*, **30**, 77 (1935).

[2] See the following reference for a survey of the systems for the high-speed compression-ignition engine.

C. B. Dicksee, Fuel Injection, *Auto. Eng.*, **25**, 91, 131 (1935).

CARBURETION AND FUEL INJECTION

Solid injection systems may have individual fuel pumps (Fig. 135) that measure and pump the fuel through the line and injection nozzle for each cylinder, or may have the pumps grouped in a multiple pump unit. In one case the pump and nozzle for each cylinder are combined into a unit, which reduces

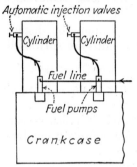

Fig. 135.—Individual-pump and injection-valve system.

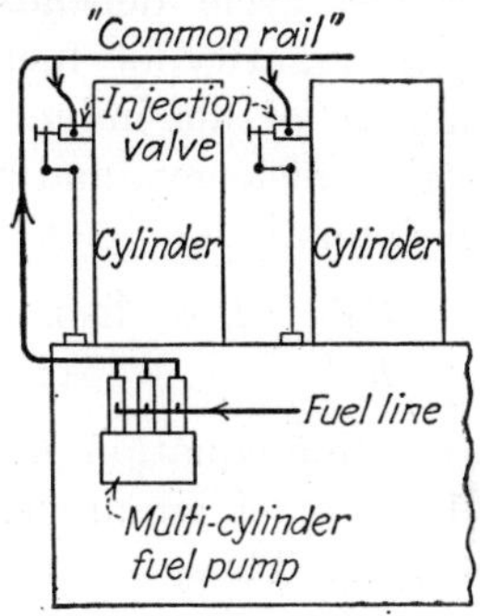

Fig. 136.—Common-rail system.

the length of connecting line from pump to nozzle to a very short passageway. The *common-rail* system (Fig. 136) employs a multicylinder pump which maintains a high pressure in a common fuel line that connects to each injection valve. All

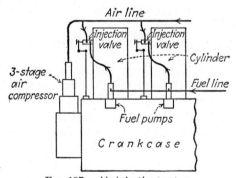

Fig. 137.—Air-injection system.

these systems employ spring-loaded injection valves which open and close either automatically under definite pressure conditions or by mechanical means.

The air-injection system (Fig. 137) requires an air compressor for supplying high-pressure air. A fuel pump meters and discharges a definite quantity of fuel into the injection valve. The injection valve is mechanically opened, and the high-pressure air drives the fuel charge and some air into the combustion chamber.

Fuel Quantity Injected.—The fuel quantity injected into each cylinder per cycle depends upon the load requirement, the maximum quantity of fuel depending upon the air inducted. The amount of air inducted depends upon the displacement, volumetric efficiency, and the air conditions surrounding the engine. Thus,

$$M = \frac{PV}{RT} = \frac{P \times \text{disp.} \times \text{vol. eff.}}{1545 \times T} \times 28.95 = \text{lb. of air.} \quad (1)$$

The reaction equation for typical fuel oil indicates that about 14 to 14.5 lb. of air are required for complete combustion of 1 lb. of fuel. Thus,

$$\text{Max. fuel requirement} = \frac{\text{air inducted}}{\text{min. air-fuel ratio}} = \text{lb. of fuel.} \quad (2)$$

Example.—Determine the maximum amount of fuel that could be injected and burned per cycle in each cylinder of an engine with a 5-in. bore and a 6-in. stroke. Vol. eff. = 0.80. Air conditions are 80°F. and 14.4 lb. in.$^{-2}$ Theoretical air-fuel ratio is 14.5:1.

$$\text{Air inducted} = \frac{14.4 \times 144 \times \pi \times \overline{2.5^2} \times 6 \times 0.80}{1545 \times 540 \times 1728} \times 28.95 =$$
$$3.93 \times 10^{-3} \text{ lb.}$$

$$\text{Max. fuel requirement} = \frac{3.93 \times 10^{-3}}{14.5} = 0.271 \times 10^{-3} \text{ lb. of fuel.}$$

Assuming an A.P.I. gravity of 30° indicates a volume of fuel of about 0.0086 in.^{3} This is slightly larger than the volume of a 0.2-in. cube of oil, which indicates the small quantity of fuel to be injected.

The volumetric efficiency of an engine usually decreases with an increase in speed. Thus, less fuel can be burned at high speeds than at low speeds. This indicates the desirability of injection systems in which the fuel injected decreases with an increase in speed.

It is practically impossible to consume all the fuel injected for maximum-load conditions unless there is an appreciable amount of *excess air*. The desirable amount of excess air at maximum load depends primarily upon combustion-chamber design and fuel-injection characteristics and varies from about 25 to 50 per cent. The lower limit for excess air should be about 10 per cent.

CARBURETION AND FUEL INJECTION

However, most engines will show smoke in the exhaust at excess air values above 10 per cent, the appearance of smoke determining the minimum desirable excess air for each engine.

The quantity of fuel is reduced for part-load conditions, and the amount of excess air increases.

Example.—Assume the amount of fuel injected in the engine in the previous example is 0.2×10^{-3} lb. per cylinder per cycle. Determine the percentage of excess air.

The correct air-fuel ratio is given as 14.5:1. Thus, the fuel now injected will require

$$14.5 \times 0.2 \times 10^{-3} = 2.9 \times 10^{-3} \text{ lb. of air.}$$

It was found in the previous example that 3.93×10^{-3} lb. of air were inducted per cylinder per cycle.

Then, $\quad$ Excess air $= \dfrac{3.93 - 2.9}{2.9} = 0.355$ or 35.5 per cent.

Rate of Injection.—The pressure-volume path of the combustion process indicates the rate at which the energy of combustion is liberated. If the assumption is made that the rate of burning and rate of injection are identical, then the rate of injection can be determined from the pressure-volume path. The time element can be determined by relating the piston position with the degrees of crank travel (see Chap. XVII).

Any increment of burning at constant-pressure results in an increment of volume change and is accompanied by a temperature rise. The increase in enthalpy of the air for the increment of temperature rise indicates the energy liberation required.

Example.—Determine the theoretical rate of fuel injection for a constant-pressure combustion process occupying 10 per cent of the stroke in an engine having a connecting-rod-crank ratio of 4.5:1. Assume T at the end of compression to be 1600°R. Compression ratio = 16:1. Use air as the medium.

V_2/V_1	Crank travel, deg.	T_2	H_{air}	ΔH	Fuel, %
1.0	0	1600	8,838	......	0
1.3	14	2080	12,710	3,872	19.2
1.6	21	2560	16,648	7,810	38.8
1.9	26	3040	20,695	11,857	58.6
2.2	30	3520	24,816	15,978	79.0
2.5	33	4000	28,981	20,143	100

Plotting the fuel injected in per cent results in the "amount" curve (Fig. 138). Plotting the slope of the amount curve results in the "rate" curve.

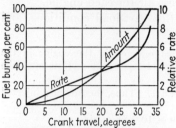

FIG. 138.—Amount and rate of fuel burned for constant-pressure combustion.

Timing Fuel Injection.—The injection of fuel should be timed so that combustion begins before a large portion of the fuel charge is injected, or detonation and combustion roughness will result. The beginning of combustion is determined by the conditions in the cylinder at the beginning of injection and by the ignition quality of the fuel.

Wilson and Rose[1] found a close relationship between the "effective" compression ratio of their test engine and the critical

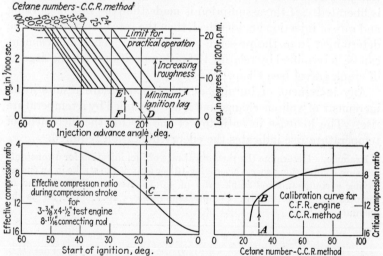

FIG. 139.—Relation between ignition quality and maximum injection advance for minimum ignition lag (see also Fig. 108). (*Wilson and Rose.*)

compression ratio for various fuels. They also found a minimum ignition lag[2,3] of about 0.001 sec. for all the fuels tested and

[1] G. C. Wilson and R. A. Rose, Behavior of High- and Low-cetane Diesel Fuels, *S.A.E. Trans.*, **32**, 343 (1937).

[2] R. F. Selden, Auto-ignition and Combustion of Diesel Fuel in a Constant-volume Bomb, *N.A.C.A. Rept.* 617 (1938).

Selden found only small decreases in ignition lag for increases in pressure

developed a chart for indicating the maximum permissible injection advance with minimum ignition lag (Fig. 139). Although this chart applies to the test engine employed, it indicates the general relationship existing between ignition quality and maximum injection advance for minimum ignition lag.

Example.—The critical compression ratio for a 30-cetane fuel is about 11:1 (*A* and *B*, Fig. 139). The effective compression ratio of 11:1 was obtained in the test engine used by Wilson and Rose at 18 deg. B.T.C. (*C* and *D*). Consequently, injection should begin at 26 deg. B.T.C. to provide for a time lag of 0.001 sec. Determine the maximum possible injection advance for 75-cetane fuel for minimum time lag.

Following the procedure indicated in the foregoing, an injection advance of 37 deg. B.T.C. could be used with a 75-cetane fuel (C.C.R. method) without appreciably affecting the ignition lag in the Wilson and Rose engine.

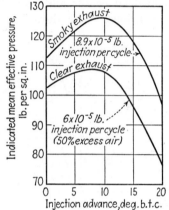

Fig. 140.—Effect of injection advance on power output. (C.F.R. Diesel; 13.5:1 c.r; 900 r.p.m.) (*Fenney*.)[1]

Although high-cetane fuels permit large injection advances, the rate and duration of injection may require a smaller injection advance to distribute the combustion process before and after top dead center, in order to obtain maximum economy (Fig. 140). Injection is always ahead of ignition by an amount equivalent to the time lag, and the effective compression ratio at the beginning of injection should be sufficiently high to result in minimum ignition lag.

Duration of Injection.—Combustion of all the fuel injected should occur, theoretically, at top dead center to obtain the maximum efficiency. Such instantaneous combustion in the compression-ignition engine would result in a very high maxi-

---

and temperature beyond those usually attained at the end of the compression process in the compression-ignition engine (see also Fig. 108).

[3] P. H. Schweitzer, Injection of Diesel Fuel into Flame Cuts Ignition Lag Only Moderately, *Auto. Ind.*, **78**, 848 (1938).

Schweitzer found an ignition lag of 0.00089 sec. for pure cetane injected into the C.F.R. Diesel engine with a compression ratio of 24:1, an intake-air temperature of 200°F., and 9 in. of mercury supercharge.

[1] Fenney, *op. cit.*

mum pressure and an infinite rate of change of pressure which would subject the engine mechanism to an enormous shock loading. A desirable rate of change of pressure with a limiting maximum pressure indicates the desirability of burning sufficient fuel at a definite rate to attain the maximum pressure at top dead center or later, and afterward at a rate designed to maintain the maximum pressure until the end of the combustion process. Thus, the combustion process should start 10 to 30 deg. before top dead center and end 20 to 30 deg. afterward for full-load conditions. The duration of injection should be from 30 to

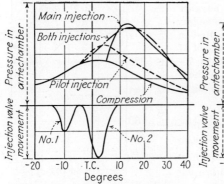

Fig. 141.—Pilot injection. Nozzle lift curve shown in lower part of diagram. (*Schweitzer*.)

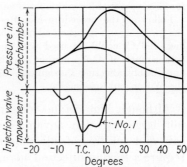

Fig. 142.—Well-tuned single injection. Nozzle lift curves shown in lower part of diagram. (*Schweitzer*.)

60 deg., the timing of the start of injection being advanced sufficiently to offset the ignition lag.

The amount of fuel is decreased for part-load conditions. This may be accomplished by reducing the rate of injection while maintaining the same duration or by maintaining the rate and decreasing the duration of injection.

The early injection of a small quantity of fuel to initiate combustion, known as "pilot" injection,[1] has been used to decrease the ignition lag of the fuel from the main injection. Pilot injection alone results in a small pressure rise (Fig. 141) whereas the main injection alone results in a knocking combustion with a rapid pressure rise. With both injections properly timed and of the desirable amounts, the combustion process is much

---

[1] D. Jäfar, Pilot Injection in Diesel Engines, *Eng.* (London), **144** (II), 417 (1937).

smoother as indicated by the low maximum rate of pressure rise.

Pilot-injection tests indicate that injection should begin at a low rate and increase rapidly after a period of time somewhat less than the minimum ignition lag (<0.001 sec). Schweitzer[1] has shown that very smooth combustion (Fig. 142) can be obtained with a single injection which began earlier than usual because of pressure-wave phenomena. The early injection of the small amount initiated combustion and reduced the lag for the remainder of the fuel. Thus, long duration of injection with controlled rates is desirable from the standpoint of smooth combustion processes.

Fuel-injection Pumps.—Fuel-injection pumps are of the plunger type (Fig. 143), the seal between the plunger and the cylinder being obtained with a very close fit over a relatively large area. The pump is usually cam driven[2] and designed to inject the fuel at a definite rate when directly connected to a fuel-injection valve.

Various methods are used to control the quantity of fuel delivered:

1. Control by varying pump stroke.
2. Control by valves:
    a. By-passing fuel delivered for part of stroke.
    b. By-passing fuel during suction stroke.

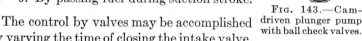
FIG. 143.—Cam-driven plunger pump with ball check valves.

The control by valves may be accomplished by varying the time of closing the intake valve, which varies the effective length of the plunger stroke. Another method is to open a by-pass port at the desired position in the plunger stroke. The effective discharge stroke begins when the plunger covers the intake and by-pass ports (Fig. 144). The effective discharge stroke ends when the scroll edge of the cutaway portion of the plunger uncovers the by-pass port. The effective

---

[1] P. H. Schweitzer, What Can Be Gained by Pilot Injection? *Auto. Ind.*, **79**, 533 (1938).

[2] The Excello pump uses a swash-plate mechanism.

stroke is varied by turning the cylinder barrel, which changes the time of uncovering the by-pass port.

The amount of fuel injected will be less than the effective plunger displacement due to leakage past the plunger and to compressibility of the fuel. The effect of compressibility can be determined from the pressure conditions and the fuel volume in the system undergoing compression. The coefficient of compressibility is

$$c = \frac{(V_1 - V_2)/V_1}{P_2 - P_1} =$$
about 90×10^{-6} (1)

with P in atmospheres.

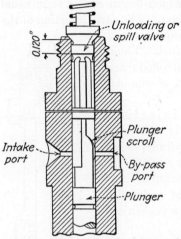

Fig. 144.—Bosch plunger, ports, and unloading valve.

Example.—The example on page 240 indicates a volume of 0.0086 in.[3] of fuel at atmospheric conditions is the maximum fuel requirement for the given engine. The fuel volume in the pump is 0.40 in.[3] The fuel line is of ⅛-in. bore and 30 in. long, the fuel volume being 0.37 in.[3] The fuel volume in the injection valve is 0.15 in.[3]

Assume an injection pressure of 2200 lb. in.$^{-2}$, and determine the pump displacement to inject the desired amount of fuel. Neglect pump leakage.

Total fuel vol. $V_1 = 0.40 + 0.37 + 0.15 = 0.92$ in.[3]

The volume change for an increase in pressure from 14.7 to 2200 lb. in.$^{-2}$ will be

$$V_1 - V_2 = cV_1(P_2 - P_1) = 90 \times 10^{-6} \times 0.92 \times \frac{2200 - 14.7}{14.7} = 0.0123 \text{ in.}^3$$

The volume to be injected will be decreased to 0.0085 in.[3] during the increase in pressure.

Effective pump disp. $= 0.0123 + 0.0085 = 0.0208$ in.[3]

The foregoing example indicates that an appreciable part of the effective plunger displacement may be required to produce injection pressures. Residual pressure in the line and injection valve from the previous injection decreases the displacement required to attain injection pressures. The residual pressure is determined from the increase in volume of the system between the discharge or "unloading" valve (Fig. 144) and the injection

CARBURETION AND FUEL INJECTION

valve while the unloading valve is closing. This increase in volume is approximately equal to the product of the lift of this valve and the port area.

The discharge may be varied somewhat by using tapered flutes and omitting the relieved section at the top of the flutes. At low pump speeds the unloading-valve lifts are increased with the tapered flutes.

Example.—Determine the required volume increase for a 300 lb. in.$^{-2}$ residual line pressure for the previous example. Total volume beyond unloading valve is 0.64 in.^{3}

$$V_1 - V_2 = cV_1(P_2 - P_1) = 90 \times 10^{-6} \times 0.64 \left(\frac{2200 - 300}{14.7} \right) = 0.0074 \text{ in.}^3$$

This indicates a valve movement of 0.2 in. for a $\frac{7}{32}$-in. valve port. Obviously, the solid cylindrical portion of the unloading valve (Fig. 144) should be less than 0.2 in.

The work of injecting the fuel may be evaluated from the P-V diagram for the injection process. The compression of the fuel (Fig. 145) may be assumed to be represented by a straight line on this diagram.

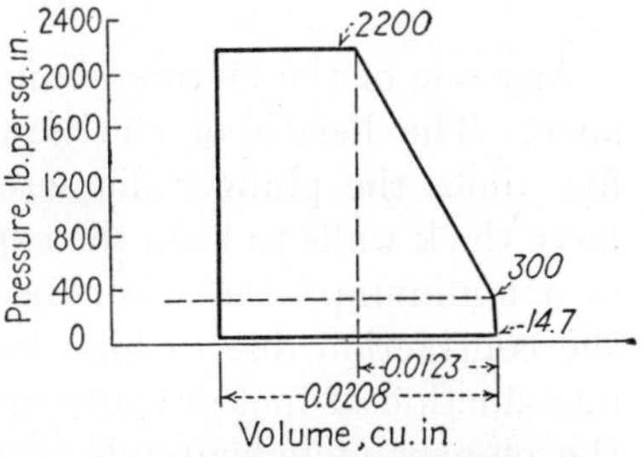

FIG. 145.—Pressure-volume diagram for fuel-injection process.

Example.—The example on page 246 resulted in an effective pump displacement of 0.0208 in.^{3} of which 0.0123 in.^{3} were required for compression. Discharge pressure is constant at 2200 lb. in.$^{-2}$ Residual pressure is 300 lb. in.$^{-2}$ Determine the work done per injection, and estimate the work of injection in percentage of engine output.

The P-V diagram (Fig. 145) shows the slight volume change in raising the pump pressure to 300 lb. in.$^{-2}$ Neglecting this volume change, results in

$$\text{Work} = 144(300 - 14.7)\frac{0.0208}{1728} + 144(2200 - 300) \left(\frac{0.0123}{2 \times 1728} + \frac{0.0208 - 0.0123}{1728} \right)$$

$$= 2.81 \text{ ft.-lb.}$$

On the assumption of a fuel rate of 0.55 lb. hp.-hr.$^{-1}$,

$$\text{Work per hp.-hr.} = \frac{0.55}{0.271 \times 10^{-3}} \times 2.81 = 5700 \text{ ft.-lb.}$$

where 0.271×10^{-3} is the weight of 0.0086 in.^{3} of fuel oil.

$$\frac{\text{Pump work}}{\text{Engine output}} = \frac{5700}{33{,}000 \times 60} = 0.0029 \text{ or } 0.29 \text{ per cent.}$$

The actual work required would be somewhat larger, because of friction losses of the pump mechanism and also because of leakage.

Leakage tests[1] have shown that plunger clearance is the most important factor. The plunger and its cylinder are lapped and have a clearance usually under 0.0001 in. The leakage with fits of this order is usually less than 0.2 per cent of the fuel injected. Leakage varies directly with the cube of the clearance, the plunger diameter, the fuel density, and the fuel pressure; it varies inversely with the fuel viscosity and the length of the lapped fit.

Thus, Leakage rate $\propto \dfrac{cl^3 d \delta P}{\mu l}$.

Any one of the factors is much less important than the clearance. The length of the lapped fit should be from three to five times the plunger diameter. The plunger cylinder should have thick walls to keep the expansion due to high fuel pressures at a minimum. Likewise, the plunger should be small to keep the contraction due to high fuel pressure at a minimum. The introduction of fuel pressure inside a hollow plunger and outside the plunger cylinder tends to maintain or reduce the clearance between the plunger and the cylinder but increases the total fuel volume and the compressibility effect of the fuel.

Plunger Displacement and Velocity Curves.—The plunger displacement at various degrees of pump rotation (Fig. 146) may be determined by clamping a dial indicator to the pump body so that it contacts the top of the plunger. The slope of the displacement curve is the velocity of the plunger.

The intake port for the Bosch pump closes at 154.5 deg., and fuel injection begins later. Maximum velocity occurs at 175 deg. at which time about 11 cm.³ of fuel have been injected. At 180 deg. about 22 cm.³ of fuel have been injected. Thus, the first 20 deg. of effective pump rotation resulted in about the same fuel injected as the next 5 deg. of rotation, indicating the effect of compressibility of the fuel as well as the rate of plunger

---

[1] A. M. Rothrock and E. T. Marsh, Effect of Viscosity on Fuel Leakage between Lapped Plungers and Sleeves and on the Discharge from a Pump Injection System, *N.A.C.A. Rept.* 477 (1934).

displacement. The volume rate of displacement of the pump plunger V_p is equal to the product of the plunger area a_p and the velocity v_p of the plunger. Thus,

$$V_p = a_p v_p = a_p \text{ in.}^2 \times v_p \frac{\text{in.}}{\text{deg.}} \times 360 \frac{\text{deg.}}{\text{rev.}} \times n \frac{\text{rev.}}{\text{min.}} \times \frac{\text{min.}}{60 \text{ sec.}}$$
$$= 6 n a_p v_p \text{ in.}^3 \text{ sec.}^{-1} \qquad (2)$$

Fuel Lines.—Fuel lines are small-bore thick-wall steel tubes. The bore is small to reduce the volume of fuel oil and com-

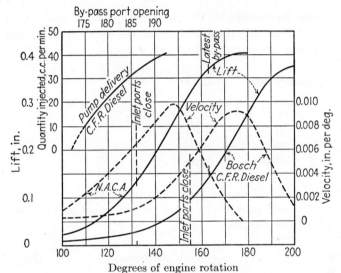

FIG. 146.—Pump delivery, plunger lift, and velocity characteristics.

pressibility effects. The use of very small bores results in high line velocities, turbulent flow, and appreciable pressure drops. Thick-wall tubes prevent appreciable line vibration and variation of line volume with pressure.

The volume of the fuel line has an effect on the plunger displacement required to attain injection pressure, and, consequently, affects the injection timing. Thus, multicylinder engines with an injection pump for each cylinder should have fuel lines of equal length and bore.

Example.—Assume a fuel line 20-in. long is substituted for the line in the example on page 246. Compare injection timing for both systems, assuming that the closing of the intake port occurs at the same crank angle.

The volume of a 20- by ⅛-in. fuel line is about 0.25 in.³ This would reduce the total fuel volume of the system to 0.80 in.³ and the required plunger displacement to attain injection pressure to 0.0107 in.³

Assuming the injection of the fuel charge of 0.0086 in.³ requires 15 deg. of crank travel, the change in required plunger displacement amounts to a crank travel of about

$$\frac{0.0123 - 0.0107}{0.0086} \times 15 = 2.8 \text{ deg.}$$

Thus, the longer line should result in injection starting about 3 deg. of crank travel later than with the shorter line.

Fuel-line Resistance.[1]—The pressure drop in the injection system will depend principally upon the type of flow, which may be either laminar or turbulent.

For laminary flow, $\quad \Delta P = \dfrac{32 v_t \mu l}{d_t^2} \quad$ (3)

where ΔP = pr. drop in line, lb. in.$^{-2}$
v_t = fuel velocity in line, in. sec.$^{-1}$
d_t = diameter of fuel line, in.
l = length of injection line, in.
μ = abs. visc. of fuel, lb. sec. in.$^{-2}$

The fuel velocity can be related to pump-plunger velocity v_p, since the volume rate of the pump, V_p, is equal to the volume rate of the injection line.

Thus, $\quad v_t \dfrac{\pi d_t^2}{4} = V_p = 6 n a_p v_p \quad$ [Eq. (2)]

or $\quad v_t = \dfrac{7.63 n a_p v_p}{d_t^2}. \quad$ (4)

Substituting this value for v_t in Eq. (3) results in

$$\Delta P = \frac{244 n a_p v_p \mu l}{d_t^4}. \quad (5)$$

Thus, the pressure drop due to frictional resistance with laminar flow will vary inversely with the fourth power of the line diameter.

The Reynolds number for a change from laminar to turbulent flow in the injection line is about 2000. For a given line

[1] Abstracted from the work of A. M. Rothrock, Hydraulics of Fuel-injection Pumps for Compression-ignition Engines, *N.A.C.A. Rept.* 396 (1931).

diameter and fuel viscosity the velocity corresponding to this number is termed the "critical" velocity v_k.

Thus,
$$v_k = 2000\frac{\mu}{d_t\rho}, \qquad (6)$$

where $\rho = \delta/g = $ lb. in.$^{-3}/32.2 \times 12$ in. sec.$^{-2}$.

Equation (4) is another expression for fuel velocity, and v can be eliminated between Eqs. (4) and (6).

Thus,
$$\frac{2000\mu}{d_t\rho} = \frac{7.63 n a_p v_p}{d_t^2},$$

or
$$d_{t_k} = \frac{n a_p v_p \rho}{262\mu}. \qquad (7)$$

For turbulent flow the resistance varies according to the relation

$$\Delta P = fl\frac{v_t^2}{2g}\frac{\delta}{d_t} = \frac{fl v_t^2 \rho}{2d_t}, \qquad (8)$$

where
$$f = 0.00714 + 0.6104\left(\frac{v_t d_t \rho}{\mu}\right)^{-0.35}. \qquad (9)^1$$

Eliminating v_t between Eqs. (4) and (8) results in

$$\Delta P = \frac{fl\rho}{2d_t} \times \frac{\overline{7.63}^2 n^2 a_p^2 v_p^2}{d_t^4} = 29.1 fl\rho \frac{n^2 a_p^2 v_p^2}{d^5}. \qquad (10)$$

This indicates that pressure drop with turbulent flow varies inversely with the fifth power of the line diameter.

The determination of the critical line diameter involves the use of the absolute viscosity, which varies with the pressure. Hersey[2] has determined the pressure-viscosity relation (Fig. 147) for Diesel oil having a Saybolt universal viscosity of 45 sec. at 80° F. and atmospheric pressure.

Example.—Determine the critical line diameter and critical fuel velocity for an injection system having a fuel pump with a plunger area of $a_p = 0.0985$ in.^{2}, a nozzle-orifice area of $a_o = 0.000314$ in.^{2}, an orifice coefficient of discharge of 0.94, $n = 750$ r.p.m., and a fuel density of 0.0307 lb. in.$^{-3}$ The fuel-pump characteristics are those of the N.A.C.A. pump (Fig. 146). Assume an average value for v_p of 0.0079 in. deg.$^{-1}$

[1] H. A. Hopf, Fluid Meters—Their Theory and Applications, *A.S.M.E. Rept.*, 2d ed. (1927).

[2] M. D. Hersey, Viscosity of Diesel Engine Fuel Oil under Pressure, *N.A.C.A. Tech. Note* 315 (1929).

Combining Eq. (11), page 256 and Eq. (2), page 249, results in

$$P_{line} - P_{cyl} = \frac{18(n a_p v_p)^2 \delta}{(a_o c)^2 g}$$

$$= \frac{18(750 \times 0.0985 \times 0.0079)^2 \times 0.0307}{(0.000314 \times 0.94)^2 \times 32.2 \times 12} = 5600 \text{ lb. in.}^{-2}$$

P_{line} is about 6000 lb. in.$^{-2}$, at which pressure the absolute viscosity is 0.13 poise (Fig. 147). Changing to the English system of units, results in

$$\mu = \frac{0.13 \text{ dyne sec. cm.}^{-2}}{981 \text{ dynes g.}^{-1}} \times \frac{(2.54 \text{ cm. in.}^{-1})^2}{454 \text{ g. lb.}^{-1}} = 1.884 \times 10^{-6} \text{ lb. sec. in.}^{-2}$$

From Eq. (7), $d_{t_k} = \dfrac{n a_p v_p \rho}{262 \mu}$

$$= \frac{750 \times 0.0985 \times 0.0079 \times 0.795 \times 10^{-4}}{262 \times 1.884 \times 10^{-6}} = 0.094 \text{ in.}$$

The solution for d_{t_k} for various pump speeds results in a maximum critical line diameter (Fig. 148) of about 0.108 at about 600 r.p.m., which should

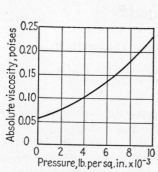 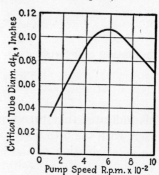

Fig. 147.—Relation of pressure to absolute viscosity of fuel oil. (45 sec. S.U. vis. at 1 atm.) (*Hersey*.)

Fig. 148.—Effect of pump speed on critical tube diameter.

be less than the diameter of line used in order to have the flow laminar throughout the speed range.

From Eq. (6), $v_k = 2000 \dfrac{1.884 \times 10^{-6}}{0.094 \times 0.795 \times 10^{-4}}$

$= 504 \text{ in. sec.}^{-1}$

The average fuel velocity in the injection line can be found from Eq. (4) by substituting the average pump-plunger velocity during the injection period. Substituting this value of average fuel velocity in Eq. (9) results in a value of f for this condition.

CARBURETION AND FUEL INJECTION

Substituting both f and the average fuel velocity in Eq. (8) results in the pressure loss for the average fuel velocity.

Example.—Determine the pressure loss for an injection system with the following conditions and data:

$n = 750$ r.p.m.; aver. $v_p = 0.0079$ in. deg.$^{-1}$; $a_p = 0.0985$ in.^{2};
$\rho = 0.795 \times 10^{-4}$ lb. sec.^{2} in.$^{-4}$; $d_t = 0.138$ in.; $\mu = 1.884 \times 10^{-6}$ lb. sec. in.$^{-2}$; and $l = 34$ in.

From Eq. (4), $v_t = \dfrac{7.63 \times 750 \times 0.0985 \times 0.0079}{0.138 \times 0.138} = 233.8$ in. sec.$^{-1}$

From Eq. (9),

$$f = 0.00714 + 0.6104\left(\dfrac{233.8 \times 0.138 \times 0.795 \times 10^{-4}}{1.884 \times 10^{-6}}\right)^{-0.35} = 0.0560.$$

Substituting these values in Eq. (8),

$$\Delta P = \dfrac{0.0560 \times 34 \times (233.8)^2 \times 0.795 \times 10^{-4}}{2 \times 0.138} = 30.0 \text{ lb. in.}^{-2}$$

If different values are substituted for d_t in the foregoing example, the pressure loss for each diameter can be determined and plotted as in Fig. 149. Also, substituting various values for d_t in Eqs. (4) and (6) provides information regarding the effect of line diameter on maximum fuel velocity and critical velocity through the injection line. These results (Fig. 149) show that the critical line diameter is 0.094 in. Above this value for d_t the critical velocity is higher than the maximum velocity, and flow through the tube will be laminar.

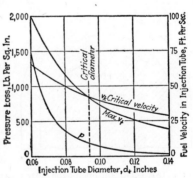

FIG. 149.—Effect of tube diameter on pressure loss and velocity.

Injection Nozzles and Valves.—Injection nozzles may be classified as the open or closed type. The *open type* has the fuel orifice, or orifices, and part of the fuel passageway open to the cylinder pressures at all times. One or more check valves (Fig. 150) may be used to prevent the flow of cylinder gases into the injection system and to prevent the "dribbling" of fuel at the orifices. The fuel is usually given a whirling motion, before flowing through the orifices, which improves the atomization. One method of producing this swirl is to have a longitudinal fuel slot in the nozzle tip which connects tangentially to holes ahead of the orifices (Fig. 151). Fairly large orifices may be used.

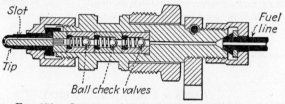

Fig. 150.—Open-type injection nozzle. (*Hesselman.*)

The injection of fuel from an open-type nozzle begins when the fuel pump raises the pressure sufficiently to unseat the check valves and ends when the fuel pressure drops below the combustion-chamber pressure. Thus, the pressure differential causing flow is low at the beginning and end of injection, and poor atomization will occur under these conditions.

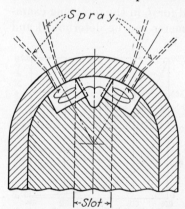

Fig. 151.—Tip of open-type nozzle. (*Hesselman.*)

The *closed-type nozzle* has a spring-loaded valve that is opened either mechanically or hydraulically. The mechanically operated valve is opened at the desired time, and fuel is sprayed through the orifices in the nozzle tip because of pressure in the *common-rail* supply for the injection valve. The amount of fuel injected may be regulated by the duration of the open period of the valve or by variation of the fuel pressure in the system. An increase in engine speed, for a given crank-angle open period and fuel pressure, decreases the time for injection and the amount of fuel injected. This must be compensated for by one of the methods indicated.

The hydraulically operated injection nozzle has a spring-loaded valve with a differential stem (Fig. 152). The fuel may be discharged through a single hole or through several orifices. Some valves have an extension known as a "pintle" (Fig. 153), which extends through the orifice when the valve is closed. The pintle nozzle may be designed to produce various spray-cone angles.

The differential-stem valve opens when the fuel pressure on the difference in cross-sectional area of the two exposed stem

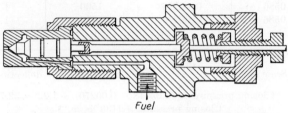

Fig. 152.—Bosch-type injection nozzle, single-hole orifice.

parts plus the gas pressure on the end of the valve exposed to the cylinder more than offsets the spring force holding the valve on its seat. As soon as the valve opens, the fuel pressure is exerted against a larger area of the valve which causes it to open appreciably. Thus, the closing pressure for this type of valve is less than the opening pressure.

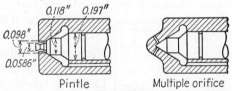

Fig. 153.—Pintle and multiple-orifice nozzle tips.

Example.—The diameters of the spring-loaded valve in Fig. 153 are 0.197, 0.118, 0.098, and 0.0586 in. Determine the required spring force with the valve seated, the valve closing pressure, and the scale of the spring if the valve opens at a pressure of 1500 lb. in.$^{-2}$ and moves 0.01 in. after opening. Gas pressure in the cylinder is 450 lb. in.$^{-2}$ at valve opening.

Dia., in.	Area, in.^{2}	Pr., lb. in.$^{-2}$	Force, lb.
0.197	0.03048	1500	45.72
0.118	0.01094	1500	−16.41
0.098	0.00754	450	3.39
		Net force to open valve	= 32.70 lb.

The fuel pressure will be exerted against all exposed portions of the valve, while open, except the end of the pintle. Any increase in combustion-chamber pressure being neglected,

Dia., in.	Area, in.^{2}	Pr., lb. in.$^{-2}$	Force, lb.
0.197	0.03048	1500	45.72
0.0586	0.00270	1500	− 4.05
0.0586	0.00270	450	1.22

Net force after valve is opened = 42.89 lb.

$$\text{Scale of spring} = \frac{42.89 - 32.70}{0.01} = 1019 \text{ lb. in.}^{-1} \text{ deflection.}$$

Also, Closing pressure × (0.03048 − 0.00270) + 1.22 = 32.70 lb.
or Closing pressure = 1133 lb. in.$^{-2}$

The volume rate of discharge through an orifice is

$$a_o v_o = a_o c \sqrt{2gh} = a_o c \sqrt{\frac{2g \, \Delta P}{\delta}} \text{ in.}^3 \text{ sec.}^{-1} \qquad (11)$$

where a_o = orifice area, in.^{2},
c = discharge coefficient,
ΔP = pressure drop through orifice, lb. in.$^{-2}$,
δ = fuel density, lb. in.$^{-3}$

Gellales[1] found discharge coefficients for fuel orifices varied from 0.65 to 0.95 depending upon the design of the orifice. Sharp-edge orifices have the lowest discharge coefficients. Enlarging the orifice entrance by making it conical in shape and rounding all sharp edges on the entrance side result in the highest coefficient of discharge. The ratio of back pressure to injection pressure has an appreciable influence on discharge coefficients except for orifices having well-rounded approaches.

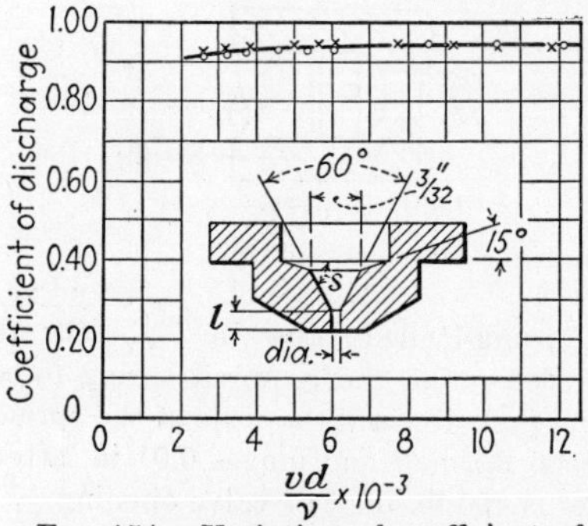

Fig. 154.—Variation of coefficient of discharge with Reynolds number. Atmospheric back pressure. (*Gellales.*)

The results of tests on various orifices of the same design show that the coefficient of discharge can be represented by a single

[1] A. G. Gellales, Coefficients of Discharge of Fuel-injection Nozzles for Compression-ignition Engines, *N.A.C.A. Rept.* 373 (1931).

curve when plotted against the Reynolds number for the flow conditions (Fig. 154). Although all these tests were made with the orifice alone instead of in an injection-valve assembly, it has been found that when a plain stem is used the discharge coefficients are the same provided that the lift of the stem is sufficient to prevent throttling of the flow at the entrance to the orifice.

Example.—Determine the orifice size to inject 0.25×10^{-3} lb. of fuel during a 20 deg. period of crank travel of an engine running at 1800 r.p.m. Injection pressure is 1500 lb. in.$^{-2}$, and compression pressure is 500 lb. in.$^{-2}$ Fuel density is 52 lb. ft.$^{-3}$ $c = 0.94$.

Duration of injection $= {}^{20}\!/_{360} \times {}^{60}\!/_{1800} = 0.00185$ sec.

0.25×10^{-3} lb. $= a_o v_o \, \delta \times $ time $= a_o c \sqrt{2g\delta \, \Delta P} \times$ time.

$$0.25 \times 10^{-3} = a_o \times 0.94 \sqrt{\frac{64.4 \times 12 \times 52(1500 - 500)}{1728}} \times 0.00185.$$

$a_o = 0.00094$ in.^{2}
$d_o = 0.035$ in. dia.

Unit Injectors.—Unit injectors combine the pump and injection nozzle. This construction (Fig. 155) eliminates the injection line and its pressure-wave phenomena which produce very erratic fuel discharges under some conditions. The plunger, in the pump illustrated, has a center passage extending from the bottom of the plunger to the metering recess. Upward motion of the plunger permits fuel to flow from the fuel cavity through the upper port and plunger passage as well as through the lower port into the plunger cylinder. Downward motion of the plunger discharges fuel through the same passageways until the plunger covers the lower port and later the upper helix covers the upper port. Rotation of the plunger by means of a rack and gear changes the position of the helix and varies the amount of fuel under the plunger at the beginning of the effective stroke. This also varies the timing of the injection period and the amount of fuel injected.

The plunger motion increases the fuel pressure, which opens the check valves and forces fuel through the small orifices in the open-type nozzle. Injection ends when the lower helix uncovers the lower port and relieves the pressure on the fuel.

With unit injectors the volume rate of displacement of the pump plunger should equal the volume rate of discharge of fuel

258 INTERNAL-COMBUSTION ENGINES

through the nozzle orifices. Equating Eq. (2), page 249, and Eq. (11), page 256, results in

$$6na_p v_p = a_o c \sqrt{\frac{2g\,\Delta P}{\delta}}, \qquad (12)$$

or

$$\Delta P = \frac{18(na_p v_p)^2 \delta}{(a_o c)^2 g}. \qquad (13)$$

For a given pump-and-nozzle combination, Eq. (13) shows that the pressure head varies directly with the square of the pump speed. Also, the rate of fuel injection varies directly with

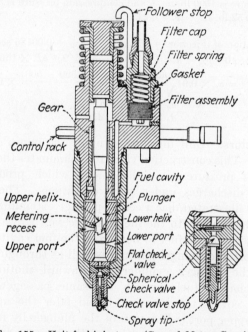

Fig. 155.—Unit fuel injector. (*General Motors Corp.*)

speed, which means that the rate per degree of crank travel should remain constant regardless of engine speed.

Tests[1] (Fig. 156) indicate that the pressure head does not increase so fast as the speed squared. This is due to the compressibility of the fuel and metal parts and to the leakage.

Another type of unit injector (Fig. 157) has an open-type nozzle with orifices connecting directly to the injector cylinder.

[1] F. G. Shoemaker, Automotive Two-cycle Diesel Engines, *S.A.E. Trans.*, **33**, 485 (1938).

CARBURETION AND FUEL INJECTION 259

A measured quantity of fuel is pumped into the injector cylinder during the upstroke of the plunger. Some of the cylinder gases are forced through the orifices into the injector during the

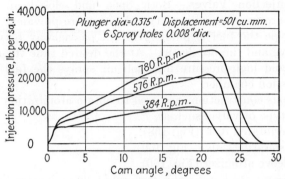

FIG. 156.—Injection pressure during injection. (*Shoemaker; unit injector, General Motors Corp.*)

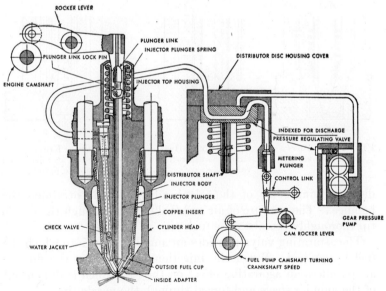

FIG. 157.—Cummins injection system.

compression stroke of the engine. During the downstroke of the plunger the fuel and these gases are injected into the cylinder.

The construction of this injector reduces the volume of the passageway between the injector plunger and the nozzle orifices to a minimum. However, the rate of fuel injection is appreciably less than the plunger-displacement rate indicates, because of the presence of cylinder gases in the injector cylinder.

Air-injection Valves.—Both open and closed air-injection valves are used, the closed type being more prevalent. In the open type the fuel is pumped into a passageway between the air valve and the nozzle tip. At the desired time the air valve opens and high-pressure injection air sprays the fuel into the combustion chamber.

Various constructions are used to obtain atomization with the closed type of valve, such as,

1. Perforated disk (Fig. 158).
2. Aspirating valve (Fig. 159).
3. Sleeve atomizer (Fig. 160).

Oil is pumped onto the perforated disks in the first type. When the valve opens, the air flows through the holes in the

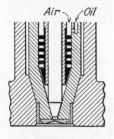 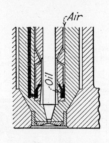 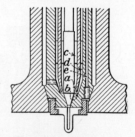

FIG. 158.—Perforated-disk valve.   FIG. 159.—Hesselman aspirating valve.   FIG. 160.—Krupp valve with sleeve atomizer and pinlike nozzle plate.

disks, driving some of the oil ahead of it and entraining the balance. The mixture of air and fuel flowing through the nozzle tip thoroughly atomizes the fuel.

The aspirating valve provides an annular space into which the fuel is pumped. When the injection valve opens, the flow of air produces an aspirating effect on the fuel which is drawn out of the annular space and forced through the nozzle tip.

The sleeve atomizer contains a sleeve which divides the space around the valve into two annular chambers. Fuel is pumped into the outer chamber (d), only a small portion entering the inner or air chamber (c) through a slot (b). As the injection valve opens, the small portion of fuel in the bottom of chamber

(c) is forced into the combustion chamber and initiates combustion. As air flow continues, it exerts an aspirating effect upon the fuel in chamber (d). The air-fuel mixture is forced through a number of holes whose axes form a cone in the "pin-like" nozzle plate.

Injection Air.—Injection-air pressures of 700 to 1000 lb. in.$^{-2}$ are required for compression pressures of 500 lb. in.$^{-2}$, in order to provide full-load fuel injection and distribution. The drop in pressure through the injection nozzle results in a very high velocity, theoretically. The restrictions provided to ensure good atomization, however, reduce the theoretical velocity materially.

Example.—Determine the theoretical air velocity for a drop in pressure from 1000 to 500 lb. in.$^{-2}$ Temperature of air is 100°F.

$$v = \sqrt{2gc_p(T_1 - T_2)}* = \sqrt{2g\frac{k}{k-1}RT_1\left[1 - \left(\frac{P_2}{P_1}\right)^{\frac{k-1}{k}}\right]}$$

$$= \sqrt{64.4 \times \frac{1.4}{1.4 - 1} \times 53.34 \times 560\left[1 - \left(\frac{530}{1000}\right)^{\frac{1.4-1}{1.4}}\right]}$$

$$= 1057 \text{ ft. sec.}^{-1}$$

The critical pressure (530) is used instead of 500 lb. in.$^{-2}$

Three-stage compressors are used, intercoolers being provided between each stage and an aftercooler after the last stage.

Injection air is about 10 to 12 per cent of the air inducted into the engine. At other than full loads, it is considered good practice to throttle the injection air, for with a small quantity of fuel there will be none left at the tip of the injection valve to initiate combustion at the next opening of the valve.

Example.—A Diesel engine has a fuel rate of 0.4 lb. of fuel per brake horsepower-hour. Determine the percentage of power consumed by the air compressor. Injection air used = 2.8 lb. lb.$^{-1}$ of fuel. Injection-air pr. = 1000 lb. in.$^{-2}$ Air temperature = 70°F.; mechanical eff. of air compressor = 0.80. $n = 1.28$.

$$W = P_1V_1\frac{3n}{n-1}\left[1 - \left(\frac{P_2}{P_1}\right)^{\frac{n-1}{3n}}\right]*$$

* See any engineering thermodynamics text.

$$W = \frac{2.8 \times 1545 \times 530}{28.95} \times \frac{3 \times 1.28}{1.28 - 1}\left[1 - \left(\frac{1000}{14.7}\right)^{0.073}\right]$$
$$= 392{,}000 \text{ ft.-lb.}$$

Input to air compressor $= \dfrac{392{,}000}{0.80} \times 0.4 = 196{,}000$ ft.-lb. (b.hp.)$^{-1}$

Power required to compress injection air $= \dfrac{196{,}000}{33{,}000 \times 60} = 0.099$ or 9.9 per cent of output of engine.

Hydraulics of Fuel-injection Systems.—The rate at which fuel will be injected into the combustion chamber of an engine depends upon the characteristics of the entire injection system. The various factors that must be considered are:
1. Pump characteristics.
2. Connecting system between the pump and nozzle.
3. Injection-valve characteristics.
4. Compressibility of the fuel.
5. Inertia of fuel column and injection-system parts.
6. Resistance to fuel flow.
7. Leakage.

The velocity v_{pl} with which the fuel enters the pump end of the fuel line, between the pump and the injection valve, depends on the volume rate of the pump and the area a_l of the fuel line. Thus, from Eq. (2), page 249,

$$v_{pl} = \frac{6na_p v_p}{a_l}. \tag{1}$$

The disturbance created by imparting a velocity to the fuel at the pump end of the line is transmitted with the acoustic velocity a to the injection-valve end of the line. This may be computed from the relation

$$a = \sqrt{\frac{kg}{\delta}} = \text{about 4800 ft. sec.}^{-1} \tag{2}$$

in which k = elastic modulus, about 280,000 lb. in.$^{-2}$ for fuel oils,
δ = density, about 0.032 lb. in.$^{-3}$
g = acceleration due to gravity.

The length l of fuel line and the acoustic velocity a provide the time element l/a for the disturbance to be transmitted from the pump to the nozzle, and vice versa.

Thus (Fig. 161), $\tan \varphi = \dfrac{\Delta l}{\Delta t} = \dfrac{l}{l/a} = \pm a,$ \qquad (3) which provides a graphical representation[1] of the transmission of a disturbance from one end of the fuel line to the other.

Any change in fuel velocity Δv at the pump end or injection-valve end of the system results in a change in pressure ΔP, the

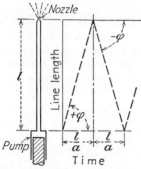

Fig. 161.—Distance-time diagram. (K. J. DeJuhasz.)

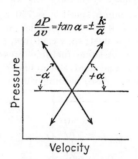

Fig. 162.—Pressure-velocity diagram. (K. J. DeJuhasz.)

change being represented graphically in Fig. 162, which is based upon the relation

$$\tan \alpha = \frac{\Delta P}{\Delta v} = \pm \frac{k}{a} = \text{about 58 lb. in.}^{-2} \text{ (ft. sec.}^{-1})^{-1}. \quad (4)$$

The effect of the various velocity disturbances transmitted to the injection valve will depend upon the rate at which fuel is being injected, which determines the fuel velocity v_{vl} at the valve end of the line.

From Eq. (11), page 256, $\quad v_{vl} = c\dfrac{a_o}{a_l}\sqrt{\dfrac{2g(P_{vl} - P_{cyl})}{\delta}}. \quad (5)*$

---

[1] The graphical analysis is based upon the work of K. J. DeJuhasz in the following references:

Graphical Analysis of Transient Phenomena in Linear Flow, *Jour. Franklin Inst.*, **223**, 463, 643, 751 (1937).

Hydraulic Phenomena in Fuel Injection Systems for Diesel Engines, *A.S.M.E. Trans.*, **Hyd-59-9**, 669 (1937).

Determination of the Rate of Discharge in Jerk-pump Fuel-injection Systems, *A.S.M.E. Trans.*, **OGP-60-2**, 127 (1938).

A mathematical analysis of the conditions existing in a fuel-injection system has been made by A. M. Rothrock, *op. cit.*

* This equation results in values for v_{vl} that should be corrected for compressibility of the fuel if high pressures are used. The correction factor is $1 - 90 \times 10^{-6}(P/14.7)$, which amounts to 0.91 for $P = 14,700$ lb. in.$^{-2}$

The velocity disturbances reflected from the injection-valve end of the line depend upon the velocity transmitted to and the velocity existing at this end of the line when the disturbance occurs.

Example.—The data for an open-nozzle fuel-injection system are as follows: pump speed $n = 750$ r.p.m.; plunger area $a_p = 0.0985$ in.^{2}; line area $a_l = 0.0150$ in.^{2}; orifice area $a_o = 0.000314$ in.^{2} Coefficient of discharge $c = 0.94$; sp. wt. of fuel, $\delta = 0.0307$ lb. in.$^{-3}$; compression pr. = 400 lb. in.$^{-2}$ Assume fuel-line pressure is 400 lb. in.$^{-2}$ at the beginning of injection and that the cylinder pressure remains at 400 lb. in.$^{-2}$ during injection. Also, assume that v_p remains constant at 0.008 in. deg.$^{-1}$ of

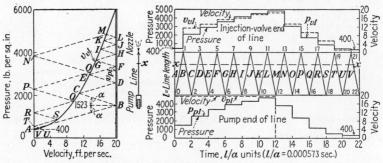

Fig. 163.—Graphical analysis of fuel-injection system. (Open nozzle—constant pump-plunger velocity.)

pump rotation. Determine the pressure-time function for both ends of the fuel line for an effective pump stroke of $12l/a$ units of time.

From Eq. (1), $\quad v_{pl} = \dfrac{6 \times 750 \times 0.0985 \times 0.008}{0.0150 \times 12} = 19.7$ ft. sec.$^{-1}$

From Eq. (5), $\quad v_{vl} = 0.94 \dfrac{0.000314}{0.0150 \times 12} \sqrt{\dfrac{64.4 \times 12(P_{vl} - 400)}{0.0307}}$

$\qquad\qquad\qquad = 0.261\sqrt{P_{vl} - 400}.$

Values of v_{vl} for various values of P_{vl} are plotted in Fig. 163. v_{pl} is constant and is indicated by a vertical line at 19.7 ft. sec.$^{-1}$ on the same diagram.

From Eqs. (2) and (4),

$$\tan \alpha = \pm \frac{k}{a} = \pm \sqrt{\frac{k\,\delta}{g}}$$

$$= \pm 12 \sqrt{\dfrac{280{,}000 \times 0.0307}{32.2 \times 12}} = 57 \text{ lb. in.}^{-2} \text{ (ft. sec.}^{-1})^{-1}.$$

The first disturbance occurs when the pump plunger imposes a velocity of 19.7 ft. sec.$^{-1}$ on the fuel at the pump end of the line, which increases the pressure from 400 lb. in.$^{-2}$ to 1523 lb. in.$^{-2}$ This is determined graphically by line AB (Fig. 163), which has a slope of 57 lb. in.$^{-2}$ (ft. sec.$^{-1}$)$^{-1}$.

The disturbance arrives at the injection valve and upon being reflected raises the pressure to that at C, since the velocity of flow v_{vl} due to this pressure is lower than the transmitted velocity. Again, reflecting to the pump end where the velocity is 19.7 ft. sec.$^{-1}$, point D indicates the pressure attained when the first disturbance is reflected from the pump end. Thus, the points A to M are determined.

The time required for a pressure wave to travel the length of the line is l/a sec. Thus, the lines 0-1, 1-2, etc. (Fig. 163) indicate the position of the wave in the line at any time. At the beginning of the process the pressure at the pump end rises abruptly from 400 to 1523 lb. in.$^{-2}$ and remains at this pressure until the reflected wave arrives at the pump at $2l/a$ units of

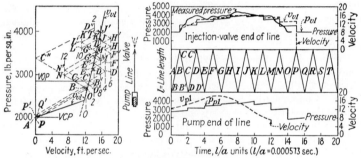

FIG. 164.—Graphical analysis of fuel-injection system. (Closed nozzle-variable pump-plunger velocity.)

time. Then, the pressure rises abruptly to above 2500 lb. in.$^{-2}$ In like manner the pressure at the injection-valve end of the line is plotted.

At $12l/a$ units of time the velocity v_{vl} drops abruptly to zero. Consequently, the line MN drawn from the velocity line v_{vl} to zero velocity indicates the decrease in pressure when the reflected wave arrives at the pump where the velocity is zero. Injection continues with decreasing pressure as indicated by the points M to V.

The pressure-time curve can be drawn for any position x in the fuel line noting that any wave coming from the pump end transmits the pressure set up at the pump, and vice versa.

Usually, pump-plunger velocity varies with time. The plot of the velocity v_{pl} on the right diagram (Fig. 164) indicates the velocities at 0, 2, 4, etc., l/a units of time. Short vertical lines labeled 0, 2, 4, etc., indicate these velocities on the left diagram.

Closed injection valves have a fixed valve-opening pressure VOP and a lower closing pressure VCP. The plot of the velocity v_{vl} has a break in it, indicating throttling of the fuel at the orifice below the break and unrestricted flow through the fuel orifice above the break. The graphical analysis for the pressure variations in a fuel-injection system with a closed type of injection valve and variable pump velocity is made in the same manner as in the foregoing example.

Example.—An injection system has a fuel pump with characteristics indicated by the N.A.C.A. curves (Fig. 146). The effective pump stroke begins at 132 deg. and ends at 164 deg. The fuel line is 34 in. long. Valve-opening pressure is 2850 lb. in.$^{-2}$ abs. and valve-closing pressure is 2000 lb. in.$^{-2}$ abs. All other data the same as in the previous example. Determine the pressure-time function for both ends of the injection system.

The velocity v_{pl} is determined at various time intervals and plotted on the right diagram (Fig. 164). The values for v_{vl} as indicated in the previous example are plotted above 2400 lb. in.$^{-2}$ on the left diagram, a straight line connecting the end of this curve with the valve-closing pressure at zero velocity.

The line AB is drawn from 2000 lb. in.$^{-2}$ pr. and zero velocity with a slope of 57 lb. in.$^{-2}$ (ft. sec.$^{-1}$)$^{-1}$ until the short vertical line labeled 0 is intersected. This vertical line is located at the velocity at the pump end at 0 time. Consequently, B indicates the pressure at the pump end at 0 time.

Since the velocity at the pump end is changing, line AB is extended to B' which indicates the pressure at the pump end of the line at the end of $2l/a$ units of time. Thus, a gradually rising pressure is noted at the pump end for the first two time units. Similarly, C and C', respectively, indicate the pressures at the injection-valve end of the line l/a units of time later, etc.

Actually, the pressure at the closed injection valve would rise to C'', if the valve does not open instantly. Thus the assumption of appreciable inertia would result in a pressure wave indicated by C'', D'', etc., from which another set of pressure curves could be obtained, the true pressure probably lying between the two limiting curves.

This analysis results in pressure curves which would be considerably smoothed in practice because of viscosity and inertia effects. The pressures existing in this system were measured by Rothrock[1] and are also plotted, good agreement being shown between theoretical and actual curves.

Graphical analysis of the characteristics of a fuel-injection system is a comparatively simple way of studying the effect of various design and operating factors. An analysis[2] of the characteristics of the injection system of the C.F.R. compression ignition engine indicated the possibility of injection-valve closure during the injection period and the tendency of the valve to vibrate. Stroboscopic analysis of the fuel spray checked the graphical analysis.

[1] Rothrock, *op. cit.*
[2] L. C. Lichty, Discussion, *A.S.M.E. Trans.*, **60**, 606 (1938).

CARBURETION AND FUEL INJECTION 267

Mechanical stroboscopes are used to determine the amount of fuel discharged during any small part of the injection period. Such experimental data may be used to construct a rate-of-injection curve which should be similar to the curve of velocity at the nozzle end of the injection system. If the effect of variations in design features and operating conditions is known, changes may be made to obtain the desired rate of injection.

Multiple-valve Injection Systems.—A single pump may be connected to several injection valves located in one cylinder. Identical injection valves should have the same injection characteristics provided that the valves are connected to the pump by identical fuel lines.[1] Any variation in fuel lines and injection valves will result in different injection characteristics.

EXERCISES

1. A 3½- by 5-in. engine with a compression ratio of 6:1 has a volumetric efficiency of 0.85 based upon atmospheric conditions of 30 in. Hg and 60°F. when operating at a speed of 2000 r.p.m. Under this condition the throttle is wide open and the intake vacuum is 2 in. Hg.

With a closed throttle and no flow into the manifold, the intake vacuum is 22 in. Hg.

The clearance gases have the following temperatures at various throttle openings and under supercharged conditions:

Manifold pr., lb. in.$^{-2}$	25	20	15	12	9	6
Temperature clearance gases, °F.	2100	1875	1500	1300	1000	650

Plot the weight of the charge inducted into the cylinder per minute, the total weight of the gases in the cylinder, and the dilution in per cent. Mol. wt. of clearance gases = 28.6.

2. Compute the size of fuel jet for a single-jet carburetor on a single-cylinder engine, with a bore of 3½ in. and stroke of 5 in., operating at a speed of 3000 r.p.m. Diameter of venturi section is 1 in. Air-fuel ratio desired is 13:1. Assume a volumetric efficiency of 85 per cent.

3. Taking values of h_w ranging from 1 to 40 in. of water for the carburetor in Prob. 2, compute the change in air-fuel ratios. Assume c_a and c_f remain constant and $h'_w = 0.35$ in.

4. Determine the air-valve openings for an auxiliary air-valve carburetor (Fig. 118) to provide the maximum economy mixture for values of h_w ranging from 1 to 40 in. of water. $A = 0.75$ in.^{2} Assume a probable rela-

---

[1] D. W. Lee and E. T. Marsh, Discharge Characteristics of a Double Injection-valve Single-pump Injection System, *N.A.C.A. Tech. Note* 600 (1937).

tion between h_w and load. Plot A_{aux} against h_w. How could such a variation in opening be obtained?

5. Work the compensating-nozzle example given in the text, page 216, for a range of $h_w = 1$ to 40 in. The desirable air-fuel ratio at 1-in. suction is 12:1 and at 40-in. suction is 16:1. Plot air-fuel ratios against load, assuming 40-in. suction is 100 per cent load. How does this compare with the maximum economy curve?

6. Determine the ratio of the diameters of the two fuel orifices in Prob. 5, assuming the same coefficients of discharge.

7. Given a carburetor with a venturi 1.5 in. in diameter, determine the areas of the two fuel jets in Probs. 5 and 6. Also compute and plot against suction head the pounds of fuel from each jet and total fuel per pound of air flowing.

8. Recompute the restricted air-bled-jet carburetor example in the text, page 218, doubling the relative size of A_b.

9. Determine the effect of restricting the air flow into the intake of the single-jet carburetor in Prob. 2.

10. Determine the air-fuel ratio at various altitudes up to 25,000 ft. for the carburetor in Prob. 2 which supplies a 13:1 ratio at sea level when the air temperature is 60°F. How much of an altitude port, compared with the venturi area, must be provided to obtain an air-fuel ratio of 13:1 at 15,000 ft. altitude?

11. A fuel spray tip penetrates the air in the combustion chamber a distance of 6 in. in 12 millisec. Combustion-chamber air density is $15 \times$ atmospheric density. Orifice diameter is 0.0135 in. Determine the time required for the spray tip to penetrate 4 in. in a combustion chamber with air density of $20 \times$ atmospheric density. Orifice diameter is 0.01 in., and injection pressure is 6000 lb. in.$^{-2}$. See Fig. 132 for initial pressure.

12. Determine the Reynolds number for both cases in Prob. 11. Sp. gr. of the fuel $= 0.85$, and $\mu = 5.21$ centipoises.

13. Estimate the maximum quantity of fuel that could be injected and burned per cycle in an 8- by 10-in. four-stroke Diesel engine.

14. Determine the rate at which fuel must be burned in a fuel-injection engine to cause a constant-temperature combustion process for 10 per cent of the stroke. The connecting rod-crank ratio is 4.5:1; the compression ratio is 15:1; temperature at the end of compression is 2000°R. Plot the results against crank travel in degrees.

15. Assume that 0.2×10^{-3} lb. of fuel are injected in the example on page 241. Determine the orifice size for an injection pressure of 20,000 lb. in.$^{-2}$ at the maximum rate of fuel injection. Combustion pr. $= 500$ lb in.$^{-2}$, $\delta = 53$ lb. ft.$^{-3}$, and engine speed $= 3000$ r.p.m. Then, compute and plot the required injection pressures at the various crank positions.

16. Determine the angle of advance for the beginning of fuel injection in Prob. 15, assuming an ignition lag of 0.001 sec. How much fuel will have been injected at top center?

17. A Diesel engine with a limited combustion pressure of 1000 lb. in.$^{-2}$ abs. uses 0.2×10^{-3} lb. of fuel per cycle. Compression pr. $= 500$ lb. in.$^{-2}$ abs. Excess air $= 25$ per cent. Determine the rate at which fuel should

CARBURETION AND FUEL INJECTION

be injected during the constant-pressure part of the combustion process. Also, solve Prob. 15 for this case.

18. Determine the work of injecting the fuel in Prob. 15.

19. Plot the pressure-time compression curve for an engine with a 15:1 compression ratio and connecting rod-crank ratio of 4.5:1. Assume

$$PV^{1.3} = \text{const. for the compression process.}$$

Sketch in a curve indicating pressure rise, starting tangent to the compression curve at 20 deg. before top center and becoming tangent to a vertical line indicating top center at a pressure of 1000 lb. in.$^{-3}$ abs. Determine the theoretical rate of injection, assuming complete combustion of $C_{12}H_{26}$. Add the rate curve from Prob. 17 to this one.

20. Determine critical line diameter, pressure loss, and fuel velocity at various pump speeds. Other data the same as in the example on page 251. Plot the results against pump speed.

21. Design a spring-loaded injection-valve stem so that the valve will open at 1000 lb. in.$^{-2}$ and move open 0.1 in. Assume the respective diameters of the valve stem, and determine the scale of the spring and the pressure at which the valve will close.

22. Determine and plot the rate of plunger movement to result in the injection pressure in Fig. 156 at a speed of 780 r.p.m.

23. Determine by graphical analysis the effect of pump speed, orifice size, and length of line upon the variation in pressure at the nozzle for the example on page 264. Assume constant pump velocity during injection.

24. Determine by graphical analysis the effect of pump speed and nozzle opening pressure upon the variation in pressure at the nozzle for the example on page 266. Assume constant pump velocity during injection.

25. Assume a check valve at the pump, in the example on page 264, that closes immediately after the pump by-pass port opens. Determine the amount of fuel injected after the check valve closes. Length of line = 30 in.

CHAPTER X

MANIFOLDS AND MIXTURE DISTRIBUTION

INTAKE MANIFOLDS

Mixture Condition and Distribution.—The fuel spray after leaving the carburetor consists of partly evaporated fuel, a mist of fine particles, and a considerable amount of heavier particles of fuel. The fuel spray becomes coarser and the proportion of heavier particles increases as the suction head and air velocity at the jets are decreased. The condition of the mixture at the engine ports depends upon the heat added between the carburetor and the ports, the velocity of flow, and the design of the intake manifold. The manifold should take the mixture supplied by the carburetor and equally distribute both the air and the fuel between the various ports. Ports must be designed to provide equal distribution between two cylinders where both are fed from a single port.

It is extremely difficult to obtain equal mixture distribution between the various cylinders of a multicylinder engine throughout the range of speed and load. Variation in compression pressure between cylinders indicates unequal air distribution which may correct or accentuate the unequal distribution of fuel. A manifold that provides equal air distribution does not necessarily provide equal fuel distribution, particularly when the mixture contains liquid-fuel particles.

Size of Manifold.—The velocity of gases in the intake manifold, during wide-open throttle conditions, depends on the displacement, r.p.m., volumetric efficiency, and size of intake manifold. Thus, for a four-stroke-cycle engine,

$$\text{Manifold velocity} = \frac{\text{disp.} \times \text{vol. eff.} \times \text{r.p.m.}}{2 \times \text{manifold cross-sectional area}}. \quad (1)$$

The minimum velocity of air that is required to entrain the heavier fuel particles in the riser section of an updraft manifold has been found to be about 40 ft. sec.$^{-1}$, which limits the size of manifold for a given engine and minimum speed.

Example.—An eight-cylinder engine having a bore of 3 in. and a stroke of 4½ in. has an intake-manifold riser with a diameter of 1½ in. Determine the minimum speed for a riser velocity of 40 ft. sec.$^{-1}$ Vol. eff. = 85 per cent.

$$\text{Engine disp.} = 8 \times 0.7854 \times \overline{3}^2 \times 4\tfrac{1}{2} = 254 \text{ in.}^3$$
$$\text{Area of riser} = 0.7854 \times \overline{1.5}^2 = 1.767 \text{ in.}^2$$
$$\text{From Eq. (1), R.p.m.} = \frac{40 \times 12 \times 60 \times 2 \times 1.767}{254 \times 0.85} = 471.$$

A downdraft manifold is not restricted to a minimum riser size since the fuel flows readily down into the manifold. Consequently, there has been a decrease in riser velocity at any given speed with the adoption of downdraft carburetion until now the average riser velocity is about 50 ft. sec.$^{-1}$ at 1000 r.p.m.[1] The course of flow may be downward all the way to the cylinder in the valve-in-head engine. However, it is usually considered desirable, with both updraft and downdraft manifolds, to have horizontal distributor sections.

A small manifold will have high mixture velocities throughout the speed range. Under accelerating conditions from low speeds, there will be less tendency for fuel to drop out of the air stream when the throttle is suddenly opened and, consequently, less accelerating fuel is required. However, the engine output at high speeds will be restricted, because of the throttling effect of the small manifold.

A large manifold is required for high outputs at high speeds, and large accelerating charges of fuel are required when accelerating from low speeds. However, the difficulties with the large manifold disappear with a dry mixture.

The mixture velocity is practically constant at any throttle position and given engine speed, the effect of the clearance gases being neglected. The expansion of the clearance gases to manifold pressure reduces not only the volume of the charge inducted but also the intake-manifold velocity as the throttle is closed. Hence, under starting and warm-up conditions the advantage of high mixture velocity is obtained with the small intake manifold.

The determination of the size of the intake manifold involves also the port and valve area. The major restriction in the entire induction system, from the carburetor to the valve area, may be in any one of its parts, and in designing for maximum output

[1] General Motors Corp.

the relation of the sizes of all of the parts must be considered. Large valves and valve areas work satisfactorily with a well-vaporized or gaseous mixture. Smaller valve areas are more satisfactory with wet mixtures, since the increased velocity of flow tends to atomize the liquid at the edge of the intake valve.

The port area must also be small enough to ensure the lifting of the liquid-fuel particles up to the valve seat in the typical L-head engine; with the valve-in-head engine, the film of fuel, flowing along the walls of the port, tends to flow by gravity to the valve seat. A manifold for a three-port engine provides a much shorter distance from the carburetor to the center port

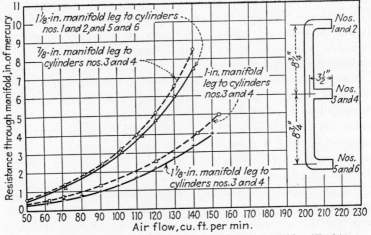

Fig. 165.—Resistance to air flow for a three-port manifold. (*Taub*.)[1]

than to either of the other ports. It is considered desirable in some cases to decrease the area of the manifold connection to the center port to produce the same resistance to flow for all three ports (Fig. 165). Although this may appear desirable from static tests, it is the belief of some that under operating conditions the ramming effect in the longer branches offsets the greater resistance.

Range of Pressure.—If the ramming effect of the flow of gases in an intake manifold is neglected, the maximum pressure that can be attained is atmospheric unless the engine is supercharged. The minimum pressure would occur while motoring

[1] A. Taub, Mixture Distribution, *S.A.E. Trans.*, **25**, 54 (1930).

the engine with the throttle closed and no fuel or air is flowing into the manifold. In a single-cylinder engine, the clearance gases at atmospheric pressure "wire-draw" into the intake manifold. Then, the gases in the cylinder and manifold are expanded during the suction stroke which reduces the manifold pressure again to the minimum. Analysis of this process[1] shows the effect of compression ratio and manifold volume (Fig. 166) and indicates pressures lower than will be attained in practice during idling conditions. Also, the late intake-valve closing and deviations from adiabatic expansion would result in higher manifold pressure.

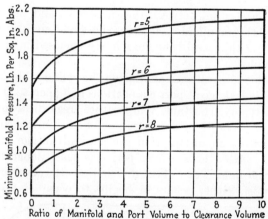

Fig. 166.—Minimum manifold pressures for single-cylinder engines.

Actually, idling-manifold pressures range from about 5 to 10 in. of mercury for multicylinder automotive engines.

Shape of Manifold.—A manifold with circular cross section offers the least resistance to flow and consequently would be the most desirable if only air or gas had to be distributed. However, the manifold must also be designed to distribute the fuel which is usually partly in liquid form. The liquid particles tend to precipitate out of the mixture stream, and the manifold should make it possible to get the precipitated fuel back into the mixture stream. The precipitated fuel will gravitate to the floor of the manifold, which, if circular in cross section, will provide a channel for the liquid stream. A flat floor for the manifold spreads the

[1] See 4th ed., p. 292.

liquid over a much larger surface and increases the possibility of evaporation and entrainment.

A rough surface in the manifold offers considerable resistance to the flow of the wall film of fuel and will tend to increase the thickness of the fuel film. On the other hand, a smooth surface reduces this tendency and is to be desired. Unevenness of the floor of the manifold provides opportunity for puddles of fuel to accumulate, which will under certain conditions flow into one of the cylinders with bad effect and consequently should be avoided. Dams may be cast integral with the floor and walls of the manifold (Fig. 178) to prevent the liquid fuel from flowing too freely to the end ports.

The riser section, between the carburetor and the distributor section of the manifold, is usually short and circular in section. It may be a thin steel tube completely enclosed in the exhaust-gas manifold.

Manifolds integral with and inside the cylinder block eliminate the possible leakage at port connections and operate at a uniform and practically constant temperature regardless of weather conditions. The possible slippage of cores during construction and the effect on distribution are a disadvantage of this construction. Also, the warm-up period is usually longer.

The Manifold Tee.—The division of the mixture for the various ports either occurs or begins to occur at the manifold

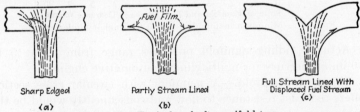

Fig. 167.—Typical updraft manifold tees.

tee. It is at this point that the heavy fuel particles, due to their inertia, become separated from the fuel stream (Fig. 167a) and impinge on the surface above or below the riser. It is extremely important that this surface be perpendicular to the axis of the riser, so that there is no tendency for the fuel to be deflected more to one branch of the manifold than to the other. Another factor that makes this part of the manifold important is the tendency of the throttle valve to deflect the fuel particles to flow more along one side of the riser than the other, which, in turn, may cause more fuel to enter one branch than the other.

Attempts to streamline the manifold tee have resulted in undesirable effects. Any streamlining that increases the area (Fig. 167*b*) reduces the velocity of flow and facilitates precipitation of the fuel particles. There is usually a film of fuel on the riser walls, and a sharp-edged entrance into the distributor section causes the wall film to be torn from the walls and reenter the mixture stream. Streamlining the tee without reduction in section (Fig. 167*c*) results in unequal fuel distribution between the branches, for the vee at the top of the riser can never equally divide the fuel since it is never uniformly distributed in the riser. Streamlining of the induction manifold is of value for sections in which flow occurs in one direction only.

Heating the Mixture.—Starting conditions require excess fuel to provide a combustible air-fuel mixture with the evapora-

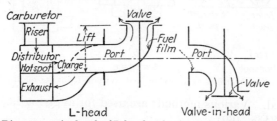

FIG. 168.—Diagrammatic sketch of L-head and valve-in-head ports and manifold.

tion of only the lighter ends of a hydrocarbon fuel. The application of heat to the manifold before starting is sometimes necessary under unusually low temperature conditions. Under starting and warm-up conditions the heavier ends of the fuel accumulate in the manifold for the L-head engine, but flow into the cylinders of the valve-in-head engine (Fig. 168). Sufficient heat should be supplied to ensure the complete evaporation of the fuel at the end of the compression stroke, which indicates the mixture need be only partly evaporated when it leaves the manifold. This should result in minimum desirable mixture temperatures and maximum power output if equal distribution is attained. Complete evaporation before the mixture reaches the manifold tee would eliminate most of the distribution difficulties but would result in higher mixture temperatures with lower volumetric efficiency and power output.

Equilibrium air-distillation data (page 131) indicate that a mixture temperature of about 100 to 110°F. is required for complete evaporation of average gasoline. Actual mixture temperature at wide-open throttle in automotive engines is about 100°F.,[1] indicating a high percentage of evaporation in the manifold although it is impossible to attain equilibrium conditions therein.

Heat is usually applied at the riser and around the tee and is usually thermostatically controlled so that the maximum possible heating effect occurs immediately after starting and under warm-up conditions. The heavy fuel particles impinge on the surface of the hot tee and are broken up and partly

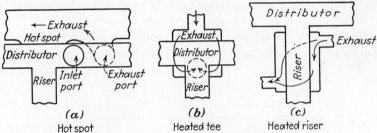

Fig. 169.—Methods of applying heat to updraft intake manifolds.

evaporated. Various methods are used for applying heat to the manifold (Fig. 169).

Intake manifolds cast integral with or bolted directly to the exhaust manifold have a "heat-flywheel" effect that causes a very appreciable rise in intake-manifold temperature upon throttling after a heavy-load run. The manifold should have low heat capacity so that its temperature will closely follow a greater or smaller application of heat. In other words, the heat-flywheel effect should be reduced to a minimum. A sheet-copper hot spot reduces the heat lag to a minimum.

Heating the air before it enters the carburetor is considered bad practice, for it increases the range of temperature through which the jets must meter the fuel and consequently increases the variations in metering characteristics. Also, heating the fuel in the carburetor bowl and passages results in some vaporization which may seriously interfere with fuel flow, and also results in the loss of the more volatile fractions of the fuel. However,

[1] General Motors Corp.

MANIFOLDS AND MIXTURE DISTRIBUTION

heating the air to prevent ice formation in the carburetor of an aircraft engine is very desirable. This may be accomplished by heating only a portion of the air with an exhaust heater (Fig. 170).

Heat should be kept from the carburetor regardless of the form of manifold heating. The heater riser is apt to be the

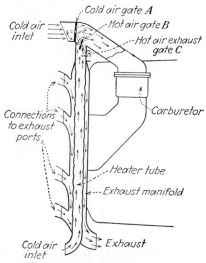

FIG. 170.—Air heater for radial aircraft engine.

worst offender, but the heat flow along the riser to the carburetor can be reduced to a minimum by the following obvious methods:

1. A short length of unheated riser between the exhaust heater and the manifold flange for the carburetor.
2. The use of thin walls for the above-mentioned length.
3. The use of a thick copper-asbestos gasket between the manifold and carburetor flanges.

Heating Surface Required.—The amount of heat that must be supplied to the mixture depends upon the following items:

1. Fuel characteristics.
2. Air-fuel ratio.
3. Percentage of evaporation desired.

The heat is supplied from the exhaust gases, and the amount Q that will be transferred depends on the following items:

1. Mean temperature difference $(T_2 - T_1)$ between exhaust products and the mixture.

2. Area A of surface transmitting the heat.
3. Heat-transfer coefficient K.
4. Time.

Thus, $\qquad Q = KA(T_2 - T_1) \text{ Time}.$ \hfill (2)

The heat-transfer coefficient K, from gas to gas, would probably be in the neighborhood of 5 to 15 B.t.u. ft.$^{-2}$ hr.$^{-1}$°F.$^{-1}$ temperature difference; for low exhaust-gas velocities the conductivity of the metal separating the two gases has very little to do with the heat transmitted. For a liquid film that is being evaporated from one of the surfaces, the heat-transfer coefficient would be higher.

The mean temperature difference $(T_2 - T_1)$ depends principally upon the exhaust-gas temperature. At any given speed the exhaust temperature varies with the power developed. Also, for any given relative power output the exhaust temperature varies with the speed. Thus, there is a wide range of exhaust temperatures depending upon speed and load

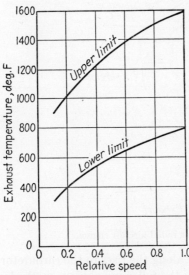

FIG. 171.—Exhaust-gas temperatures.

(Fig. 171). The greatest demand for heat occurs at low speeds with the throttle in a position to result in less than full-load condition.

Example.—Determine the heating surface required for a 200-in.^{3} engine, for 75 per cent evaporation; speed 600 r.p.m. with open throttle. Assume a volumetric efficiency of 85 per cent and a manifold mixture pressure of 14 lb. in.$^{-2}$ abs.

$$\text{Disp. per hr.} = \frac{200 \times 600 \times 60}{2 \times 1728} = 2083 \text{ ft.}^3$$

Air inducted per hr. $= 0.85 \times 2083 = 1771$ ft.^{3}

About one-sixtieth of this air volume would be fuel vapor if the fuel were evaporated. The mixture temperature for 75 per cent evaporation would be about 85°F. and the heat required would be about 200 B.t.u. lb.$^{-1}$ of fuel

(Table 13, p. 155). Assuming 3.5 ft.³ lb.⁻¹ of fuel vapor, and mean exhaust-gas and mixture temperatures of 900°F. (Fig. 171) and 75°F, resp., result in

$$A = \frac{Q}{K(T_2 - T_1)} = \frac{(200 \div 3.5)(1771 \div 60)}{10(900 - 75)} = 0.21 \text{ ft.}^2$$

This example indicates that 1 in.² of heating surface is required for about 7 in.³ of piston displacement to approach a condition of 75 per cent evaporation in the intake manifold.

Considerable heat would be received by radiation from the exhaust pipe, also by conduction at the cylinder connections, which would reduce the heating-surface requirement. However, equilibrium-air-distillation computations result in minimum

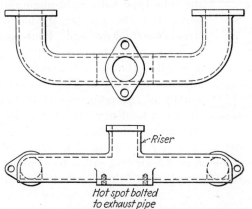

Fig. 172.—Conventional 2-port manifold. (Flat floor and hot spot.)

heat requirements which should be less than those required in practice to attain a given mixture condition.

Two- and Four-cylinder Manifolds.—A manifold for a two- or four-cylinder engine is usually of the two-port type (Fig. 172). This type of manifold offers the same resistance to flow for each cylinder, having the same length and bends to any cylinder. The tee should have the same abruptness as all other tees, and the manifold may be streamlined (Fig. 172) or of the "rake" type (Fig. 174). Each pair of cylinders in the four-cylinder engine has a Siamese intake port (Fig. 173), and consequently the branch should have a short section (d) pointing directly at the port, to direct the mixture flow before entering the port. The manifold for the two-cylinder engine can be streamlined up to the port since only one cylinder is fed through each port.

The two-port manifold for a four-cylinder engine with the usual firing order (1-3-4-2) has reversal of flow from one branch of the manifold to the other after the induction period for cylinders 1 and 4, but not after induction periods for cylinders 2 and 3. This tends to produce unequal mixture distribution and has led to the development of a four-port multiple manifold with separate branches between each port and the tee.

Six-cylinder Manifolds.—The most common six-cylinder manifold is the three-port type in which each port feeds two adjacent cylinders (Fig. 165). The manifold may be of the rake type with right-angle outlets to the ports, but the greater tendency is toward streamline construction. Four-port manifolds have the outer branches feeding the end cylinders while the inner branches each feed two adjacent cylinders. The six-port manifold is usually of the rake type (Fig. 174) although there is a tendency toward multiple manifolding through providing branches which extend from each port to or nearly to the tee (Fig. 175). Such designs will reduce a spread of 2.5 air-fuel ratios to a spread of about 1 air-fuel ratio between cylinders if the distribution is fairly uniform at the tee.

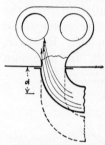

Fig. 173.—Siamese port with two types of approaches. (Broken lines indicate better practice.)

Fig. 174.—Rake-type manifold.

The reversals of flow across the tee for the conventional and rake-type six-cylinder manifold are equally spaced and occur three times each revolution when either one of the customary firing orders is used. This reversal of flow can be eliminated by the use of two carburetors, each feeding one-half of the manifold which consists of two separate parts, one part and its carburetor supplying cylinders 1-2-3, etc.

The pressure differences (Fig. 176) between the riser and cylinders for a typical three-port manifold indicate that exhaust gases flow from the cylinder into the intake manifold during the first part of intake-valve opening. There is no assurance that

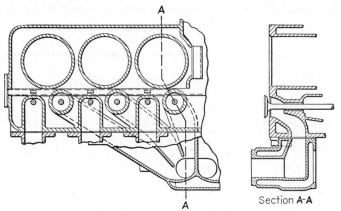

Fig. 175.—Multiple manifolding. (*White Motor Co.*)[1]

the same amount of exhaust gases will be returned to the same cylinder. Flow from the cylinder into the manifold also occurs during the last part of the valve-open period, indicating that better charging would be accomplished at this speed with valve closing very shortly after outer center.

The pressure differences (Fig. 177) between various positions in the manifold and the port of a charging cylinder indicate the complex nature of the pressure conditions in a manifold.

Eight-cylinder Manifolds.—Eight-cylinder in-line engine manifolds are usually of the four-port type (Fig. 178). The rake-type manifold has large variations in distances from the carburetor to the various cylinders.

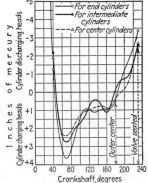

Fig. 176.—Effect of "event order" on pressure difference between riser and cylinders for a typical 3-port manifold. (*Tice*)[2]

Equal distances to all cylinders are obtained by providing two-port manifolds for cylinders 1 to 4 and for cylinders 5 to 8,

---

[1] F. S. Baster, Why Not 125 B.M.E.P. in an L-head Truck Engine, *S.A.E. Trans.*, **34**, 72 (1939).

[2] P. S. Tice, Good Manifold Design Based on Specific Natures of Flow, *Auto. Ind.*, **77**, 416, 585 (1934).

and then providing each of these manifolds with mixture from another two-port manifold (Fig. 179). Obviously, all three parts of this manifold may be in the same plane or otherwise, as shown.

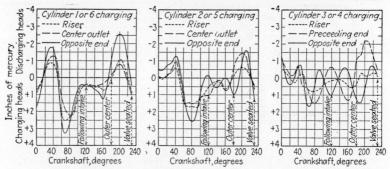

Fig. 177.—Pressure differences between positions indicated and port of charging cylinder in a typical 3-port manifold. (*Tice*.)[1]

Either updraft or downdraft carburetion may be used with the customary hot spot at the tee. This type of manifold provides two places where the wall film will be torn off the walls and impinged against a surface perpendicular to the line of flow. Any

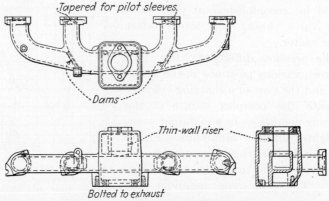

Fig. 178.—Four-port manifold with dams.

particles in the mixture stream will also impinge against the same surfaces.

Dual manifolds are quite commonly used with eight-cylinder in-line engines, the central four cylinders being treated as one system and the outer four cylinders as another system. Crossing

[1] Tice, *op. cit.*

the two manifolds (Fig. 180) provides more nearly equal distances to the various cylinders.

V-type engines may use dual manifolding, half of the cylinders in each bank being fed by one section of the manifold.

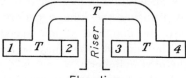

Fig. 179.—Manifold with equal distances to all ports.

Liquid Flow at Elbow.—Flow through an elbow results in higher pressures at the larger radius of the elbow. Tice[1] has found that this pressure difference causes flow from the outer to the inner radius of the elbow and that

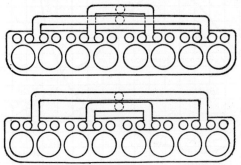

Fig. 180.—Two types of dual manifolding.

a pool of liquid fuel tends to accumulate on the manifold wall beyond the inner radius (Fig. 181). This pool of liquid will have eddy currents as indicated and will be maintained by the deposition of entrained fuel particles that are prevented from striking and clinging to the outer wall by the flow toward the low-pressure area.

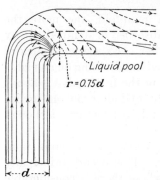

Fig. 181.—Flow through elbow.

Intermittent flow causes collapse of the liquid pool and eddy, but the net result of flow around an elbow is to displace the fuel to one side of the elbow and cause unequal distribution between two cylinders with Siamese ports.

Valve Overlap and Blowback.—The average duration of inlet-valve opening is about 220 deg. The effect of closing an

[1] Tice, *op. cit.*

intake valve after bottom dead center is to reduce the charge at low speeds and to increase it to a maximum at some higher speed by utilizing the kinetic energy of the charge. The blowback into the manifold at low speeds sets up pulsations (Fig. 182) in the manifold that may be very disturbing to the charging of the next cylinder.

With a four-cylinder engine there is about 40 deg. overlapping of inlet valves, with a six-cylinder engine about 100 deg., and with an eight-cylinder engine about 130 deg. In general, the

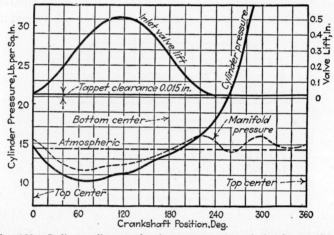

Fig. 182.—Indicator diagram showing pressure waves in intake manifold.[1]

more cylinders that connect into a manifold, the more the blowback from one cylinder will be absorbed or damped out by the induction of the charge into another. Experience indicates that, with a four-cylinder engine, or dual manifolding on an eight-cylinder engine, a 40-deg. late closing of the inlet valve is about as late as is satisfactory. With eight cylinders on one manifold a 50-deg. closing has been found satisfactory, although later closings are sometimes used.

Another difficulty is that valve overlap accentuates non-uniform fuel distribution in the riser. This occurs principally with six- and eight-cylinder engines where there is considerable flow in both directions from the tee at the same time. If most of the fuel is flowing along one side of the riser, most of the fuel will be inducted into the branch which takes off on that side.

[1] *Air Service Eng. Div. Rept.* 2608, p. 160.

Manifolds for Inertia Supercharging.—The periodic opening and closing of an intake valve and the accompanying flow into the cylinder may set up a vibratory or kinetic-energy effect which under certain conditions results in supercharging the cylinder. A smooth, straight pipe attached to the intake port of a cylinder is the most desirable construction.

The length of pipe that will produce resonance between the vibrations in the air column in the pipe and engine speed depends on the velocity of sound in air, which is about 1130 ft. sec.$^{-1}$, and on the engine r.p.m.

$$\text{Thus,} \qquad l \propto \frac{1}{\text{r.p.m.}} \qquad (3)$$

which indicates the desirable length of pipe for resonance and possible supercharging decreases with engine speed.

The kinetic energy of the column of air in motion in the pipe under any condition depends on the mass in the pipe and the air velocity squared.

$$\text{Thus,} \qquad KE \propto Mw^2 \propto d^2l \times \frac{1}{d^4} \propto \frac{l}{d^2} \qquad (4)$$

where d = the pipe diameter.

This indicates that the longer the pipe and the smaller its diameter the greater will be the supercharging effect. However, friction will affect the flow and impose limits on pipe length and diameter beyond which no beneficial effect will occur. The friction effect per unit quantity of fluid is directly proportional to pipe length and kinetic energy, and inversely proportional to the diameter since the larger the diameter the smaller will be the rubbing surface.

$$\text{Thus,} \qquad \text{Friction effect} \propto l \times \frac{1}{d^4} \times \frac{1}{d} \propto \frac{l}{d^5} \qquad (5)$$

from which it might be expected that variation in pipe diameter would have a very appreciable effect.

Dennison[1] found that compression pressure went through a maximum when varying only the length of induction pipe (Fig. 183, curves 1 and 2). He also found an appreciable

---

[1] E. S. Dennison, Inertia Supercharging of Engine Cylinders, *A.S.M.E. Trans.*, **O.G.P. 55**, 53 (1932).

gain in compression pressure, eliminating the effect of speed when an induction pipe of desired diameter was used (curves 3 and 4). Methods[1,2] have been developed for computing the pressure in the

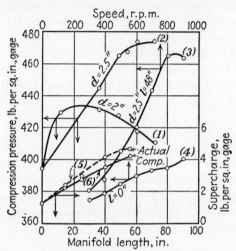

Fig. 183.—Effect of manifold diameter, length, and engine r.p.m. on supercharging. (*Dennison.*)

Curves (1), (2), (3), and (4) for 8¼- by 12-in. engine at 800 r.p.m.; curves (5) and (6) for 9- by 12-in. engine at 1000 r.p.m. $d = 3''$.

cylinder at intake-valve closure which checked fairly well with actual results (curves 5 and 6). (See also Valve, Piston, and Flow Relations on page 324.)

EXHAUST MANIFOLDS

Back Pressure.—The exhaust system should be designed to maintain the back pressure at nearly atmospheric pressure. The m.e.p. is reduced 1 lb. for each pound the back pressure is above atmospheric pressure, owing directly to the extra work done by the piston during the exhaust stroke. There is another loss due to the effect of back pressure on the clearance gases left in the cylinder at the closing of the exhaust valve (see page 154). Tests (Fig. 184) indicate a decrease of about 11 per cent in power for an increase of 4.4 lb. in.$^{-2}$ in back pressure. This indicates about 1.5 per cent decrease in power per 1 lb. increase

---

[1] Dennison, *op. cit.*

[2] Hans List, Increasing the Volumetric Efficiency of Diesel Engines by Intake Pipes, *N.A.C.A. Tech. Memo.* 700 (1933).

in back pressure, due to the effect of back pressure on amount of clearance gases and volumetric efficiency.

Valve Overlapping.—An exhaust valve is usually open for a period of about 200 to 225 deg. during a cycle of two revolutions. Obviously, only three cylinders with firing orders equally spaced can exhaust into one manifold without one exhaust interfering with another. In a four-cylinder engine the exhaust valve for

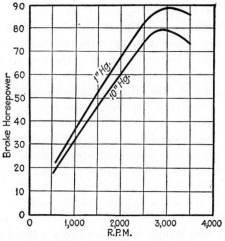

FIG. 184.—Effect of exhaust back pressure on power output. (From tests on a six-cylinder 3½- by 5-in. engine.)

cylinder 2 closes shortly after the exhaust valve for cylinder 1 opens. In a six-cylinder engine with one of the customary firing orders the exhaust valve for cylinder 1 usually closes just before the one in cylinder 3 opens, and cylinder 3 closes while piston 5 is on the exhaust stroke. In an eight-cylinder engine with cylinders in line and with the firing order of 1-6-2-5-8-3-7-4, one exhaust valve is opening while another is closing and a third is somewhere in between. Whenever there is overlapping of the exhaust valves, there is a tendency toward higher back pressure.

Types of Exhaust Manifolds.—The ideal exhaust manifold has separate leads from each cylinder which may be streamlined at some distance from the cylinder block into a common or group exhaust system. Such a system is usually found on racing cars where the maximum output is desired. Next, in order of preference, would be a system which divides the exhaust manifold into various sections so that two cylinders do not exhaust into the same branch at the same time. This design complicates the exhaust system for a four-cylinder engine but is readily adapted to the six-cylinder engine since the exhaust ports from the first three cylinders can be connected to one branch and the last three to another branch, both of which may be streamlined into a common exhaust line.

A V-eight is naturally adapted for two separate exhaust systems, each one having the disadvantages of a four-cylinder manifold. The "straight"-eight with the customary firing order does not lend itself to the grouping of the first four and the last four cylinders into separate sections. Connecting cylinders 3, 4, 5, and 6 together to exhaust into one section duplicates typical four-cylinder conditions, which is also true if cylinders 1, 2, 7, and 8 are connected together. Considering all the factors, including the space available under the hood of a motorcar, it appears that the best compromise would be to have short individual stacks from each cylinder, streamlined as much as possible into a common exhaust header.

In all cases the exhaust manifold should be designed for ample capacity with no restrictions to build up back pressure. Restriction to flow should be reduced to a minimum by streamlining all sections and by using large radius bends wherever possible.

Exhaust-manifold Expansion.—The temperature of the exhaust manifold is subject to wide variations, which range from atmospheric temperature to red heat. This results in considerable expansion and contraction in length and must be provided for, or enormous strains will be set up in the manifold, studs, and cylinder block. In general, the shorter the exhaust stacks connecting the exhaust ports to the main exhaust header, the greater is the possibility of strain in the manifold.

In some cases the exhaust manifold is built in two sections and an expansion slip joint is used to connect the two sections. The joint requires some form of packing that need be replaced at infrequent intervals.

The tendency of cast iron to "grow" when subjected to high temperatures has led to the development of a cast iron that has a small growth characteristic. A material with this characteristic should be used, particularly in manifolds not provided with expansion joints.

Individual Exhaust Pipes.—Single exhaust pipes can be used to obtain optimum performance. The opening of the exhaust valve starts a pressure wave which travels with the velocity of sound along the exhaust pipe. This wave is reflected at the open end of the pipe and returns to the exhaust valve or port where it sets up a pressure disturbance. Whitefield[1] gives the relationship

$$\text{Sound velocity} = 49\sqrt{T°} \qquad (6)$$

where $T = °R$.

He found exhaust-pipe lengths of 38, 64, and 86 ft. resulted in increased output from a two-stroke engine operating at 362 r.p.m. For a temperature of 870°R., the time period for the 38-ft. exhaust pipe is equivalent to 115 deg. of crank travel, which is about the angle during which the exhaust port is open. Obviously, the inertia and friction effects have an influence in fixing the other optimum lengths of exhaust pipe.

MIXTURE DISTRIBUTION

Mixture-distribution Criterion.—The problem of mixture distribution between cylinders is eliminated in the single-cylinder engine which receives all the air-fuel mixture leaving the carburetor. Single-cylinder-engine performance is affected principally by air-fuel ratio and fuel distribution in the combustion chamber, other operating conditions being optimum. On the assumption that desirable fuel distribution occurs, performance of a single-cylinder engine with various air-fuel ratios establishes

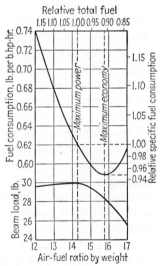

Fig. 185.—Single-cylinder performance at constant speed. Gasoline fuel.

[1] K. C. Whitefield, Improving Diesel-engine Operation by the Selection of a Proper Exhaust-pipe Length, *A.S.M.E. Paper* (unpublished).

290   INTERNAL-COMBUSTION ENGINES

a criterion of performance which can be used to indicate the perfection of mixture distribution in a multicylinder engine.

Wide-open-throttle runs at constant speed on a single-cylinder engine indicate air-fuel ratios for maximum power and maximum economy (Fig. 185) which depend upon the fuel used. Such tests on various single-cylinder engines, gasoline being used as the fuel, result in air-fuel ratios of about 14.3 for the leanest value for maximum power and about 15.7 for maximum economy. The obtaining of these two air-fuel ratios for these points in a multicylinder-engine test using gasoline is indicative of excellent mixture distribution.

The ratio between the specific fuel consumption at maximum economy and at the leanest mixture for maximum power for a single-cylinder engine is about 0.95, while this ratio based on total fuel consumption is about 0.92.

Multicylinder Performance.—If each cylinder of a multicylinder engine received the same quantity of fuel and air, the performance would be similar to that of a single-cylinder engine. However, it is very difficult to obtain equal distribution at any one speed, let alone all speeds.

The worst distribution situation, from a fuel-loss standpoint, is to have one weak cylinder while all the others are receiving uniform mixtures. In this case the mixture must be enriched in order to make the lean cylinder fire regularly and increase its power. This results in an overrich mixture for all but the one cylinder. With three lean cylinders, there will be three rich cylinders in a six-cylinder engine. Thus, the extra fuel would be three-fifths of the extra fuel required with one weak cylinder, whereas with five lean cylinders out of the six, it would be only one-fifth of the extra fuel with one weak cylinder.

The maximum power of the engine will be attained when a slight increase of mixture richness causes a loss of power of the rich cylinders that is greater than the gain in power of the lean cylinders.

Example.—Five cylinders of an engine are supplied with a 14:1 mixture when the sixth cylinder receives a 17:1 mixture. Determine the maximum-power air-fuel ratio for the engine and the loss of fuel due to unequal distribution. Assume that the mixture is enriched in each cylinder by relatively the same amount. Then, using the beam load as in Fig. 185, the maximum-power air-fuel ratio is determined as follows:

MANIFOLDS AND MIXTURE DISTRIBUTION

A/F ratio, 1 cyl.	Rel. fuel consumption	A/F ratio, 5 cyl.	Beam load, lb.		
			1 cyl.	5 cyl.	Total
17.0	1.000	14.00	25.6	149.7	175.3
16.5	1.030	13.60	26.9	149.5	176.4
16.0	1.063	13.18	27.9	149.2	177.1
15.5	1.097	12.75	28.8	148.9	177.7
15.0	1.133	12.36	29.5	148.5	178.0
14.5	1.172	11.95	29.9	148.0	177.9
14.0	1.214	11.52	29.9	147.5	177.4

On the assumption of 1 lb. of air per cylinder,

$$\text{Original air-fuel ratio} = \frac{6}{\frac{5}{14} + \frac{1}{17}} = \frac{6}{0.416} = 14.42.$$

$$\text{Max.-power air-fuel ratio} = \frac{6}{\frac{5}{12.36} + \frac{1}{15}} = \frac{6}{0.472} = 12.71.$$

With perfect distribution the fuel consumption at maximum power would be $6 \div 14.3 = 0.420$ lb. per 6 lb. of air.

$$\text{Fuel wasted} = \frac{0.472 - 0.420}{0.420} = 0.124 \text{ or } 12.4 \text{ per cent.}$$

If the air-fuel ratio supplied to an engine for maximum power is not much richer than that required for maximum power in a single cylinder, it is quite apparent that the mixture ratio in the leanest cylinder is almost the maximum-power ratio and that the distribution is fairly good.

Distribution of Heavy Fuel Fractions.—Nearly perfect distribution can be obtained when gaseous fuels are used. This indicates that most of the distribution difficulty is due to heavy ends of the fuel which are not evaporated in the manifold tee. Thus, the leanest mixture provided any cylinder may be due to receiving only its portion of the evaporated light ends of the fuel, F_l. Other cylinders will receive in addition various portions of the liquid heavy ends of the fuel, nF_h, where n may vary from zero to two or three times the normal portion, F_h, for perfect distribution. Thus, for any cylinder,

$$\text{Air-fuel ratio} = \frac{\text{air/cyl.}}{F_l + nF_h}, \qquad (7)$$

from which some measure of the unequal distribution of heavy fractions may be obtained.

Example.—Determine the distribution of the heavy ends of a fuel for the following distribution. Assume that the leanest cylinder receives only the evaporated lighter fractions of the fuel.

The leanest air-fuel ratio is 16.01. On the assumption of 1 lb. of fuel per cylinder,

$$F_l = \frac{1}{16.01} = 0.0625 \text{ lb. of fuel.}$$

Cyl.	1	2	3	4	5	6	7	8
Distribution in terms of air-fuel ratio.	14.59	16.01	13.29	14.13	12.63	11.56	15.27	12.37
Total fuel/lb. air.	0.0685	0.0625	0.0752	0.0708	0.0791	0.0865	0.0655	0.0808
Heavy fractions, nF_h.	0.0060	......	0.0127	0.0083	0.0166	0.0240	0.0030	0.0183
$n = 8nF_h/\Sigma nF_h$.	0.54	0	1.14	0.75	1.50	2.16	0.27	1.65

The values for n represent the relative distribution of the heavier unvaporized fuel fractions, cylinder 6 receiving 2.16 times its normal proportion of this part of the fuel.

The foregoing example indicates that 85 per cent of the fuel is evaporated by hot-spot heating. This appears to be fairly high for wide-open-throttle low-speed accelerating conditions under which the data were obtained, and indicates that some of the unevaporated fuel may have been distributed to the leanest cylinder.

Distribution from Exhaust-gas Analysis.—An analysis of the products of combustion provides information regarding the air-fuel ratio when the chemical composition of the fuel is known. The usual procedure is to analyze the exhaust products from each cylinder and compute the air-fuel ratio by the use of the various stoichiometric relationships. This method has been simplified (see Exhaust-gas Analysis, page 158) for hydrocarbon fuels, with H/C ratios that lie within the usual range, so that only CO_2 and O_2 content need be determined. Various gas analyzers have been developed that may be calibrated for the given fuel and give reliable indications of the air-fuel ratio.

Samples of exhaust products obtained from each exhaust port or even close to the exhaust valve are contaminated by the products from other cylinders. The use of separate exhaust stacks changes the charging effect for each cylinder and removes the hot spot, both of which seriously affect the distribution characteristics.

Sampling the exhaust products in the cylinder during a portion of the exhaust stroke is the only reliable method for determining mixture distribution by exhaust-gas analysis. A solenoid-operated sampling valve in combination with a spark plug[1]

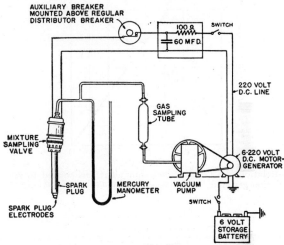

FIG. 186.—Mixture sampling valve and auxiliary apparatus. (*Bartholomew, Chalk, and Brewster.*)

makes possible such analysis. This sampling valve with auxiliary apparatus (Fig. 186) was used for the determination of mixture distribution in an automobile engine on the road, the results in one case being given in the preceding example.

A study[2] of mixture distribution in a nine-cylinder radial aircraft engine, with individual exhaust pipes, was made from CO_2 analyses only, after determining the relationship between CO_2, CO, equivalent fuel wasted, and air-fuel ratio (Fig. 187). The results obtained with constant-power output and speed but varying the specific fuel consumption (Fig. 188) indicate a variation in mixture ratio between cylinders of about 4 per cent. This study also indicated that the temperature of the

[1] E. Bartholomew, H. Chalk, and B. Brewster, Carburetion, Manifolding, and Fuel Antiknock Value, *S.A.E. Trans.*, **33**, 141 (1938).

[2] H. C. Gerrish and F. Voss, Mixture Distribution in a Single-row Radial Engine, *N.A.C.A. Tech. Note* 583 (1936).

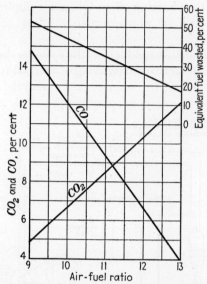

Fig. 187.—Relation of CO, CO_2, and equivalent fuel wasted to air-fuel ratio. (Aviation gasoline. Army spec. Y-3557-G.) (*Gerrish and Voss.*)[1]

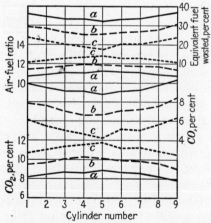

Fig. 188.—Effect of specific fuel consumption on distribution. (1800 r.p.m., 422 to 429 i.hp. $a = 0.567$, $b = 0.515$, and $c = 0.467$ lb./i.hp.-hr.) (*Gerrish and Voss.*)[1]

---

[1] Gerrish and Voss, *op. cit.*

rear spark-plug boss was not a good indicator of mixture ratio in the cylinder for ratios from 10:1 to 13:1.

The effect of speed and load on mixture distribution was studied by Blackwood, Lewis, and Kass[1] (Table 19) who analyzed exhaust-gas samples taken at a point ½-in. from the exhaust valve of each cylinder.

TABLE 19.—SPREAD, IN AIR-FUEL RATIOS, BETWEEN MAXIMUM AND MINIMUM AIR-FUEL RATIO IN THE INDIVIDUAL CYLINDERS
(Regular-grade gasoline, eight-cylinder in-line valve-in-head engine)

M.p.h.	10	15	20	30	40
Road load	...	3.5	3.0	1.5	2.6
2% grade	2.5	...	...	...	1.6
4% grade	4.1	3.3	2.8	1.0	0.8
6% grade	4.1	2.6	1.1	1.1	0.9
8% grade	3.9	3.1	2.2	1.2	0.7
10% grade	2.0	1.4	0.8	0.3	2.7
Full load	3.0	6.6	6.9	5.9	5.6

They found a distribution spread of 7.1 air-fuel ratios for regular-grade gasoline, 4.7 for aviation gasoline, and 0.9 for butane. Thus, distribution can be improved by increasing the volatility of the fuel.

Distribution from Temperature Indication.—A study of the theoretical temperatures of combustion for various air-fuel ratios indicates that combustion temperatures attain a maximum value at slightly less than the correct air-fuel ratio and decrease both with richer or leaner mixtures. This characteristic is made use of in air-cooled aircraft engines by observing the temperatures indicated by thermocouples located in the cylinder head or in the spark-plug gasket. Variation of the air-fuel ratio in a single-cylinder engine indicates a definite relationship between the mixture ratios for maximum economy, maximum head temperature, and maximum power (Fig. 189). The results (Fig. 190) from the seven cylinders of a radial aircraft engine indicate maxima at various fuel flows, the relation of these maxima indicating the mixture distribution between the various cylinders.

Temperatures of the center electrodes of spark plugs (Fig. 191) are used for indicating mixture distribution in a water-

[1] A. J. Blackwood, O. G. Lewis, and C. B. Kass, Antiknock and Mixture Distribution Problem in Multicylinder Engines, *Auto. Ind.*, **79**, 821 (1938).

cooled engine. More than three times the variation in temperature is obtained with the usual range of air-fuel ratios as compared

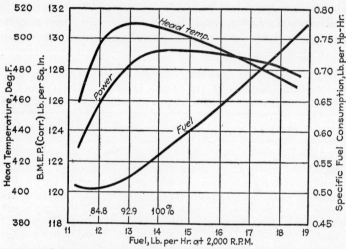

Fig. 189.—Variation of cylinder-head temperature with fuel consumption. (*DuBois*.)

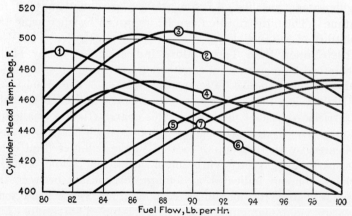

Fig. 190.—Cylinder-head temperatures on a seven-cylinder, radial, air-cooled engine. (*DuBois*.)

with cylinder-head temperature of an air-cooled cylinder. Obviously, the greater the range of temperature the more desirable is the use of this means for air-fuel-ratio evaluation.

[1] R. DuBois, Production and Overhaul Testing Combined with Research Laboratory, *Auto. Ind.*, **64,** 723 (1931).

The results of tests (Fig. 192) on an overhead-valve engine with cylindrical combustion chamber, in which spark-plug temperatures were determined, indicator cards taken, and gas analyses made, show that a definite relation exists between the various factors charted. The power curve peaks at the richest mixture, the temperature curve peaks at a leaner mixture, and the CO_2 curve peaks at the leanest mixture of the three. It was found for this engine that the temperature and CO_2 peaks occurred at fuel consumptions (based upon pounds per hour) of 96 and 92 per cent, respectively, of the fuel consumption for the maximum power. This relation between the maximum values for power, temperature, and CO_2 is a function of combustion-chamber design and spark-plug location.

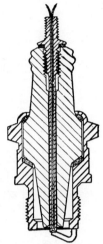

Fig. 191.—Spark plug with thermocouple in center electrode. (*A. C. Spark Plug Co.*)

The results of tests (Fig. 193) on one cylinder of a six-cylinder engine, having a combustion chamber of uniform height over both valves and over all of the piston, show that the temperature peak occurs at 90 per cent of the fuel consumption for the maximum power. This series of tests was made by removing pistons, rods, and tappets as well as blocking the intake-manifold passages of five of the cylinders.

The mixture distribution in a multicylinder engine (Fig. 194) is determined at a given speed by varying the mixture ratio through a range large enough to indicate the various peaks. The peak values of the indicated m.e.p. curves vary considerably and are due to variations in charge mass. A test of compression pressures showed the same relative variations. The percentage of CO_2 in the exhaust gases at the various peaks of the i.m.e.p. curves lies in the range from about 10.5 to 11.5. This corresponds to an air-fuel-ratio range of about 11.7 to 12.5.

The total fuel consumption at the maximum beam load is about 30 lb. hr.$^{-1}$ If all the cylinders received the same air-fuel ratio, the temperature curves would all peak at about 0.90×30 or 27 lb. hr.$^{-1}$ Those cylinders that peak below 27 lb. hr.$^{-1}$ are receiving more than their share of fuel and thus are running richer than those peaking above 27 lb. hr.$^{-1}$ Dividing the total fuel consumption for each of the temperature peaks by the fuel consumption at which all would peak if provided with the same air-fuel ratio indicates the relative mixture ratio. Thus, cylinder 3 is 11 per cent too lean, cylinders 1, 2, and 4 are 4 per cent too rich, cylinder 6 is 7 per cent too rich, and

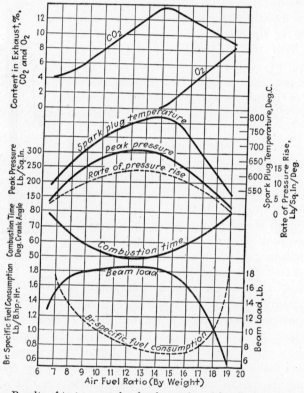

Fig. 192.—Results of tests on overhead-valve engine with cylindrical combustion chamber firing on the exhaust side. (*Rabezzana and Kalmar.*)[1]

cylinder 5 is 13 per cent too rich, when the engine is provided with the maximum power mixture. The air-fuel ratios are computed on the basis of a 13:1 air-fuel ratio, providing the maximum power for any cylinder receiving this mixture ratio.

The actual fuel consumption for each cylinder was evaluated from a determination of the air-fuel ratio and the air consumption

[1] H. Rabezzana, and S. Kalmar, Mixture Distribution in Cylinders Studied by Measuring Spark Plug Temperatures, *Auto. Ind.*, **66**, 13 (1932).

for each cylinder. The air-fuel ratio was determined from an exhaust-gas analysis. The total air consumption was determined and then divided by 6 and multiplied by a *charge factor* for each cylinder. The charge factor is the ratio of the cold compression

Cylinder	1	2	3	4	5	6
Temperature-peak fuel consumption	26	26	30	26	23.5	25
Relation to max.-power fuel consumption	0.96	0.96	1.11	0.96	0.87	0.93
Air-fuel ratios	12.5	12.5	14.4	12.5	11.3	12.1

to the average compression of all cylinders at the same speeds. Dividing the pounds of air per hour by the air-fuel ratio gives the

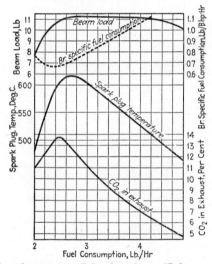

FIG. 193.—Results of test on an L-head engine. (*Rabezzana and Kalmar.*)[1]

pounds of fuel per hour. Results of these tests and computations indicate that the air-fuel ratios are about 1 ratio lower than those in the foregoing tabulation. The computed air-fuel ratios would check the foregoing if a maximum-power single-cylinder air-fuel ratio of about 12:1 had been used. However, this is considerably lower than is reported on a number of single-cylinder tests.

[1] Rabezzana and Kalmar, *op. cit.*

300 INTERNAL-COMBUSTION ENGINES

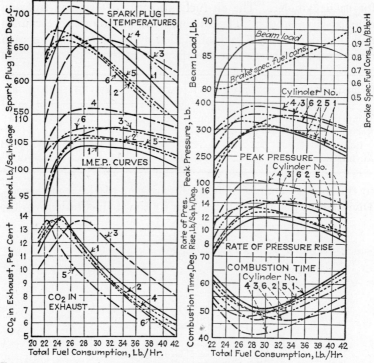

Fig. 194.—Results of test on a six-cylinder engine. (*Rabezzana and Kalmar.*)[1]

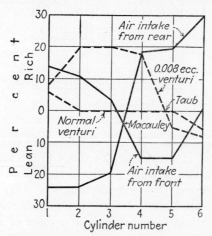

Fig. 195.—Effect of direction of air intake and eccentric venturi on distribution.

[1] Rabezzana and Kalmar, *op. cit.*

Carburetor Design and Construction.—Mixture distribution has been found to be peculiarly sensitive to direction from which the air approaches the carburetor. Macauley[1] found the distribution reversed for a reversal of direction of air intake (Fig. 195). Distribution is sensitive to the location of the fuel jet with regard to the center of the venturi. Taub[2] found that an eccentric venturi (Fig. 195) changed the distribution characteristics. The data reported for the eccentric venturi indicate an increase in fuel flow also, since the average of the points is not zero.

The butterfly throttle valve is an unusually bad offender, often causing the fuel to flow along one side of the riser and to flow more readily into the manifold branch which connects to that side.

EXERCISES

1. Discuss the change in intake-manifold velocities with various throttle openings and constant engine speed for the following conditions:
   a. No clearance volume.
   b. Clearance gases at intake-manifold pressure when the piston is beginning the suction stroke.
   c. Actual conditions.

2. Determine the minimum possible inlet-manifold pressure while motoring a four-cylinder engine that has a manifold and port volume ten times the clearance volume in one cylinder. The connecting-rod-crank ratio is 4:1. The inlet valves open at 0 deg. and close at 45 deg. past bottom dead center. The compression ratio is 6:1.

Note.—The minimum pressure will occur when one piston is on bottom center on the suction stroke. Immediately afterward, the next inlet valve opens and raises the pressure in the manifold and first cylinder. Then the first cylinder decreases the total volume, while the second increases it until the second piston is at 45 deg. after top dead center. Determine the pressure at this point. Then the inlet valve in the first cylinder closes and changes the volume to be expanded by the second cylinder. When the second piston reaches bottom dead center, the pressure should be the same as was assumed to exist when the first piston was in this position.

3. Determine the riser heating surface required for a 200-in.³ engine at various speeds for 50 per cent evaporation of the fuel, with wide-open throttle. Assume a volumetric efficiency of 0.85.

4. An engine develops 1000 hp. at a given speed with a back pressure of 4 in. Hg. If the back pressure is reduced to 2 in. Hg, how much power should the engine develop?

[1] J. B. Macauley, Discussion, *S.A.E. Trans.*, **24**, 109 (1929).
[2] Alex Taub, Motor-car Engines in England, *S.A.E. Trans.*, **33**, 229 (1938).

302   INTERNAL-COMBUSTION ENGINES

5. Approximately how far apart will the two end studs for fastening the exhaust manifold be located on an eight-cylinder 3.5-in.-bore engine, having cylinder walls ¼ in. thick and with spaces of ¼ in. between the cylinders of an L-head engine.

6. Under some conditions of operation the exhaust manifold reaches a dull red heat. Approximately how much expansion of the manifold in Prob. 5 would occur in going from a cold manifold to a hot one? Explain the effect of heating the block from the same cold condition up to a temperature of 150°F.

7. Determine the variation in exhaust-pipe lengths with speed when designed to prevent the reflected wave from returning to the exhaust port before exhaust-valve closure. Exhaust valve opens at 45 deg. before bottom center and closes at top center. Assume exhaust-gas temperatures vary as indicated by the maximum curve in Fig. 171. Plot the results.

8. Three cylinders are supplied with a 14:1 mixture and the other three with a 17:1 mixture. Determine the maximum-power-mixture ratio for the engine and the loss of fuel due to unequal distribution. Use Fig. 185.

9. The conditions are the same as in Prob. 8 except that 1 cylinder receives a 14:1 mixture and 5 cylinders receive a 17:1 mixture. Compare results with the example on page 290 and with Prob. 8.

10. Estimate the percentage of CO_2, CO, and O_2 in the exhaust from each cylinder having the distribution indicated in the example on page 292.

11. A nine-cylinder radial aircraft engine has a CO_2 content of the gases of the individual exhaust stacks of 4.8, 5.75, 6.7, 7.5, etc., per cent for cylinders 1, 2, 3, 4, etc., respectively (Fig. 187). Assume cylinder 8 receives a mixture with no unevaporated heavy ends of fuel. Determine the distribution of the unevaporated heavy ends.

12. One cylinder of a three-cylinder radial engine is receiving 11.5 lb. of fuel per hour. The two other cylinders are receiving a maximum-power mixture of 14 lb. of fuel per hour. Determine the b.m.e.p. of the engine. Increase the total fuel consumption, and determine maximum power and specific fuel consumption of the engine. Use the data in Fig. 189.

13. For the conditions in the preceding problem, how much excess fuel is used when the engine develops maximum power? Note that with perfect distribution the engine develops maximum power with a total fuel consumption of $3 \times 14 = 42$ lb. hr.$^{-1}$.

14. Assuming that maximum engine power occurred at 92 lb. of fuel per hour (Fig. 190), which represents an engine air-fuel ratio of 13:1, determine the air-fuel ratio in each of the cylinders.

15. Determine the excess fuel in Prob. 14 and the ideal air-fuel ratio.

CHAPTER XI

VALVES AND VALVE MECHANISMS

Requirements.—The operation of the internal-combustion engine necessitates the admission to, the trapping in, and the exhausting of the working medium from the engine cylinder, all of which is accomplished with valves and valve mechanisms. The valves must open and close at definite crank angles and with a minimum of noise and wear. The valves should open sufficiently to permit unrestricted breathing of the engine if maximum power output is desired. Also, the valves must seat tightly and prevent the escape of gases during the working process.

The valves should be made of material that will not be seriously affected by the various temperature, erosive, and corrosive conditions.

Size of Valve Ports.—The port diameter is fixed primarily by the size of the cylinder bore and the distance between cylinder centers. Obviously, the larger the bore, the larger the ports, which, for a given displacement per minute, indicates a decided advantage in breathing capacity for the large-bore short-stroke engine.

The mean velocity of gases through ports depends upon piston and port diameters, or areas, and the mean piston speed.

Thus,
$$v_p = \left(\frac{d_P}{d_p}\right)^2 v_P = \left(\frac{a_P}{a_p}\right) v_P \tag{1}$$

where subscripts P and p denote piston and port, respectively.

The mean gas velocity varies from 75 ft. sec.$^{-1}$ for small stationary engines to 120 ft. sec.$^{-1}$ for large stationary engines. For automotive and aircraft engines, mean velocities of 150 to 250 ft. sec.$^{-1}$ are used. Good design usually results in mean gas velocities between 120 and 160 ft. sec.$^{-1}$

For a port one-half the piston diameter, the velocity through the port is approximately four times the piston speed.

Example.—Determine the mean velocity through an intake port in a 3½- by 5-in. engine running at 3600 r.p.m. Port diameter = 1.75 in.

$$\text{Mean vel. } v_p = \left(\frac{3.5}{1.75}\right)^2 \left(\frac{2 \times 5 \times 3{,}600}{12 \times 60}\right)$$
$$= 200 \text{ ft. per sec.}$$

The actual velocity of the piston for each stroke varies from 0, at top dead center, to a maximum and back to 0 again, at bottom dead center. At the maximum the velocity is slightly higher than the crankpin velocity. The relation between crankpin velocity and mean velocity of the piston is

$$\frac{\pi \times \text{stroke}}{2 \times \text{stroke}} = 1.57.$$

Thus, at about half stroke the instantaneous velocity of the piston is about 57 per cent higher than the mean speed. This variation in piston velocity has a direct effect on the velocity of gases through the port, the change in gas velocity lagging behind the change in piston velocity.

The demand for higher outputs which require better cylinder charging and better cooling of the exhaust valves has resulted in the use of exhaust ports 20 to 30 per cent smaller than the intake ports.

Valve-seat Angles, Lifts, and Areas.—Valve seats may be either flat or beveled. The beveled valve presents a conical seating surface which is self-centering. For this reason the bevel seat is almost universally used. The flat valve ($\alpha = 0$, Fig. 196; also Fig. 198) provides an opening that is directly proportional to the valve lift. The area of valve opening is equal to

$$A = \pi d_p l \tag{2}$$

where d_p = diameter of the port, in.,
l = lift of the valve, in.

For a beveled or conical valve seat, the valve opening is equal to the area of a frustum of a cone (Fig. (196) having the two radii r (one-half the port diameter) and r', and the slant height s.

Thus, $s = l \cos \alpha$, $\quad x = s \sin \alpha$, $\quad r' = r + x$.

Also, the area of the frustum is $\pi(r + r')s$, from which

$$\text{Valve opening area} = \pi(2r + l \cos \alpha \sin \alpha)l \cos \alpha. \tag{3}$$

The value of the valve opening from Eq. (3) approaches that of Eq. (2) as the seat angle approaches zero. Flat seats appear to be desirable owing to the smaller lifts required and to certain advantages in maintaining a tight seat. However, the charging ability is a maximum for valves with seat angles of about 30 deg.

Example.—Determine the valve lift to make valve and port areas equal when a 45-deg. bevel seat is used. Area of port = area of valve opening.

$$\frac{\pi d_p^2}{4} = \pi(d_p + l \cos \alpha \sin \alpha) l \cos \alpha.$$

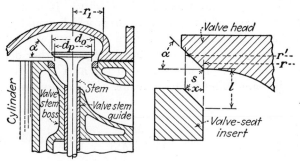

Fig. 196.—Beveled valve, and seat insert.

With $\alpha = 45$ deg., $\sin \alpha = \cos \alpha = 0.707$, and the above relation becomes

$$d_p^2 - 2.828\ d_p l - 1.414 l^2 = 0.$$

The solution of this equation results in a lift-diameter relation of

$$\frac{l}{d_p} = 0.307.$$

This lift-diameter relation is apparently the maximum desirable, since larger lifts would make the port the major restriction. However, with a design having a poor coefficient of discharge at this lift, it might be advisable to increase the lift and reduce the resistance at this point. Also, since flow must occur through the valve while it is opening and closing, a lift beyond the so-called maximum desirable provides a period during which the valve does not restrict the flow. The foregoing example does not take into account the valve-stem cross section which reduces the port area and would decrease the above lift-diameter relation.

The radius r_1 of the pocket in which the valve is located (Fig. 196) should be large enough to provide an unrestricted flow passage around the valve head.

Thus, $$0.7854[(2r_1)^2 - d_o^2] \geqq 0.7854d_p^2. \qquad (4)$$

This is usually satisfied when

$$r_1 \geqq 0.8d_p. \qquad (5)$$

It is generally the case, however, that the form of the combustion chamber furnishes more space than is required.

Flow through Ports and Valve Openings.—The coefficient of discharge, c, for an orifice is defined as the ratio of the mean actual velocity to the theoretical velocity. It can also be defined as the ratio of an equivalent area, which would theoretically discharge the same quantity, to the actual area through which the given quantity is discharged. Still another definition is the ratio of the equivalent area to the area of a flat-seated valve of the same dimensions and lift. The equivalent area is equal to

$$A = \frac{M}{18.3\sqrt{h_w d}}, \qquad \text{[From Eq. (4), p. 208]}$$

and the area of valve opening for a flat seat is

$$A_f = \pi d_p l.$$

Thus, $$c = 144\frac{A}{A_f} = \frac{7.86M}{\pi d_p l \sqrt{h_w d}}. \qquad (6)$$

The results of various experiments with approximately the same h_w were compared[1] on this basis and indicate a decrease in the coefficient of discharge with an increase in the l/d_p ratio (Fig. 197).

Coefficients of discharge for orifices vary from about 0.6 for a sharp-edge orifice to about 0.98 for a well-rounded or streamlined orifice. Also, the velocity of flow through an orifice increases with pressure drop until the lower pressure is about 0.53 of the higher pressure. Consequently, the following methods should be used to attain maximum flow with minimum pressure drop:

1. Use a small seat angle, and obtain maximum opening with a given lift.

[1] E. S. Dennison, T. C. Kuchler, and D. W. Smith, Experiments on the Flow of Air through Engine Valves, *A.S.M.E. Trans.*, **OGP, 53,** 6 (1931).

2. Round all the entering and leaving edges of the valve and seat, particularly the entering edges.

3. Use large and smooth passageways to and from the valves.

4. The pressure drop should never exceed the critical value which is usually never approached with good design. (Critical pressure is 0.53 of initial pressure.)

Multiple Valves.—When two valves are used instead of one, the combined area of the two should at least equal the area of

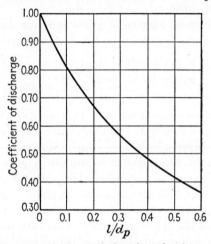

Fig. 197.—Coefficients of discharge for flow through valve openings. Approximate values for valves having 30° and 45° seats. (*Dennison Kuchler, and Smith.*)[1]

opening of the one valve. If a given lift-diameter ratio of k, and equal valve-seat angles in both cases, are assumed,

$$l = kd_p$$

and $\alpha_1 = \alpha_2 = \alpha.$

Then, from Eq. (2), using d_{p_1} for the single valve and d_{p_2} for the two valves, results in

$$\pi(d_{p_1} + kd_{p_1} \cos \alpha \sin \alpha)kd_{p_1} \cos \alpha = 2\pi(d_{p_2} + kd_{p_2} \cos \alpha \sin \alpha)kd_{p_2} \cos \alpha,$$

from which $\qquad d_{p_2} = 0.707 d_{p_1}.\qquad (7)$

If the same lift is used in both cases,

$$\pi(d_{p_1} + l \cos \alpha \sin \alpha)l \cos \alpha = 2\pi(d_{p_2} + l \cos \alpha \sin \alpha)l \cos \alpha,$$

from which $\qquad d_{p_2} = \dfrac{d_{p_1} - l \cos \alpha \sin \alpha}{2}.\qquad (8)$

---

[1] Dennison, Kuchler, and Smith, *op. cit.*

This indicates a valve size for the two valves of less than one-half the size of the single valve. This would be true for lifts for the smaller valve that did not exceed the maximum desirable value.

There are two principal reasons for using multiple valves:
1. Lower valve weight.
2. More valve seat and better cooling.

Lower weights reduce the forces to which the cams and valve springs are subjected.

The ratio between lengths of valve seat for dual- and single-valve constructions for the same lift-diameter ratio is as follows:

$$\text{Ratio} = \frac{2 \times \text{perimeter of dual valves}}{\text{perimeter of single valve}} = \frac{2 \times 0.707\pi d_1}{\pi d_1}$$
$$= 1.414.$$

With the same width of valve seat this means 41.4 per cent more surface through which heat from the valves can be dissipated.

Valve-seat Inserts.[1]—Valves are ordinarily seated in the cylinder block or head which is usually a gray-iron mixture. This material forms a very desirable seat for car engines or comparatively light- or medium-duty engines. However, under heavy duty the cast-iron seat for the exhaust valve "picks up" and erodes, and the valve gradually sinks into the block. This erosion of the valve seat decreases the tappet clearance; and, if not offset by frequent adjustment of the tappets, the valve will be held off the seat, and burning and warping will result. Aluminum cylinder heads used on aircraft engines must be provided with valve inserts, since the aluminum is too soft to stand this service.

Valve-seat wear, which includes pounding, "pickup," and erosion, may be reduced to a minimum with valve-seat inserts.

[1] The reader is referred to the following articles from which considerable material has been obtained:

H. L. Horning, Successful Valve-design Considerations, *S.A.E. Jour.*, **28**, 224 (1933).

J. Geschelin, No More Valve Grinding, *Auto. Ind.*, **68**, 188 (1933); Valve-seat Insert Practice Is Nearing Standardization Stage, *Auto. Ind.*, **70**, 644 (1934).

R. Jardine and R. S. Jardine, Designing Valves and Related Parts for Maximum Service, *S.A.E. Trans.*, **30**, 268 (1935).

A. T. Colwell, Wear Reduction of Valves and Valve Gear, *S.A.E. Trans.*, **43**, 366, (1938).

VALVES AND VALVE MECHANISMS

Hard materials are used most extensively because of low cost. The soft materials cost more and require wider valve seats, but they can be reamed whereas the hard materials must be ground. Various materials are used which range from stellite to cast iron and include high-speed steels, monel metal, and various alloys and bronzes. Stellite is used for the face of an insert made of steel.

The valve-seat insert must fit tightly against the cylinder or head material in which it is fastened with a press and roll fit, a shrink fit, or a screwed and shrink fit. The insert should have the same coefficient of expansion as the adjacent material which should be enlarged in section because of stresses set up by the seat inserts.

Valve seats should be narrow to obtain high unit pressures and tight seating, but wide to obtain effective cooling of the valve head. Wide seats trap more hard carbon particles between the valve and seat, which tends toward holding the valve open and results in burned and warped valves. Thus, there is an optimum valve-seat width, depending upon materials and conditions, that will result in the most satisfactory service. Soft materials for valve seats and high-output engines necessitate the use of wide valve seats for exhaust valves (Table 20).

TABLE 20.—EXHAUST-VALVE SEATS

Port diameters, in.	Seat width, in.	Port diameters, in.	Seat width, in.
$\frac{7}{8}$ to $1\frac{1}{8}$	$\frac{1}{32}$ to $\frac{5}{64}$	$3\frac{3}{8}$ to $3\frac{5}{8}$	$\frac{3}{32}$ to $\frac{1}{4}$
$1\frac{1}{4}$ to $1\frac{7}{8}$	$\frac{1}{32}$ to $\frac{3}{32}$	$3\frac{3}{4}$ to 4	$\frac{1}{8}$ to $\frac{9}{32}$
2 to $2\frac{1}{4}$	$\frac{1}{16}$ to $\frac{1}{8}$	$4\frac{1}{8}$ to $4\frac{3}{8}$	$\frac{1}{8}$ to $\frac{5}{16}$
$2\frac{3}{8}$ to $2\frac{7}{8}$	$\frac{1}{16}$ to $\frac{5}{32}$	$4\frac{1}{2}$ to $4\frac{3}{4}$	$\frac{5}{32}$ to $1\frac{1}{32}$
3 to $3\frac{1}{4}$	$\frac{1}{16}$ to $\frac{7}{32}$	$4\frac{7}{8}$ to 5	$\frac{5}{32}$ to $\frac{3}{8}$

Minimum values are for hard materials or low-output engines.

Poppet-valve Design and Materials.[1]—The elementary poppet valve consists of a head and stem (Fig. 198). The head is subjected to a uniform gas-pressure loading, which depends upon the maximum combustion pressure, and to a concentrated load at the center due to the valve-spring force when the valve

[1] See footnote, p. 308.

is closed. An additional load is impressed on the valve head depending upon the seating velocity, which should be low.

The stress due to uniform loading is

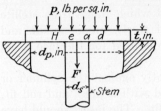

FIG. 198.—Elementary flat-seat poppet valve.

$$s = \frac{p}{4}\left(\frac{d_p}{t}\right)^2 \quad (9)$$

where s = the stress, lb. in.$^{-2}$,
p = max. combustion pr., lb. in.$^{-2}$,
d_p = port diameter, in.,
t = head thickness, in.

Example.—Determine the stress in a flat valve head having a diameter of 2 in. and a thickness of ¼ in. The maximum pressure in the combustion chamber is 500 lb. in.$^{-2}$

Substituting in Eq. (9) results in

$$s = \frac{500}{4}\left(\frac{2}{0.25}\right)^2 = 8000 \text{ lb. in.}^{-2}$$

The stress due to the valve-spring load is approximately

$$s = 1.4\left(1 - \frac{2}{3}\frac{d_s}{d_p}\right)\frac{F}{t^2} \quad (10)$$

where d_s = stem diameter, in.,
F = spring force, lb.

Example.—Determine the stress in the preceding example with a stem diameter of ½ in. The spring force is 150 lb.

Substituting in Eq. (10) results in

$$s = 1.4\left(1 - \frac{1}{6}\right)\frac{150}{0.25^2} = 2800 \text{ lb. in.}^{-2}$$

The maximum total combined stress is 10,800 lb. in.$^{-2}$ which is fairly high for some materials and conditions of operation. However, a very large fillet between the stem and head and an increase in head thickness at the center reduce the stress appreciably.

Heat flows into both valves from the hot gases in the combustion chamber and particularly into the exhaust valve during the exhaust process. During this process the hot gases moving at fairly high velocity contact the valve seat and part of the stem. Consequently, it is axiomatic that the valve shape be designed to absorb the least heat and to dissipate the most (Fig. 199). All edges should be well rounded and should not extend over the edge of the valve seat. The valve should be symmetrical to prevent unequal expansion and warping, which indicates the elimination of screw-driver slots or spanner holes in the valve head.

Inlet valves run much cooler than exhaust valves and are made of such materials as carbon steel, nickel steel, chrome nickel and chrome molybdenum alloys, which may be hardened and will withstand repeated high stresses. High-output engines require materials for inlet valves which will retain their strength under moderately high-temperature conditions and resist rusting and corrosion. These materials lie between those already mentioned for average service and those required for exhaust valves in high-output engines.

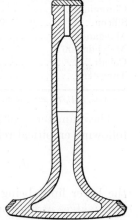

Fig. 199.—Well-designed aircraft-engine exhaust valve.

Exhaust valves must be made of materials that will maintain their strength at high temperatures and offer resistance to air hardening, deformation, oxidation, wear, pitting, and etching. The austenitic types of nickel chromium steels provide the greatest strength and most resistance to etching. However, the standard chromium steels are usually sufficiently strong, and the chromium, the silicon, and the nickel contents provide excellent resistance to hot oxidation. Metals having critical and hardening temperatures of about 1800°F., which is well above the maximum operating temperature, are resistant to deformation and air hardening. Pitting and wear normally occur more readily with the softer austenitic steels, but stellited seat surfaces provide better resistance against these effects than the hardenable chromium steels.

The valve stem must have sufficient section so that the stresses set up by opening and closing the valve at high speed will not produce excessive stress at operating temperatures. Also, it must present sufficient surface to the valve-stem guide so that considerable heat can be transferred from the valve stem.

Hollow valve stems are used to reduce the weight of the valve (Fig. 199), the variation in stem section providing a well-designed path for heat flow down the valve stem. The hollow stem is partly filled with a salt which melts at valve-stem temperatures and carries the heat from the head toward the end of the valve stem.

TABLE 21.—COMPOSITIONS OF FIVE VALVE STEELS*

Carbon	0.30 to 0.45	<0.50	1.00	0.2 to 0.4	1.2 to 1.5
Nickel	....	25 to 31	....	7 to 9	0.65
Chromium	8.5 to 12	11 to 15	20	11.5 to 14	11.5 to 14
Silicon	>0.03	1 to 2	0.5	0.65	0.65
Manganese	....	1 to 2	0.36	<0.40	<0.40
Molybdenum	0.7 to 1.3	....	1	0.45 to 0.95	0.45 to 0.95
Cobalt	....	....	2–3	2.5 to 3.5	2.5 to 3.5
Tungsten	....	....	2.5		

* *Auto. Ind.*, **79**, 224 (1938).

The diameter d_s of the valve stem may be determined from the following empirical relation:

$$d_s = \frac{d_p}{8} + \frac{1}{4} \text{ in. to } \frac{d_p}{8} + \frac{7}{16} \text{ in.,}$$

the lower limit being for small valves and the upper limit for large valves or high-output conditions.

Some valves and stems are filled with copper to facilitate the transfer of heat from the head to the stem, the conductivity of copper being about fifteen times that of average valve material. Such construction lowers the head temperature 100 to 200°F. but raises the valve-stem temperature and makes the cooling of the valve-stem guides a more important problem.

Valve stems are copperplated to a depth of about 0.0002 in. to reduce the initial wear. Chromium plating of the valve stem also reduces wear, and in some cases the valve stem has been nitrided with good effect.

Valve-stem Guides and Bosses.[1]—Good valve seating is dependent upon the alignment maintained by the valve-stem guide. The inherent unsymmetrical construction (Fig. 196) of the port passageway with regard to the seat in the block or head makes distortion appear to be inevitable, particularly for the exhaust valve. Sweeping or streamlined passageways with

[1] See footnote, p. 308.

bosses fairly high in the passageway and with the guides not extending beyond the boss are desirable.

The valve-stem boss should be well cooled, and the fit between the guide and the boss should be excellent. The length of the valve-stem guide does not depend directly upon the valve size but more upon the design of the engine. Short guides may be used where no side thrust is given the valve stem. Longer guides will be required where short levers or rocker arms act directly on the valve stems. The clearance between the valve stem and its guide should be about $\frac{1}{100}$ of the valve-stem diameter. This may be reduced appreciably when an excellent

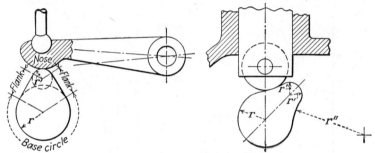

Fig. 200.—Tangential cam and pivoted follower.   Fig. 201.—Concave-flank cam and roller follower.

finish is given both the stem and guide surfaces, and results in improved performance.

Bronze guides with the soft austenitic stems are used with good success in aircraft engines, whereas hardened valve stems wear much faster. In general, hard-soft combinations usually result in less wear for the hard material and more wear for the soft material. A hard stem and hard guide, with excellent surface finish, small clearance, good alignment, and thorough lubrication, would be the most desirable condition.

Cams and Followers.—Valves are usually operated by means of cams of which there are three general types, namely, the tangential, concave-flank, and convex-flank cams.

The tangential cam is constructed of a base circle, a nose circle, and two flanks tangential to both circles (Fig. 200).

The concave-flank cam (Fig. 201) and the convex-flank cam (Fig. 202) have flanks that are arcs of circles tangent to both the base and the nose circles.

There are three principal types of cam follower, known as the "roller" (Fig. 201), "pivoted" (Fig. 200), and "mushroom" followers (Fig. 202). Clearance must be provided between the valve stem and the cam follower, or other intervening mechanism, to allow for valve-stem expansion and sinking of the valve seat. Adjustment is provided for reseating or regrinding of the valve.

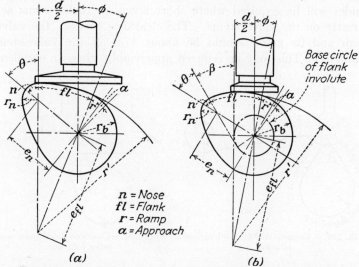

n = Nose
fl = Flank
r = Ramp
α = Approach

Fig. 202.—Convex-flank cams with mushroom followers. (*Bouvy.*)[1]

An adjustable tappet (Fig. 203) is used to obtain the desired clearance, any variation from which changes the valve timing.

It is customary to have the tappet strike the valve stem and the valve strike the valve seat during engine operation with a low constant velocity (about 0.0002 in. deg.$^{-1}$ of crank angle as a maximum) which reduces the noise and hammering effect to a minimum. Thus, cams are designed with ramps to provide this characteristically slow valve opening and closing for a crank angle of from 50 to 100 deg. in both cases.

The ramp for opening the valve is not required with hydraulic cam followers (Fig. 204) which automatically maintain zero clearance and eliminate opening noise. The tappet is held against the valve stem by a comparatively light plunger spring.

[1] C. H. Bouvy, Triple-curve Cam Gives Maximum Lift Where Space Puts Limit on Tappet Head Diameter, *Auto. Ind.*, **70**, 740 (1934).

Motion of the follower is transmitted to the plunger by oil in the oil chamber. During follower motion the oil is under considerable pressure; a small amount leaks past the plunger but is replenished by oil which flows from the oil duct into the follower and through the ball check valve into the oil chamber.

Valve-lift Diagrams.[1]—The lift of a valve, the velocity, and the acceleration at any instant depend upon the design of the cam and cam follower and any intervening valve mechanism.

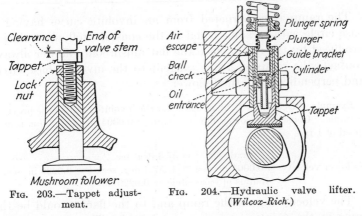

Fig. 203.—Tappet adjustment.

Fig. 204.—Hydraulic valve lifter. (*Wilcox-Rich.*)

The motion can be divided into two parts, the opening and the closing, each of which is usually symmetrical but opposite to the other. Each part can also be further divided into the motion due to the ramp, flank, and half of the nose of the cam.

If a constant acceleration for the first part of the ramp is assumed, the time required for the attainment of the constant velocity desired for the major portion of the ramp is given by the relation

$$\text{Vel.} = \text{acceleration} \times \text{time} \qquad (11)$$

and the lift obtained during this time is given by the relation

$$\text{Lift} = \frac{\text{acceleration} \times \overline{\text{time}}^2}{2} \qquad (12)$$

Example.—It is desired to have a cam ramp that will give the follower a constant velocity 0.0004 in. deg.$^{-1}$ of cam angle for a cam speed of 1 rad. sec.$^{-1}$ The first part of the ramp is to provide a constant acceleration of

[1] Based on the work of Bouvy, *op. cit.*

0.0001 in. deg.$^{-2}$ Determine the time required to attain the desired velocity and the lift per degree during this time.

From Eq. (11),   Time $= \dfrac{0.0004 \text{ in. deg.}^{-1}}{0.0001 \text{ in. deg.}^{-2}} = 4$ deg.

From Eq. (12),   Lift $= 0.00005$ in. deg.$^{-2} \times$ deg.^{2}

Time, deg	1	2	3	4
Lift, in	0.00005	0.0002	0.00045	0.0008

The ramp is constructed from an involute curve having a base circle with a radius equal to the constant follower velocity produced by the ramp, since the flat mushroom follower always contacts the ramp on a line tangent to the involute base circle and perpendicular to the follower.[1]

Example.—Determine the base circle for the involute ramp in the previous example. Constant velocity due to ramp $= 0.0004$ in. deg.$^{-1}$ for a cam speed of 1 rad. sec.$^{-1}$

1 rad. sec.$^{-1} = 57.3$ deg. sec.$^{-1}$

Follower velocity $= 0.0004$ in. deg.$^{-1} \times 57.3$ deg. sec.$^{-1} = 0.0229$ in. sec.$^{-1}$

Radius of base circle $= 0.0229$ in.

The velocity due to the ramp and to the flank should be the same at the point where they are joined. The motion of the follower while in contact with the flank arc during the angle ϕ (Fig. 202) depends on the eccentricity of the flank arc compared with the base circle of the flank arc.

Thus, $$\text{Lift} = e_{fl}(1 - \cos \phi). \tag{13}$$

$$\text{Vel.} = \frac{d(\text{lift})}{d(\text{time})} = e_{fl} \sin \phi \frac{d\phi}{dt} \tag{14}$$

or $$v_{fl} = e_{fl}\omega \sin \phi. \tag{15}$$

Also, $$\text{Acceleration} = a_{fl} = \frac{dv_{fl}}{dt} = e_{fl}\omega^2 \cos \phi. \tag{16}$$

The maximum value of ϕ is determined by d and e_{fl} (Fig. 202).

Thus, $$\sin \phi_{max} = \frac{d/2}{e_{fl}}. \tag{17}$$

---

[1] An angular cam velocity of 1 rad. sec.$^{-1}$ results in a lift in 1 sec. equal to the radius of the involute base circle from which the ramp profile is developed.

Example.—Determine the angle between the beginning of the flank at its base circle and the point at which the follower velocity would be equal to that given it by the ramp in the previous example. Also, determine ϕ_{max}, maximum v_{fl}, maximum a_{fl}, and total lift due to the flank. $e_{fl} = 3.5$ in.

From Eq. (15), the angle during which the latter part of the ramp overlaps the first part of the flank is

$$\phi_{overlap} = \sin^{-1}\frac{v_{fl}}{e_{fl}\omega} = \sin^{-1}\frac{0.0229}{3.5 \times 1} = 22'.$$

Neglecting this angle during which the ramp is moving the follower, and using a value of 0.70 for $d/2$, results in

$$\phi_{max} = \sin^{-1}\frac{d/2}{e_{fl}} = \sin^{-1}\frac{0.70}{3.5} = 11°32' \qquad \text{[Eq. (17)]}$$

$$\phi_{max} - \phi_{overlap} = 11°32' - 22' = 11°10'.$$

From Eq. (15),

$$\text{Max. } v_{fl} = e_{fl}\omega \sin \phi_{max} = 3.5 \times 1 \times \frac{0.70}{3.5} = 0.70 \text{ in. sec.}^{-1}$$

From Eq. (13),

$$\text{Max. lift}_{fl} = e_{fl}(1 - \cos \phi) = 3.5(1 - 0.9798) = 0.0707 \text{ in.}$$

From Eq. (16),

$$a_{fl} \text{ at } \phi_{max} = e_{fl}\omega^2 \cos \phi_{max} = 3.5 \times 1^2 \times 0.9798 = 3.43 \text{ in. sec.}^{-2}$$

Max. a_{fl} occurs with $\phi = 0$ and amounts to e_{fl} or 3.5 in. sec.$^{-2}$ for $\omega = 1$ rad. sec.$^{-1}$

The motion of the follower while in contact with the nose depends upon the nose eccentricity e_n and on the angle θ, during which half of the nose arc is in contact with the follower. Referring to Fig. 202a, and noting that angle θ is measured from the tip, results in

$$-\text{Lift} = e_n(1 - \cos \theta), \qquad (18)$$

$$\text{Vel.} = \frac{d(-\text{lift})}{d(\text{time})} = e_n \sin \theta \frac{d\theta}{dt}, \qquad (19)$$

or

$$v_n = e_n\omega \sin \theta. \qquad (20)$$

$$\text{Acceleration} = a_n = \frac{dv_n}{dt} = e_n\omega^2 \cos \theta. \qquad (21)$$

Also, from Fig. 202a,

$$\sin \theta_{max} = \frac{d/2}{e_n}. \qquad (22)$$

318   INTERNAL-COMBUSTION ENGINES

Example.—Determine the maximum cam-nose angle, θ_{max}, lift, acceleration, and velocity characteristics for $e_n = 1$ in., $d/2 = 0.7$ in., and $\omega = 1$ rad. sec.$^{-1}$

From Eq. (22), $\theta_{max} = \sin^{-1}\dfrac{d/2}{e_n} = \sin^{-1}\dfrac{0.7}{1} = 44°26'.$

From Eq. (18),

$$-\text{Lift} = e_n(1 - \cos\theta) = 1(1 - 0.7141) = 0.2859 \text{ in.}$$

At $\theta = 0$, the tip of the cam is contacting the follower and the velocity is zero.

From Eq. (21), $a_n = e_n\omega^2 \cos\theta = 1 \times 1^2 \times \cos\theta,$

which indicates values of a_n of 1 and 0.7141 in. sec.$^{-2}$ at 0° and 44° 26', respectively. These values are negative compared with those of the flank since the angle is measured from the tip of the cam.

Values obtained from the foregoing examples (curves I, Fig. 205)[1] show the typical characteristics for this type of cam and follower. Curves III show the effect of reducing the size of the cam follower, and curves IV show the additional effect of reducing the flank eccentricity.

The total lift of the follower (Fig. 202) is

$$\text{Total lift} = e_n + r_n - r_b. \tag{23}$$

The assumption of the nose-arc radius r_n fixes the base-circle radius r_b for a given lift and nose-arc eccentricity.

Example.—Determine the base-circle radius for the cam in the preceding examples, for a value of $r_n = 0.1$ in. Total ramp and approach angle = 24°.
From the example, page 315.

Approach + ramp lift = $0.00080 + 20 \times 0.0004 = 0.00880$ in.

From the example, page 317,   Total flank lift = 0.07070 in.

However, zero flank angle begins 22' before the end of the ramp. During this angle either the ramp or the flank would lift the follower the same

---

[1] Since the velocity curve is determined by differentiating the lift curve with respect to time,

$$\text{Lift} = \int \text{vel.} \times dt,$$

or the area under the velocity curve up to any given crank angle indicates the lift of the valve. A similar relationship exists between the area under the acceleration curve and the velocity at any angle. The velocity and acceleration curves have the wrong sign as plotted in the right half of the diagram.

amount. This will be

$$2\tfrac{2}{60} \times 0.0004 \text{ or } 0.00015 \text{ in.}$$

From the example, page 318, Total nose lift = 0.2859 in. Substituting in Eq. (23) results in

$$0.0088 + 0.0707 - 0.0002 + 0.2859 = 1.0 + 0.1 - r_b,$$
$$r_b = 0.7348 \text{ in.}$$

The angle during which the valve is open for a given cam and follower depends upon the clearance between the tappet and

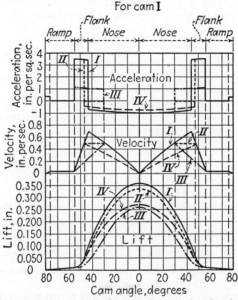

FIG. 205.—Lift, velocity, and acceleration curves for flat followers. Cam velocity = 1 radian sec.$^{-1}$ Acceleration and velocity are negative during the closing period.) (*Curves II, III, and IV, from the work of C. H. Bouvy.*)[1]

Cam	e_n	e_{fl}	$d/2$
I	1	3.5	0.70
II	1	3.5	0.50
III	1	1.14	0.50
IV	0.675	3.5	0.50

valve stem. The larger the clearance the later the valve opens and the earlier it closes, or the smaller the angle of opening. However, the clearance must be kept within the range that will

[1] Bouvy, *op. cit.*

cause the valve stem to be struck by the follower while in contact with the ramp. Changes in temperature and deformation of the various parts appreciably affect these events. The use of materials with coefficients of thermal expansion that reduce the variation in tappet clearances to a minimum is desirable.

Example.—Determine the angle during which the valve is open from the data in the preceding examples for tappet clearances of 0, 0.005, and 0.010 in. With zero clearance,

$$\text{Total angle} = 2[24° + 11°10' + 44°26'] = 159°12'.$$

Clearance of 0.005 in. will be taken up by the ramp in

$$4° + \frac{0.005 - 0.0008}{0.0004} = 14.5°.$$

This indicates a 29-deg. crank-angle later opening and earlier closing of the valve than with zero clearance.

A clearance of 0.010 in. is more than the ramp lift, which indicates that the flank will be moving the follower when the clearance is taken up. The flank angle at which the lift equals the clearance is,

$$(1 - \cos \phi) = \frac{0.010 - 0.0088 + 0.0002}{3.5} = 0.00037, \quad [\text{Eq. (13)}]$$

$$\phi = 1°33'.$$

There are included in the ramp angle 22' of this angle. Hence, the total required angle is $24° + 1°33' - 22' = 25°11'$. The follower velocity at the striking of the valve will be

$$v_{f1} = e_{f1}\omega \sin \phi = 3.5 \times 1 \times 0.0232 = 0.08 \text{ in. sec.}^{-1} \quad [\text{Eq. (15)}]$$

This is 3.6 times higher than the velocity due to the ramp and is undesirable.

The total valve-open cam angles are 159.2, 130.2, and 108.8 deg. for clearances of 0, 0.005, and 0.010 in., respectively.

Bouvy[1] proposed using a triple cam profile, in addition to the ramp, to provide maximum lift with limited follower size (Fig. 202b). This is accomplished by using an involute profile between the flank and nose arcs. Curve II (Fig. 205) shows a reduction of almost 30 per cent in maximum velocity with a reduction of only 11 per cent in lift with a reduction of about 30 per cent in follower size.

The total cam angle exclusive of the ramps is

$$2\alpha = 2(\phi + \beta + \theta) \quad \text{(Fig. 202b).} \tag{24}$$

[1] Bouvy, *op. cit.*

Obviously, to increase β requires a reduction in ϕ or θ or both. θ is fixed by the nose radius of curvature which should not be below 0.05 in. The flank may be designed to give a constant maximum acceleration which depends upon the loading that the mechanism can stand. This would increase the velocity and lift. A master cam can be designed from the lift curve by graphically constructing the cam to give the desired lift at each degree of cam rotation.

It is sometimes desirable to have the valve remain in wide-open or maximum-lift position for a given period of time. This is accomplished by having a cam nose (Fig. 206), made up of an arc of a circle with its center of curvature at the center of the cam base circle. This arc and the flank are joined by an arc of small radius which is tangent to both. During the period of dwell, the acceleration and velocity of follower and valve are zero.

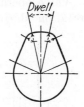

Fig. 206.—Cam with dwell period.

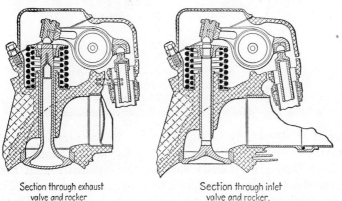

Section through exhaust valve and rocker.

Section through inlet valve and rocker.

Fig. 207.—Overhead valve mechanism. (*Pratt and Whitney.*)

For overhead valves, or any type with rocker arms (Fig. 207) the leverage ratio must be taken into account in drawing lift diagrams.

Camshafts.—Cams are forged or cast integral with the camshaft (Fig. 208) which should be designed for rigidity and supported to reduce deflections and vibration to a minimum. Accurately designed and constructed cams are of little value unless the desirable relationship with the follower is maintained.

INTERNAL-COMBUSTION ENGINES

The size of the cams limits the size of the shaft which should be provided with liberal bearing surfaces to reduce wear to a minimum. Usually two to four cams are located between bearings, depending on the length of the camshaft.

Valve operation twists the camshaft against the direction of rotation during the opening period, and vice versa. This

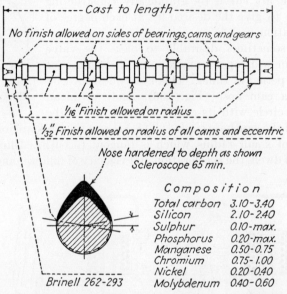

Fig. 208.—Finish and hardness details of cast camshaft.[1]

reversal of stress may be offset by the use of properly located "dummy" cams and tappets. A six-cylinder camshaft has six inlet (and six exhaust) cams uniformly distributed about the camshaft circle which tends to eliminate the reversal effect at the driving end but does not eliminate local stress reversals that tend to set up torsional vibrations.

Small single-acting engines usually have the camshaft parallel to the crankshaft. The camshaft is driven from the crankshaft, usually by gears or "silent" chain. The overhead camshaft may be driven by a shaft with two sets of bevel or other types

[1] D. J. Vail, The Development of "Proferall" Cast Camshafts, *S.A.E. Trans.*, **31**, 288 (1936).

F. J. Walls, Cast Camshafts and Crankshafts, *S.A.E. Trans.*, **32**, 284 (1937).

of gears. When the camshaft is at right angles to the crankshaft, either bevel or helical gears are used. In all cases the camshaft for four-cycle engines is driven at one-half the crankshaft speed.

The use of two sets of gears makes possible the "hunting-tooth" system which permits the gears to shift their relative position while maintaining the desired relationship between the crankshaft and camshaft. This distributes the wear evenly over all the gear teeth.

Example.—A camshaft is driven from a crankshaft through a pair of bevel gears with 24 and 25 teeth, respectively, and another pair of bevel gears with 25 and 48 teeth, respectively.

$$\text{The overall ratio} = \frac{24}{25} \times \frac{25}{48} = \frac{1}{2}.$$

A similar effect is usually obtained with silent chain and sprockets by choosing a chain with a number of links which is not a multiple of the number of teeth of either sprocket. Thus, a system with two sprockets with 10 and 20 teeth and a chain with 101 links will produce the "hunting-tooth" effect.

Valve Timing.—Theoretically, the intake valve should be opened at top center and closed at bottom center, whereas the exhaust valve should open at bottom center and close at top center. This timing would be fairly satisfactory for very slow-speed engines, but with an increase in speed it was found desirable to advance the time of opening of the exhaust valve and retard the time of closing for the intake valve, in order to reduce the work of exhaust and to induct the maximum amount of air or charge. Thus, a given valve timing results in maximum compression pressure at a definite speed (Fig. 318).

Late closing of the intake valve at low speeds permits some of the charge to escape from the cylinder during the compression stroke. This reduces the compression pressure and the tendency to detonate that is usually characteristic at low engine speed.

Overlapping of the closing of the exhaust and the opening of the intake valve makes possible the partial scavenging of the clearance space provided that exhaust conditions are favorable. Under throttled conditions, appreciable valve overlap permits exhaust products to be drawn into the cylinder and increase the dilution of the charge.

Valve timings are based upon definite tappet clearances with a hot engine. Automotive practice varies from 0.006 to 0.016 in.

TABLE 22.—VALVE-TIMING DATA*

Engines	Intake valve		Exhaust valve	
	Opens at deg. B.T.C.†	Closes at deg. A.B.C.†	Opens at deg. B.B.C.†	Closes at deg. A.T.C.†
Chevrolet............	9	29	52	1 B.T.C.
Ford................	0	44	48	6
Plymouth............	6 A.T.C.	46	42	8
Amer. auto. practice..	15 B.T.C. to 16 A.T.C.	29 to 71	40 to 57	1 B.T.C. to 25 A.T.C.
Average values.......	5	47	48	9
World auto. Diesel and other heavy oil engines.	30 B.T.C. to 10 A.T.C.	19 to 60	15 to 64	25 B.T.C. to 20 A.T.C.

* Data obtained from *Auto. Ind.*, **78**, 730 (1938); **79**, 598 (1938).
† A.T.C.—after top center. B.T.C.—before top center. B.B.C.—before bottom center.

clearance for intake valves and from 0.009 to 0.016 in. clearance for exhaust valves. Any variation in clearance will affect the valve timing.

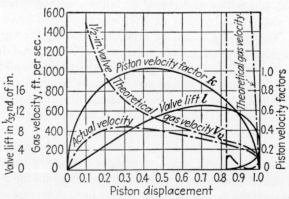

FIG. 209.—Valve lift, piston velocity, and theoretical gas velocity. Gas velocity based upon 3½- by 5-in. cylinder, 3½ to 1 connecting-rod-crank ratio, and 3000 r.p.m. Valve opens at top center and closes at 40° after bottom center.

Valve, Piston, and Flow Relations.—The piston velocity (Fig. 209) varies from zero at one crank center to a maximum somewhere near the middle of the stroke and back to zero at the other crank center. During this period the valve opening varies from zero to the maximum and back toward zero again in the case of the inlet valve. On the assumption that the volume displaced by the piston is equal to the volume passing through the valve opening,

$$A_p v_p = A_v v_v, \qquad (25)$$

$$\frac{0.7854 \overline{D}^2}{144} \times v_c k = \frac{\pi d l}{144} v_v,$$

$$v_v = 0.25 \left(\frac{\overline{D}^2}{d}\right) \frac{v_c}{l} k, \qquad (26)$$

where d and D = diameters of the valve and piston, resp.,

v_p, v_c, and v_v = vel., ft. sec.$^{-1}$ of the piston, of the crank, and of the gas through the valve, resp.,

k = vel. factor relating piston vel. to crank vel.

Applying the foregoing relationships to a given set of engine conditions results in high gas velocities through the intake valve during the early part of the induction stroke (Fig. 209). This theoretical velocity decreases to zero at outer dead center, reverses in direction and approaches infinity as the valve closes. The curve of actual velocity is at zero velocity at the beginning of the stroke and drops to zero at valve closure.

The actual instantaneous rate of flow may be determined from Eq. (6), page 306.

Thus, $$M = \frac{\pi d_p l c}{7.86 m} \sqrt{h_w d} \qquad (27)$$

where M = mols sec.$^{-1}$,

m = mol. wt. of the fluid flowing,

d_p and l = port dia. and lift, resp., in.,

h_w = head, in. of water,

d = density, lb. ft.$^{-3}$

Plotting M against crank angle or time results in an area between any two crank angles ($\theta_2 - \theta_1$) that represents the amount of fluid which has flowed into the cylinder during that increment of time. Adding the increase in quantity to that already in the cylinder results in the total quantity in the cylinder at the second angle. The characteristic equation is used to determine the temperature of the fluid in the cylinder, and this is checked by applying the energy equation to the process between the two crank angles. The energy equation for the process is

$$(M_cE_c + M_{m_1}E_{m_1})_1 + (M_{m_2} - M_{m_1})(E_m + PV_m)_a =$$
$$(M_cE_c + M_{m_2}E_{m_2})_2 + W_{out} \quad (28)$$

where m and c = mixture and clearance gases, resp.,
      1 and 2 = piston positions,
           a = the condition of the mixture in the port passageway.

Example.—It has been found by calculations not shown that the condition in the cylinder, at $\theta = 30°$, for 3.43×10^{-6} mol of clearance gas and 0.46×10^{-6} mol of correct mixture of octane and air are 1650°R. and 11.8 lb. in.$^{-2}$ abs. $M_{30°} = 2.36 \times 10^{-3}$ mol sec.$^{-1}$. Manifold conditions are 13.2 lb. in.$^{-2}$ abs. and 580°R. Determine the condition in the cylinder at $\theta = 60°$ for an engine speed of 3000 r.p.m.

At $\theta = 60°$, $l = 0.248$ in., $d_p = 1.375$, and $c = 0.68$ from Fig. 197. Also, $V_{30} = 10.46$ in.$^{-3}$ and $V_{60} = 17.83$ in.^{3}

Assume cyl. pr. at $\theta = 60°$,	10.2	11.2	12.2
M at 30°,	2.36×10^{-3}	2.36×10^{-3}	2.36×10^{-3}
M at 60°, Eq. (27),	6.65×10^{-3}	5.36×10^{-3}	3.98×10^{-3}
M aver. for period,	4.51×10^{-3}	3.86×10^{-3}	3.17×10^{-3}
M_m added during period,	7.52×10^{-6}	6.43×10^{-6}	5.28×10^{-6}
M_{m_1} at 30°,	0.46×10^{-6}	0.46×10^{-6}	0.46×10^{-6}
M_c,	3.43×10^{-6}	3.43×10^{-6}	3.43×10^{-6}
Total mols at 60°, M_2,	11.41×10^{-6}	10.32×10^{-6}	9.17×10^{-6}
$T_{60} = PV/MR$, °R.,	860	1043	1282

Then, the terms in the energy equation are evaluated as follows:

M_cE_c at 1650°R. (Table III, Appendix),	23.06×10^{-3}	23.06×10^{-3}	23.06×10^{-3}
$M_{m_1}E_{m_1}$ at 1650°R.,	3.32×10^{-3}	3.32×10^{-3}	3.32×10^{-3}
$(M_{m_2} - M_{m_1})E_m$ at 580°R.,	2.50×10^{-3}	2.14×10^{-3}	1.75×10^{-3}
$(M_{m_2} - M_{m_1})PV_m$ at 580°R.,	8.66×10^{-3}	7.40×10^{-3}	6.08×10^{-3}
Left side of Eq. (28),	37.54×10^{-3}	35.92×10^{-3}	34.21×10^{-3}
M_cE_c at T_{60},	6.44×10^{-3}	10.07×10^{-3}	15.00×10^{-3}
$M_{m_2}E_{m_2}$ at T_{60},	15.61×10^{-3}	21.23×10^{-3}	26.58×10^{-3}
$W_{out} = P_{ave} \Delta V$,	8.70×10^{-3}	9.10×10^{-3}	9.48×10^{-3}
Right side of Eq. (28),	30.75×10^{-3}	40.40×10^{-3}	51.06×10^{-3}

Plotting the values for both sides of the energy equation against the assumed pressures results in $P_{60} = 10.8$ lb. in.$^{-2}$ abs., $T_{60} = 965°$R., $M_{60} = 5.85 \times 10^{-6}$ mol sec.$^{-1}$, and $(M_m)_{60} = 7.32 \times 10^{-6}$ mol.

The foregoing method applied to the entire induction period provides information regarding the suction pressure (Fig. 210) and charge inducted. Variations in valve-opening and -closing characteristics may be studied to determine the most desirable valve characteristics.

The induction strokes of the various cylinders in a multi-cylinder engine have an effect on the pressure difference between the port and the cylinder (Fig. 211).

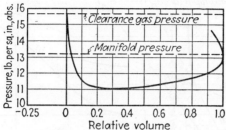

FIG. 210.—Computed cylinder pressure during induction at 3000 r.p.m. (cl. gas expanded in cyl. to man. pr.)

Valve-spring Forces and Deflections.—Valve springs are required for maintaining contact between the cams and the valves and intermediate mechanisms during periods of negative acceleration (Fig. 205). The required spring force depends upon the mass of the moving parts and the acceleration desired and may be determined graphically by laying off lengths of lines equivalent to product of mass and acceleration ma at various valve-lift positions (Fig. 212). Joining the ends of these lines results in a curve AB indicating the variation of spring force with lift.

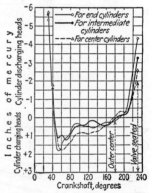

FIG. 211.—Pressure differences between port and cylinder. (*Tice.*)[1]

Valve springs are usually made of round carbon-steel wire, wound into cylindrical, helical shapes, and have straight-line relations between force and deflection. The spring characteristic $O'P'$ (Fig. 212) would have no initial force, also no initial deflection, but would have a very high maximum force. It indicates a short stiff spring. Spring characteristic $O''P''$ indicates the lowest maximum force but an enormous initial deflection.

A spring with characteristics indicated by OP would be more desirable than either of the two limiting cases. Actually, the

[1] Tice, *op. cit.*

328 INTERNAL-COMBUSTION ENGINES

spring should exert enough force to overcome the frictional resistance of the parts, as well as to provide the desired accelera-

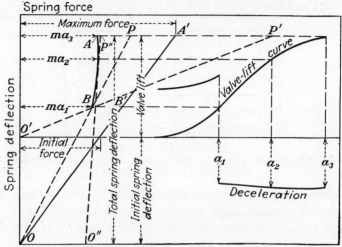

FIG. 212.—Determination of static valve-spring forces. ($O''P''$ has a different spring deflection scale with a zero valve much below the diagram.)

tion. Thus, OA' would be satisfactory, indicating an excess force of BB' where spring force is first required and an excess of AA' at maximum lift. The maximum force divided by the total spring deflection will give the scale of the spring.

Valve-spring Design.—There are various combinations of coil-spring mean diameter ($2r$) and wire dimension (b or d) (Fig. 213) which will satisfy a given spring requirement. However, the shearing stress S increases as the ratio of the mean coil diameter to the wire size decreases. Wahl[1] introduced the stress multiplication factor K (Fig. 214) into the customary coil-spring formulae which follow:

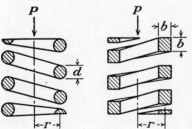

FIG. 213.—Helical springs with round and square wire.

$$\text{Round Wire} \qquad \text{Square Wire}$$
$$f = \frac{64Pr^3n}{Gd^4}, \qquad f = \frac{44.5Pr^3n}{Gb^4}, \qquad (29)$$

[1] A. M. Wahl, Helical Compression and Tension Springs, *A.S.M.E. Trans.*, **57**, A-35 (1935).

$$P = \frac{\pi d^3 S}{16 r K}, \qquad P = \frac{S b^3}{4.8 r K}, \tag{30}$$

where f = spring deflection, in.,
P = load on the spring, lb.,
G = torsion modulus, lb. in.$^{-2}$,
 = 11,400,000 for steel and 6,000,000 for phosphor bronze,
n = number of active turns,
r, b, and d = coil and spring-wire dimensions, in.

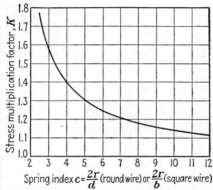

Fig. 214.—Stress multiplication factor K. Helical tension or compression springs. (A. M. Wahl.)[1]

Example.—Design a valve spring for a valve having a lift of 0.4 in. The desirable spring forces with the valve closed and open are 80 lb. and 120 lb., respectively.

The spring scale is

$$\frac{120 - 80}{0.4} = 100 \text{ lb. in.}^{-1} \text{ deflection.}$$

Total deflection = $^{120}\!/_{100}$ = 1.20 in.

On the assumption that S has a value of 60,000 lb. in.$^{-2}$, C_1 (Fig. 215) is 20. At this value, various combinations of wire and coil diameter are determined. Thus:

	1	1½	2	3
Coil dia. D, in.,				
Wire dia. d, in.,	0.175	0.200	0.218	0.255
C_2 (Fig. 216),	0.008	0.019	0.032	0.072
Active coils, n,	25.0	10.5	6.2	2.8
$(D - d)/d = 2r/d$,	4.7	6.5	8.2	10.8

[1] Wahl, *op. cit.*,

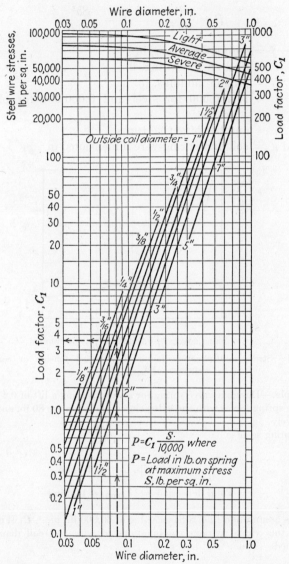

Fig. 215.—Chart for spring load factor. Round wire. (*Wahl.*)[1]

[1] Wahl, *op. cit.*

Good practice indicates that the ratio $(D - d)/d$ should lie in the range of 6 to 8, which indicates a spring with an outside coil diameter of 1½ to 2 in.

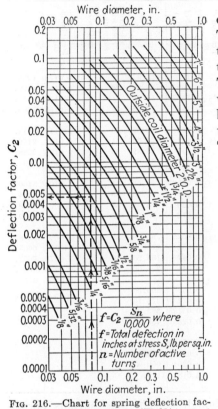

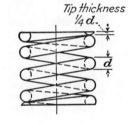

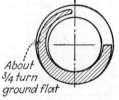

It is customary to add one coil to each end of the spring. These two coils are assumed to be inactive and include the ground ends (Fig. 217). The free length is computed with an assumed distance between coils of $\frac{1}{16}$ in. when the full load is on the spring. The ratio $L_c/(2r)$ should be about 2:1, where L_c is the length compressed.

FIG. 216.—Chart for spring deflection factor. Round wire. (*Wahl.*)[1]

FIG. 217.—Compression spring. Ends set up and ground.

Example.—The desirable spring in the previous example would have a wire size of 0.207 in. diameter (No. 5, Washburn and Moen, Table 23). Fixing the active coils at 10 makes $r = 0.694$ in., from Eq. (29). Also, $2r/d = 6.7$.

$$\text{Free length} = 12(0.207 + 0.063) + 1.2 = 4.4 \text{ in.}$$

$$\frac{L_c}{2r} = \frac{4.4 - 1.2}{1.39} = 2.3.$$

$$\text{Wire length} = 2\pi r n = 2 \times 3.1416 \times 0.694 \times 12 = 52.3 \text{ in}$$

$$\text{Wt.} = \frac{52.3}{12 \times 8.375} = 0.498 \text{ lb. (Table 23).}$$

[1] Wahl, *op. cit.*

The actual spring would weigh less because of the ground ends.

TABLE 23.—SIZE AND WEIGHT OF STEEL SPRING WIRE

Washburn & Moen wire gage no.	Fraction of in.	Decimal of in.	Ft. lb.$^{-1}$	Washburn & Moen wire gage no.	Fraction of in.	Decimal of in.	Ft. lb.$^{-1}$	Washburn & Moen wire gage no.	Fraction of in.	Decimal of in.	Ft. lb.$^{-1}$
	1/2	0.5000	1.500	7		0.177	11.97	27		0.0173	1253.
7-0		0.490	1.562	8		0.162	14.29	28		0.0162	1429.
	15/32	0.46875	1.706		5/32	0.15625	15.36		1/64	0.0156	1547.
6-0		0.462	1.760	9		0.148	17.05	29		0.0150	1666.
	7/16	0.4375	1.959	10		0.135	20.57	30		0.0140	1913.
5-0		0.431	2.023		1/8	0.125	24.00	31		0.0132	2152.
	13/32	0.40625	2.272	11		0.120	25.82	32		0.0128	2288.
4-0		0.394	2.418	12		0.105	33.69	33		0.0118	2693.
	3/8	0.3750	2.666		3/32	0.09375	42.66	34		0.0104	3466.
3-0		0.3629	2.853	13		0.092	44.78	35		0.0095	4154.
	11/32	0.34375	3.173	14		0.080	58.58	36		0.009	4629.
2-0		0.331	3.422	15		0.072	72.32	37		0.0085	5189.
	5/16	0.3125	3.839	16	1/16	0.0625	95.98	38		0.008	5858.
0		0.307	3.991	17		0.054	128.60	39		0.0075	6665.
1		0.283	4.681	18		0.047	166.20	40		0.007	7652.
	9/32	0.28125	4.740	19		0.041	223.00	41		0.0066	8607.
								42		0.0062	9753.
2		0.263	5.441	20		0.035	309.60				
	1/4	0.250	5.999	21		0.032	373.10				
3		0.244	6.313		1/32	0.03125	383.90				
4		0.225	7.386	22		0.0286	458.40				
	7/32	0.21875	7.835	23		0.0258	563.3				
5		0.207	8.750	24		0.0230	708.7				
6		0.192	10.17	25		0.0204	900.9				
	3/16	0.1875	10.66	26		0.0181	1144.				

Valve-spring Surge.—Although considerable surplus spring force is usually provided, it is a common experience to have the cam follower leave the cam at apparently critical camshaft speeds. Under this condition the center coils of the spring may be seen to surge back and forth. The surging offsets enough of the spring force to allow the follower to leave the cam. This occurrence is associated with the natural frequency of vibration of the spring.

The natural frequency in vibrations per second for a coiled spring is given by the expression

$$\text{Frequency} = \frac{1}{2\pi}\sqrt{\frac{g}{\text{static deflection}}}. \tag{31}$$

The static deflection is the ratio of the acting weight of the spring to the force required to produce unit deflection.

$$\text{Thus,} \quad \text{Static deflection} = \frac{M \text{ lb.}}{P/f \text{ lb. in.}^{-1}} = \frac{Mf}{P} \text{ in.} \quad (32)$$

With one end of the spring fixed, the active weight is about one-half the spring weight. With both ends fixed, which is the case for vibration of the center coil, the active weight is still about one-half the spring weight. Also, when both ends are fixed, it will require four times as much force to move the center coil a given distance as when one end is fixed, for the coils on one side are being compressed while the others are being expanded. Thus, for surging of the center coil,

$$\text{Frequency} = \frac{1}{2\pi}\sqrt{\frac{8gP}{Mf}} = 8.85\sqrt{\frac{P}{Mf}} \text{ vibrations sec.}^{-1} \quad (33)$$

Substituting for M its equivalent $\pi d^2/4 \times 2\pi rn \times 0.28$, where 0.28 is the density of 1 in.³ of spring metal, and for P/f its value from Eq. (29), Eq. (33) becomes

$$\text{Frequency} = 0.94 \frac{d}{r^2 n}\sqrt{G} \text{ vibrations sec.}^{-1} \quad (34)$$

The use of this equation results in frequencies about 4 per cent higher than that given by Ricardo[1] and about 10 per cent lower than that given by Swan.[2]

Example.—Determine the natural frequency of spring surge for the spring designed in the previous example. $d = 0.207$ in., $r = 0.694$ in., and $n = 12$.

$$\text{Frequency} = 0.94 \frac{0.207}{0.694^2 \times 12}\sqrt{11{,}400{,}000} = 113.5 \text{ vibrations sec.}^{-1}$$
$$= 6810 \text{ vibrations min.}^{-1}$$

An engine running at 1362 r.p.m. will have a camshaft speed of 681 r.p.m. or one-tenth of the natural frequency of spring in the preceding example. Thus, the tenth harmonic of the valve-lift curve may set the spring to surging. The valve-lift curve

---

[1] H. R. Ricardo, "The Internal-combustion Engine," Vol. 2, p. 210, D. Van Nostrand Company, Inc., New York, 1922.

[2] A. Swan, Valve Springs, *Auto. Eng.*, **16**, 290 (1926).

may be analyzed for the amplitude of the various harmonics in the following manner:[1]

Plotting the valve-lift curve with maximum lift at zero degrees (Fig. 218) makes the maximum value of all harmonics occur at 0, 360, 720 deg. etc. The equation for the valve lift can then be written

$$l = a_0 + a_1 \cos \theta + a_2 \cos 2\theta + \cdots + a_n \cos n\theta \tag{35}$$

where a_0 = mean valve lift for 360 deg.
a_1 = maximum value of the first harmonic, etc.,
θ = cam angle measured from maximum lift,
n = highest harmonic (a whole number) to be considered.

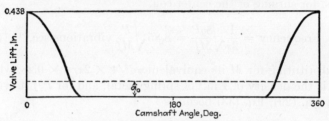

Fig. 218.—Valve-lift curve for harmonic analysis.

The value for a_0 can be found by planimetering[2] the area beneath the valve-lift curve and dividing this area by the length of the plot for 360 deg.

The value of a_n is found by determining the height of the valve-lift curve at intervals of $360/n$ deg., starting with the height at zero degrees. Thus, n heights are determined, added together, and divided by n. Subtracting a_0 from this average gives a_n and the higher harmonics that are multiples thereof, as a_{30}, a_{60}, and a_{90}. However, in general, the higher the harmonic the smaller its magnitude, and so the higher harmonics can safely be neglected. When the lower harmonics are to be determined, it is necessary to subtract the higher multiples; thus, a_{30} is subtracted to obtain a_{15}; a_{20} and a_{30} are subtracted to obtain a_{10}; etc. This is shown in the computations for five harmonics (p. 335), most of the values for lifts of zero having been omitted.

[1] F. Jehle and W. R. Spiller, Idiosyncrasies of Valve Mechanisms and Their Causes, *S.A.E. Trans.*, **24**, 197 (1929).

[2] Planimetering of the area under the valve-lift curve can be eliminated as explained in the following reference:

W. R. Spiller, Simplifying Steps in Harmonic Analysis, *Machine Design*, **4**, 26 (1932).

Sample Computation of Harmonic Amplitudes

18th harmonic		17th harmonic		16th harmonic		15th harmonic		14th harmonic	
Angle, deg.	Lift, in.	Angle, deg.	Lift, in.	Angle, deg.	Lift, in.	Angle, deg.	Lift, in.	Angle, deg.	Lift, in.
0	0.438	0	0.438	0	0.438	0	0.438	0	0.438
20	0.362	21.2	0.353	22.5	0.342	24	0.329	25.7	0.312
40	0.145	42.4	0.113	45.0	0.074	48	0.038	51.4	0.012
60	0	63.6	0	67.5	0	72	0	77.1	0
220	0	233.0	0	247.5	0	264	0	282.9	0
240	0	254.1	0	270.0	0	288	0	308.6	0.012
260	0	275.3	0	295.5	0	312	0.038	334.3	0.312
280	0	296.5	0	315.0	0.074	336	0.329	Total..	1.086
300	0	317.6	0.113	337.5	0.342	Total..	1.172	Ave.....	0.0776
320	0.145	338.8	0.353	Total..	1.270	Ave.....	0.0782	$a_0 = 0.0796$	
340	0.362	Total..	1.370	Ave.....	0.0794	$a_0 = 0.0796$		$a_{14} + a_{28} =$	
Total..	1.452	Ave....	0.0805	$a_0 = 0.0796$		$a_{15} + a_{30} =$			-0.0020
Ave.....	0.0807	$a_0 = 0.0796$					-0.0014	$a_{28} = -0.0001$	
$a_0 = 0.0796$						$a_{30} = 0.0001$			
$a_{18} = +0.0011$		$a_{17} = +0.0009$		$a_{16} = -0.0002$		$a_{15} = -0.0015$		$a_{14} = -0.0019$	

The natural frequency of the spring should be high enough so that at maximum engine speed there is no chance for resonance with any harmonic of appreciable amplitude, say 1 per cent of the valve lift or above. Thus, if the tenth harmonic is the highest one to have an amplitude of 1 per cent of the valve lift, the spring frequency should be 11 times maximum camshaft speed.

Example.—Given the following amplitudes in inches for the given harmonics:

1st............	+0.1580.	7th............	−0.01246.
2d.............	+0.1100.	8th............	−0.00610.
3d.............	+0.0650.	9th............	+0.00111.
4th............	+0.0300.	10th...........	+0.00492.
5th............	+0.00215.	11th...........	+0.00389.
6th............	−0.01154.	12th...........	+0.00085.

Determine the desirable spring frequency for an engine speed of 4000 r.p.m. Camshaft speed is 2000 r.p.m. $a_0 = 0.08$.

By adding $a_0 = 0.08$ to the sum of the first five harmonics, it is obvious that the maximum valve lift is about 0.45 in., making the tenth or eleventh harmonic of appreciable value.

Min. spring frequency $= 11 \times 2000/60 = 367$ vibrations sec.$^{-1}$

Jehle and Spiller[1] also derived a relation for the maximum amplitude of spring vibration which can be written as follows:

$$\text{Max. amplitude} \propto \left(\frac{d^3}{rn}a_n\right) \tag{36}$$

where a_n = amplitude of the lowest harmonic that can come into resonance with the spring. It is obvious that the wire should be the smallest permissible and that there should be a large number of coils of large coil diameter for minimum amplitude of spring vibration.

Sleeve Valves.—Rotary, disk, and sleeve valves have been used in internal-combustion engines. All three types of valves have very desirable port-opening and port-closing characteristics and are quiet and positive in action at all speeds. Sealing, lubrication, and scoring difficulties have prevented the rotary and disk valves from coming into general usage. However, both single- and double-sleeve valves have given satisfactory performance.

The sleeve valve is well adapted for high-output engines since the valve is well cooled and no hot spots develop to limit compression ratio or cause preignition. The double-sleeve valve, used extensively in the "Silent Knight" engines in this country and abroad, required an eccentric drive to give each sleeve a reciprocating motion, the eccentrics being out of phase with each other to provide the desired valve timing.[2] The inherently high friction loss due to the large bearing surfaces of the two sleeves, the difficulty in controlling lubrication, and the cost apparently led to its disuse.

A single-sleeve valve[3] requires a reciprocating and oscillating motion which is usually obtained with a spherical sliding bearing in an extension of the sleeve. The axis of the crank is perpendicular to the cylinder axis (Fig. 219) and runs at half crankshaft speed. The spherical bearing slides back and forth on the crank, because of the oscillation of the sleeve. The reciprocating

---

[1] Jehle and Spiller, *op. cit.*

[2] See 4th ed. for valve diagrams.

[3] W. A. Frederick, The Single Sleeve Valve Engine, *S.A.E. Trans.*, **22**, 102 (1927).

A. H. R. Fedden, The Single Sleeve as a Valve Mechanism for the Aircraft Engine, *S.A.E. Trans.*, **33**, 349 (1938).

motion imparted to the sleeve is equal to the diameter of the crank circle A, while the oscillating motion B depends on this diameter and the distance between the center of the spherical bearing and the cylinder axis. Thus, the actual motion of the sleeve is elliptical with the major axis of the ellipse parallel to the cylinder axis.

If C is the distance from the cylinder axis to the center of the spherical bearing (Fig. 219), the angle of oscillation can be determined from the following trigonometric relationships:

$$\sin \frac{\alpha}{2} = \frac{A}{2} \div C, \quad (37)$$

$$\alpha = 2 \sin^{-1} \frac{A}{2C}. \quad (38)$$

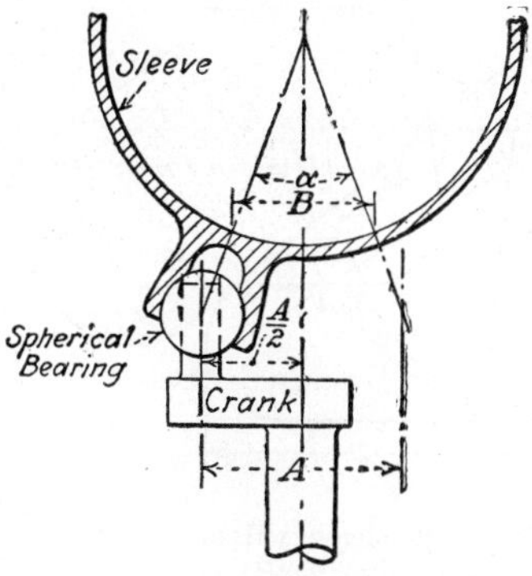

Fig. 219.—Drive for single-sleeve valve.

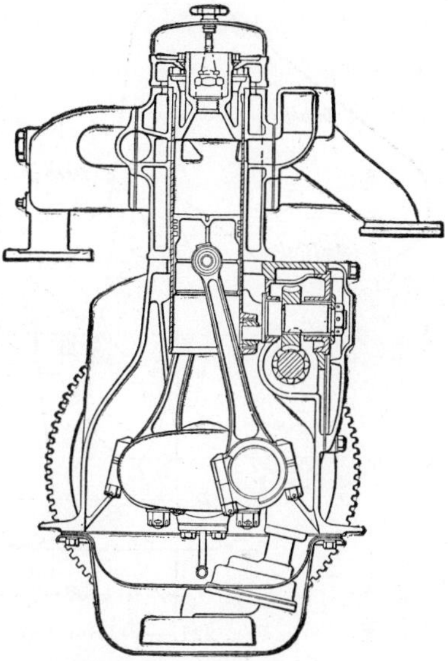

Fig. 220.—Cross-sectional view of single-sleeve-valve engine. (*Frederick*.)[1]

It is desirable to have both intake and exhaust ports located at the same level in the cylinder walls, which makes possible the choice of having the ports opened and closed during the lower or upper part of the valve motion. The use of the lower part of the motion for this purpose permits the ports in the sleeves to be located between the cylinder head and wall during compression and expansion (Fig. 220).

The valve action can be analyzed by laying out the ports in a plane surface and giving the ports in the sleeve the elliptical movement mentioned before. The right-hand side of the dia-

[1] Frederick, *op. cit.*

338   INTERNAL-COMBUSTION ENGINES

gram (Fig. 221) relates to the exhaust port and the left-hand side to the intake port. The exhaust port opens at position OA'' and closes at OA, the angle of opening being 115 deg. of travel which is equivalent to 230 deg. of crank travel. Halfway between opening and closing positions the sleeve port is shown registering with the exhaust port, the lighter dotted area indicating the port opening. The port opening for any position of the valve can be determined by planimetering the port-opening area.

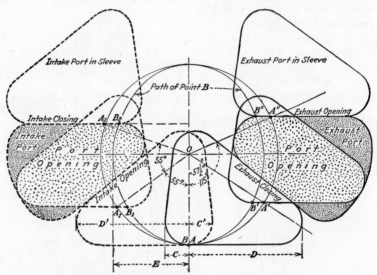

Fig. 221.—Valve diagram for single-sleeve valve.

The ports are designed to provide quick opening for the exhaust ports and quick closing for the intake ports as shown in Fig. 221.

A double port in the sleeve can be made by combining the intake and exhaust ports at the bottom of the diagram. Only one pair of ports can be operated, however, by a double port in the sleeve without complicating the exhaust- and intake-manifold systems. As arranged, the exhaust port would close as the intake port opens.

The number of ports that can be used depends upon the oscillation and the length of the sleeve port, provided that the wall port is no longer. In the diagram, $2E$ represents the oscillation and $C + D$ the length of port. If both intake and exhaust ports are of the same length, the maximum number of ports will be equal to the circumference of the sleeve divided by

$(C + D + 2E + \text{seal})$, where the seal or cover is the **margin** allowed between the sleeve port and the adjacent wall port. The seal should not be less than 0.05 in.

The sleeve is subjected to the maximum pressure and should be designed to prevent seizure due to expansion.

Example.—Determine the clearance between a steel sleeve and the cylinder to prevent seizure in an engine having a maximum pressure of 700 lb. in.$^{-2}$ abs. Sleeve thickness = $\frac{1}{8}$ in. Engine bore = 3.5 in. Modulus of elasticity = 30×10^6 lb. in.$^{-2}$

$$\text{Tension stress in material} = \frac{700 \times 3.5}{2 \times \frac{1}{8}} = 9800 \text{ lb. in.}^{-2}$$

$$\text{Strain, length in.}^{-1} = \frac{9800}{30{,}000{,}000} = 0.00033 \text{ in. in.}^{-1}$$

$$\text{Increase in diameter} = 3.5 \times 0.00033 = 0.0012 \text{ in.}$$

The clearance should be somewhat larger than this figure.

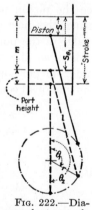

Fig. 222.—Diagram for two-cycle port.

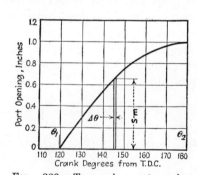

Fig. 223.—Two-cycle port-opening curve.

Ports for Two-cycle Engines.—The ports for the two-cycle engine are usually opened and closed by the movement of the piston. The amount of port opening at any crank angle can be determined by laying out the position of the top edge of the piston with regard to the port (Fig. 222). It is obvious that the first part of the port to be opened is the last to close and consequently remains open longer than any other portion of the port.

For a given pressure difference the amount of charge that will flow through a port depends upon the area of the port and the coefficient of flow through the port. Since the area of the port is continually varying, the mean port area for the port-open period is required. This can be found by plotting (Fig. 223) the port opening as the ordinate and the corresponding crank angle as the abscissa. Integrating the area beneath this curve and the crank-angle axis and dividing by the length of the diagram result in the mean port opening, from which the mean port area can be obtained.

The distance (Fig. 222) the piston has traveled from top dead center at any given angle θ is given by the expression[1]

$$s = r(1 - \cos\theta) + \frac{r^2}{4l}(1 - \cos 2\theta) \tag{39}$$

where r = crank radius,
l = connecting-rod length, center to center.

The amount of port opening at any crank angle depends upon the height of the port and the distance the piston must move from top dead center before uncovering the port. If the distance from top dead center to the top edge of the port is represented by m, then, when s is greater than m, the port opening will be $s - m$ (Fig. 222). The area under the port-opening curve from θ_1 the opening angle to θ_2, the bottom dead center, is equal to

$$\text{Area} = \int_{\theta_1}^{\theta_2} (s - m)\, d\theta$$

or, $\quad \text{Area} = k_1(\theta_2 - \theta_1) + r \sin\theta_1 + \dfrac{r^2}{8l}\sin 2\theta_1 \tag{40}$

where $k_1 = \left(r + \dfrac{r^2}{4l} - m\right) \div 57.3$ deg. rad.$^{-1}$,

θ = deg.,

and $\sin\theta_2$ and $\sin 2\theta_2$ are both equal to zero since $\theta_2 = 180$ deg.

The mean port opening is obtained by dividing the result from Eq. (40) by $(\theta_2 - \theta_1)$ deg. and multiplying by 57.3 deg. rad.$^{-1}$, since the dimensions of the equation are inch-radians. The solution is made for the opening period, since the closing period is exactly the same and results in the same mean height of port opening.

[1] See the chapter on Mechanics of Principal Moving Parts.

VALVES AND VALVE MECHANISMS

Example.—Determine the mean port opening for an intake port 1 in. high in a 3½- by 5-in. cylinder of a two-stroke-cycle engine. The connecting-rod-crank ratio is 3.5:1.

At θ_1, $\quad s = m = 5 - 1 = 4$ in.

Substituting in Eq. (39) results in

$$4 = 2.5(1 - \cos \theta_1) + \frac{\overline{2.5}^2}{4 \times 3.5 \times 2.5}(1 - \cos 2\theta_1)$$

the solution of which results in $\theta_1 = 119.5$ deg.

Then, $\quad \theta_2 - \theta_1 = 60.5$ deg.,

$$k_1 = \left(2.5 + \frac{6.25}{35} - 4\right) \div 57.3 = -0.0231,$$

$\sin 119.5$ deg. $= \cos 29.5$ deg. $= 0.8704,$
$\sin 239$ deg. $= -\sin 59$ deg. $= -0.8572.$

Substituting in Eq. (40) results in

$$\text{Area} = -0.0231 \times 60.5 + 2.5 \times 0.8704 - \frac{6.25}{70} \times 0.8572$$

$$= 0.703 \text{ in.-rad.}$$
$$= 0.703 \times 57.3 = 40.3 \text{ in. deg.}$$

$$\text{Mean port opening} = \frac{40.3 \text{ in. deg.}}{60.5 \text{ deg.}} = 0.67 \text{ in.}$$

The mean pressure required to transfer a given charge through an intake port depends upon the mean area of the port, the time it is open, and the coefficient of discharge. The mean area is the product of the mean port opening and the circumferential length of the port. The time depends upon the total angle through which the port is open and the engine speed. The coefficient of discharge will depend upon the approach to the port and will be in the range of 0.6 to 0.8.

Example.—Determine the mean pressure difference required to transfer a charge equal to the displacement of the engine in the preceding example at atmospheric pressure and 140°F. The engine speed is 3000 r.p.m.; circumferential length of port is 4 in.

The displacement of a 3½- by 5-in. cylinder is 48.1 in.[3] At a pressure of 14.7 lb. and a temperature of 140°F., this volume of charge will have a weight about the same as air.

Thus, $\quad M = \dfrac{144 \times 14.7 \times 48.1}{53.34 \times 600 \times 1728} = 1.84 \times 10^{-3}$ lb.

The mean port area is

$$0.67 \times 4 = 2.68 \text{ in.}^2 \text{ or } 1.86 \times 10^{-2} \text{ ft.}^2$$

The density of the charge will be

$$\frac{1728}{48.1} \times 1.84 \times 10^{-3} = 0.066 \text{ lb. ft.}^{-3}$$

The time of port opening is equal to

$$\frac{121 \text{ deg.}}{360 \text{ deg. rev.}^{-1}} \times \frac{1}{50 \text{ rev. sec.}^{-1}} = 0.67 \times 10^{-2} \text{ sec.}$$

The mean rate at which the charge must flow into the cylinder is

$$1.84 \times 10^{-3} \text{ lb. per } 0.67 \times 10^{-2} \text{ sec., or } 0.274 \text{ lb. sec.}^{-1}$$

Substituting the foregoing data in Eq. (4), page 208, results in

$$0.274 = 18.3 \times 1.86 \times 10^{-2} \times 0.70 \sqrt{h_w \times 0.066},$$
$$h_w = 20.0 \text{ in. of water,}$$
$$= 0.72 \text{ lb. in.}^{-2}$$

Two cycle engines with exhaust valves in the cylinder head may have intake ports extending completely around the cylinder

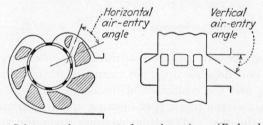

Fig. 224.—Inlet ports for a two-stroke-cycle engine. (*Earl and Duttee.*)[1]

(Fig. 224). The best entrance angle varies from 45 to 60 deg. from radial, the lower values applying to higher engine speeds and lower scavenging air pressures. The vertical entry angle should be zero. The air flow for the design shown varied directly with flow area and scavenging pressure and inversely with the engine speed.[1]

EXERCISES

1. Determine and plot the relation between the mean gas velocity through the intake port and the stroke-bore relation for a displacement of 50 in.[3]

[1] S. L. Earle and F. J. Duttee, Effect of Air-entry Angle on Performance of a Two-stroke Cycle Compression-ignition Engine, *N.A.C.A. Tech. Note* 610 (1937).

VALVES AND VALVE MECHANISMS 343

in a four-cycle engine running at 4000 r.p.m. Assume the port diameter is one-half the cylinder bore in each case.

2. Work Prob. 1 using a mean piston speed, in all cases, of 3000 ft. min.$^{-1}$

3. Determine and plot the relation between lift and valve-seat angle for the same valve opening and for various valve-seat angles between 0 and 45 deg. Assume the lift for a 0-deg. seat to be one-fifth the port diameter.

4. Determine the effect on the lift-diameter ratio for a valve to make port and valve areas equal for valve-seat angles between 0 and 45 deg.

5. Assume a valve-stem diameter of $\frac{1}{4}d_p$, and determine the effect on maximum desirable lift-diameter ratio for a 45-deg. valve seat with various valve diameters.

6. Determine the port diameters for two valves to replace one valve, both having the same lift. Do this for seat angles from 0 to 45 deg.

7. Determine the thickness of a single 2-in. valve and of the two valves to replace it, keeping the stress at 6000 lb. in.$^{-2}$ Gas pressure is 500 lb.in.$^{-2}$ Assume the valves have a 45-deg. seat.

8. Assume that the spring required for the 2-in. valve has an initial compression of 100 lb. The stem diameter is $\frac{1}{2}$ in. Determine the stress in the valve head in Prob. 7, due to spring force. Then determine the stem diameter for the same stress in the valve head for the multiple valves, assuming that the spring force for each valve is directly proportional to the valve weight. Stem length is 6 in.

9. Draw a valve-lift diagram with constant acceleration. Positive acceleration 40 per cent of the time. (Suggestion. What are the relations between acceleration, velocity, distance, and time?)

10. Design a cam and follower for a total lift of 0.35 in. Tappet clearance = 0.007 in. $e_{fl} = 3e_n = 6(d/2)$. Use an involute profile for joining the flank and nose arcs. No ramps.

11. Determine the velocity with which the cam follower, in Prob. 10, strikes the valve stem at an engine speed of 4000 r.p.m.

12. Design a ramp having a desirable velocity for quiet tappet action, having an approach giving constant acceleration, and reaching the desired ramp velocity in 5 deg. of cam travel.

13. Combine the results of the design in Probs. 10 and 12.

14. Plot the piston-velocity factor and theoretical gas velocity for a $3\frac{1}{2}$- by 5-in. engine, 4-to-1 connecting-rod to crank ratio, running at 3000 r.p.m. Valve size = 2 in. Use a valve lift similar to that in Fig. 209 with 10 deg. later opening and closing. (See the chapter on Mechanics of Principal Moving Parts for piston-velocity factor.)

15. Compute and plot the pressure in the cylinder during the induction period. There is no induction pipe but a well-rounded entrance to intake port. Clearance-gas pressure is atmospheric at top center. A 2-in. valve opens at top center and follows lift curve I, Fig. 205. Use a $3\frac{1}{2}$- by 5-in cylinder. Engine speed = 4200 r.p.m.

16. Design a valve spring for a 2.5-in. valve having a lift of 0.5 in. Weight of valve = 1 lb. The desirable spring forces with the valve closed and open = 65 lb. and 110 lb., resp.

17. Determine the natural frequency of spring surge for the spring designed in Prob. 16.

18. Redesign the foregoing spring so as to increase the surge frequency, keeping in mind the relation given by Eq. (36).

19. Determine graphically and plot the valve-lift curve for a tangential cam having a base-circle diameter of $1\frac{1}{4}$ in., a nose radius of $\frac{1}{4}$ in., and an angle, from opening to closing, of 120 deg. Use a mushroom follower.

20. Determine the maximum amplitudes of the ninth, tenth, and eleventh harmonics for the cam in Prob. 19. This requires the determination of the eighteenth, twentieth, twenty-second, twenty-seventh, and thirtieth harmonics. What percentage of total lift are the amplitudes of each of the three harmonics determined?

21. Scale the single-sleeve valve diagram (Fig. 221), and plot port opening vs. crank angle. Design the sleeve ports for maximum port opening, and plot the results.

22. Determine the mean port opening for an intake port $\frac{1}{2}$ in. high in a 2- by 3-in. two-cycle engine cylinder. The connecting-rod to crank ratio is 3:1. Circumferential length of port = 2 in. Engine speed = 6000 r.p.m.

23. Determine the mean pressure difference required to transfer a charge equal to the displacement of the engine in Prob. 22. Assume charge pressure and temperature in the cylinder to be 14.7 lb. in.$^{-2}$ and 140°F., respectively. Coefficient of discharge = 0.70.

CHAPTER XII

IGNITION OF THE CHARGE

Ignition.—Early internal-combustion engines used flames for igniting the combustible charge in the engine cylinder. These engines ran very slowly, had little or no compression, and required mechanical devices for exposing the charge to the flame at the desired time.

The charge can be ignited by contact with a red-hot piece of metal such as uncooled spots or bulbs as part of the combustion-chamber walls. These bulbs were heated with flame until red hot, which permitted starting the engine. Engine operation maintained the high temperature of the bulb, which ignited the charge. This type of ignition system can be satisfactory only for a limited range of speed and load in a fuel-injection engine.

A low-tension current flowing through a coil of wire around a soft-iron core tends to continue to flow after opening the circuit and produces a low-tension arc across the gap in the circuit. Locating the breaker points inside the combustion chamber provided another means for igniting the charge. The high current consumption and the packing gland required for the operation of the breaker points inside the cylinder are the principal disadvantages of this system which is adapted only to slow-speed low-compression engines.

The development of the high-compression engine resulted in the adoption of a system producing a high-tension spark across a short fixed gap in the combustion chamber for the ignition of the charge. Obviously, a combustible mixture must exist between the discharge gap when the spark occurs or ignition of the charge will not result.

The flow of electrons during the electrical discharge activates and ignites the mixture in the path of the discharge, and the combustion process is started. The comparatively small conflagration nucleus loses appreciable energy by heat transfer to the cooler surroundings, which exerts a retarding influence

346   INTERNAL-COMBUSTION ENGINES

on the activating of the surrounding mixture and the propagation of flame. The variation in the rate at which flame develops, after electrical discharge at the spark-plug gap, results in undesirable variations in combustion time and indicates the desirability of large spark-plug gaps for initiating a comparatively large reaction.

High-tension Ignition Systems.—The source of energy for a high-tension ignition system is an electric generator, battery, or magneto. The generator and battery supply direct current at 6 or 12 volts potential, while the magneto supplies alternating current with peak voltages somewhat higher. Regardless of

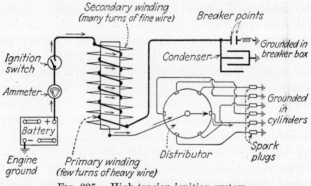

FIG. 225.—High-tension ignition system.

energy source, the high-tension ignition system has a primary circuit for the low-voltage current and a secondary circuit for the high-voltage current.

The Primary Circuit.—This circuit consists of the battery, ammeter, switch, primary coil winding, breaker points, and condenser (Fig. 225). Closing the ignition switch and the breaker points causes current to flow through the primary circuit and builds up a magnetic field (Fig. 226a) which extends through and around the soft-iron coil core. A cam driven usually at half engine speed opens the breaker points and also permits them to close.

The breaking of the primary circuit, by the opening of the breaker points, causes the magnetic field to start to collapse (Fig. 226b). The collapsing magnetic field induces current which continues to flow in the same direction in the primary circuit and charges the condenser plates. The condenser builds up a potential opposing flow and soon discharges back through the primary circuit. This causes a sudden collapse of the remaining magnetic field, and a high voltage is induced in the

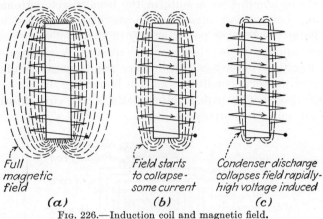

Fig. 226.—Induction coil and magnetic field.

secondary coil winding which is also cut by the collapsing magnetic field.

The condenser not only assists in the collapse of the magnetic field but prevents arcing at the breaker points by providing a

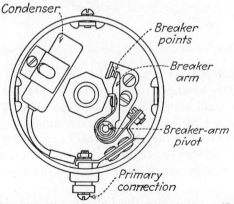

Fig. 227.—Breaker mechanism for eight-cylinder engine. (*Delco-Remy*.)

place for the current to flow. Too large or too small condenser capacity causes excessive pitting of the breaker points, the hole forming on the positive point with capacities too large. Condensers for automotive systems range from 0.15 to 0.45 mf.

capacity. Obviously, a short-circuited or open-circuited condenser will cause ignition failure.

Current should flow through the closed primary circuit at the rate of about 4 to 5 amp. in the usual 6-volt automotive system. Low-current readings indicate high resistance which may be at the breaker points or any of the connections of the various apparatus in the primary circuit. A voltage drop exceeding 0.1 volt across the breaker points usually indicates an unsatisfactory condition, and a knowledge of the usual voltage drop across various other parts of the circuit forms a basis for indicating circuit condition.

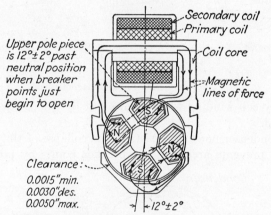

FIG. 228.—Four-pole rotating-magnet magneto. (*Scintilla*.)

The breaker points are adjusted for a maximum gap of about 0.020 in. The breaker arm supports one breaker point. The arm may be rigid and pivoted at one end (Fig. 227) or flexible and clamped in position. In either case, sufficient spring tension should be provided to make the breaker arm follow the cam and not chatter. Otherwise, the breaker points are not closed sufficiently long at high engine speed to permit the desirable build-up of current.

The magneto eliminates the battery and includes in its construction the balance of the apparatus required in both the low-tension and the high-tension circuits. The magneto may have either a rotating coil and stationary permanent magnets or rotating magnets and a stationary coil (Fig. 228). The relative movement between the primary coil winding and the electromagnets induces an alternating current in the primary circuit, the breaking of which induces a high-voltage current in the secondary as in the battery system. Obviously, the breaking of the circuit and motion of the rotor must be timed so that the desirable current is flowing in the primary circuit.

The Secondary Circuit.—This circuit consists of the secondary coil winding, the lead to the distributor rotor, the distributor, the spark-plug leads, and the spark plugs. The secondary coil winding consists of many turns of fine well-insulated wire, the insulation between adjacent wires offering appreciably greater resistance to "shorting" than the balance of the secondary circuit which includes the spark-plug gap.

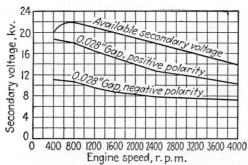

FIG. 229.—Effect of engine speed and polarity of spark-plug center electrode. (*Critchfield.*)[1]

The voltage attained in the secondary winding is higher, and the leakage from the various parts of the circuit is less, when cool than when the circuit is hot. The available voltage and that required to cause the spark to jump the spark-plug gap depend upon various factors.[1,2] The available voltage decreases with an increase in engine speed, and the voltage required is lower when the insulated center electrode of the spark plug has a negative polarity (Fig. 229).

The lead to the distributor and the leads from the distributor to the spark plugs are well insulated to prevent short-circuiting. The distributor rotor may or may not contact the connections

---

[1] R. M. Critchfield, Effect of Application on Maintenance of Automotive Electrical Equipment, *S.A.E. Trans.*, **33**, 403 (1938).

[2] L. H. Middleton, "The Physics of Ignition." The Electric Auto-lite Company, 1938.

for the spark-plug leads; if it does not, another air gap besides that of the spark plug is introduced into the circuit.

Spark Plugs.—A spark plug consists of two electrodes, one usually grounded through the shell of the plug and the other well insulated with porcelain or mica (Figs. 230 and 231). The center electrode and insulator are exposed to the combustion process and vary considerably in temperature depending upon the conditions of operation. Heat flow occurs from the center electrode and insulator to the spark-plug shell, which contacts the cooled cylinder head, and to the end of the spark plug exposed to the atmosphere. A cool plug has a short path for heat flow,

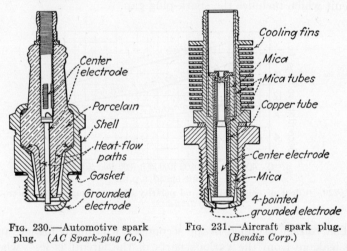

Fig. 230.—Automotive spark plug. (*AC Spark-plug Co.*)

Fig. 231.—Aircraft spark plug. (*Bendix Corp.*)

relatively large areas through which heat transfer occurs, and the minimum of exposure for the center electrode and insulator.

The small plug has less exposed surface and runs cooler than the large plug. Increases in engine output have necessitated the development of smaller plugs until now $\frac{7}{8}$-in. 18-, 14-, and 10-mm. plugs are available.

Rabezzana[1] has found that the preignition temperature is about 650°F. above the temperature required for burning any carbon deposited on the insulator.

Thus, the spark-plug center electrode should run below 1500°F. under the most severe conditions or preignition will result, and

---

[1] Hector Rabezzana, Trend toward Metric Spark Plugs Due to Higher Compression of Engines, *Auto. Ind.*, **59**, 900 (1928).

it should run at least part of the time above 850°F. to eliminate carbon deposits formed during light or idling loads.

Any appreciable gas leakage past the center electrode or past any of the gaskets is very undesirable and soon results in an overheated plug and the accompanying effects of preignition.

Lean mixtures may be more readily ignited with hot plugs and large gaps. Gaps vary from 0.015 to 0.045 in., the larger gaps requiring the higher voltage for sparking (Fig. 232). Some evidence[1] indicates that it is not the size of the gap that is the important factor so much as the energy involved in the discharge

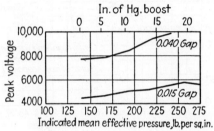

FIG. 232.—Effect of indicated m.e.p. and spark gap on peak voltage required. (*Tognola and De Chard.*)[2]

across the gap; the larger the energy discharge, the larger is the initial conflagration. Also, consecutive cycles show less variation with large spark gaps, indicating more uniform beginning of combustion following spark discharge.[3]

For good part-throttle performance the plug points should extend beyond the surface of the combustion chamber and never be located in a pocket unswept by the fresh charge. Extending the electrodes into the combustion chamber is not permissible in high-output engines because of the greater exposure and higher temperatures of the center electrode. The most important test[4] for a spark plug for high-output engines is the evaluation of its ability to operate at high temperatures without causing preignition.

[1] Alex Taub, Fuel Consumption Problems, *S.A.E. Trans.*, **31**, 66 (1936).

[2] T. Tognola and A. W. De Chard, Spark Plug Endurance Tests, *Auto. Ind.*, **78**, 87 (1938).

[3] Hector Rabezzana and Stephen Kalmar, Fuel Economy, *Auto. Eng.*, **27**, 260 (1937).

[4] A. L. Beall and L. M. Townsend, Hi-duty Spark-plug Testing, *S.A.E. Trans.*, **35**, 465 (1938).

Timing the Ignition.—The desirable ignition timing depends upon various factors that affect the rate of burning or the time available for the combustion process. In addition the detonation characteristics of the fuel must be considered. For automotive vehicles, it is desirable to have the maximum-economy

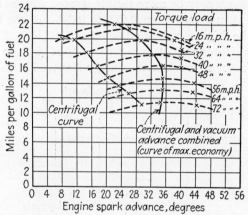

FIG. 233.—Effect of spark advance on fuel consumption at various speeds at level-road conditions. (*Critchfield.*)[1]

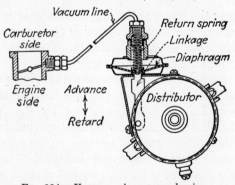

FIG. 234.—Vacuum advance mechanism.

spark advance for level-road operation. A series of runs at level-road conditions and at various speeds with a range of spark advance indicates the desirable spark advance for this operating condition (Fig. 233).

Full-load operation requires less spark advance than level-road operation but requires an increase in spark advance with

[1] R. M. Critchfield, Modern Automotive Electrical Equipment, *S.A.E. Trans.*, **32**, 358 (1937).

speed. A centrifugal device in the distributor head is used to obtain the desirable full-load spark advance. Closing the throttle to level-road condition decreases the intake-manifold pressure which acts on a diaphragm-linkage mechanism and advances the spark to the desired level-road advance (Fig. 234).

Idling requires a retarded spark, which is accomplished by having the vacuum-line connection located on the carburetor side of the throttle plate when closed. This causes the available vacuum at road torque to pass through a maximum (Fig. 235)

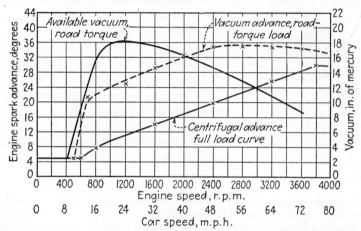

FIG. 235.—Typical spark-advance curves and available manifold vacuum at road load. (*Critchfield.*)[1]

and fall rapidly toward zero as the throttle is closed. Thus, the spark advance at idling speed is the same for idling and full-load conditions. The spark advance is also the same for top-speed road load and full-load condition, which are the same condition.

The octane number of a fuel must be high enough to prevent serious detonation; otherwise optimum spark advance cannot be used at wide-open throttle (Fig. 95). Also, the desirable spark advance will be different for each cylinder, owing to difference in conditions, and in addition the variation in cam contour will result in different spark advances for the various cylinders. Thus, the contour of the shaded area (Fig. 236) represents the actual spark advance obtained by Blackwood, Lewis, and Kass,

[1] Critchfield, *op. cit.*

on an eight-cylinder engine, and the contour of the unshaded area represents the spark advance to give incipient detonation in each cylinder. Rotating the shaded area 45 deg. at a time will show the given arrangement to be the best with the smallest maximum difference between the two contours measured at the cylinder numbers.

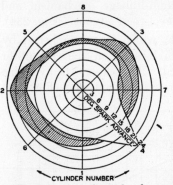

FIG. 236.—Actual spark advance and spark advance for incipient detonation with a given fuel. (*Blackwood, Lewis, and Kass*.)[1]

Firing Order for Multicylinder Engines.—The number of possibilities of firing orders depends upon the number of cylinders and throws of the crankshaft. It is desirable to have the power impulses equally spaced and this has led to certain conventional arrangements of crankshaft throws (Table 34). A greater variation in firing order is possible with the four-stroke-cycle engine than with the two-stroke-cycle engine which fires every time a piston is at top-center position.

Example.—Determine the possible firing orders for a conventional six-cylinder crankshaft with cranks spaced at 120 deg. as indicated in Table 34.

Assuming clockwise rotation of the crankshaft, cylinder 1 is in the firing position. It will be followed by cylinder 2 or 5 and then by cylinder 3 or 4. Thus, the possible firing orders are:

$$1\text{-}2\text{-}3\text{-}6\text{-}5\text{-}4$$
$$1\text{-}2\text{-}4\text{-}6\text{-}5\text{-}3$$
$$1\text{-}5\text{-}4\text{-}6\text{-}2\text{-}3$$
$$1\text{-}5\text{-}3\text{-}6\text{-}2\text{-}4$$

The last firing order is usually preferred, since no two consecutive explosions occur in adjacent cylinders.

EXERCISES

1. Determine the possible firing orders for the crank and cylinder arrangements shown in Table 34.

2. Discuss the desirability of the various firing orders in Prob. 1.

[1] Blackwood, Lewis, and Kass, *op. cit.*

CHAPTER XIII

COMBUSTION-CHAMBER AND CYLINDER-HEAD DESIGN

SPARK-IGNITION ENGINES

Conditions at Ignition.—A definite air-fuel mixture is inducted into the engine cylinder during the suction process. This entering charge mixes with the hot clearance gases and contacts the hot combustion chamber, cylinder, and piston, with resulting heat transfer to the charge. The compression process also increases the energy of the mixture so that the unvaporized portion of the fuel entering the cylinder will probably be vaporized before the beginning of combustion.

A method has been developed for determining the conditions at the end of the suction process (page 151), and the condition at the beginning of combustion can be determined by the use of one of the compression charts (A, B, and C, Appendix).

The Combustion Process.—The charge is ignited by means of an electric spark, and combustion begins at this point and spreads in all directions from the spark-plug gap until stopped by coming in contact with the combustion-chamber walls. The flame is propagated through the combustion chamber by the spreading of the reaction from one fuel molecule to another and by the expansion of the burned fraction of the charge which advances the flame front while compressing the unburned fraction.

Actually, combustion begins before the piston is at top center and ends with the piston moving away from top center. Thus work is done on both the burned and unburned fractions during the first part of the combustion process and by both fractions during the latter part of the process.

The burned fraction of the charge loses heat to the walls from the time it is burned until exhausted. The expansion of the burned fraction compresses the unburned fraction and raises its temperature so that it is losing heat also to the walls, but at a much slower rate than the burned fraction.

The burning of a given fraction a (Fig. 237) of the charge increases its volume from V_0 to V'_0, its pressure from P_1 to P'_0, and its temperature from T_1 to T'_0, which will later be shown to be a mean temperature. While the fraction a burns, it expands and does work indicated by the area beneath the path 1–2, which is also equal to the work of compressing the unburned fraction from V_1 to V'_1. The pressures on both sides of the flame front are assumed to be equal. Applying the energy equation to the part a and neglecting the motion of the piston result in

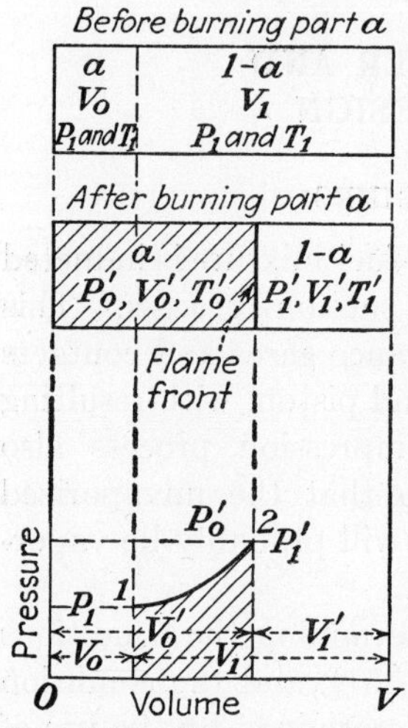

$$(E_1 + C_1)_a = (E_2 + C_2)_a + (W_{out})_a + (Q_{out})_a. \quad (1)$$

The term $(E_1 + C_1)_a$ represents the internal and chemical energy associated with the part a of the mixture at the condition before burning. $(Q_{out})_a$ represents the heat transferred from the burned fraction to the cylinder walls during the process.

Fig. 237.—The combustion process.

Also,

$$(W_{out})_a = (W_{in})_{1-a} = (E_2 - E_1)_{1-a} + (Q_{out})_{1-a}. \quad (2)$$

An assumption of the flame-front position, V'_0, fixes V'_1. The adiabatic compression of the unburned fraction from V_1 to V'_1 determines P'_1, T'_1, and the work term. An assumption of Q permits the evaluation of the term $(E_2 + C_2)_a$ from Eq. (1). At $P'_0 = P'_1$ and $(E_2 + C_2)_a$ the value of T'_0 and V'_0 can be obtained from one of the combustion charts (A, B, and C, Appendix). The correct solution is indicated when the determined V'_0 checks the assumed value.

Example.—A correct mixture of octane and air at $P_1 = 100$ lb. in.$^{-2}$ abs. and $T_1 = 1000°$R. is ignited in a constant-volume combustion chamber. After 25 per cent of the mixture burns with 10 per cent heat loss, determine the flame-front position and condition of the burned and unburned fractions.

On the assumption that the fraction to be burned contains 1 lb. of air,

$$(E_1 + C_1)_a = 100 + 1278 = 1378 \text{ B.t.u. (Chart } B, \text{ Appendix).}$$

Also, $\qquad (Q_{out})_a = 0.10 \times 1278 = 127.8$ B.t.u.

The unburned fraction will contain 3 lb. of air, and

$$(E_1)_{1-a} = 3 \times 100 = 300 \text{ B.t.u. (Chart } B)$$
and $V_0 + V_1 = 3.8 + 3 \times 3.8 = 15.2 \text{ ft.}^3$ and $V_1 = 11.4 \text{ ft.}^3$

Assume V'_0,	8	9	10
$V'_1 = (V_0 + V_1) - V'_0$,	7.2	6.2	5.2
$V'_1/3$,	2.40	2.07	1.73
$(E_2)_{1-a}$ (Chart B, Appendix),	405	444	489
$P'_1 = P'_0$ (Chart B, Appendix),	185	225	280
$(E_2 - E_1)_{1-a}$,	105	144	189
$(E_2 + C_2)_a$, from Eq. (1),	1145	1106	1061
T'_0 (Chart E, Appendix),	4480	4400	4300
V'_0 (Chart E, Appendix),	9.8	7.9	6.3

Plotting the determined V'_0 values against the assumed values and other results indicates a solution at $V'_0 = 8.6 \text{ ft.}^3$, $T'_0 = 4435°\text{R.}$, $P'_0 = 207 \text{ lb. in.}^{-2}$ abs. Also, from Chart B, $T'_1 = 1190°\text{R.}$

Then, $\dfrac{V'_0}{V_0 + V_1} = \dfrac{8.6}{15.2} = 0.57$, the flame-front position,

which is the part of the combustion chamber occupied by the burned fraction

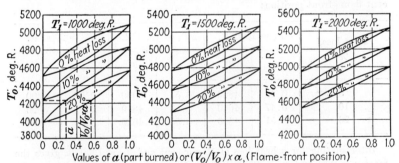

Fig. 238.—Temperature of burned fraction and flame-front position. ($P_1 = 100 \text{ lb./sq. in. abs.}$)

Burned Fraction of the Charge.—The results of similar computations[1] show that the combustion of an infinitesimally small part of the mixture ($a = 0$; Fig. 238) results in a high temperature T'_0 for the burned fraction. The part burned and the flame-front position are identical at $a = 0$ and $a = 1$, but in between the flame-front position is appreciably greater than the part burned. This indicates the effect of the expansion of the burned fraction on the flame-front position.

[1] L. C. Lichty, R. R. Faller, and M. F. Smith, A Study of the Combustion Process, *S.A.E. Trans.*, **27**, 101 (1932).

Heat loss reduces the temperature of the burned part. It also reduces its volume (Fig. 239). An increase in initial temperature, T_1 also reduces the volume of the burned fraction. The results for any intermediate condition of T_1 and heat loss Q may be obtained by interpolation.

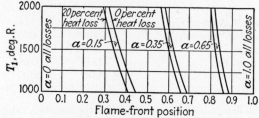

Fig. 239.—Flame-front positions.

Example.—Determine the percentage of charge mass that has been burned and its temperature when the volume of the burned fraction is 75 per cent of the combustion-chamber volume. Assume $T_1 = 1200°R.$ and a heat loss of 15 per cent.

From Fig. 238 the following values of a and T'_0 were obtained for 75 per cent flame-front position:

Heat loss, %	$T_1 = 1000°R.$		$T_1 = 1500°R.$		$T_1 = 2000°R.$	
	a	T'_0	a	T'_0	a	T'_0
0	0.42	4860	0.51	5090	0.58	5300
10	0.47	4640	0.53	4880	0.59	5090
20	0.51	4360	0.55	4620	0.60	4860

Interpolating for a heat loss of 15 per cent and $T_1 = 1200°R.$ gives the mass $a = 0.51$ and $T'_0 = 4600°R.$

Unburned Fraction of the Charge.—The temperature rise of the unburned $(T'_1 - T_1)$ for any burned fraction is practically independent of initial temperature but is decreased with heat loss from the burned fraction (Fig. 240). The maximum temperature attainable by the unburned fraction for all conditions is approximately 500°F. higher than the initial temperature at which the combustion process begins, 50 per cent of this rise being attained when 35 per cent of the charge has been burned.

The pressure of the unburned fraction, which is the pressure in the combustion chamber, depends upon the initial temperature, T_1, initial pressure P_1, and the temperature rise $T'_1 - T_1$ (Fig. 240).

COMBUSTION-CHAMBER AND CYLINDER-HEAD DESIGN

Example.—Determine the pressure in the combustion chamber when the volume of the burned fraction is 75 per cent of the total volume, with $T_1 = 1200°R.$, $P_1 = 100$ lb. in.$^{-2}$ abs., and a heat loss of 15 per cent.

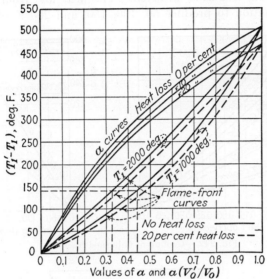

Fig. 240.—Temperature rise of unburned fraction.

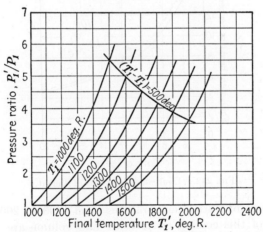

Fig. 241.—Pressure-temperature relationship for the unburned fraction.

From the example on page 358, it was found that $a = 0.51$ for the preceding conditions. From the a curves in Fig. 240, the temperature rise of the unburned fraction is found to be 310°F., making $T_1' = 1510°R.$

From the curve for $T_1 = 1200°R.$, in Fig. 241, the pressure ratio is found to be 2.65. The pressure is 2.65×100 or 265 lb. in.$^{-2}$ abs.

Cooling the Unburned Fraction.—The heat loss from the unburned fraction to produce 50, 100, 150, and 200°F. of cooling

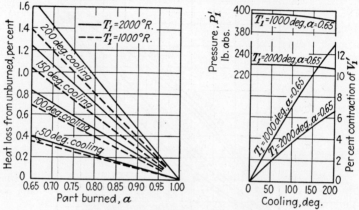

FIG. 242.—Cooling of unburned fraction.

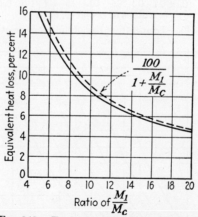

FIG. 243.—Equivalent heat loss of diluent.

for the range of initial temperatures from 1000 to 2000°R. and the effect of this cooling on pressure and volume are shown in Fig. 242. The effect on pressure is slight for $a = 0.65$, and for values of $a > 0.65$ the effect would be less than for $a = 0.65$. Also, T'_0 is not appreciably affected by the range of cooling indicated.

COMBUSTION-CHAMBER AND CYLINDER-HEAD DESIGN

The unburned fraction contracts when cooled and increases the volume of the burned fraction. The contraction for values of a greater than 0.65 is only slightly larger than for a equal to 0.65.

Effect of Diluent.—The effect of clearance-gas dilution on the combustion process can be treated as a heat loss since the clearance gases reduce the temperatures attained. The computed effect of clearance-gas dilution (Fig. 243) is almost the same as obtained from the assumption that the equivalent heat loss is equal to the percentage of clearance gases in the total change in the cylinder.

Thus,

$$\text{Equiv. loss} = \frac{100}{1 + \dfrac{M_1}{M_c}} \quad (3)$$

where M_1 = mols of inducted charge,
M_c = mols of clearance gases.

Both curves extend to 100 per cent heat loss for $M_1/M_c = 0$ and to 0 per cent heat loss when this ratio is infinite.

Effect of Initial Pressure.—The foregoing illustrations are based upon an initial pressure of 100 lb. in.$^{-2}$ abs. Higher initial pressures result in more complete combustion and slightly higher values for T'_0. Practically, the effect of initial pressure may be neglected except in the computation of pressure. Thus, the illustrations apply to combustion processes with initial pressures from less to more than 100 lb. in.$^{-2}$ abs.

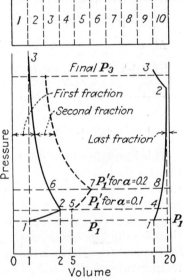

FIG. 244.—The combustion process by parts.

Variation of Temperature of Burned Fraction.—Hopkinson[1] was the first to call attention to an inherent variation in the temperature of the burned fraction in the constant-volume process. The first fraction of the charge to burn (Fig. 244)

[1] D. Clerk "The Gas Petrol and Oil Engine," p. 191, John Wiley & Sons, Inc., 1909.

expands and compresses the unburned fractions from 1 to 2. This displaces the second fraction, to the position 2–5, which burns and compresses the first burned fraction from 2 to 6, and the remaining unburned fractions from 5 to 7.

The last fraction of the charge after being compressed to 2 burns and expands along the line 2–3. In the meantime, the first burned fraction has been compressed along its 2–3 line.

Applying the energy equation, and noting that for equal fractions $(E_1 + C_1)$ is the same for all fractions, results in

$$[(E_1 + C_1) - {}_1W_2 + {}_2W_3]_{1st\ fraction} = (E_3 + C_3)_{1st\ fraction}. \quad (4)$$

Also,

$$[(E_1 + C_1) + {}_1W_2 - {}_2W_3]_{last\ fraction} = (E_3 + C_3)_{last\ fraction}. \quad (5)$$

On the assumption that $V_3 = V_1$ for both fractions,

$$_2W_3 > {}_1W_2$$

and $\quad (E_3 + C_3)_{1st\ fraction} > (E_3 + C_3)_{last\ fraction}. \quad (6)$

This inequality indicates that T_3 is higher for the first fraction than for the last fraction of the charge to burn. This also indicates that P_3 is higher for the first fraction than for the last fraction if each occupies the same volume. The inequality in pressure is impossible; consequently, V_3 for the first fraction must be larger than for the last fraction, which also indicates that T_3 is higher for the first fraction than for the last fraction to burn. Thus, the temperature of the products at the end of the constant-volume combustion process is highest at the spark plug and lowest at the opposite end of the combustion chamber.

Example.—Determine the final temperature of the first and last fractions of the charge to burn in a constant-volume combustion of a correct mixture of octane and air. Initial conditions are 100 lb. in.$^{-2}$ abs. and 1000°R. Divide the charge into 10 equal fractions with 1 lb. of air in each fraction. Assume no heat loss.

$$V_1 = 3.8 \text{ ft.}^3 \text{ (Chart } B, \text{ Appendix)}.$$

For the burned fraction, $T_2 = T'_0$ (Fig. 238) = 4620°R., for $a = 0.1$.
For the unburned fraction, $T_2 = T'_1$ (Fig. 240) = 1085°R., and

$$P_2 = P'_1(\text{Fig. 241}) = 135 \text{ lb. in.}^{-2} \text{ abs.,}$$

for $a = 0.1$. This is also the pressure of the burned fraction for this condition.

COMBUSTION-CHAMBER AND CYLINDER-HEAD DESIGN

Compressing the burned fraction adiabatically from $P_2 = 135$ lb. in.$^{-2}$ abs. and $T_2 = 4620°$R. to its original volume before burning, 3.8 ft.^{3}, (Chart E, Appendix) results in $T_3 = 5620°$R. and $P_3 = 620$ lb. in.$^{-2}$ abs.

Also, for the unburned fraction, $T_2 = T_2'$ (Fig. 240) = $1475°$R. and $P_2 = P_2'$ (Fig. 241) = 520 lb. in.$^{-2}$ abs., for $a = 0.9$.

At this condition,

$$E_2 + C_2 = 211 + 1278 = 1489 \text{ B.t.u.} \text{ (Chart } B, \text{ Appendix).}$$

The mean temperature of the burned fraction for $a = 0.9$ is $5070°$R. and the flame-front position is 0.97 (Fig. 238). Compressing each of the nine fractions burned, since the charts are designed for 0.1 of the entire charge in this problem, from $5070°$R. and $0.97(10V_1)/9$ or 4.1 ft.^{3} to V_1 or 3.8 ft.^{3} (Chart E, Appendix) requires a total work amount, supplied by the expansion of the last 0.1 to burn, of

$$(_2W_3)_{last\ fraction} = 9(1455 - 1420) = 315 \text{ B.t.u.}$$

Then, for the last fraction, $(E_3 + C_3) = (E_2 + C_2) - {_2W_3}$
$$= 1489 - 315 = 1174 \text{ B.t.u.}$$

From Chart E, Appendix, at $V_1 = 3.8$ ft.^{3} and $(E_3 + C_3) = 1174$ B.t.u., $T_3 = 4600°$R. and $P_3 = 490$ lb. in.$^{-2}$ abs.

Obviously, the two values for P_3 should be the same, which is found to occur at $T_3 = 5530°$R. and $V_3 = 4.3$ ft.^{3} for the first fraction; also, $T_3 = 4690°$R. and $V_3 = 3.5$ ft.^{3} for the last fraction when $P = 545$ lb. in.$^{-2}$ abs.

Effect of Piston Motion during Combustion.

Movement of the piston during the first part of the combustion process decreases the volume and consequently results in a faster pressure rise than would occur with constant-volume combustion (Fig. 245). After top center the movement of the piston reduces the rate of pressure rise until in most cases the combustion process ends with the pressure decreasing.

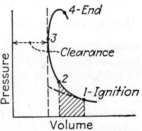

Fig. 245.—Analysis of part of the actual combustion process.

Applying the energy equation to the combustion process from the point of ignition to any point 2 (Fig. 245) results in

$$(E_1 + C_1) + {_1W_2} = (E_2 + C_2) + Q_{out}. \tag{7}$$

$(E_1 + C_1)$ is known from the conditions at ignition, $_1W_2$ can be obtained from the work area on the P-V diagram, and Q_{out} may be estimated, all of which provides a solution for $(E_2 + C_2)$ and from this solution the fraction burned may be determined.

Example.—The correct mixture of octane and air is inducted into an engine. The clearance gases amount to one-ninth of the weight of the charge inducted. The conditions at ignition are 90 lb. in.$^{-2}$ abs. and 1000°R. Combustion proceeds and the piston moves until the pressure is 180 lb. in.$^{-2}$ abs. and the volume is eight-tenths of the volume at ignition. Determine the amount of charge burned, assuming an adiabatic process.

Assume the total charge in the cylinder contained originally 1 lb. of air. Then, from Chart B (Appendix) the volume at point 1 (Fig. 245) is very nearly 4.2 ft.^{3}

Then, $\qquad V_2 = 0.8 V_1 = 0.8 \times 4.2 = 3.4$ ft.^{3}

$$\int_1^2 P \, dV = 19 \text{ B.t.u., work in, from area on Fig. 245.}$$

$E_1 + C_1 = 100 + 0.9 \times 1278 = 1250$ B.t.u. (Chart B, Appendix).
Then, $\qquad E_2 + C_2 = 1250 + 19 = 1269$ B.t.u.

Assume that a represents the fraction of the total charge that has been subjected to the combustion of the mixture at point 2. Then, at P_2 and S_1 (Chart B, Appendix),

E_2 for the unburned fraction $= (1 - a)140$ B.t.u.
V_2 for the unburned fraction $= (1 - a)2.5$ ft.^{3}
C_2 for the unburned fraction $= (1 - a)(0.9 \times 1278)$.
$\qquad (E_2 + C_2)_{unb} = (1 - a)1290$ B.t.u.

For the burned fraction,

Assume T_0' at point 2,	4200	4400	4600
V_{burned} at P_2 and T_0',	$9.7a$	$10.0a$	$10.4a$
$V_2 = 3.4$ ft.$^3 = V_b + (1 - a)2.5$ ft.,			
	$2.5 + 7.2a$	$2.5 + 7.5a$	$2.5 + 7.9a$
a,	0.125	0.120	0.114
$(E_2 + C_2)_b$ at P_2 and T_0' (Chart E, Appendix),	$1022a$	$1112a$	$1210a$
or	128	133	138
$(E_2 + C_2)_{unb} = (1 - a)1290$,	1129	1135	1143
$E_2 + C_2$,	1257	1268	1281

Plotting the values of $E_2 + C_2$ against T_0' indicates a solution for $E_2 + C_2$ at 1269 B.t.u. at $T_0' = 4410$°R. and $a = 0.120$.

The estimate of the initial temperature and heat transfer introduces an error in the foregoing solution when applied to the actual engine process. However, Rassweiler and Withrow[1] have developed the following simple approximate analysis which does not require a knowledge of the initial temperature.

[1] G. M. Rassweiler and L. Withrow, Motion Pictures of Engine Flames Correlated with Pressure Cards, *S.A.E. Trans.*, **33**, 185 (1938).

Assume that the compression of the unburned fraction follows the path $PV^n = $ const. Thus, with reference to Fig. 237,

$$V_1 = V_1' \left(\frac{P_1'}{P_1}\right)^{1/n}. \tag{8}$$

The unburned fraction of the mixture at any instant is

$$\frac{V_1}{V_i} = \frac{V_1'}{V_i}\left(\frac{P_1'}{P_i}\right)^{1/n} = 1 - a \tag{9}$$

where subscript $i = $ conditions at ignition.

At the end of combustion the burned fraction will be compressed into some volume V_0.* Assume that the compression of the burned fraction is also described by the path $PV^n = $ const.

Then,
$$V_0 = V_0'\left(\frac{P_1'}{P_e}\right)^{1/n} \tag{10}$$

where subscript $e = $ end of combustion.

Then,
$$\frac{V_0}{V_e} = \frac{V_0'}{V_e}\left(\frac{P_1'}{P_e}\right)^{1/n} = a. \tag{11}$$

Also, from Eq. (9),

$$1 - \frac{V_1}{V_i} = 1 - \frac{V_1'}{V_i}\left(\frac{P_1'}{P_i}\right)^{1/n} = 1 - \left(\frac{V - V_0'}{V_i}\right)\left(\frac{P_1'}{P_i}\right)^{1/n} = a. \tag{12}$$

Substituting the value for V_0' from Eq. (11) into Eq. (12), P for P_1', and solving, result in

$$\text{Wt. fraction burned} = \frac{\left(\frac{P}{P_i}\right)^{1/n}\left(\frac{V}{V_i}\right) - 1}{\left(\frac{P_e}{P_i}\right)^{1/n}\left(\frac{V_e}{V_i}\right) - 1} = a. \tag{13}$$

Also, equating Eqs. (11) and (12), substituting P for P_1', and solving for V_0'/V, result in

$$\text{Vol. fraction burned} = \frac{\left(\frac{P_i}{P}\right)^{1/n}\left(\frac{V_i}{V}\right) - 1}{\left(\frac{P_i}{P_e}\right)^{1/n}\left(\frac{V_i}{V_e}\right) - 1} = a\frac{V_0'}{V}. \tag{14}$$

* Actually, this volume will vary somewhat from the volume indicated in Fig. 237.

Equations (13) and (14) require a knowledge only of the total pressure in and the volume of the combustion chamber at the given instant, at ignition, and at the end of combustion for the evaluation of the fraction burned and the flame-front position at the given instant.

Example.—The conditions at ignition are $P_i = 82$ lb. in.$^{-2}$ abs. and $V_i = 10.2$ in.^{3} At the end of combustion, $P_e = 319$ lb. in.$^{-2}$ abs. and $V_e = 9.8$ in.^{3} Determine the fraction burned and flame-front position when $P = 249$ lb. in.$^{-2}$ abs. and $V = 8.4$ in.^{3} Assume $n = 1.33$.

Substituting in Eq. (13) results in
$$a = \frac{\left(\frac{249}{82}\right)^{1/1.33}\left(\frac{8.40}{10.2}\right) - 1}{\left(\frac{319}{82}\right)^{1/1.33}\left(\frac{9.8}{10.2}\right) - 1} = 0.54.$$

Substituting in Eq. (14) yields
$$a\frac{V_0'}{V_0} = \frac{\left(\frac{82}{249}\right)^{1/1.33}\left(\frac{10.2}{8.40}\right) - 1}{\left(\frac{82}{319}\right)^{1/1.33}\left(\frac{10.2}{9.8}\right) - 1} = 0.76.$$

These values are obtained from Fig. 238 at $T_1 = 1000°$R. and 20 per cent heat loss. This heat loss appears high but might be expected since the movement of the piston before top center (ignition to point 11, Fig. 246) causes a displacement of gases into the valve chamber of the L-head engine, retards the flame progress, and gives the appearance of heat loss.

Rassweiler and Withrow developed a method for separating the increments of pressure rise into the part due to piston motion and that due to combustion.[1] The pressure rise is assumed to take place in steps (Fig. 246): first, pressure rise due to movement of the piston only; then, pressure rise due to constant-volume combustion. If the change in volume between any two points is known, the pressure rise due to piston movement may be computed from the relation $PV^n = $ const. and the pressure rise due to combustion may be determined by the difference between the total rise and that due to piston movement. The pressure rise due to combustion occurs at different volumes. Consequently, the equivalent pressure must be determined at some fixed volume, such as, at the point of ignition, to be indicative of the fraction burned.

For a given initial T_1, the burning of a given fraction will result in an increase in pressure from P_1 to P_2 equivalent to that of

[1] Rassweiler and Withrow, *op. cit.*

heating the entire mixture to T_2. This equivalent heating effect will be the same regardless of the volume of the chamber. Thus, for any chamber volume V_1,

$$\frac{P_2 - P_1}{P_1} = \frac{T_2 - T_1}{T_1} \qquad (15)$$

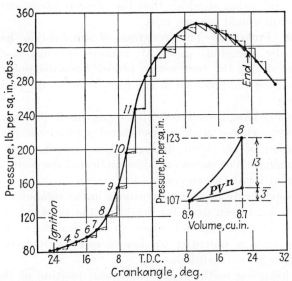

Fig. 246.—Pressure-time diagram of combustion process. (*Rassweiler and Withrow.*)[1]

and comparing with the relation for any other volume V_2, results in

$$\frac{(P_2 - P_1)_{V_2}}{(P_2 - P_1)_{V_1}} = \frac{(P_1)_{V_2}}{(P_1)_{V_1}} = \frac{V_1}{V_2} \qquad (16)$$

which may be used to determine the equivalent pressure rise due to combustion.

Example.—Determine the equivalent pressure rise due to combustion between points 7 and 8, Fig. 246.

$$P_2 = P_1 \left(\frac{V_1}{V_2}\right)^n = 107 \left(\frac{8.9}{8.7}\right)^{1.33} = 110 \text{ lb. in.}^{-2} \text{ abs.}$$

Pressure increment due to piston motion = $110 - 107 = 3$ lb. in.$^{-2}$
Pressure increment due to combustion = $123 - 110 = 13$ lb. in.$^{-2}$

[1] Rassweiler and Withrow, *op. cit.*

Correcting this increment to the equivalent value for the combustion-chamber volume at ignition, 10.2 in.³, results in

$$(P_2 - P_1)_{V_2} = (P_2 - P_1)_{V_1}\frac{V_1}{V_2} = 13 \times \frac{8.7}{10.2} = 11 \text{ lb. in.}^{-2}$$

The foregoing analysis provides a method for determining the end of combustion at which time the pressure increment due to combustion should become zero.

Flame Progress.—The progress of combustion has been studied by sampling gases at various combustion-chamber positions during the combustion process,[1] the disappearance of O_2 indicating the presence of the flame front at the position for sampling.

Cylinders have been fitted with many small quartz windows which were viewed with a mechanical stroboscope permitting observation of the head during a given small crank angle.[2] The illumination of various windows indicated the position of the flame front for the given crank angle.

Spark-plug gaps at various positions in the combustion-chamber wall have been used for detecting the presence of flame.[3] Flame ionizes the gap and causes an electrical discharge across the gap from a potential too low to cause a discharge under other than flame conditions.

The foregoing methods give the mean position of the flame for a series of processes and indicate that the flame progresses radially from the spark-plug gap in all possible directions. The flame-front appears to be somewhat spherical but varies from this shape because of mass movement of the gases set up during the induction period and due to variation in mixture condition throughout the combustion chamber (Fig. 247). An increase in speed altered the diagram somewhat in detail, but the general pattern was the same. The crank angle for a given flame travel

[1] L. Withrow, W. G. Lovell, and T. A. Boyd, Following Combustion in the Gasoline Engine by Chemical Means, *Ind. Eng. Chem.*, **22**, 945 (1930).

[2] C. F. Marvin, Jr., and R. D. Best, Flame Movement and Pressure Development in the Engine Cylinder, *N.A.C.A. Rept.* 399 (1931).

[3] D. MacKenzie and R. K. Honaman, The Velocity of Flame Propagation in Engine Cylinders, *S.A.E. Trans.*, **15**, 299 (1920).

K. Schnauffer, Engine-cylinder Flame Propagation Studied by New Methods, *S.A.E. Trans.*, **29**, 17 (1934).

H. Rabezzana and S. Kalmar, Factors Controlling Engine Combustion, *Auto. Ind.*, **72**, 324, 354, 394 (1935).

COMBUSTION-CHAMBER AND CYLINDER-HEAD DESIGN 369

is about the same at 600 and 1200 r.p.m., indicating an increase in flame speed with engine speed. Central location of the spark plug shortened the flame path, and decreased the crank

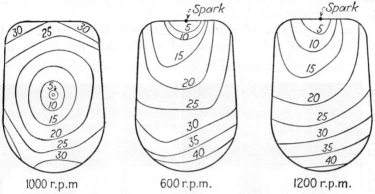

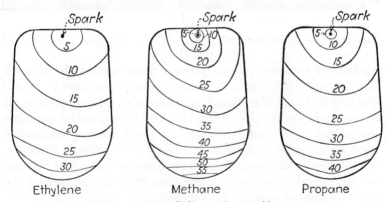

Fig. 247.—Flame-progress diagrams. Numbers indicate degrees after occurrence of spark. (*Marvin and Best.*)[1]

angle required for the combustion process. The flame front for a rapid-burning fuel traverses the combustion chamber in less time than is required for a slow-burning fuel.

A series of photographs (Fig. 248) of a single combustion process shows the progress of a rather ragged flame front which

[1] Marvin and Best, *op. cit.*

proceeds in all directions from the spark-plug gap until it reaches the combustion-chamber walls. The ragged flame front appears to be the effect of turbulence and nonuniformity of mixture since combustion experiments under quiescent and uniform conditions result in nearly spherical flame fronts.[1]

For the purpose of combustion analysis, it has been found desirable[2,3] to reduce the actual combustion process to an

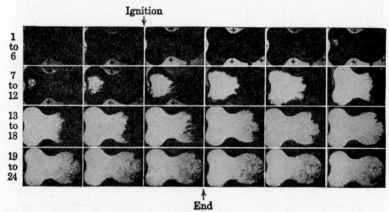

FIG. 248.—Flame progress during nonknocking combustion. (*Rassweiler and Withrow*.)[2]

equivalent constant-volume process, inasmuch as the time of ignition and the subsequent motion of the piston have an appreciable effect on the actual pressure rise. Consequently, it appears that the relative effect of changes in combustion-chamber design and spark-plug location on pressure rise may be studied from an assumption of constant-volume combustion. However, any change in design that influences the turbulence or mixture condition may accentuate or offset an effect anticipated by the constant-volume combustion analysis.

Flame-front Area.—One of the principal factors that influence the rate of energy liberation is the area of the flame front. The combustion starts from a small nucleus and has a rapidly increasing area of flame front until it is restricted in various places by contact with the combustion-chamber wall which

---

[1] F. W. Stevens, *N.A.C.A. Repts.* 176, 280, 305, 337, 372 (1923–1930).
E. F. Fiock and C. H. Roeder, *N.A.C.A. Repts.* 531, 532 (1935).
[2] Rassweiler and Withrow, *op. cit.*
[3] Rabezzana and Kalmar, *op. cit.*

COMBUSTION-CHAMBER AND CYLINDER-HEAD DESIGN

ends the progress in that direction. Obviously, the area of the flame front depends upon the design of the combustion chamber and the location of the spark plug or plugs.

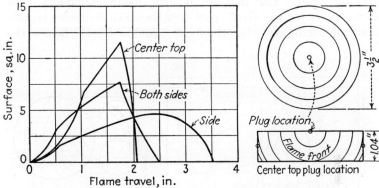

Fig. 249.—Flame-front areas, for a cylindrical combustion chamber.

A spherical combustion chamber, ignited at the center, would have the maximum flame-front area for a given flame travel. If fired from both sides or from only one side of the sphere, the flame travel would be increased and the flame-front area decreased for a given flame travel.

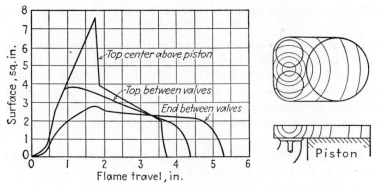

Fig. 250.—Flame-front areas, for a flat L-head combustion chamber.

The flat cylindrical combustion chamber (Fig. 249) presents several possible plug locations which result in appreciably different diagrams of flame-front area. An analysis of a flat L-head combustion chamber (Fig. 250) with the same volume results in appreciably less flame-front area and longer flame travel for equivalent plug locations.

The Ricardo high-turbulence head (Fig. 251) provides a compact combustion chamber near the valves, but extends into the small clearance space between part of the piston and the cylinder head. The flame-front area increases rapidly with flame travel for the given spark-plug location, decreases rapidly from the maximum, when the flame front reaches the floor of the

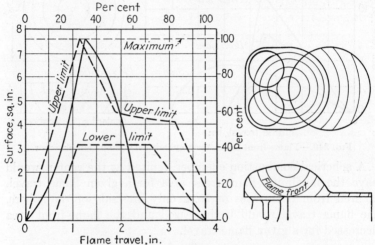

Fig. 251.—Flame-front areas, Ricardo high-turbulence head. (*Dashed limiting lines according to Aske.*)[1]

valve chamber, to a rather low value when the flame enters the small clearance space, and decreases to zero at end of flame travel.

Aske[1] has determined limits (Fig. 251) between which desirable combustion processes are usually obtained. The upper limit indicates the percentage of maximum flame-front area at various percentages of flame travel below which smooth combustion ordinarily results.

A simple method for determining flame-front areas is to determine the rate of change of combustion-chamber volume per a small increment of flame travel. Thus,

$$\text{Flame-front area} = \frac{\Delta \text{ vol.}}{\Delta \text{ flame travel}} = \frac{\text{in.}^3}{\text{in.}} = \text{in.}^2 \quad (17)$$

A cast is made of the combustion chamber, a molten waxlike substance being used. Increments may then be removed from

[1] I. E. Aske, private communication, 1938.

the cast by means of a spherical cutter, the cast being mounted with the spark plug at the spherical center. The weight of the remainder indicates the volume removed for the given distance increment, from which the flame-front area is obtained.

Flame Velocity and Rate of Reaction.—The flame velocity may be divided into two components:

1. The rate at which the flame progresses into the unburned portion relative to the unburned portion.

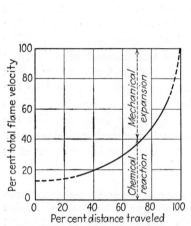

FIG. 252.—Division of total flame velocity. (*From data in Univ. of Ill., Eng. Exp. Sta., Bull.* 157, 1926.)

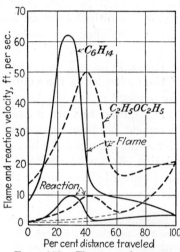

FIG. 253.—Flame and reaction velocity in constant-volume chamber. $C_2H_5OC_2H_5 + 85$ per cent required air.[1] $C_6H_{14} + 88$ per cent required air.[2]

2. The rate at which the burned portion expands and moves the flame front.

Rosecrans[1] developed a method for dividing flame velocity in a constant-volume process into these two parts, the results of which indicate (Fig. 252) that the reaction velocity is a small part of the flame velocity at the beginning of the process but is the flame velocity at the end of the process. The application of this analysis to several constant-volume-combustion experi-

[1] C. Z. Rosecrans, An Investigation of the Mechanism of Explosive Reactions, *Univ. Ill., Eng. Expt. Sta., Bull.* 157 (1926).

[2] M. R. Duchene, Contribution to the Study of Normal Burning in Gaseous Carbureted Mixtures, *N.A.C.A. Tech. Memo.* 548 (1930).

ments results in reaction velocities which vary with the flame-front position (Fig. 253). The reaction velocities for the two different mixtures pass through a maximum at about one-third of the flame travel and then again increase with flame travel.

Analysis for Relative Rate of Pressure Rise.—The rate at which energy is liberated in the combustion process depends upon the flame-front area, the concentration of the unburned mixture at the flame front, and the rate of reaction relative to the unburned portion. The area of the flame front, at any position, multiplied by an increment of distance traveled per increment of time represents the volume of charge burned in the increment of time. The energy liberated by the combustion of this volume will depend principally upon the concentration of the molecules in the volume, and this depends directly upon the pressure and inversely upon the absolute temperature. The time required to burn an increment of charge is equal to the increment of distance divided by the rate of reaction. Rates of reaction are not known for gasoline-air mixtures, but the relative effect with different combustion-chamber designs may be obtained by assuming a rate of reaction which is constant or which varies with flame-front position.

Example.—Determine the relative mass rate of reaction and rate of pressure rise for the cylindrical valve-in-head type of combustion chamber with the spark plug located at the side. Engine dimensions, 3.5 by 5 in.; combustion-chamber volume, 10 in.[3]; height of combustion chamber, 1.04 in. Assume no heat loss during the process, a constant rate of reaction, and an initial temperature of 1165°R. for the octane-air mixture.

The flame-front areas for given flame positions, measured from the plug location to the flame front, were computed for a layout similar to Fig. 249 but for the side location of the spark plug. The volume behind the flame front was computed from the same diagram.

The mass burned was determined from the volume of the burned fraction and the initial temperature (Fig. 240). The volume of the burned fraction is the factor $a(V_0'/V_0)$ which is the abscissa for the flame-front curves.

The temperature rise of the unburned fraction added to T_1 gives the value of T_1'.

The pressure ratio is determined from Fig. 241.

The product of the flame-front area, the pressure ratio, and the reciprocal of the temperature ratio results in a number that is a measure of the rate of mass or molecular reaction for a constant linear rate of reaction. The values for these computations are given in the table at the top of page 375.

The mass rate and mass are plotted against length of flame travel in **Fig. 254.** Although the mass rate is simply a relative figure, it represents

COMBUSTION-CHAMBER AND CYLINDER-HEAD DESIGN

Flame position	Flame travel, in.	Flame-front area, in.²	Vol. of burned fraction		Mass burned $a \times 100$	T_1' °R.	P_1'/P_1	T_1'/T_1	Mass rate of reaction
			In.³	$a(V_0'/V_0)$					
1	0.125	0.095	0.0042	0.0004	0.02	1165+	1.000+	1.000	0.095
2	0.250	0.38	0.032	0.0032	0.16	1166	1.005+	0.999	0.382
3	0.520	1.55	0.26	0.026	1.32	1175	1.04	0.992	1.60
4	0.875	2.62	0.94	0.094	3.5	1188	1.10	0.982	2.83
5	1.250	3.27	1.99	0.199	7.5	1225	1.24	0.952	3.86
6	1.750	3.82	3.92	0.392	14	1293	1.55	0.902	5.34
7	2.500	4.07	6.82	0.682	36	1435	2.40	0.812	7.94
8	3.000	3.41	8.83	0.883	67	1575	3.55	0.740	8.97
9	3.375	2.58	9.53	0.953	87	1640	4.20	0.711	7.71
10	3.500	1.92	9.84	0.984	95	1660	4.35	0.702	5.87
11	3.540	0	10.00	1.000	100	1675	4.50	0.696	0

mass per unit of time. Hence, dividing the mass burned between any two positions of the flame front, not far apart, by the mean mass rate of reaction between the two positions will give a relative figure for the time required for the flame front to move from one position to the other. The computations for the time for the flame to reach the various positions are given in the following table, and the results plotted as the time curve.

Position	100 × mass burned between positions	Mean rate of reaction	Relative time between positions	Total time to reach position	Time, %
0– 1	0.02	0.04	0.50	0.50	2.76
1– 2	0.14	0.20	0.70	1.20	6.64
2– 3	1.16	1.00	1.16	2.36	13.1
3– 4	2.18	2.21	0.99	3.35	18.5
4– 5	4.0	3.35	1.19	4.54	25.1
5– 6	6.5	4.60	1.41	5.95	32.9
6– 7	22.0	6.64	3.32	9.27	51.3
7– 8	31.0	8.46	3.67	12.94	71.6
8– 9	20.0	8.60	2.33	15.27	84.4
9–10	8.0	7.00	1.14	16.41	90.8
10–11	5.0	3.00	1.67	18.08	100.0

The relation between the mass burned and the percentage of combustion time, plotted in the right-hand side of Fig. 254, was obtained by plotting values of time and mass given by the curves in the left-hand side of the same illustration. The pressure-ratio curve was plotted from the data in the first tabulation.

The pressure-time curve developed in the foregoing analysis may be compared with similar curves obtained with variations in the combustion-chamber design and location of spark plug.

Turbulence.—The time required for the combustion process depends considerably upon the turbulence or state of agitation of the mixture. The greater the turbulence the faster the rate of burning, the slowest rate being obtained with a stagnant mixture.

An increase in engine speed increases the intake velocity. However, this mass velocity is fairly well damped out when the flame sweeps through the combustion chamber as evidenced by the flame-front diagrams at various speeds. Thus, the

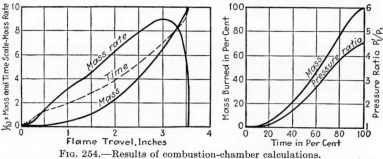

Fig. 254.—Results of combustion-chamber calculations.

flame velocity is not aided appreciably by the mass movement of the gases, although it is increased appreciably with an increase in engine speed.

The entering mixture strikes the cylinder head, cylinder walls, and piston head, which must produce innumerable wave effects similar to that occasioned by throwing a handful of pebbles into a pool. These wave effects have less chance to die out at high speeds before combustion occurs and apparently enlarge the effective flame-front area and speed up the reaction.

High-turbulence heads have a small clearance between part of the head and the piston when at top center. As the piston approaches top center, some of the gases in the small clearance space are projected violently into the main part of the combustion chamber. This produces a turbulent effect which speeds up the reaction.

A turbulent-mixture condition tends to prevent a stagnant-mixture film in contact with the chamber walls and, hence, results in more complete combustion. However, this movement of gases increases the heat transfer.

Combustion Roughness.—An increase in mixture turbulence may result in a roughness or harshness of engine operation, which is due principally to a too rapid increase in the rate of pressure rise during the combustion process. Janeway[1] has called attention to the well-known fact that any force P impressed upon an elastic structure, such as the engine crankshaft and crankcase,

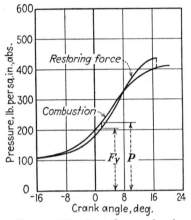

FIG. 255.—Smooth combustion. 1600 r.p.m., 16° spark advance, and 5.25 to 1 compression ratio. (*Janeway*.)

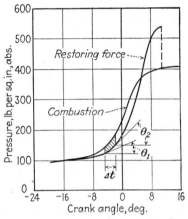

FIG. 256.—Rough combustion. 1600 r.p.m., 16° spark advance, and 5 to 1 compression ratio. (*Janeway*.)

will result in a deflection, and a restoring force will be exerted by the structure in direct proportion to the deflection. If the impressed force is increased at an appreciable rate, the deflection of the structure will lag behind the impressed force and consequently there will be a difference between the impressed and restoring forces, which sets up a velocity of deflection of the structure. This force differential provides a given momentum during the accelerating period of deflection, which will be equal in amount to the decrease in momentum when the velocity of deflection decreases from the maximum value to zero, at which condition the structure will be at the point of maximum deflection and will exert the maximum restoring force.

The restoring force F_y (Fig. 255) is lower than the combustion pressure during the first part of the combustion process and above it during the latter part.

[1] R. N. Janeway, Combustion Control by Cylinder-head Design, *S.A.E. Trans.*, **24**, 139 (1929).

Since force equals mass times acceleration, and acceleration is dv/dt,

$$dv = \frac{(P - Fy)\,dt}{M} \qquad (18)$$

where v = vel. of the deflecting mass, in. sec.$^{-1}$,
M = deflecting mass, slugs in.$^{-2}$ of piston area,
dt = duration of force differential, $P - Fy$, sec.,
y = deflection, in.,
F = force required to produce unit deflection, lb. in.$^{-2}$ in.$^{-1}$

If dv is integrated between the limits v_1 and v_2, and the integration of the right-hand side between t_1 and t_2 is indicated, Eq. (18) becomes

$$v_2 - v_1 = \int_{t_1}^{t_2} \frac{P - Fy}{M}\,dt. \qquad (19)$$

The slope of the curve of the restoring force, $\tan \theta$, is indicated by

$$\frac{d(Fy)}{dt} = F\frac{dy}{dt} = Fv = \tan\theta \qquad (20)$$

or
$$v = \frac{\tan\theta}{F}$$

and
$$v_2 - v_1 = \frac{\tan\theta_2 - \tan\theta_1}{F}. \qquad (21)$$

Eliminating $v_2 - v_1$ between Eqs. (19) and (21) results in

$$\tan\theta_2 - \tan\theta_1 = \frac{F}{M}\int_{t_1}^{t_2}(P - Fy)\,dt. \qquad (22)$$

The factor $\int_{t_1}^{t_2}(P - Fy)\,dt$ represents the area between the P and Fy curves. Consequently, the Fy curve is constructed so that between any two points the change in the tangent of the Fy curve is equal to the area between the two curves, multiplied by the constant F/M (Fig. 256).

Obviously, a rough combustion process will have larger areas between the curves than those for a smooth process, the upper area always being equal to the lower area when the Fy curve reaches its maximum value.

Janeway's analysis of a smooth and a rough combustion process (Figs. 255 and 256) resulted in the following values, on the assumption of $F = 80,000$ lb. in.$^{-2}$ in.$^{-1}$ deflection, and $M = 1$ slug.

Item	Smooth	Rough	Relative
Max. rate of pr. rise, lb. in.$^{-2}$ sec.$^{-1}$..	162,300	344,000	2.18
Max. rate of change of restoring force, lb. in.$^{-2}$ sec.$^{-1}$............	188,370	465,000	2.47
Max. accel. in pr. rise, lb. in.$^{-2}$ sec.$^{-2}$..	93,000,000	567,000,000	6.1
Max. accel. in restoring force, lb. in.$^{-2}$ sec.$^{-2}$......................	149,000,000	605,000,000	4.05
Relative K.E. of deflecting mass....	$(1)^2$	$(2.47)^2$	6.10
Max. restoring force, lb. in.$^{-2}$.......	439	543.7	
Max. pr. lb. in.$^{-2}$..................	410	410	
Shock factor, %..................	7	33	4.7

Note.—Shock factor $= \dfrac{\text{Max. } Fy - \text{Max. } P}{\text{Max. } P} \times 100.$

These data and illustrations indicate that smooth combustion will be obtained with comparatively low maximum rates of pressure rise occurring at about half pressure rise or earlier.

Ricardo[1] has found, experimentally, that a pressure rise with a maximum rate of 30 lb. in.$^{-2}$ deg.$^{-1}$ of crank travel produces the maximum power from an engine. When the rate of pressure rise increases to a rate above 35 lb. in.$^{-2}$ deg.$^{-1}$, the turbulence required to produce this rate increases the heat loss and causes a loss in power. In addition the engine becomes rough at the higher rates. Ricardo has also shown that, if the pressure rise is increased at a uniform rate instead of abruptly, high rates of pressure rise may be attained without appreciable roughness.

Whatmough[2] developed the noncompact, nonturbulent, streamlined combustion chamber (Fig. 257) to smooth out the combustion process. Considerable clearance is provided around the valves, and the cylinder-head walls are curved to provide easy flow into and out of the cylinder. The transfer passage between the main part of the combustion chamber and cylinder is large and streamlined so as not to restrict the flow.

[1] H. R. Ricardo, "The High-speed Internal-combustion Engine," p. 108, D. Van Nostrand Company, Inc., New York, 1931.

[2] W. A. Whatmough, Combustion-chamber Design in Theory and Practice, *S.A.E. Trans.*, **24**, 115 (1929).

The Transfer Passage.—High-turbulent or nonturbulent combustion chambers for L-head engines must have a transfer passage from the part of the combustion chamber over the valves to the part over the piston. The amount of close clearance over the piston, compression ratio, valve dimensions, etc., are factors that determine the area of the transfer passage. Jardine[1] found that best results were obtained when the transfer passage had an area in square inches equivalent to 5 to 7 per cent of the piston displacement in cubic inches. The height of transfer passage is measured from the valve-chamber floor to the cylinder head above the edge of the cylinder bore on the transverse center line. The length of the passage may be more or less than the cylinder bore, depending on the combustion-chamber design, volume of combustion-chamber, etc.

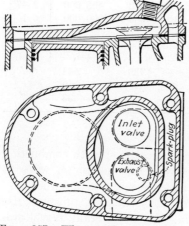

Fig. 257.—Whatmough streamlined nonturbulent combustion chamber.

Clearance above Piston.—The small clearance between part of the piston and cylinder head (Fig. 258) serves to create considerable turbulence at the end of the compression stroke and also provides a comparatively cool pocket for the last part of the charge to be burned. This permits the use of a higher compression ratio without detonation for combustion chambers incorporating this feature. This type of pocket provides a large surface for the volume contained and facilitates heat transfer from the unburned gases which are being compressed and heated during the combustion process.

Ricardo[2] found that a decrease in head clearance in the high-turbulence head from 0.25 to 0.15 in. permitted only a small increase in the highest useful compression ratio. Decreasing the clearance from 0.15 to 0.10 in. allowed the highest useful

---

[1] Frank Jardine, Tests Show How to Design Aluminum Cylinder Heads for Optimum Results, *Auto. Ind.*, **71**, 24 (1934).

[2] Ricardo, *op. cit.*

COMBUSTION-CHAMBER AND CYLINDER-HEAD DESIGN 381

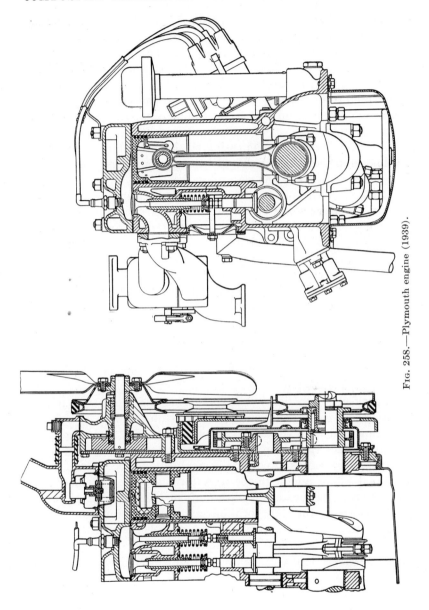

Fig. 258.—Plymouth engine (1939).

compression ratio to be raised from 5.75 to 6.6, and decreasing the clearance to 0.05 raised the ratio to above 7.

Janeway[1] found that an increase in the head clearance of a single-cylinder engine resulted in an increase in detonation. He found a small increase in detonation when the clearance was increased from $\frac{1}{32}$ to $\frac{1}{16}$ to $\frac{3}{32}$ in. and concluded that a clearance of anything up to $\frac{3}{32}$ in. would result in practically the least detonation.

Jardine[2] found that increasing the head clearance resulted in a loss of power, the minimum head clearance used being that provided by the customary head gasket.

Not all of the piston can have a close clearance since there must be a passageway from the cylinder to the main part of the combustion chamber. Consequently, designs vary from small areas to a maximum of about 50 per cent of the piston area.

The amount of charge in the close-clearance space over the piston when the flame reaches it depends upon ignition timing and rate of combustion. Normally, the flame should arrive at this position very shortly after top center. Thus, the position of the piston must be considered in determining the volume of the close-clearance space and the total combustion-chamber volume for any flame-front position. The part unburned at this time $(1 - a)$ may be determined from the volume of the burned $(aV_0'/V_0$, Fig. 240).

Spark-plug Location.—The spark-plug location is very important; there must be a combustible mixture between the gap at ignition under all conditions of operation; also, the spark-plug location fixes the variation in flame-front area, length of flame travel, and time required for the combustion process.

Good idling and high-speed performance are obtained with the spark plug located near the intake valve. High output is obtained with the plug location giving minimum flame travel. Detonation characteristics require a plug location near the hot exhaust valve, and smooth combustion requires a plug location that produces desirable variation in flame-front area and volume of the burned fraction.

---

[1] Janeway, *op. cit.*
[2] Jardine, *op. cit.*

COMBUSTION-CHAMBER AND CYLINDER-HEAD DESIGN   383

Good practice[1] in standard L-head designs is to locate the spark plug between the valves and in the plane of valve centers and at a distance from the exhaust-valve center of 30 per cent of distance between the valves.

Combustion Chambers for Maximum Output.—The power output of a given engine can be increased, as far as combustion-chamber design is concerned, by the following methods:

1. Decreasing the combustion-chamber volume.
2. Providing larger valve and port areas.
3. Streamlining the combustion chamber.
4. Decreasing combustion time with more compact combustion chambers.

Any change in design that improves the antidetonation characteristics of a combustion chamber permits the use of a higher compression ratio which should result in higher output and efficiency.

Larger valves and ports will permit the charge to flow into the cylinder with less pressure drop and consequently will increase the volumetric efficiency. This means more charge per stroke and a proportionate increase in the power output.

Streamlining the combustion chamber is a refinement that further reduces the drop in pressure necessary to induct the charge as well as eject the products of combustion during the exhaust stroke. Both this factor and the larger valves not only increase the power output over the normal speed range, but also raise the speed at which the power curve peaks and thus still further increase the power due to increasing the speed and displacement per minute.

Compact combustion chambers reduce the length of flame travel to a minimum. With a given turbulence, this reduces the time of combustion and minimizes the inherent combustion-time loss (Chap. VI).

Locating the valves of an engine in a single line restricts the valve size to some function of the distance between cylinder centers. Consequently, engines of high output usually have intake valves in one line and exhaust valves in another (Fig. 259). The valve-in-head type of combustion chamber is usually

[1] Bohn Aluminum and Brass Corp., Combustion-Chamber Design, *Power Plant Eng.*, **39**, 169 (1935).

"domed" (Fig. 259a), and for maximum output two inlet and two exhaust valves are used per cylinder. The spark plug is usually located in the top of the dome which makes a compact combustion chamber. The flame-front area increases rapidly with this type of chamber, the length of flame travel is short, and the combustion rate is very rapid.

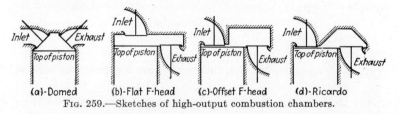

Fig. 259.—Sketches of high-output combustion chambers.

The F-head type (Fig. 259b) has the exhaust valve located at the side of the cylinder and the inlet valve in the head. The charge is fired near the hot exhaust valve, and the flame travels toward the cool inlet valve. The compact, offset, F-head type (Fig. 259c) has small clearance between the piston and the part of the head containing the cool inlet valve. Thus, the last

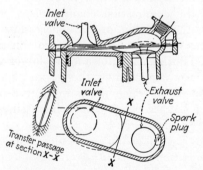

Fig. 260.—Streamlined offset F-head chamber. (*Whatmough.*)[1]

part of the charge to burn is in a cool space of such shape that the flame-front area is reduced near the end of the combustion process.

Ricardo[2] has made use of the domed combustion chamber with the F-head valve arrangement. Placing the spark plug in the top of the dome (Fig. 259d) provides the shortest flame

[1] Whatmough, *op. cit.*
[2] Ricardo, *op. cit.*

COMBUSTION-CHAMBER AND CYLINDER-HEAD DESIGN 385

travel and might result in roughness under certain conditions of operation.

The offset F-head chamber may be streamlined (Fig. 260) and have the spark plug located for long flame travel to obtain smoothness.

The typical valve-in-head engine with its inherently high turbulence was at one time conceded to be the most powerful type of engine. The advent of the high-turbulence L-head combustion chamber increased the performance of this type of engine to equal or better than

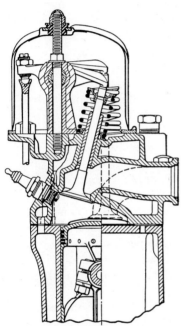

Fig. 261.—Chevrolet combustion chamber (1939).

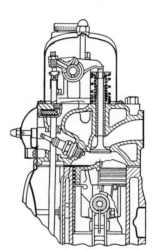

Fig. 262.—Buick combustion chamber (1939).

that of the valve-in-head engine. However, late designs of valve-in-head engines have increased turbulence and volume control. The Chevrolet engine (Fig. 261) has the inlet valve vertical and close to the piston on top center, the exhaust valve being inclined in a raised portion of the combustion chamber. The Buick engine (Fig. 262) has a piston head with an irregular shape that causes the piston head to approach one side of the combustion-chamber wall more closely than the other. The spark plug in both cases is located in a position to promote rapid burning of the first part of the charge.

Combustion-chamber Material.[1]—Tests show that the use of materials of high heat conductivity for the combustion-chamber walls permits higher compression and greater output without detonation. Heat transfer from the inflamed charge is theoretically undesirable since it reduces the thermal efficiency of the engine process. However, heat transfer from the last fraction of the charge to burn may sufficiently influence the detonating characteristics so that there may be more gain obtained from increased compression ratio than loss resulting from the higher heat transfer from the inflamed fraction of the charge.

The use of a material of high heat conductivity or of combustion chambers of improved design may not result in better performance if such parts as the spark plug, exhaust valve, or piston head are operating at such conditions as to prevent the use of higher compression ratio.

Aluminum alloy heads are used extensively in automobile engines and practically exclusively in aircraft engines of appreciable size. Some heads are built of copper or copper alloy, and others have copper inserts located to contact the last part of the charge to burn. The ability of these materials to conduct heat away from a hot spot and the higher specific heats appear in some cases to account for the improved performance.

The combustion-chamber surface should be smooth but not necessarily highly polished, since a thin coating of carbon will soon be deposited on the surface.

The ribs in the water-jacket space of the head should be designed to act as beams supporting the combustion load. Consequently, these ribs should tie in with the material around the head bolts. The bolts in the middle of the head between the two center cylinders should have little clearance, thus locating the head. Appreciable clearance should be allowed the other bolts to permit longitudinal expansion and growth. Bolt

---

[1] F. F. Kishline, Aluminum Cylinder Heads, *S.A.E. Trans.*, **28,** 121 (1933).
H. W. Gillette, Copper-alloy Cylinder Heads, *Auto. Ind.*, **71,** 473 (1934).
H. W. Ristine, Copper Cylinder Heads, *Auto. Ind.*, **71,** 593 (1934).
I. E. Aske, Iron and Copper Bonded in Composite Cylinder Heads, *Auto Ind.*, **73,** 724 (1935).
D. E. Anderson, Let's Take a Step Forward in Engine Design, *Auto. Ind.*, **76,** 498 (1937).

bosses for aluminum-alloy heads should have larger areas of contact for the washers than are used with cast-iron heads.

Combustion-chamber-design Summary.—Many factors must be considered in obtaining a desirable combustion process under various conditions of engine operation. These factors have been summarized diagrammatically by Rabezzana and Kalmar[1] in Fig. 263, which indicates the necessity of both analytical and experimental work to obtain the desired results.

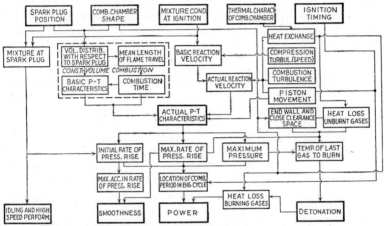

FIG. 263.—Relation of the various factors affecting the actual combustion process. (*Rabezzana and Kalmar.*)

Combustion-chamber Efficiency.—It is customary to base the efficiency of an engine upon the heating value of the fuel supplied the engine, although it is impossible to liberate all of this energy unless an excess of oxygen is supplied. The fact that the power output does not change appreciably as the mixture is made somewhat richer than the maximum-power mixture (Fig. 185) indicates that the energy liberation per unit quantity of air is practically constant for this range of mixtures. Thus the air consumption of an engine may be used as a basis for determining combustion-chamber efficiency when rich mixtures are supplied.

The amount of energy liberated per pound of air for the last three hydrocarbons in Table VIII (Appendix) is about 1380 B.t.u. This may be used as the specific energy input and the thermal

[1] Rabezzana and Kalmar, *op. cit.*

efficiency based upon energy liberated by the air determined as follows:

$$\text{Eff.} = \frac{\text{output}}{\text{input}} = \frac{\text{hp.} \times 2544 \text{ B.t.u. hp.-hr.}^{-1}}{\text{lb. air hr.}^{-1} \times 1380 \text{ B.t.u. lb.}^{-1}}$$

$$= 1.84 \frac{\text{hp.}}{\text{lb. air hr.}^{-1}}. \quad (23)$$

This method of obtaining thermal efficiency, suggested by Ricardo, depends upon providing a rich mixture so that practically all the oxygen in the air enters into the reaction. An engine that shows a high efficiency on this basis has good combustion-chamber design. If fuel-economy tests result in low thermal efficiencies, however, the trouble is with the carburetion and distribution systems. To quote from Ricardo[1]:

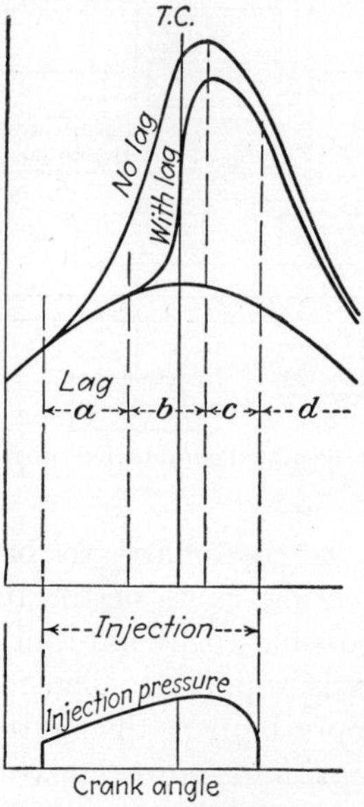

If an engine consumes its air efficiently, then it is an efficient engine and to render it economical in fuel is a question solely of carburetion and distribution. If its air consumption is heavy, then no amount of finessing with carburetor adjustments or distribution design will render it efficient.

FIG. 264.—The compression-ignition process.

FUEL-INJECTION ENGINES

The Combustion Process.[2]—The fuel is injected into the combustion chamber when the piston is near top center, at which time the work of compression has increased the air temperature sufficiently to autoignite the fuel. Combustion of the fuel with no lag would make possible the controlling of the combustion process by the rate of injection. However, ignition lag causes a delay (a, Fig. 264), and considerable fuel may be injected before combustion begins to affect the pressure in the combustion chamber. Rapid combustion of most of the fuel in the chamber

---

[1] Ricardo, *op. cit.*
[2] H. R. Ricardo, Combustion in Diesel Engines, *Auto. Eng.*, **20**, 151 (1930).
G. D. Boerlage and J. J. Broeze, *Chem. Reviews*, **22**, 61 (1938).

COMBUSTION-CHAMBER AND CYLINDER-HEAD DESIGN

follows the development of flame (*b*, Fig. 264), the rise in temperature during this period of the process reducing the time lag to a minimum, and combustion proceeds at about the rate of injection during the third period (*c*, Fig. 264). The combustion process should end very shortly after the third period; but because of poor distribution of the fuel particles, combustion continues during part of the remainder of the expansion stroke.

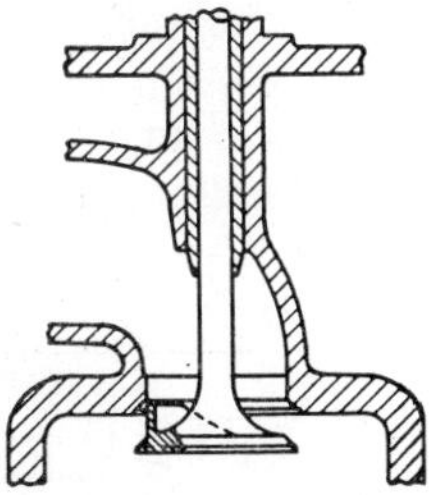

The fuel-injection compression-ignition engine has no flame front proceeding through the combustion chamber from a given starting point as has the spark-ignition engine. Consequently, the design of combustion chambers for fuel-injection engines depends principally upon the spray form of the injected fuel, the degree of atomization, and the means for obtaining the desired air agitation and control of the rate of combustion. The first two items are dealt with in Chap. IX.

Fig. 265.—Shrouded intake valve.

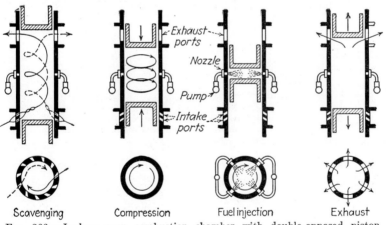

Fig. 266.—Junkers open combustion chamber with double-opposed piston engine.

Air Swirl and Turbulence.—Either the fuel must be injected into all parts of the combustion chamber, or the air must be moved to and through the fuel spray which has a definite shape and position in the chamber. Air is given a "swirling" motion by shrouding the intake valve (Fig. 265), but this reduces the flow area and is not desirable for high-speed engines. A two-cycle engine and a sleeve-valve engine may use tangential ports (Fig. 266) to provide this motion.

Turbulence may be obtained by the use of small clearance between the cylinder head and part of the piston when at top center. This construction displaces air from the small clearance space and results in the same type of agitation as in the high-

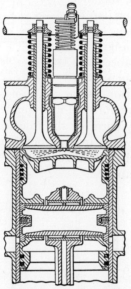

Fig. 267.—General Motors open combustion chamber. Air swirl produced by intake ports.

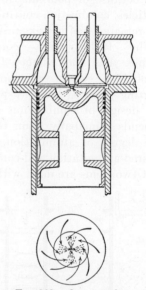

Fig. 268.—Open combustion chamber—high turbulence and some swirl.

turbulence spark-ignition combustion chamber and may be used with all types of combustion chambers.

Injection with high-pressure air may provide the desirable mixing and turbulence. However, solid injection is now almost universally used.

Compression Ratio.—The conditions at the time of fuel injection should be such that ignition of the charge readily occurs under cold starting conditions. The compression ratio required depends upon the size, speed, and type of combustion chamber used with a given engine.

Auxiliary ignition devices such as glow plugs (Fig. 269) eliminate starting difficulties and permit the use of lower com-

COMBUSTION-CHAMBER AND CYLINDER-HEAD DESIGN 391

pression ratios under operating conditions. The use of some form of spark ignition under operating conditions, as in the Hesselman engine, or the use of a hot tube, bulb, or plate, permits the use of compression ratios ranging from 6:1 to 10:1. Engines with moderate compression, ranging from 10:1 to 14:1, usually require some form of ignition for starting but operate without auxiliary ignition devices after warming up.

High-compression engines have compression ratios ranging from about 14:1 to 16:1 and require no auxiliary ignition devices. A compression ratio of 15:1 usually results in compression pressures and temperatures of about 500 lb. in.$^{-2}$ abs.

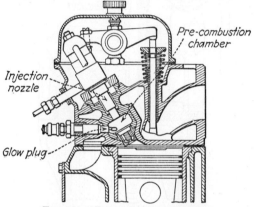

Fig. 269.—Precombustion chamber.

and 1200°R., respectively, depending upon initial conditions, heat loss, and leakage during compression.

Open Combustion Chambers.—Open or direct-injection combustion chambers (Figs. 266 to 268) usually have a shape to conform to the spray form. Air swirl is usually employed to complete the mixing of the air and fuel. High injection pressures are required for producing the desired atomization.

An open-chamber design with high turbulence (Fig. 268) may or may not have appreciable swirl depending upon the intake-port design or the shrouding of the valve. The chamber may have various shapes, and the injection nozzle may be offset from the center of the chamber.

Precombustion Chambers.—These chambers are divided into two parts, one between the piston and the cylinder head, the other in the cylinder head (Fig. 269). Comparatively small passageways connect the two chambers. Fuel is injected into the precombustion chamber, and under full-load conditions sufficient air for complete combustion is not present in this chamber which is usually the smaller of the two chambers. Partial combustion of the fuel discharges the burning mixture through the small passageways into the air in the various parts of the second chamber where the combustion is completed. This type of combustion chamber produces a smooth combustion process but has high fluid-friction and heat-transfer losses.

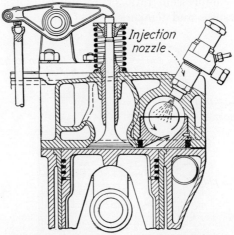

Fig. 270.—Antechamber with high turbulence.

Antecombustion Chambers.—A modification of the precombustion chamber has resulted in the antechamber type with a passageway offering a much milder restriction and with the volume of the chamber over the piston reduced to a minimum. Thus, sufficient air is usually present in the antechamber to burn completely all but the overload quantities of fuel injected.

The small clearance over the piston produces a high tangential velocity in the antechamber (Fig. 270); this distributes the fuel injected radially into the chamber and provides high turbulence, which promotes rapid combustion. Part of the antechamber containing the transfer passage is usually not well cooled, which provides a regenerative action; the antechamber receives heat from the combustion products and gives up heat to the entering air.

The velocity through the transfer passage depends upon the piston area and velocity, the area of transfer passage, and the

COMBUSTION-CHAMBER AND CYLINDER-HEAD DESIGN

relation between the volume of the antechamber and the volume in the cylinder above the piston. Actually, a pressure difference must exist to cause flow, but a fair approximation of the velocity may be made by assuming equal pressures. Then, any change in volume due to piston displacement may be assumed to be proportionately divided between the two chambers. Thus,

$$\text{Vel.} = \frac{\text{vol. transferred}}{\text{area of passage}} = \frac{V_{ac}}{V_c A_t} \times \frac{A_p \, ds}{d\theta} = \frac{\text{in.}}{\text{rad.}} \quad (24)$$

where V_{ac} = vol of antechamber, in.^{3},
A_p = area of piston, in.^{2},
s = distance of piston from top center,
$V_c = A_p s + V_{cl}$, in.^{3},
V_{cl} = cl. vol. not including precombustion chamber volume,
A_t = area of transfer passage, in.^{2},
θ = crank angle, rad.

Substituting for V_c its equal as defined and introducing the necessary constants result in Eq. (24) becoming

$$\text{Vel.} = \frac{V_{ac} A_p}{A_t} \times \frac{ds/d\theta}{A_p s + V_{cl}} \times \frac{\pi \text{r.p.m.}}{360} = \text{ft. sec.}^{-1} \quad (25)$$

Making use of Eq. (3), Chap. XVII, results in

$$\text{Vel.} = \frac{V_{ac} A_p}{114.6 A_t} \times \frac{\left(r \sin \theta + \frac{r^2}{2l} \sin 2\theta\right) \times \text{r.p.m.}}{A_p r \left[(1 - \cos \theta) + \frac{r^2}{4l}(1 - \cos 2\theta)\right] + V_{cl}}. \quad (26)$$

Example.—Given an engine with a piston area of 15 in.^{2} and a stroke of 5 in. A 16:1 compression ratio indicates a total clearance volume of 5 in.^{3} of which 4 in.^{3} are in the antechamber and 1 in.^{3} is in the cylinder. The transfer passage is ½ in. × ½ in. The length of the connecting rod is 10 in. Determine the theoretical velocity through the transfer port when the crank is 45 deg. from top center and the engine is running at 2000 r.p.m.

Substituting in Eq. (26) results in

$$\text{Vel.} = \frac{4 \times 15}{114.6 \times 0.5 \times 0.5} \times \frac{\left(2.5 \times 0.707 + \frac{6.25}{20} \times 1\right) \times 2000}{15 \times 2.5 \left[(1 - 0.707) + \frac{2.5}{40}(1 - 0)\right] + 1}$$

$$= 162 \text{ ft. sec.}^{-1}$$

One design (Fig. 271) restricts the transfer passage as the piston approaches top center, thus obtaining a higher velocity at top center than with a passage of constant area. High fluid-friction loss is inherent with high fluid velocity. Consequently, engine performance must be improved to justify any change in design that increases fluid velocities through the transfer port.

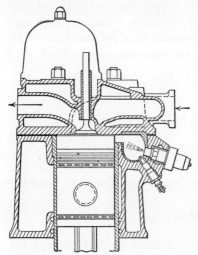

Fig. 271.—Hercules combustion chamber with variable antechamber orifice.

Air-cell Combustion Chambers.—These combustion chambers (Figs. 272 and 273) are divided into two compartments with a restricted passage between them. In some cases (Fig. 272) the fuel is injected toward the transfer passage, but little if any fuel is supposed to enter the outer compartment or air cell. Some combustion takes place in the open part of the combustion chamber, and combustion is completed on the down stroke of the piston while air is discharged from the air cell into the partly burned mixture.

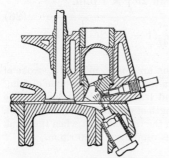

Fig. 272.—Air-cell combustion chamber with glow plug.

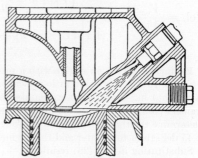

Fig. 273.—M.A.N. air-cell combustion chamber.

The Lanova combustion chamber (Fig. 274) has two air chambers. Fuel is injected across the main chamber, some entering and igniting in the first auxiliary chamber. This

COMBUSTION-CHAMBER AND CYLINDER-HEAD DESIGN 395

raises the pressure in the auxiliary chambers, and the burning mixture is discharged into the main chamber which is designed to produce a double swirl. The air in the second auxiliary chamber acts the same as in the typical air cell. The use of the two auxiliary chambers and a starting valve that can shut off

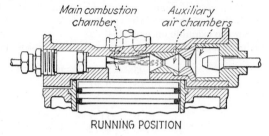

RUNNING POSITION

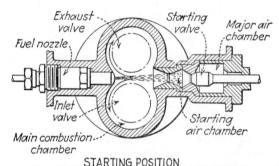

STARTING POSITION

Fig. 274.—Lanova air-cell combustion chamber.

the main air chamber makes possible a higher compression ratio for starting.

EXERCISES

1. Determine the compression pressures, in pounds per square inch gage, for compression ratios from 4.5:1 to 8:1, for suction pressures of 13.0, 14.7, and 16.0 lb. in.$^{-2}$ abs. and a suction temperature of 150°F. Plot compression pressures against compression ratios. Use Chart B (Appendix).

2. Determine the temperature of the burned and the flame-front position when 30 per cent of the mass of the charge has been burned and a heat loss of 7 per cent has occurred. $T_1 = 1200°R$.

3. Determine the pressure in the combustion chamber for the conditions in Prob. 2. The compression ratio is 6:1 and the suction pressure is 12 lb. in.$^{-2}$ abs.

4. Plot the pressure in the combustion chamber against mass burned and against flame-front position. Compression pressure is 100 lb. in.$^{-2}$ gage and compression temperature 1100°R. Use Figs. 240 and 241.

5. Plot mass burned against volume of the burned fraction for the conditions in Fig. 240.

6. Determine the final temperature for adiabatic combustion of the first $\frac{1}{100}$ and last $\frac{1}{100}$ of the charge to burn. Initial conditions of the correct octane-air mixture are 100 lb. in.$^{-2}$ abs. and 1000°R. Assume constant-volume combustion.

7. Determine the amount of charge burned at point 3, Fig. 245. The pressure at 3 is 375 lb. in.$^{-2}$ abs., and the volume is $0.7V_1$. Other conditions are the same as in the example on page 364.

8. Given the following data regarding the combustion process in Fig. 246:

θ, deg.,	-25	-23.2	-20.8	-18.4	-16.0
V, in.^{3},	10.15	9.91	9.61	9.34	9.11
P, lb. in.$^{-2}$ abs.,	82	85	89	93	97
θ, deg.,	-13.6	-11.2	-8.8	-6.4	-4.0
V, in.^{3},	8.90	8.72	8.58	8.47	8.40
P, lb. in.$^{-2}$ abs.,	107	123	155	197	249
θ, deg.,	-1.6	$+0.8$	3.2	5.6	8.0	10.4
V, in.^{3},	8.38	8.38	8.39	8.44	8.54	8.67
P, lb. in.$^{-2}$ abs.,	288	309	319	334	344	348
θ, deg.,	12.8	15.2	17.6	20.0	22.4	24.8
V, in.^{3},	8.84	9.04	9.26	9.52	9.81	10.13
P, lb. in.$^{-2}$ abs.,	346	340	335	327	319	305

Determine the end of the combustion process, the part burned for each interval, and the equivalent pressure rise due to combustion for each interval. Compare the two sets of results on a percentage basis.

9. A spherical combustion chamber has a volume of 10 in.^{3} Determine and plot the flame-front area vs. flame travel for a spark-plug location at the center, at the chamber wall, and with two spark plugs located diametrically opposite and in the chamber wall.

10. A cylinder having a compression ratio of 5.81:1 has a combustion chamber in the form of a spherical segment of which the top of the piston forms the base of the segment. The bore of the cylinder is 3.5 in., and the stroke is 5 in. Suction temperature and pressure are 200°F. and 14.7 lb. in.$^{-2}$ abs., respectively. Use the correct mixture of octane and air.

a. Determine the combustion-chamber volume, the height of the spherical segment, and the spherical radius.

b. Determine the area of the flame front for a spark plug centrally located in the top of the head, for various lengths of flame travel.

c. Plot the flame-front area against the length of flame travel.

d. Determine the volume of the burned portion for various positions of the flame front.

e. Determine the mass burned and the temperature of the unburned portion for the flame-front positions determined in (*d*).

f. Determine the pressure ratio and temperature ratio for the various flame positions.

g. Determine the theoretical mass rate of reaction for the various flame positions.

h. Plot both the mass rate of reaction and the mass against the flame travel in inches.

i. Determine the relative time between various flame positions and the time in per cent to reach the various positions, and plot the latter against the flame travel in inches.

j. Plot mass and pressure ratio against time in per cent and compare with results in Fig. 254, which are for the same conditions but different combustion-chamber shape and spark-plug location.

11. A $3\frac{1}{2}$- by 5-in. engine with a compression ratio of 5.81 has a clearance space of $\frac{1}{16}$ in. covering half the piston area. The flame reaches this space at 0, 5, 10, and 15 deg. after top dead center under various conditions. The connecting-rod-crank ratio is 5:1. Determine the volume and mass of the charge in the small clearance space when the flame reaches it. Also, plot mass in the small clearance space vs. crank angle when flame reaches the space. Heat loss from the burned fraction is 10 per cent.

12. A test on an engine resulted in a fuel rate of 1 lb. of fuel hp.-hr.$^{-1}$ The measured air consumption was 10 lb. of air hp.-hr.$^{-1}$ Determine the efficiency based upon the fuel consumption and air consumption. What causes the high fuel rate?

13. Determine and plot the velocity through the transfer passage to the precombustion chamber for various crank angles on the compression stroke. Use the data from the example on page 393, with the exception of the transfer port which is 1 by 0.5 in.

14. Assume a Hercules combustion chamber with the transfer passage 1 in. wide and 0.5 in. high. The piston is 0.1 in. below the top of the passage when on top center. Modify Eq. (26) to fit this case, and determine and plot the velocity through the passage for the same conditions as in Prob. 13.

CHAPTER XIV

ENGINE LUBRICATION

Mechanics of Fluid Lubrication.—An engine has various parts that are in motion and must be guided and kept in definite relationship with other parts. This indicates more or less contact between surfaces moving relative to each other, which is minimized by the introduction of an oil film between the surfaces. The oil film adheres to both surfaces and offers resistance to the shearing of the film when relative motion of the surfaces occurs.

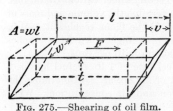

Fig. 275.—Shearing of oil film.

The force per unit area of oil film, F/A (Fig. 275), required to move one of the surfaces relative to the other is directly proportional to the rate of shear v/t and the viscosity μ.

Thus,
$$\frac{F}{A} = \mu\left(\frac{v}{t}\right) = \mu\left(\frac{dv}{dt}\right). \tag{1}$$

Equation (1) defines the *absolute viscosity* μ as the force per unit area required to move a surface at unit velocity when it is separated by a fluid film of unit thickness from a stationary surface. The unit of absolute viscosity is the poise.

Thus,
$$\begin{aligned}
1 \text{ poise} &= 1 \text{ dyne cm.}^{-2} \text{ (cm. sec.}^{-1})^{-1} \text{ cm.} \\
&= 1 \text{ dyne sec. cm.}^{-2} \\
&= 2.09 \times 10^{-3} \text{ lb. sec. ft.}^{-2}
\end{aligned} \tag{2}$$

The application of a load or force to the moving surface, some component of which is perpendicular to the direction of motion and toward the stationary surface, will force the fluid film out of the space between the two surfaces and result in metal-to-metal contact. A reversal of this force will increase the clearance space and permit a thicker oil film. Thus, a film of oil may be maintained between two surfaces if the time required to force the film out of the space between the two surfaces is greater than the time between load reversals, provided that a supply of lubricant is available to fill the clearance space when this is increasing.

If one surface is inclined at a small angle to the other surface (Fig. 276), the relative motion of the surfaces will develop an oil-film pressure that will support a load. On the assumption that the bearing surfaces are of infinite width, which is equivalent

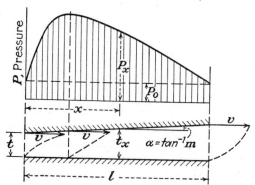

Fig. 276.—Development of oil-film pressure with inclined surfaces.

to the assumption of no end leakage of the oil, and that the oil adheres to the surfaces,

$$P = P_o + \frac{6\mu vm}{(2t + ml)} \cdot \frac{lx - x^2}{(t + mx)^2} \qquad (3)*$$

where P = pr., lb. ft.$^{-2}$ at a distance x, ft.,
P_o = external pr. on oil film,
v = vel., ft. sec.$^{-1}$ of surface motion,
t = smallest clearance, ft.,
l = surface length, ft.,
m = tangent of angle of inclination.

The solution of Eq. (3) for a given set of data (see Piston-ring Lubrication) provides information regarding the pressure distribution (Fig. 276). The pressure is a maximum near the position of smallest clearance.

The area beneath the total-pressure curve is a measure of the load that can be supported by the oil film, provided that the

* See R. A. Dodge and M. J. Thompson, "Fluid Mechanics," p. 462, McGraw-Hill Book Company, Inc., New York, 1937.

external pressure P_o does not also act downward on the upper surface. Equation (3) indicates that an increase in loading decreases the clearance between the surfaces. Increasing the load sufficiently will cause the surface irregularities to break through the oil film and come in contact with each other. Thus, a high degree of surface smoothness[1] is desirable to minimize wear under thin-film conditions.

Surface irregularities are measured with a "profilometer" in microinches (millionths of an inch). Some surfaces in regular production have average irregularities as low as 1 to 2 microinches.

The *coefficient of friction* is defined as the ratio between the resistance due to friction in the direction of motion and the load carried normal to the plane of motion. This coefficient varies appreciably with the type of bearing and its lubrication (Table 24).

TABLE 24.—COEFFICIENTS OF FRICTION*

Type of bearing	Range	Average
Unlubricated or very poorly lubricated	0.10 to 0.40	0.160
Semilubricated	0.01 to 0.10	0.030
Well lubricated	0.002 to 0.010	0.006
Roller bearings	0.002 to 0.007	0.005
Ball bearings	0.001 to 0.003	0.002

* T. C. Thomsen, "The Practice of Lubrication," McGraw-Hill Book Company, Inc., New York, 1926.

Operating Conditions.—The operating conditions of the lubricated parts in an internal-combustion engine vary considerably. The cylinder surfaces are subjected to flame temperatures. The top of the piston is subjected to the same condition, which results in high temperatures for the ring section and upper part of the piston. The pistons exert low to moderate pressures against the cylinder wall, reversing the direction at least several times during each cycle. The mean piston speed is high. Thus, there must be supplied to the cylinder wall a lubricant satisfactory under conditions of high temperature, high speed, and various pressures.

[1] E. J. Abbott and F. A. Firestone, New Profilometer Measures Roughness of Finely Finished and Ground Surfaces, *Auto. Ind.*, **69**, 204 (1933).

ENGINE LUBRICATION

The lubricant on the cylinder wall must also perform the function of sealing the piston, so that only a small percentage of the gases in the combustion chamber can escape past the piston and rings.

The conditions for the piston pin are high pressures and low speed, the motion being oscillatory through an angle of about 30 to 45 deg. The temperatures are moderately high.

The crankshaft and crankpin bearings operate at high speeds and moderate pressures if properly balanced and if light reciprocating parts are used. The temperatures are comparatively low particularly if considerable oil is circulated through the bearings and if the oil temperature is controlled.

The camshaft and valve-mechanism bearings are subject to moderate loads and fairly low speeds. The temperatures are low except in the ends of the valve-stem guides.

These operating conditions can be summarized as follows:

Bearing	Motion	Condition		
		Pressure	Temperature	Speed
Pistons................	Reciprocating	Moderate	High	High
Piston pins............	Oscillating	High	Moderate	Low
Crank and crankpin....	Rotating	Moderate to high	Moderately low	High
Camshaft and valve mechanism.	Rotating and reciprocating	Low	Low	Low
Exhaust-valve stems...	Reciprocating	Low	High	Low

A lubricant of high viscosity is required to sustain a heavy load in a given bearing (Eq. 3). An increase in temperature reduces the viscosity of a lubricant. Thus, high-temperature conditions reduce the load-carrying properties of lubricants. An increase in speed increases the pressure developed in an oil film, and consequently the higher the speed the lower the viscosity required to sustain a given bearing loading. Viscosity is apparently an important property.

The conditions in the internal-combustion engine are so varied that one lubricant could scarcely be expected to produce ideal lubrication conditions in all the bearings. However, satisfactory results in a well-designed engine are obtained with a lubricant having all the desirable properties in varying degrees.

Properties of Lubricants.—Lubricants obtained by the refining of petroleum are used in internal-combustion engines. Not long ago it was thought that such lubricants should not include any addition agents not derived from petroleum. However, the use of other addition agents to improve the properties of a lubricant is now common practice.

The customary laboratory inspection provides data concerning various properties, the more common of which and their significance are as follows:[1]

The gravity, degrees A.P.I., of a lubricant is an index of its specific gravity. It is of little significance as an index of the quality of a lubricant.

The *flash point* of an oil is the temperature at which a flash appears on the oil surface when a test flame is applied under specified test conditions. It is a rough indication of the tendency to vaporize.

The *fire point* is the temperature at which the oil ignites and burns for at least 5 sec. under specified test conditions. It is of little significance.

The *pour point* is determined by noting the temperature at which the oil will not flow when under specified test conditions. It indicates the temperature below which oil will not flow to the oil-pump suction line. If dewaxed, the paraffin-base oils may have a lower pour point than asphalt-base oils.

The *viscosity* of an oil is determined by the time required for a given quantity to flow through a capillary tube under specified conditions. For all but very small capillary tubes,

$$\text{Kin. visc.} = \nu = \frac{\mu}{\delta} = kt \text{ stokes} \qquad (4)$$

where μ = abs. visc., poises,
δ = mass density, g. cm.$^{-3}$ (sp. g. with c. g. s. units),
k = const. for the viscometer,
t = time for flow, sec.

[1] The comments regarding the properties of lubricants are based upon the following reports:
A.S.T.M. Standards on Petroleum Products and Lubricants, Issued Annually. *A.S.T.M. Com. D-2 Report*, The Significance of Tests of Petroleum Products, 1934.

The relation between kinematic viscosity and Saybolt universal viscosity is

$$\text{Kin. visc. (centistokes)} = 0.226t - \frac{195}{t} \qquad (5)$$

when $t = 100$ sec. Saybolt universal or less,

$$\text{Kin. visc. (centistokes)} = 0.220t - \frac{135}{t} \qquad (6)$$

when $t = 100$ sec. Saybolt universal or more.

Example.—The kinematic viscosity of a lubricant at 100°F. is 40 centistokes. Determine the Saybolt universal and absolute viscosity. Specific gravity at 100°F. is 0.85.

From Eq. (4), $\mu = 0.85 \times 40 = 34$ centipoises.

From Eq. (6) $40 = 0.220t - \dfrac{135}{t}$

or $t = 185$ sec. Saybolt universal visc.

The *carbon residue* is determined (*Conradson* carbon test) by evaporating, under specified test conditions, a known weight of oil and weighing the residue. It is not a sure indication of the relative carbon deposits that may form in an engine. Paraffin-base oils usually show a higher Conradson carbon value than do the asphalt-base oils.

S.A.E. Viscosity Classification.—The S.A.E. has adopted a series of numbers (Table 25) that classify engine lubricants in ranges of viscosity at one temperature only.

TABLE 25.—VISCOSITY RANGES FOR S.A.E. VISCOSITY NUMBERS

S.A.E. visc. no.	Visc. range, Saybolt universal sec.			
	At 130°F.		At 210°F.	
	Min.	Max.	Min.	Max.
10	90	Less than 120		
20	120	Less than 185		
30	185	Less than 255		
40	255	....	...	Less than 80
50	...	....	80	Less than 105
60	...	....	105	Less than 125
70	...	....	125	Less than 150

NOTE.—In the case of prediluted oils, S.A.E. viscosity numbers by which the oils are classified shall be determined by the viscosity of the undiluted oils.

404   INTERNAL-COMBUSTION ENGINES

The viscosity of lubricants varies with pressure, temperature, the base of the oil, and also with addition agents. This has led to the development of the A.S.T.M. viscosity-temperature chart (Fig. 277) on which the variation of viscosity with temperature appears as a straight line which makes extrapolation possible in the range between the cloud point[1] and the initial boiling point.

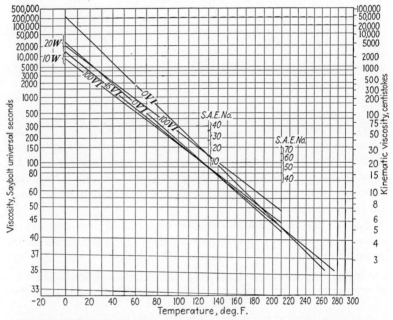

Fig. 277.—A.S.T.M. tentative viscosity-temperature chart for liquid petroleum products (D 341-37 T). Modified by omitting some lines and scale values, and adding the kinematic-viscosity scale.

The plot of the viscosities of the two lubricants from different base stocks shows that the viscosity classification is no indication of the viscosity at temperatures other than the given temperature for the classification. This has led to the adoption of the W classification for winter lubricants by practically all the automobile manufacturers in order to obtain satisfactory starting characteristics.

[1] The temperature at which wax and other substances begin to crystallize out or separate from solution is the cloud point.

Visc. no.	Saybolt universal visc., sec. at 0°F.	
	Min.	Max.
10-W	6,000	12,000
20-W	12,000	50,000

Though an engine requires a low-viscosity oil at starting temperatures, there is a minimum viscosity for operating conditions below which the lubrication film will be broken through and excessive wear or destruction of the bearings may result. The viscosity of a lubricant at operating conditions depends upon the viscosity at any given temperature, and the viscosity-temperature characteristic.

Viscosity Index.—The viscosity index is a measure of the change of viscosity with temperature of an oil compared to two reference oils having the same viscosity at 210°F., one of naphthenic base and the other of paraffinic base. If A, B, and x are the viscosities in Saybolt Universal seconds at 100°F. for the naphthenic, paraffinic, and unknown oils, respectively,

$$\text{Visc. index (V.I.)} = 100 \frac{A - x}{A - B}. \qquad (7)*$$

Thus, the naphthenic reference oil has a V.I. of 0, whereas the paraffinic reference oil has a V.I. of 100. Evaluation of V.I. is facilitated by the use of a diagram (Fig. 278) based on the foregoing relation and reference oils.

Example.—The viscosity of a lubricant is 400 and 55 Saybolt universal sec. at 100 and 210°F., respectively. Determine the V.I. of the lubricant.

The intersection of the two lines representing the given viscosities occurs at a V.I. of 78 (Fig. 278).

Processes have been developed that result in high V.I. oils until now lubricants are available with V.I. values ranging from about 0 to well over 100. Three oils having the same low-temperature viscosity for equal cold-starting characteristics may have the minimum desirable viscosity at different operating temperatures. Thus, the low V.I. oils are as satisfactory as

---

* E. W. Dean and G. H. B. Davis, Viscosity Variation of Oils with Temperature, *Chem. and Met. Eng.*, **36**, 618 (1929).

high V.I. oils from a viscosity standpoint, under milder operating conditions. The low V.I. oils might be unsatisfactory at severe operating conditions because of lower viscosity.

Oil Stability.—Lubricating oils are subjected to high temperatures in internal-combustion engines, particularly the oil in the

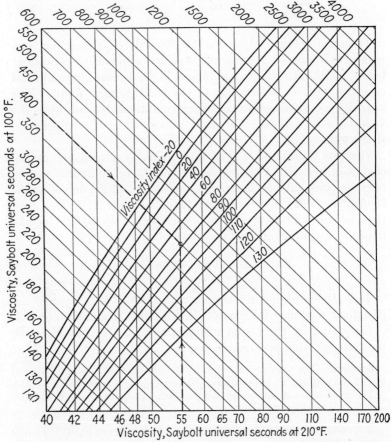

Fig. 278.—Viscosity-index chart. (*Dean and Davis Scale, Union Oil Co. of Cal.*)

"ring belt" and that which contacts the underside of the piston head. High temperatures promote oil decomposition, which is accelerated because of the presence of oxygen. This results in sludge and acid formations in the crankcase, gumming and sticking of the rings, and carbon formation on the underside of the piston.

Lubricating oils have a natural high stability or resistance to change under engine operating conditions that may be improved with some addition agents. However, an addition agent that may improve stability may lower the value of another property and make the oil less desirable for some types of service.

The determination of the stability of engine oils under actual engine conditions appears to be the best method for rating oils for this characteristic. However, variation in engine and operating conditions will change the stability rating of lubricants.

The cost and time required to run engine stability tests have led to the development of accelerated laboratory methods, such as the Sligh and the Indiana oxidation stability tests. Such tests[1] expose the surface of the oil under definite pressure and temperature conditions to air or oxygen, or aerate the oil under specified conditions. At present the correlation with engine tests is not entirely satisfactory.

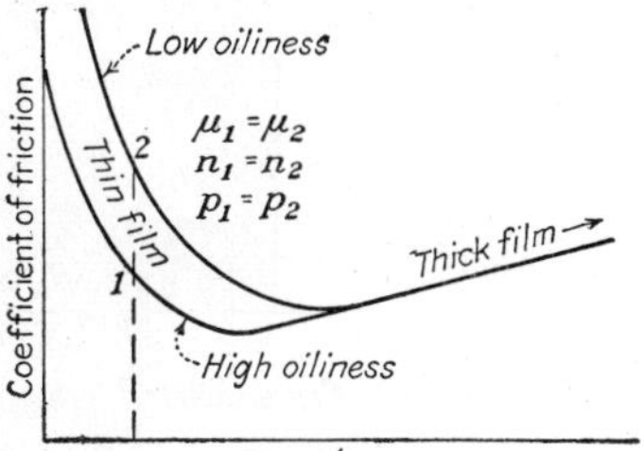

Thin-film Lubrication.—Thin-film lubrication occurs when the conditions of bearing operation reduce the thickness of the oil film sufficiently to make the rubbing of the bearing surfaces a near possibility. The term "high-film strength" is given to oils that offer high resistance to the incipient scarring or scuffing of the bearing surfaces under such operating conditions.[2]

FIG. 279.—Thin-film friction and oiliness.

The coefficient of friction for a journal bearing passes through a minimum when plotted against the variable $\mu n/p$ (Fig. 279). Thin-film lubrication exists to the left of this minimum where the coefficient of friction increases with a decrease in $\mu n/p$. Two lubricants should have the same friction coefficient in the same bearing with the same $\mu n/p$ condition and thick-film lubrication but may have different coefficients of friction (Fig. 279) under thin-film conditions at the same viscosity μ, load p, and speed n. This difference indicates the oiliness property[3] of the oil.

[1] D. P. Barnard, E. R. Barnard, T. H. Rogers, B. H. Shoemaker, and R. E. Wilkin, Sludge Formation in Motor Oils; *S.A.E. Trans.*, **29**, 167 (1934).

[2] The Timken, Almen, and S.A.E. machines are used for the relative determination of film strength.

[3] M. D. Hersey, Logic of Oiliness, *Mech. Eng.*, **55**, 561 (1933).

Lubrication Systems.—Small and low-duty engines may have individual oil supplies for each bearing, in which case the oil is usually wasted after escaping from the bearings. Bearings in a heavy-duty engine require a continuous oil supply usually under appreciable pressure, which accounts for the various self-contained oiling systems.

The *splash system* pumps oil from the sump to troughs, into which the connecting rods dip or splash and force oil into the

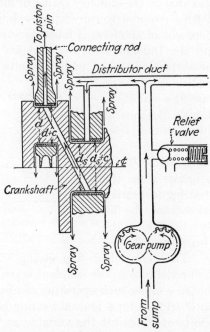

Fig. 280.—Diagrammatic sketch of full force-feed lubrication system (c = diametral clearance).

connecting-rod bearings. Splash and spray from the connecting rods distribute oil to the cylinders and various other bearings, troughs and grooves being provided to direct the oil to the bearings. The oil leaks out of the various bearings and returns to the sump.

The *force-feed system* supplies a large quantity of oil under considerable pressure to the main- and connecting-rod bearings. A gear pump (Fig. 280) takes the oil from the sump and delivers it to a distributor duct which connects with all the main bearings.

The crankshaft is drilled to provide an oil passage to the connecting rod which is also drilled to provide an oil passage to the piston pin. Considerable oil escapes through the clearance spaces at the ends of the main and connecting-rod bearings, some of which is sprayed on the cylinder walls. Under cold-starting conditions an appreciable amount of time is usually required before sufficient oil reaches the cylinder wall, since the oil flows slowly through the small crankpin-bearing clearances from which it is sprayed on to the walls. One method of oiling cylinders, under starting conditions, consists of having a hole in the rear side of the upper half of the crankpin bearing. This hole registers with the oilhole in the crankpin at such a position that oil will be projected onto the cylinder wall at low speeds. The position of the projected oil stream with regard to the cylinder bore varies with speed so that at high speed no oil will reach the cylinder bore from this source.

A spring-loaded pressure-regulating valve maintains the desired pressure by bleeding some oil from the system. All the oil that leaks from the bearings returns to the sump and is recirculated. Oil pressure is no indication of the quantity of oil being pumped, for with small bearing clearances not much oil can flow and the pressure would be high. Also, with larger clearances and oil of high viscosity, the quantity flowing would be small and the pressure would be high. Finally, with large clearance and oil of low viscosity, the quantity flowing may be large and the pressure low.

Large aircraft engines operate with a dry crankcase system. This necessitates the use of a scavenging pump for removing the oil from the crankcase and pumping it to the oil reservoir. The oil reservoir serves as a sump from which the oil is pumped into the force-feed lubrication system.

The quantity of oil that must be circulated to maintain the desired oil pressure at the various bearings depends principally upon the viscosity of the oil and the bearing clearances, both radial and end.

The effect of centrifugal force must be carefully considered in engines with large crankshaft bearings, particularly at high speeds and in cases where the main bearings are large in diameter. In some cases the pressure in the crankshaft duct may offset the oil pressure sufficiently to produce disastrous results.

The expression for centrifugal force in pounds is

$$F = 0.0003410 wrn^2 \tag{8}$$

where w = wt. of the body, lb.,
r = radius of the curved path, ft.,
n = r.p.m.

Any increment of length of duct Δr (Fig. 281) will contain a weight of oil w equal to $A \Delta r \delta$, where A is the area of cross section of duct and δ is the density of the oil. The centrifugal force of this element, which may be considered at any radius r from the center of rotation, is

$$\Delta F = 0.0003410 A \, \delta n^2 r \, \Delta r,$$

or
$$F = 0.0003410 A \, \delta n^2 \int_{r=0}^{r=r} r \, dr$$
$$= 0.0001705 A \, \delta n^2 r^2 \tag{9}$$

where r = max. length of oil duct measured radially from the shaft center.

Example.—Determine the oil pressure set up by centrifugal force in the oil duct in the main bearing part of the crankshaft which has a diameter of 6 in. Speed = 3600 r.p.m. Oil density = 56 lb. ft.$^{-3}$

Substituting the values given into Eq. (9), using a duct cross section of 1 in.^{2}, results in

$$F = 0.0001705 \times \tfrac{1}{144} \times 56 \times \overline{3600}^2 \times \overline{0.25}^2$$
$$= 54 \text{ lb. in.}^{-2}$$

At a speed of 1800 r.p.m., or with a 1½-in. radius, the maximum pressure would be one-fourth as much, or 13.5 lb. in.$^{-2}$

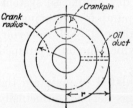

Fig. 281.—Oil duct into center of crankshaft of barrel-type crankcase.

Oil Cooling.—The temperature to which an oil will rise during engine operation depends upon the conditions of operation. In any case, the temperature will rise until the heat transmitted from hotter parts and that generated by friction in the fluid films balance the heat transferred from the oil to the cooler engine parts.

Low-output engines usually have sufficient crankcase surface from which heat can be transferred to the surrounding air to prevent the oil from rising to an undesirable temperature. High-output engines have a comparatively small crankcase area, and in such engines the oil temperature may continue to rise and reduce the viscosity until the connecting-rod bearings are destroyed. Obviously, it is desirable to prevent the oil temperature from rising sufficiently to reduce the viscosity to the danger point for the conditions of operation. The danger-point viscosity varies considerably but is probably in the neighborhood of 30 to 40 Saybolt universal sec.

Oil coolers are constructed of radiator sections and use air or water for the cooling medium.

Oil Consumption.—There are four ways in which oil can be consumed in an engine in good condition:

1. Burning in the combustion chamber.
2. Loss as oil mist or vapor from the breather or in the exhaust gases.
3. Leakage where the crankshaft or other shafts protrude from the crankcase.
4. Decomposition of the oil.

The amount of oil that will be consumed in the combustion chamber depends upon the amount pumped into it by the piston and rings. This depends primarily upon the amount of oil thrown on the cylinder walls, the design of the pistons and rings, and their relation to each other and to the cylinder bores. Some oil consumption is desirable or the upper end of the cylinder bores would run dry and excessive wear occur. High oil consumption usually results in appreciable carbon formation, fouling of plugs, and smoky exhaust. Also, any wear of pistons, rings, and cylinders after running in or any "out-of-roundness" will increase the oil consumption. Thus, it appears impossible to obtain the desirable amount of lubrication at the upper end of the cylinder bores under all conditions of speed and load with the customary lubrication systems, for this part of the bores is fairly dry at low speed and oversupplied with oil at high speeds if average oil consumption is attained.

Oil having high-volatility characteristics will cause appreciable oil consumption under high-temperature operating conditions. Crankcase breathing will cause the loss of oil vapor and particles in suspension in the crankcase.

Leakage should be negligible with good construction, tight joints, and oil slingers which throw the oil off the rotating shafts and prevent leakage.

The oil consumption due to decomposition is usually negligible. However, under some conditions of operation, decomposition may account for an appreciable part of the oil consumption.

The principal factors affecting oil consumption in a well-designed engine that is not appreciably worn are

1. Engine speed.
2. Engine output.
3. Oil viscosity.
4. Quantity of oil circulated.

Everett and Stewart[1] investigated the effect of various factors on oil consumption in three Dodge engines (1933 models),

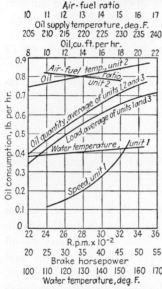

Fig. 282.—Effect of variables on oil consumption. (*Everett and Stewart.*)

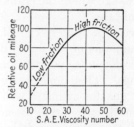

Fig. 283.—Relative effect of oil viscosity on oil consumption. (*The Texas Co.*)[2]

arbitrarily taking 3250 r.p.m., 48 b.hp., 230°F. oil supply, 16 ft.³ of oil circulation per hour, 14:1 air-fuel ratio, and 160°F. outlet-water temperature as the standard conditions and varying one condition at a time to determine its effect (Fig. 282). All investigators have found that increasing the engine speed at high speeds has more effect on oil consumption than any other factor. More oil is thrown on the cylinder walls, and the piston rings are less effective in controlling the quantity of oil that reaches the combustion chamber at high speeds.

Tests indicate that oil mileage is a maximum in a given automobile engine with an oil of definite viscosity and that lower mileage is obtained with both higher- and lower-viscosity oils (Fig. 283). Apparently, the high-viscosity oils provide thicker

[1] H. A. Everett and F. C. Stewart, Performance Tests of Lubricating Oils in Automobile Engines, *Penn. State Coll., Bull.* 44 (1935).

[2] The Texas Co., *Lubrication*, **24**, 55 (1938).

oil films on the cylinder walls, the thickness becoming greater than that which results in best control. Also, the high-viscosity oils drain off the cylinder walls at a slower rate. The low-viscosity oils flow very readily and are more easily pumped to the combustion chamber. Oil control is usually so satisfactory in automotive engines that the lowest viscosity oils are used for easy-starting and low-friction characteristics.

Oil consumption in automotive engines may vary from 200 to 2000 miles per quart of oil depending upon relative fits of the various parts and the conditions of operation. Oil consumption of aircraft engines under normal operating conditions should be in the range between 0.01 and 0.02 lb. of oil per brake horse-

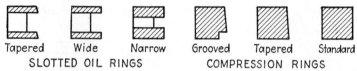

Fig. 284.—Sections of piston rings. More severe oil scraping toward the left.

power-hour. The high-load conditions in the aircraft engine make the use of the higher viscosity oils desirable.

Oil consumption is controlled principally by the use of slotted oil rings (Fig. 284) having relieved outer surfaces that increase the unit pressure of the ring against the cylinder wall. The higher the unit pressure and the larger the slot, the greater will be the capacity of the ring for reducing the oil film and consumption. Less severe oil rings have smaller slots and more area to contact the cylinder wall. Scraper compression rings have the lower edge relieved, and standard compression rings are square in cross section. Compression rings may be tapered several degrees on the outer surface to cause the lower edge to present a "cutting" edge to the oil film. Such rings seat rapidly and reduce the initial oil consumption.

High ring pressure results in severe scraping action and requires stiff ring construction. This has led to the use of rings with fairly low pressures backed by spring-steel expanders to increase the ring pressure. Ring wear reduces the ring pressure due to the ring but does not appreciably lower the pressure due to the expander.

Wear.—Wear in an engine may be caused by three different processes, erosion, abrasion, and corrosion. Erosion is the wearing away or shearing off of the high spots of bearing surfaces that break through the oil film. Abrasion is the scratching of bearing surfaces by foreign particles in an oil film that are smaller than the film thickness. Corrosion is the result of chemical reaction between acids, formed from the products of combustion or by the decomposition of the oil, and the bearing surfaces.

Wear of crankshaft and connecting-rod bearings depends upon the loading, the distortion and deflection of the bearing

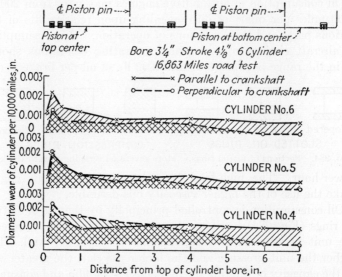

Fig. 285.—Cylinder wear—normal road operation with air cleaner. Basis 10,000 miles. (*Sparrow & Scherger.*)

surfaces, the quantity of oil supplied, and the ability of the oil film to support the load.

High piston-ring pressure causes rings to break through the oil film, particularly at the top of the stroke where the piston stops and the film is later subjected to flame temperatures. Any distortion of the bores, pistons, and rings usually increases the pressure at some point, which increases the wear.

Normal-cylinder-wear data (Fig. 285) show more wear at the upper end of ring travel.[1] The effect of the addition of an abrasive, by removing the air cleaner and taking the air from

[1] S. W. Sparrow and T. A. Scherger, Cylinder Wear, Where, and Why, *S.A.E. Trans.*, **31**, 117 (1936).

M. M. Roensch, Observations on Cylinder-bore Wear, *S.A.E. Trans.*, **32**, 89 (1937).

beneath the running board while traveling on a sandy road, is to increase the wear enormously and carry it farther down the bore (Fig. 286). Ricardo[1] suggested that corrosive action of the gases on the upper end of the cylinder bore, which may be dry or covered only with a thin oil film, accounts for some of the excessive wear at this position. He noted that the loss of material from the cylinder liner was much more than from the rings and

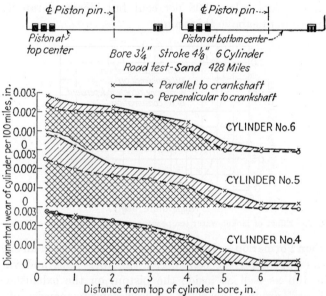

FIG. 286.—Cylinder wear—abnormal road operation. Basis 100 miles. (*Sparrow & Scherger.*)[2]

that the wear at this position was high even with materials known to be highly resistant to abrasive action.

Operation with low jacket-water temperature or during the warming-up period following cold starting is responsible for rapid cylinder wear. The low cylinder-wall temperature permits the condensation of water vapor; this combines with carbon, nitrogen, and sulphur compounds to form carbonic, nitric, and

---

[1] H. R. Ricardo, Some Notes and Observations on Petrol and Diesel Engines, *Inst. Auto. Eng. Proc.*, **27,** 434 (1933).

[2] Sparrow and Scherger, *op. cit.*

sulphuric acids, which cause wear by corrosion. Williams[1] has shown that low jacket temperature greatly increases the wear (Fig. 287). He divides the temperature range into two parts, one below 185°F. where wear is dependent on oil consumption, which determines film thickness, and the other above 185°F. where wear is largely independent of oil consumption. Apparently, the increase in wear in region A must be due principally to corrosion. He also found that wear was low for sulphur contents of fuel below 0.08 per cent but increased rapidly for sulphur contents above this value.

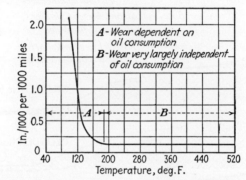

FIG. 287.—Effect of jacket-water temperature on cylinder wear. (*Williams.*)

Carbon Formation.—During the normal operation of an engine, deposits appear on the combustion-chamber walls and valves, on the top as well as inside the piston, on the exhaust-valve stems, and around the rings. These deposits consist of dirt from the air, carbon formed by the destructive distillation of the oil, carbon formed by the incomplete combustion of the fuel, and products of oxidation such as lacquers.

The carbon formed from the lubricating oil appears to be independent of the oil consumption, although the general belief is probably the exact opposite. It has been found[2] that the volatility characteristics are a better guide as to carbon residue than the Conradson carbon-residue test. Thus, if the distillation curve for the heavier ends runs high, the carbon deposition will be high.

[1] C. G. Williams, Cylinder Wear, *Auto. Eng.*, **23**, 259 (1933).

[2] W. H. Bahlke, D. P. Barnard, J. O. Eisinger, and O. Fitzsimons, Factors Controlling Engine-carbon Formation, *S.A.E. Trans.*, **26**, 373 (1931).

Carbon deposits tend to insulate the combustion-chamber wall and raise the inside surface temperature. Sooty surfaces of an aluminum head tend to cause more heat loss from the charge than do clean surfaces, apparently because of better heat-absorption characteristics of a black body. The high temperatures of the inside surface of heavy carbon deposits result in some oxidation under heavy-load conditions so that the thickness of the deposit tends to reach a maximum with use. Engines having compression ratios near the highest useful ratio will detonate with considerable severity as the deposit forms, because of the higher inside temperatures and the increase in compression ratio.

Carbon accumulation around the rings may restrict their operation until they cease to function properly and finally become stuck in the grooves. The oil consumption and blowby increase as the ring operation becomes restricted. The blowby of hot gases aggravates the condition which finally results in ring failure. The formation of carbon in the channels of the oil rings restricts the flow of oil off the walls through the rings and piston and back to the crankcase.

Oil Dilution.—Considerable quantities of liquid gasoline are drawn into the cylinders during starting and warm-up conditions. This is a result of the volatility characteristics of the fuel. The unvaporized gasoline tends to work down past the pistons, in a vertical engine, and not only wash the oil off the cylinder walls but also dilute the oil supply in the crankcase. This changes the viscosity characteristics and usually results in undue wear, if dilution is considerable.

The amount of dilution depends primarily on the oil temperature, being high when the oil temperature is low, and vice versa. Thus, the dilution tends to reach a maximum which is determined by the operating conditions. In some cases, it is desirable to predilute the lubricant so that when the oil is changed the engine does not have to deal with an oil more viscous than that which will exist after a few cold starts.

Journal Bearings.—A shaft that comes to rest in a bearing forces out part of the film of oil and settles to the bottom of the bearing. The eccentric position of the shaft inclines the surfaces with regard to each other. At the beginning of rotation the shaft tends to roll up one side of the bearing, but the wedge-shaped oil film acts similarly to that of the wedge-shaped film between two plain surfaces, and an oil-film pressure is developed that carries the load.

Under operating conditions the shaft is displaced to the opposite side from the one up which it tends to roll during starting. The oil-film pressure is a maximum near the point of smallest clearance and varies theoretically in the manner indicated in Fig. 288, based upon the action of an inclined surface on an oil film with no end leakage. The resultant of the oil-film pressure passes through the shaft center at a slight angle to that of the load. This displaces the shaft to the left in the illustration until the resultant of the load and oil-film pressure are balanced by the tangential shearing forces at the surface of the shaft. The load varies in amount and direction with crankshaft and crankpin bearings, which varies the position and amount of the minimum clearance as well as the oil-film pressure during each cycle.

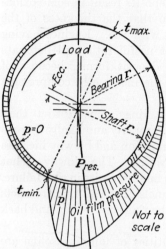

FIG. 288.—Theoretical oil-film pressure in a journal bearing.

The coefficient of friction f has been found to be a straight-line function of the factor $\mu n/p$ (Fig. 289) except at low values of this factor which indicate thin-film lubrication.

Thus, with thick-film lubrication,

$$f = a + k_1 \frac{\mu n}{p} \tag{10}$$

where a = correction factor for length-diameter, l/d, ratio.

The factor a is practically constant at about 0.002 for l/d ratios greater than 0.75 but increases rapidly at lower ratios (Fig. 290).

The value of k_1, Eq. (10), or the slope of the straight part of the curve (Fig. 289) depends upon the bearing clearance. However, dividing the factor $\mu n/p$ by bearing clearance c and plotting f vs. $\mu n/pc$ causes all points to fall on the same line. Also,

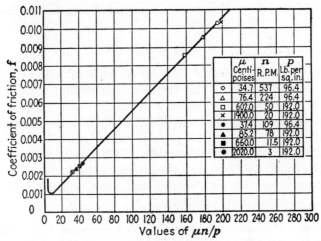

Fig. 289.—Variation of journal-bearing friction coefficient with $\mu n/p$. (*Dickinson and Bridgeman.*)[1]

when the bearing diameter is introduced, Eq. (10) becomes

$$f = a + k_2\left(\frac{\mu n}{p} \cdot \frac{d}{c}\right) \qquad (11)[1]$$

where $k_2 = 4.73 \times 10^{-8}$ when μ = abs. visc., centipoises,
 n = r.p.m.,
 p = load, lb. in.$^{-2}$ of projected bearing area.

Equation (11) indicates that, for a given set of conditions, a change to a lower viscosity oil or a lower speed decreases the coefficient of friction as long as thick-film lubrication is maintained. In either case the oil-film temperature is lowered, which increases the viscosity and tends to restore the original value of the factor $\mu n/p$.

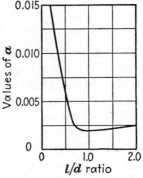

Fig. 290.—Effect of l/d ratio on a. (*McKee.*)[2]

[1] H. C. Dickinson and O. C. Bridgeman, Fundamentals of Automotive Lubrication, *S.A.E. Trans.*, **27**, 278 (1932).

[2] S. A. McKee, Journal-bearing Design as Related to Maximum Loads, Speeds and Operating Temperatures, *Nat. Bur. Standards, Jour. Research*, **19**, 457 (1937).

The friction force F per unit of projected area is the product of f and p.

Thus, from Eq. (11), $F = ap + k_2\mu n \dfrac{d}{c}.$ (12)

Equation (12) indicates that an increase in p immediately increases the friction. This increases the oil-film temperature and lowers the viscosity, which reduces the value of the second factor and tends to offset the effect of the increase in the first factor. However, the factor $\mu n/p$ is lowered, and the bearing is forced toward thin-film lubrication.[1] Obviously, $\mu n/p$ should be high enough (>20) so that any increase in load with the lowest operating values for oil viscosity and speed will not force the bearing into thin-film conditions.

Example.—Determine the permissible oil temperature for a value of $\mu n/p$ of 20. $p = 4000$ lb. in.$^{-2}$, $n = 1000$ r.p.m. $\mu = 45$ Saybolt universal sec. at 210°F., V.I. of oil $= 100$, and sp. gr. $= 0.87$.

$$\frac{\mu n}{p} = \frac{\mu \times 1000}{4000} = 20,$$

$$\mu = 80 \text{ centipoises.}$$

$$\text{Kin. visc.} = \frac{\mu}{\text{sp. g.}} = 80 \div 0.87 = 92 \text{ centistokes.}$$

From Eq. (6),   Visc. $= 418$ Saybolt universal sec.
From Fig. 277, for 100 V.I. oil, $t = 68°$F.

The temperature of the oil film can be only a few degrees above the bearing temperature, hence the temperature obtained in the foregoing manner is very nearly the permissible bearing temperature.

Heat Generation and Bearing Temperature.—The heat generated in a bearing is equal to the friction force times the distance through which it acts. The force is $pfdl$ based on the projected bearing area, dl, and the distance moved per unit of time is πdn. Thus, the heat generated is

$$Q = (pfdl) \times (\pi dn) = \pi d^2 nplf \text{ in. lb. min.}^{-1} \quad (13)$$

*Combining Eqs. (11) and (13) and substituting values for the constants result in

---

[1] This is also indicated by Eq. (3), page 399.

ENGINE LUBRICATION

$$Q = (pn)(\pi d^2 l)\left[0.002 + 4.73 \times 10^{-8}\left(\frac{\mu n}{p}\right)\left(\frac{d}{c}\right)\right]. \quad (14)^1$$

This equation may be written as follows:

$$Q = (pv)(dl)\left[0.002 + 4.73 \times 10^{-8}\left(\frac{\mu n}{p}\right)\left(\frac{d}{c}\right)\right] \quad (15)^1$$

where $v = \pi dn$, the peripheral veloc. of the bearing surface.

Thus, for a given bearing the heat generated or power dissipated by friction depends both upon the pv factor and upon the $\mu n/p$ factor. The effect of the various factors upon the heat generated can be determined from Eq. (15).

Example.—Determine the value of $\mu n/p$ so that both factors will be equally important. Assume a clearance of 0.001 in. in.$^{-1}$ of bearing diameter. Equal importance between the two factors will exist when

$$4.73 \times 10^{-8}\left(\frac{\mu n}{p}\right)\left(\frac{1}{0.001}\right) = 0.002,$$

$$\frac{\mu n}{p} = 42.$$

When the factor $(\mu n/p)$ is much below 42 for this case, pv will be the controlling factor, and vice versa. In the latter case, for a given bearing, Q will be approximately proportional to $\mu n v$ or its equivalent μn^2.

The heat dissipated from a bearing must equal the heat generated by friction when the bearing and oil film reach an equilibrium temperature. The heat dissipation from a bearing depends upon the heat-transfer coefficient k_3 between the bearing and the surrounding atmosphere, the area of exposed surface, A, and some power of the temperature difference ΔT. Thus, McKee[2] uses the following relation:

$$Q = k_3(k_4\pi\, dl)\Delta T^{1.3} \text{ B.t.u. min.}^{-1} \quad (16)$$

where k_4 = ratio of exposed surface to the working area of the bearing.

Combining Eqs. (14) and (16) results in

$$pnd\left[a + k_2\left(\frac{\mu n}{p}\right)\left(\frac{d}{c}\right)\right] = 778 \times 12 k_3 k_4\, \Delta T^{1.3}. \quad (17)$$

[1] Dickinson and Bridgeman, *op. cit.*
[2] McKee, *op. cit.*

422   INTERNAL-COMBUSTION ENGINES

For any given bearing, the substitution of values for $\mu n/p$ and ΔT results in the expression

$$pn = \text{const.} \tag{18}$$

Example.—The dimensions of a bearing are $d = 3$ in., $l = 2.5$ in., and c, the diametrical clearance, $= 0.003$ in. Heat-dissipation conditions are such that $k_3 k_4 = 0.0012$. Min. allowable $\mu n/p = 30$. Atmospheric temperature surrounding bearing $= 150°$F. Max. allowable bearing temperature $= 250°$F. Determine the limiting values of p and n for the given conditions for various viscosities.

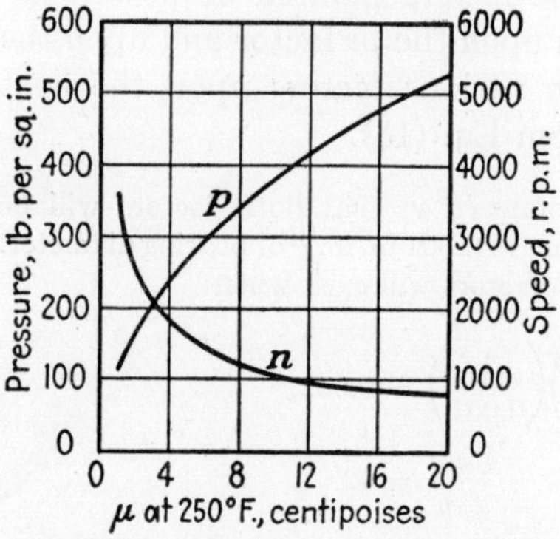

Substituting in Eq. (17) results in

$$3pn(0.002 + 4.73 \times 10^{-8} \times 30 \times 1000) = 778 \times 12 \times 0.0012 \times (100)^{1.3}$$

or $pn = 435{,}000$ lb. in.$^{-2}$ r.p.m.

Combining this relation with $\mu n/p = 30$ results in

$$p = 120\sqrt{\mu}$$

and

$$n = 3610\sqrt{\mu^{-1}}.$$

Fig. 291.—Effect of viscosity on load and speed for assumed limiting bearing conditions. (*McKee.*)[1]

Solving these relations for various viscosity values at the bearing temperature of 250°F. results in the limiting values (Fig. 291) for load and speed for the given conditions.

The foregoing analysis assumes that the bearing dissipates all the heat generated by friction. Most of the energy may remain in the oil which rises in temperature in passing through the bearings, particularly in high-speed engines. The amount of energy increase in the oil is

$$Q_{oil} = Mc(T_{bear.} - T_{oil}) \tag{19}$$

where $c =$ about 0.47 B.t.u. lb.$^{-1}$ deg.$^{-1}$ rise in oil temperature. The value of Eq. (19) divided by $\pi d l$ should be added to the right side of Eq. (17) and may be considered the controlling factor in high-output engines with oil coolers.

Example.—Assume oil enters the bearing at 230°F. in the previous example. The quantity of oil passing through the bearing is about 1.6 lb. min.$^{-1}$ if the specific gravity is about 0.9. Determine the effect of energy increase of the oil on the value of pn in Eq. (17).

[1] McKee, *op. cit.*

Substituting in Eq. (19) and dividing by πdl result in

$$\frac{Q_{oil}}{\pi dl} = \frac{1.6 \times 0.47 \times 20 \times 778 \times 12}{\pi \times 3 \times 2.5} = 596{,}000 \text{ in. lb. in.}^{-2}\text{ min.}^{-1}$$

Adding this value to the right side of Eq. (17), in the previous example, results in

$$pn = 435{,}000 + 58{,}100{,}000.$$

The factor for energy increase of the oil is many times that of the factor for heat dissipation from the bearing to the surrounding atmosphere. Obviously, a cool and comparatively large supply of oil is a most important factor affecting permissible bearing speeds and loads.

Oil Flow.—The flow of oil through a tube is given by the relation[1]

$$V = \frac{\pi}{8}\left(\frac{\Delta P}{l}\right)\frac{r^4}{\mu} \tag{20}$$

where V = quantity of oil, in.^{3} sec.$^{-1}$,
ΔP = pr. drop, lb. in.$^{-2}$,
l = tube length, in.,
r = tube radius, in.,
μ = abs. visc., lb. sec. in.$^{-2}$

Example.—An oil duct 0.25 in. in diameter and 12 in. long connects an oil manifold and one of the engine bearings. The quantity of oil flowing to the bearing is 2 in.^{3} sec.$^{-1}$ The oil viscosity is 20 centipoises. Determine the pressure drop in the oil duct.

Substituting in Eq. (20) results in

$$\Delta P = \frac{2 \times 8 \times 12 \times 0.20 \times 2.09 \times 10^{-3}}{\pi \times 0.125^4 \times 144} = 0.73 \text{ lb. in.}^{-2}$$

The axial flow of oil through a bearing, having a central oil groove completely around the bearing, is given by the approximate relation[2]

$$V = \frac{\pi r c^3 \Delta P}{48 \mu l}\left[1 + 6\left(\frac{\text{ecc.}}{c}\right)^2\right] \tag{21}$$

where r = bearing radius, in.,
c = diametral clearance, in.,
and other quantities the same as in Eq. (20).

[1] M. D. Hersey, Theory of Lubrication, p. 28, John Wiley & Sons, Inc., New York, 1936.

[2] E. S. Dennison, Film-lubrication Theory and Engine-bearing Design, *A.S.M.E. Trans.*, **58**, 25 (1935).

A bearing with no load has an eccentricity of zero whereas an actual bearing may have an eccentricity approaching $c/2$ as the limit when thin-film conditions exist and the destruction of the bearing is imminent. Thus, the bracketed term will range from > 1 to < 2.5 for thick-film lubrication.

Example.—Determine the range of pressure that may be required to flow the oil through the bearing in the examples on page 422. $\mu = 4$ centipoises. Assume a central oil groove with a width of 0.1 in.

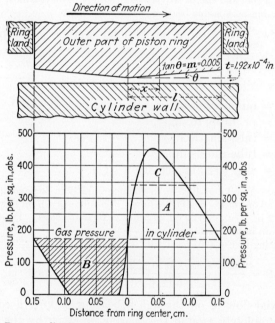

Fig. 292.—Pressure distribution in oil film between piston ring and cylinder wall.

Substituting the values from the previous examples in Eq. (21), results in

$$\Delta P = \frac{0.00025 \times 1728 \times 48 \times 0.04 \times 2.09 \times 10^{-3} \times (1.25 - 0.05)}{\pi \times 1.5 \times (0.003)^3 \times [> 1 \text{ to } < 2.5] \times 144}$$

or ΔP is > 47 and < 117 lb. in.$^{-2}$

Note that half the quantity of oil flows each way from the groove.

Piston-ring Lubrication.—Piston rings are usually designed to have surfaces parallel with the cylinder walls. Wear between the rings and cylinder walls produces a beveled surface on the ring[1] (Fig. 292) with a high point about in the center or below the center of the surface of the ring. Thus, it may be assumed that a ring, after running in, presents surfaces that are inclined to the cylinder wall. An oil-film pressure above that due to the gas pressure exerted on the oil film will be developed under the leading ring surface and below that due to gas pressure under the trailing ring surface when the ring is moving in either direction.

The pressure developed under the leading surface of the ring is given by Eq. (3).

Example.—The conditions of piston-ring lubrication examined by Castleman[1] were $\mu = 0.1$ poise, $v = 39.4$ ft. sec.$^{-1}$, $m = 0.005$, $t = 1.92 \times 10^{-4}$ in., $l = 0.059$ in., and the gas pr. = 170 lb. in.$^{-2}$ abs. at this instant. Determine the oil-film pressure for the given conditions.

Substituting in Eq. (3) results in

$$P = 170 + \frac{6 \times 0.1 \times 2.09 \times 10^{-3} \times 39.4 \times 12 \times 0.005}{144 (2 \times 1.92 \times 10^{-4} + 0.005 \times 0.059)} \times \frac{0.059x - x^2}{(1.92 \times 10^{-4} + 0.005x)^2}$$

$= 453$ lb. in.$^{-2}$ when $x = 0.04$ cm.

Substituting values for x between the limits $x = 0$ and $x = l = 0.059$ in. results in data from which the pressure-distribution curve (Fig. 292) was plotted.

The pressure distribution for one ring surface would be the "mirror image" of the other if both surfaces are equal in length and symmetrical with regard to the cylinder surface. However, the pressure cannot be below absolute zero in the oil film under the trailing edge if oils are assumed to have no absolute tension.

Castleman[1] noted that the pressure tending to expand the ring is the difference between the gas pressure and the oil-film pressure in the film under the trailing surface, whereas the pressure tending to compress the ring is the similar difference acting upon the leading surface. Thus, the net pressure tending to compress the ring is the difference between pressures indicated by areas $(A + C)$ and B (Fig. 292) which amounts to that indicated by area C. This must equal the force exerted by the ring if equilibrium conditions exist.

Example.—Determine the net force tending to compress the ring in the previous example. Cyl. diameter = 3.25 in.

[1] R. A. Castleman, Jr., A Hydrodynamical Theory of Piston-ring Lubrication, *Physics*, **7**, 364 (1936).

Integrating area C (Fig. 292) and dividing by the length of the surface, $2l$, results in a mean pressure of 18.7 lb. in.$^{-2}$

The pressure per square inch of circumference of the ring is

$$18.7 \times 0.30 \div 2.54 = 2.21 \text{ lb. in.}^{-1}$$

The ring tension required to balance this pressure is

$$3.25 \times 2.21 = 7.18 \text{ lb.}$$

Obviously, if the ring tension is different, the minimum oil-film thickness and pressure distribution would be changed.

EXERCISES

1. Determine the value of 1 poise in pound seconds per square foot.

2. An oil has a specific gravity of 0.9 and a viscosity of 40 Saybolt universal sec. at 250°F. Determine the kinematic and absolute viscosity at this temperature.

3. Determine the possible variation in V.I. of oils having a viscosity classification of S.A.E. 10 and a viscosity range of 5000 to 10,000 Saybolt universal sec. at 0°F.

4. Determine the possible viscosity range in Saybolt universal seconds at 250°F. for the oils referred to in Prob. 3.

5. An oil lead from an oil pump feeds nine bearings in an eight-cylinder engine. On the assumption of a diameter-clearance ratio of 1000, with a 3-in. crankshaft, what size oil line is required to equal the areas of flow out of the bearings? Assume that the oil flows into the center of each bearing, completely around it, and out at both ends.

6. In the example, page 410, dealing with centrifugal force, assume that the 6-in. bearing section of the crankshaft has a 3-in. hole located at the center of the shaft. Determine the minimum pressure required to force oil into the center hole for a speed of 3600 r.p.m.

7. In a given crankshaft 25 lb. in.$^{-2}$ oil pressure is required at the center of a crankshaft in order to flow the desired quantity of oil into the shaft. What is the maximum oil pressure at the oilhole in the crankpin bearing which has a diameter of 2 in.? The crank radius is $4\frac{1}{4}$ in., and the oilhole position in the crankpin is 30 deg. from the line connecting the crankpin and crankshaft centers and is located in the quadrant nearer the crankshaft.

8. Determine the oil-film thickness that must be evaporated or burned from the cylinder walls per cycle of an eight-cylinder, 3- by 4-in. engine to account for an oil consumption of 1 qt. per 1000 miles. The engine speed is 3600 r.p.m. at a car speed of 60 m.p.h.

9. A $3\frac{1}{2}$- by 5-in. valve-in-head engine has a flat cylindrical combustion chamber of such size that the compression is 5:1. What uniform thickness of carbon deposit is required to increase the compression ratio to 5.1:1? To 5.2:1?

10. Determine the equilibrium bearing temperature for the bearing referred to in the example on page 422 at various speeds for a constant value of $pn = 435{,}000$ lb. in.$^{-2}$ r.p.m. and an oil having a V.I. of 80 and a viscosity of 55 Saybolt universal sec. at 210°F.

11. Solve Prob. 10, assuming 0.03 ft.^{3} of oil is supplied per minute at a temperature 10°F. below the equilibrium temperature.

12. Determine the effect of variation of angle of inclination of the surfaces of a piston ring on oil-film pressure and pressure distribution. Use the data in the example on page 425.

13. Determine the net force tending to compress the piston ring in Prob. 12.

14. A 16-cyl. V-type engine has connecting-rod bearings 2.0 in. in diameter by 1.75 in. in length with a diametrical clearance of 0.0015. There are eight main bearings 2.5 in. in diameter by 1.063 in. in length and 1 main bearing 2.5 in. by 2.156 in., all with a diametrical clearance of 0.0015 in. Cylinder bore is 3.25 in., and stroke is 3.25 in. Design the oil manifold and feed lines to the main bearings so that the oil pressure at the bearings will be 50 lb. in.$^{-2}$ gage. Estimate the quantity of oil that must be circulated per hour to maintain this pressure at an engine speed of 4000 r.p.m. Assume that oil temperature is 250°F. and viscosity is 40 Saybolt universal sec.

CHAPTER XV
ENGINE COOLING

Heat Transfer.—Heat transfer occurs when a temperature difference exists. The combustion process results in high temperature differences and causes considerable heat transfer from the gases to the surrounding metal walls. The shearing of the oil films separating the bearing surfaces transforms available energy into internal energy of the oil film. This increases the temperature of the oil film and results in heat transfer from the oil to the bearing surfaces. Hence, the cylinder walls and the lubricant must be cooled to maintain safe operating temperatures.

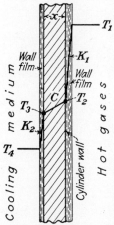

Fig. 293.—Heat transfer from gases to cooling medium.

Heat is transferred by radiation, convection, and conduction from the gases to the cylinder walls, by conduction through the walls, and by radiation, convection, and conduction to the cooling medium. Stagnant-fluid films against the cylinder wall offer high resistance to heat transfer and cause a high temperature drop between the gases and the wall (Fig. 293) and between the wall and a gaseous cooling fluid. A liquid coolant offers less resistance and lowers the temperature drop from the wall to the coolant. The metal wall offers low resistance to heat transfer and has a low temperature gradient.

The amount of heat transferred depends upon the temperature difference, the area of the surface A through which heat is transferred, a heat-transfer coefficient K or C, which depends upon the media and their condition, and the time Δt. Thus, from the gases to the cylinder wall,

$$Q = K_1 A_1 (T_1 - T_2)\, \Delta t. \qquad (1)$$

Through the wall, $\quad Q = \dfrac{C}{x} A_m (T_2 - T_3)\, \Delta t \qquad (2)$

where C = conductivity,
x = thickness of the metal wall.

From the wall to the cooling medium,

$$Q = K_2 A_2 (T_3 - T_4) \Delta t. \qquad (3)$$

Also, the heat transferred from the gases to the cooling fluid is

$$Q = K A_m (T_1 - T_4) \Delta t \qquad (4)$$

where K = overall heat-transfer coefficient.

The heat transferred is the same in all cases and the relation between the various coefficients, on the assumption that A_1 equals A_2 equals A_m, is

$$K = \frac{1}{\frac{1}{K_1} + \frac{x}{C} + \frac{1}{K_2}}. \qquad (5)$$

Gas-temperature Variation.—The temperature of the gases in the cylinder varies appreciably during the different processes.

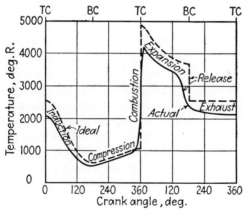

FIG. 294.—Gas-temperature variation. (5.3 to 1 comp. ratio; C_8H_{18} + 59.5 Air + clearance gas.)

The temperature at the beginning of the induction process (Fig. 294) is that of the clearance gases. The temperature falls rapidly as the cool mixture is inducted. The temperature rises during the compression process and is elevated to the maximum temperature by the combustion process. The expansion process decreases the temperature, and release very rapidly drops the temperature of the gases remaining in the cylinder. There is some drop in temperature of the gases in the cylinder during the exhaust process in the actual engine, but none in the ideal case without heat transfer.

Surface Variation.—The amount of combustion-chamber surface varies with the type of combustion chamber for a given clearance volume. This surface and the area of the top of the piston are fixed for a given engine. However, movement of the piston exposes the cylinder bore and increases the exposed surface until bottom center is reached. The variation in cylinder-bore surface exposed with crank angle depends on the bore diameter, the length of stroke, and the length of the connecting rod. Thus, the surface exposed to the gases in the cylinder varies from a minimum at top center (Fig. 295) to a maximum at bottom center.

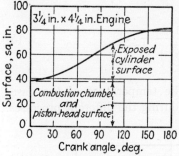

Fig. 295.—Surface variation.

Heat-transfer Integration Areas.—Janeway[1] plotted the product of temperature difference and exposed surface against the crank angle. Following his method, this has been done by assuming instantaneous combustion, by using temperatures corrected for heat transfer[2] (Fig. 294), with cylinder-wall temperatures of 300°F. for wide-open-throttle conditions and 200°F. for idling conditions, and for the exposed surface as given in Fig. 295, all of which results in the curves in Fig. 296. The area between each curve and its zero line represents the product of surface area, temperature difference, and time, and has been termed "heat-transfer integration area" by Janeway.[1]

---

[1] R. N. Janeway, Quantitative Analysis of Heat Transfer in Engines, *S.A.E. Trans.*, **33**, 371 (1938).

[2] Heat transfer lowers the gas temperature. Consequently, an end temperature and path must be assumed for each process such that the heat transfer, the work done, and the energy in the gases at the beginning and end of the process satisfy the energy equation. The assumption of instantaneous combustion eliminates heat loss during this process. However, this is offset by the higher temperatures and amounts to including the heat loss during combustion with that of the expansion process, the net result being a slightly high heat loss.

Heat-transfer Coefficients.—Janeway[1] has evaluated the mean coefficients of heat transfer for the expansion process based on test results on a 3¼- by 4¼ in. L-head engine with a

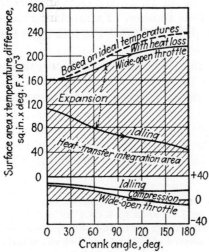

FIG. 296.—Heat-transfer integration areas.

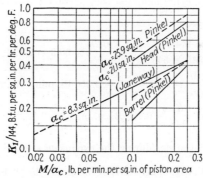

FIG. 297.—Heat-transfer coefficients K_1 from gases to walls. (*Janeway;* water-cooled automobile engine. *Pinkel;* air-cooled aircraft engine.

5.4:1 compression ratio, on the analytical separation of the heat losses for the various processes, and on the heat-transfer integration method previously outlined. His results have been plotted (Fig. 297) against weight of charge per unit time per unit piston area.

[1] Janeway, *op. cit.*

Pinkel[1] computed the heat-transfer coefficients (Fig. 297) from the gases to the cylinder barrel and cylinder head from experimental data on heat loss from the two parts. Variation of the amount of cooling air varied the barrel and head temperatures as well as the heat transfer. Extrapolating these temperatures to zero heat transfer indicated the mean gas temperature used in his computations.

These heat-transfer coefficients apply to the turbulent conditions existing in the engine for which the values were determined. Consequently, the same turbulence is assumed when these coefficients are used.

By Janeway's method, the heat transfer is determined by multiplying the heat-transfer integration area by the heat-transfer coefficient for the given conditions of speed and charge density.

Example.—The area under the expansion curve for wide-open-throttle conditions (Fig. 296) amounts to 3.42×10^7 in.^{2}, °F., deg. crank angle. Determine the heat transfer per second at wide-open throttle with a $3\frac{1}{4}$-by $4\frac{1}{4}$-in. engine at a speed of 3600 r.p.m. Vol. eff. = 0.75.

Piston displacement per cylinder is 35.3 in.^{3}

$$\text{Vol. disp. cyl.}^{-1} \text{ min.}^{-1} = 35.3 \times 3600 \div 2 = 63{,}500 \text{ in.}^3$$

Mol. wt. of charge is about 30.3.

Hence, $\dfrac{14.7 \times 144 \times 63{,}500 \times 0.75}{1545 \times 540 \times 1728} \times 30.3 = 2.12$ lb. min.$^{-1}$

$$\frac{M}{a_c} = 2.12 \div 8.3 = 0.256 \text{ lb. min.}^{-1} \text{ in.}^{-2}$$

From Fig. 297, $\dfrac{K_1}{144} = 0.435$ B.t.u. in.$^{-2}$ hr.$^{-1}$ °F.$^{-1}$

Then, the heat transfer during the expansion stroke is

$$\frac{3.42 \times 10^7 \text{ in.}^2 \text{ °F. deg. angle} \times 0.435 \text{ B.t.u. in.}^{-2} \text{ hr.}^{-1} \text{ °F.}^{-1}}{3600 \text{ sec. hr.}^{-1} \times 360 \text{ deg. rev.}^{-1} \times 60 \text{ rev. sec.}^{-1}} = 0.191 \text{ B.t.u.}$$

$$0.191 \times 6\% = 5.73 \text{ B.t.u. sec.}^{-1}$$

Janeway[2] has called attention to the fact that, although the pressure is much higher during expansion than during compression, the charge density is the same at any piston position

---

[1] B. Pinkel, Heat-transfer Processes in Air-cooled Engine Cylinders, *N.A.C.A.Rept.* 612 (1938).

[2] Janeway, *op. cit.*

for both strokes. The charge density is obviously the highest when the piston is at top center and should result in the highest heat-transfer coefficient. The charge density is lowest after release has permitted much of the charge to escape.

Thus, the same mean heat-transfer coefficient may be used for compression and expansion, and the coefficients for exhaust and induction may be evaluated on a density basis. The heat transfer for compression and expansion only, at road-load conditions, in percentage of energy supplied, computed in the foregoing manner, varies from less than 10 per cent at wide-open conditions and high speed to about 60 per cent at idle conditions (Fig. 298). This heat loss directly affects the thermal efficiency of the process; the heat loss during release, exhaust, and induction, though appreciable in quantity, has an indirect and small effect on thermal efficiency.

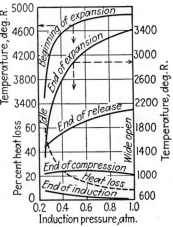

Fig. 298.—Gas temperatures at road-load conditions. Supercharged to 1 atm. at wide-open throttle. Heat transfer during compression and expansion.

Heat Transfer through Cylinder Walls.—The net heat transferred to the walls must be transferred through the walls if the wall temperature is to remain constant. The conductivity of metals varies appreciably (Table 26), aluminum and copper, which may be used for cylinder heads, having high conductivity.

TABLE 26.—CONDUCTIVITY OF METALS
(B.t.u. ft.$^{-2}$ hr.$^{-1}$ °F.$^{-1}$ ft.

Material	Conductivity
Cast iron	27
Steel	26
Aluminum	119
Copper	220

The thickness of the cylinder walls varies from about $3/16$ in. to more than 1 in. depending on cylinder size. The temperature difference for a given heat transfer can be obtained from Eq. (2).

Example.—Assume 8 per cent of the energy input into a $3\frac{1}{4}$- by $4\frac{1}{4}$-in. engine is transferred as heat through the cast-iron cylinder walls which are

¼ in. thick. Determine the temperature drop through the cylinder wall. Engine speed = 3600 r.p.m. Vol. eff. = 0.80.

The volume of charge is

$$\left(\frac{\pi \times \overline{3\frac{1}{4}}^2}{4} \times 4\frac{1}{4} \div 1728\right) \times 1800 \times 0.80 = 29.4 \text{ ft.}^3 \text{ min.}^{-1}$$

From Table VIII (Appendix), the energy supplied is 100 B.t.u. ft.$^{-3}$ of octane-air mixture. Thus,

$$\text{Heat transfer} = \frac{0.08 \times 29.4 \times 100 \times 60}{\pi \times 3\frac{1}{4} \times 4\frac{1}{4} \div (2 \times 144)} = 93{,}500 \text{ B.t.u. ft.}^{-2} \text{ hr.}^{-1}.$$

Substituting in Eq. (2) results in

$$(T_2 - T_3) = \frac{Q}{(C/x)A} = \frac{93{,}500}{27 \times 4 \times 12 \times 1} = 72.2°\text{F. temp. diff.}$$

More heat is transferred through a unit area of cylinder-head surface than through the same area of cylinder surface since the cylinder surface is exposed to the gases only half the time. Also, a scrubbing action is set up by the combustion process which pushes the burned gases back from the flame front. Consequently, it appears that parts of the cylinder head would transfer, on the average, two to three times as much heat per unit area as the cylinder walls. This indicates a temperature difference through certain parts of the cylinder-head wall of well over 100°F. The use of aluminum-alloy or copper-alloy heads should reduce this temperature difference 60 to 80°F.

An increase in wall thickness increases the temperature difference. Also, an increase in the outside-wall temperature will raise the inside-wall temperature the same amount. The outside-wall temperature depends upon the resistance to heat transfer from the wall to the cooling medium.

Heat Transfer to Cooling Medium.—The cooling medium is usually water or air. The resistance to heat transfer from the cylinder wall to a liquid is very low particularly with fairly high velocities of flow. Thus, heat-transfer coefficients are high (Fig. 299) and low temperature differences are found between the cylinder wall and liquid coolant.

Example.—Determine the temperature drop from the cylinder wall to cooling water having a velocity of 1.5 ft. sec.$^{-1}$ Use the data given in the previous example.

From Fig. 299, $K_2 = 1010$ B.t.u. ft.$^{-2}$ hr.$^{-1}$ °F.$^{-1}$

Substituting in Eq. (3) results in

$$T_3 - T_4 = \frac{93{,}500}{1010 \times 1} = 92.5°\text{F. temp. diff.}$$

On the assumption of a mean water temperature around the cylinders of 150°, the mean temperature of the inside cylinder wall will be

$$150 + 92.5 + 72.2 = 314.7°\text{F}.$$

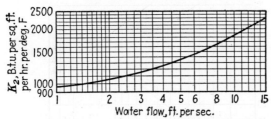

Fig. 299.—Heat-transfer coefficients K_2 from metal to water.

Fig. 300.—Section of cylinder and fins.

A liquid cooling system for a high-duty engine should be designed to direct the liquid with considerable velocity at the walls which absorb large quantities of heat so that high rates of heat transfer will result and minimum inside-surface temperatures will be maintained. The exhaust-valve seat and port, as well as that section of the cylinder head which is subjected to the combustion scrubbing action, and that part around the spark plug, should receive special consideration.

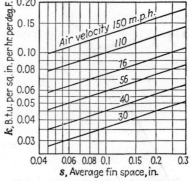

Fig. 301.—Local heat-transfer coefficients k (unbaffled cylinder). (*Brevoort.*)[1]

Heat Transfer from Finned Metal Cylinders.—An engine cylinder can be air-cooled only by the use of extended surface or fins. The air flows through the channels between the fins (Fig. 300) and transfers heat from the surface to the air stream. Tests[1] on unbaffled finned cylinders having an outside diameter

[1] M. J. Brevoort, Principles Involved in the Cooling of a Finned and Baffled Cylinder, *N.A.C.A. Tech. Note* 655 (1938).

of 5.81 in. showed that fin spacing and air-stream velocity to which the cylinder was subjected were the two principal variables (Fig. 301). The local heat-transfer coefficient can be increased appreciably by the use of baffles which force the air to flow through the channels between the fins.

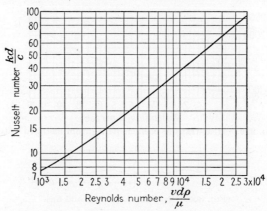

FIG. 302.—Nusselt's number vs. Reynolds' number for cylinder fins with baffles. (*Brevoort*.)

Brevoort[1] has developed a relationship (Fig. 302) between the Reynolds number $vd\rho/\mu$, describing the flow conditions, and the Nusselt number kd/c, describing the heat-transfer conditions. In these relationships, v = vel., ft. sec.$^{-1}$,

d = hydraulic dia., $4 \times$ hyd. radius, ft.,
ρ = density of air $\div g$, lb. ft.$^{-4}$ sec.^{2},
μ = abs. visc., lb. sec. ft.$^{-2}$,
k = local heat-transfer coefficient, B.t.u. ft.$^{-2}$ hr.$^{-1}$ °F.$^{-1}$,
c = conductivity of air, B.t.u. ft.$^{-1}$ hr.$^{-1}$ °F.$^{-1}$

Example.—Determine the local heat-transfer coefficient k for an air velocity of 100 m.p.h. between the fins, a fin spacing s of 0.06 in., and a fin width w of 1 in. Fins are baffled. Air conditions are 80°F. and 14.7 lb. in.$^{-2}$

$$\mu = 0.000165 + 2.5 \times 10^{-7} t \quad (Marks'\ Handbook)$$
$$= 1.85 \times 10^{-4}\ \text{poise}$$
$$= 3.87 \times 10^{-7}\ \text{lb. sec. ft.}^{-2}$$

$$\rho = \frac{d_a}{g} = \frac{14.7 \times 144}{53.4 \times 540 \times 32.2} = 2.28 \times 10^{-3}\ \text{lb. ft.}^{-4}\ \text{sec.}^2$$

[1] Brevoort, *op. cit.*

ENGINE COOLING

Hyd. dia. $= \dfrac{4 \times \text{area}}{\text{perimeter}} = \dfrac{4 \times 1 \times 0.06}{2.12} = 0.113$ in. or 9.42×10^{-3} ft.

The air velocity of 100 m.p.h. is equivalent to 146.6 ft. sec.$^{-1}$

Then, Reynolds No. $= \dfrac{146.6 \times 9.42 \times 10^{-3} \times 2.28 \times 10^{-3}}{3.87 \times 10^{-7}} = 8140.$

From Fig. 302, $\dfrac{kd}{c} = 32.5.$

Conductivity of air at 80° $= 13.5 \times 10^{-3}$ B.t.u. ft.$^{-1}$ hr.$^{-1}$ °F.$^{-1}$

Then, $k = \dfrac{32.5 \times 13.5 \times 10^{-3}}{9.42 \times 10^{-3}} = 46.6$ B.t.u. ft.$^{-2}$ hr.$^{-1}$ °F.$^{-1}$

Biermann and Pinkel[1] developed a relation between the local heat-transfer coefficient k and the surface coefficient K_2 based upon the outside area of the cylinder to which the fins are attached.

Thus,
$$K_2 = \frac{k/144}{s+t}\left[\frac{2}{a}\left(1 + \frac{w}{2r_b}\right)\tanh aw + s\right] \qquad (6)$$

and
$$a = \sqrt{k/72c_m t}, \qquad (7)$$

where $K_2 =$ B.t.u. hr.$^{-1}$ °F.$^{-1}$ in.$^{-2}$ of base surface,

$c_m =$ conductivity of the metal, B.t.u. in.$^{-1}$ hr.$^{-1}$ °F.$^{-1}$
   (2.17 for steel and 7.66 for aluminum Y alloy),

and s, t, w, and r_b (Fig. 300) are in inches.

Example.—Determine the optimum thickness of the steel fins for the preceding example. $2r_b = 5.81$ in.

Substituting in Eq. (7) results in $a = \sqrt{\dfrac{46.6}{72 \times 2.17 t}} = 0.546\sqrt{t^{-1}}.$

Substituting in Eq. (6) results in

$$K_2 = \frac{0.324}{0.06 + t}[3.66\sqrt{t}(1.172)\tanh 0.546\sqrt{t^{-1}} + 0.06].$$

Substituting various values of t results in the maximum value of $K_2 = 2.96$ when $t = 0.048$ in.

The heat transferred from an 8-in. finned section of the cylinder barrel for an assumed mean temperature difference of 200°F. is

$$Q = 2.96 \times \pi \times 5.81 \times 8 \times 200 = 86.5 \times 10^3 \text{ B.t.u. hr.}^{-1}$$

In a similar manner, optimum fin thickness may be obtained for various spacing and width arrangements.

---
[1] A. E. Biermann and B. Pinkel, Heat Transfer from Finned Metal Cylinders in an air stream, *N.A.C.A. Rept.* 488 (1934).

The weight of the metal fins per unit base area of outside cylinder surface is given by the relation

$$M = d_m \frac{wt}{s+t}\left(1 + \frac{w}{2r_b}\right) \qquad (8)$$

where d_m = density of metal, lb. in.$^{-3}$ (0.282 for steel and 0.101 for aluminum Y alloy).

The solution of Eq. (6) for a given fin thickness and quantity of metal in the fins per square inch of cylinder barrel surface for various fin spacings results in the optimum fin spacing and maximum heat-transfer coefficient for these conditions. Other maxima can be obtained for other fin-thickness values which

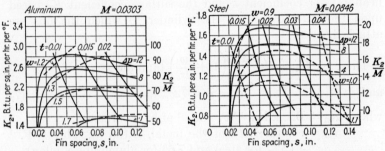

Fig. 303.—Optimum dimensions when fin spacing is specified. Applicable only to Biermann's test cylinder and baffling arrangement. (*Biermann.*)

may be plotted against fin thickness. A family of curves may be plotted in a similar manner for other air velocities.

The same procedure may be followed, using the same quantity of metal and a given fin spacing, and the optimum thickness determined, etc., the result being another family of curves.

Biermann[1] has done this for various pressure drops in inches of water instead of air velocity (Fig. 303) for both aluminum and steel fins. Similar diagrams with fin thickness as the specified variable would complement these diagrams.

Example.—Determine the fin dimensions for steel fins, giving a maximum K_2 value for $\Delta p = 4$ in. of water, fin wt. = 0.0846 lb. in.$^{-2}$ of cylinder-wall area, and with t and s not less than 0.025 and 0.10 in., respectively.

From a chart similar to Fig. 303, with t as the specified variable, at $t = 0.025$ and $\Delta p = 4$, it would be found that s would be less than 0.10 in. Referring to Fig. 303, at $s = 0.10$ and $\Delta p = 4$, $t = 0.032$, $w = 1.03$, and $K_2 = 1.26$. The fin dimensions meet the specified requirements.

[1] A. E. Biermann, The Design of Metal Fins for Air-cooled Engines, *S.A.E. Trans.*, **32**, 388 (1937).

Figure 303 shows that the optimum thickness of aluminum fins is much less, and the value of K_2 is much higher than the corresponding values for steel fins.

Power Required for Air Cooling.—The flow work done on the air at the entrance to the cylinder baffle minus the flow work done on the air at the exit of the baffle is a measure of the power required to flow the air over the fin surface.

Thus,
$$\text{Power} = \frac{\Delta P \times V}{\text{Const.}} \qquad (9)$$

where V = vol. of air flowing per unit of time.

Tests[1] on a finned 4.66-in. cylinder show the effect of small space between the fins on the pressure drop required to produce a given velocity (Fig. 304). The slope of these lines indicates that the pressure drop varies directly with the air velocity raised to the 1.25 power for $s = 0.02$ in. and to the 1.9 power for $s = 0.20$ in. Thus, doubling the velocity increases the power requirement 5 to 7.5 times, depending on the fin spacing.

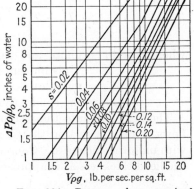

Fig. 304.—Pressure drop required for air flow through cooling-fin spaces. The cylinder is completely baffled except for air entrance and exit. (*Biermann*.)

Lowering the air density, indicated by the mass density ρ, (ρ_0 = standard atmospheric mass density) lowers the mass rate of flow for a given pressure drop. The air velocity is theoretically proportional to $d^{-1/2}$ where d is the air density. The mass flow is proportional to the velocity and density or proportional to $d^{1/2}$. Hence, lowering the air density lowers the mass rate of flow although the velocity increases for a given pressure drop.

Heat Transfer in Piston Head.—The heat transferred from the cylinder gases to the piston head is transferred through the head to the ring section and then to the cylinder walls, except that which is transferred through the head to the crankcase

---

[1] Biermann, *op. cit.*

gases and oil sprayed against the piston head. The heat transferred from the gases at temperature T_g, to any central portion of the piston head, at temperature T_p, and bounded by a circle of radius r (Fig. 305) will equal the heat transferred through the annular section Δr if none is transferred through the piston head.

Thus,

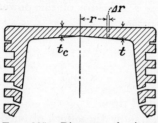

Fig. 305.—Diagram of piston head for heat flow.

$$K_1 \pi r^2 (T_g - T_p) = -\frac{c_m}{dr} 2\pi r t \, dT \quad (10)$$

$$-dT = \frac{rK_1(T_g - T_p)}{2c_m t} dr. \quad (11)$$

Integrating between the limits $r = 0$ and $r = r$, and $T = T_c$ and $T = T$,

$$T_c - T = \frac{r^2 K_1 (T_g - T_p)}{4 c_m t} \quad (12)$$

where T and r = temp. and radius at any distance from the piston center,

T_c = temp. at the center of the piston,

c_m = the conductivity of the metal.

The heat flow into the piston is a function of r^2 and the heat flows through a section that is a function of r for a piston head with uniform thickness. Consequently, it is desirable to increase the head thickness as it approaches the ring section.

By assuming $\quad t = t_c + ar \quad (13)$

and by substituting in Eq. (11) and integrating,

$$T_c - T = \frac{K_1(T_g - T_p)}{2c_m}\left(\frac{r}{a} - \frac{t_c}{a^2} \log_e \frac{t_c + ar}{t_c}\right). \quad (14)$$

Example.—Assume a temperature of 600°F. for T_c, and $t_c = 0.15$ in. $K_1 = 0.3$ B.t.u. in.$^{-2}$ hr.$^{-1}$ °F.$^{-1}$ The mean T_g is 1200°F.; $c_m = 2.27$ B.t.u. in.$^{-1}$ hr.$^{-1}$ °F.$^{-1}$ Determine the thickness at a radius of 1 in. from the piston center required to maintain a temperature of 500°F. at this position. The mean temperature of this section of the piston head, T_p, was found to be 533°F., a constant temperature gradient being assumed. By eliminating a between Eqs. (13) and (14), substituting the various values, and solving, $t = 0.26$. Substituting this value in Eq. (13) results in $a = 0.11$.

Equation (14) indicates that the use of a material having a high conductivity and a thicker piston head will reduce the temperature difference between the center and edge of the piston.

Piston temperatures are higher with the larger bore engines and increase with an increase in load (Fig. 306).

Baffling.—The coefficient of heat transfer is obtained when the air is flowing in the fin spaces with a definite velocity. This velocity is dependent upon the available pressure drop, the baffling, and the fin spacing. The type of baffling used in Biermann's experiments[1] (Fig. 307a) would have an overall pressure drop of

$$\Delta P = P_1 - P_2. \quad (15)$$

The ideal case would be negligible kinetic energy at entrance and exit positions, with a well-rounded entrance to reduce the entrance loss and a well-designed exit section (Fig. 307c) to transform the velocity head into pressure head. Brevoort[2] has found that the exit duct should be as long as possible and that the radius at the baffle exit should be about ¾ in. A small radius is required to maintain air flow against the rear of the cylinder.

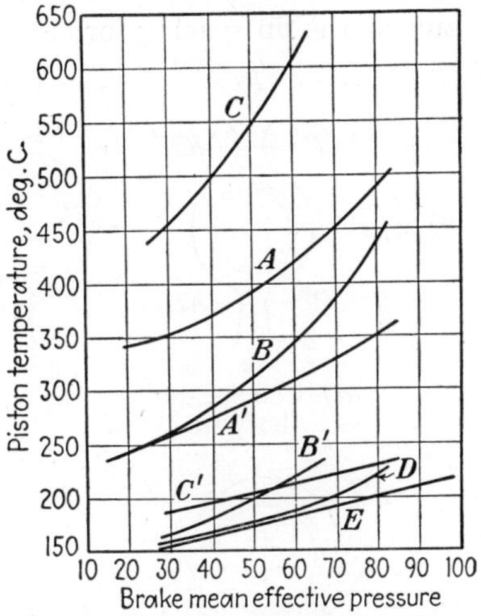

Fig. 306.—Piston temperatures vs. brake mean effective pressure. (*Baker.*)[3]

A and A', center and edge temperatures for 12-in. cast-iron piston No. 2; mean cylinder temperature, 42 deg. C.
B and B', center and edge temperatures for 5⅝ in. Mirrlees-Ricardo cast-iron piston; mean cylinder temperature, 90 deg. C.
C and C', center and edge temperatures for 14-in. cast-iron piston (*Mucklow*).
D, center temperature, Mirrlees-Ricardo "L8" alloy piston (No. 1).
E, center temperature, "J.A.P." petrol engine running at 3000 r.p.m.

The flow should be turbulent for optimum heat transfer, and the pressure drop dp in the channels between the fins with close-fitting baffles should be similar to that in pipes. Glauert[4]

[1] Biermann, *op. cit.*
[2] M. J. Brevoort, Energy Loss, Velocity Distribution, and Temperature Distribution for a Baffled Cylinder Model, *N.A.C.A. Tech. Note* 620 (1937).
[3] H. W. Baker, Piston Temperatures in a Sleeve-valve Oil Engine, *I.M.E. Proc.*, **135**, 35 (1937).
[4] H. Glauert, "The Elements of Aerofoil and Airscrew Theory," p. 107, Cambridge University Press, London, 1930.

gives the relationship

$$\frac{dp}{dl} = \text{const.} \frac{1}{(R)^{1/4}} \frac{\text{K.E.}}{d} \qquad (16)$$

where R is the Reynolds number, and l and d are length and diameter of pipe. The pipe diameter compares with the hydraulic diameter of the flow channel between fins, which is nearly the same as the fin spacing for close spacing.

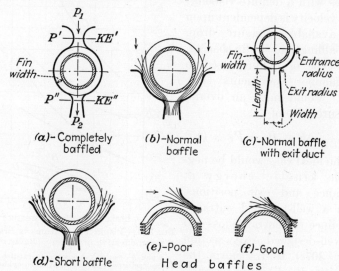

FIG. 307.—Baffles for air-cooled cylinders.

An examination of Eq. (16) discloses the fact that

$$\frac{dp}{dl} \propto \frac{v^{1.75}}{d^{1.25}} \qquad (17)$$

where v = air vel. Thus, decreasing the baffle length l while maintaining the same pressure drop and air velocity permits decreasing the space between the fins, which increases the surface.

Example.—Determine the permissible decrease in fin spacing by halving the baffle length while maintaining the same pressure drop and air velocity.

$$\frac{(dp/dl)_2}{(dp/dl)_1} = \frac{2}{1} = \left(\frac{d_1}{d_2}\right)^{1.25}$$

or

$$\frac{d_2}{d_1} = 0.57.$$

ENGINE COOLING

This increases the fin surface about 74 per cent for a given fin width.

The relationship between the Nusselt number N and the Reynolds number R in the turbulent region (Fig. 302) is

$$NR^{-0.78} = \text{const.} \tag{18}$$

Substituting the equivalents for N and R results in

$$k \propto \frac{v^{0.78}}{d^{0.22}} \tag{19}$$

which indicates that hydraulic diameter or fin spacing compared with the air velocity has only a small effect on the heat-transfer coefficient.

Example.—Determine the effect on the heat-transfer coefficient k of reducing the baffle length in the previous example.

Since the air velocity is held constant,

$$\frac{k_2}{k_1} = \left(\frac{d_1}{d_2}\right)^{0.22} = \left(\frac{1}{0.57}\right)^{0.22} = 1.13.$$

Thus, halving the baffle length while maintaining constant-pressure drop and velocity permits an increase in surface of 74 per cent and causes an increase in heat-transfer coefficient of 13 per cent, the combined result being an increase in heat transfer of 97 per cent. Thus, halving the baffle length for these conditions practically doubled the heat transfer per unit of base area.

The foregoing example illustrates the possibility of much more effective cooling with a number of baffles. However, the power required for cooling increases with the number of baffles. Pinkel[1] has found that the indicated horsepower of an engine may be increased as the value of $K_2^{1.5}$ increases with practically no increase in head temperature.

Thus,
$$\frac{\text{I.hp.}_2}{\text{I.hp.}_1} = \frac{(K_2)_2^{1.5}}{(K_2)_1^{1.5}}. \tag{20}$$

Example.—Assuming that doubling the number of baffles doubles the heat transfer for the conditions indicated in the previous example, and also doubles the power required for cooling, determine the relative increase of the possible indicated horsepower and the power required for cooling.

$$\frac{\text{I.hp.}_2}{\text{I.hp.}_1} = (2)^{1.5} = 2.83.$$

[1] Pinkel, *op. cit.*

Thus, there is 83 per cent more gain in indicated horsepower than in power required for cooling. Obviously, power or blower cooling is desirable for high outputs.

Fins on the rear of cylinder heads require baffling for adequate cooling. Such baffles should follow closely the contour of the edges of the fins near the rear of the cylinder (Fig. 307f) and force the air to flow through the entire channel where the high cooling effect is desired. Baffles that are not closely fitted to the fins (Fig. 307e) permit much of the air flow to short-circuit the fin and result in low cooling effect near the base of the fins.

The temperature of the air if confined to the spaces between the fins will rise as the air flows around the cylinder. This causes the cylinder temperature to be higher in the rear than in front. This effect can be minimized by introducing cool air into the air stream in the fin spaces, which is accomplished by allowing considerable clearance between the baffles and fins near the entrance and gradually reducing the clearance to zero near the rear end of the cylinder.

Piston Cooling.—Most of the heat transferred to the piston head of a small engine is transferred to the ring belt and through the rings to the cylinder. Some heat is transferred from the piston skirt to the oil film and then to the cylinder. The rest of the heat is transferred to the cool entering charge and also to the air and oil spray in the crankcase.

The amount of heat that a piston can absorb depends on its area or diameter squared, and the amount transferred through the head to the ring belt depends on the area of head metal at various radii from the piston center. Even though the head thickness is increased directly with the bore, making the temperature drop per inch of length a constant, the increased length through which heat is transferred results in higher piston-head temperatures with larger pistons. Thus, large-bore engines usually require piston cooling either with an oil spray directed at the underside of the piston head or direct water-cooling (Fig. 308) as in the case of a double-acting engine. Water connections are made to the crosshead with telescopic joints, the inner or moving pipe being of bronze. A generous stuffing box must be provided and air chambers used to prevent surges of water due to the pumping action. The water pressure should be at least 40 lb. in.$^{-2}$, and where this is not available a triplex pump driven from the main engine may be used to force the water through the piston-cooling system.

Valve Cooling.—The intake valve is cooled by the comparatively cool mixture that contacts it during the induction process. The exhaust valve not only is subjected to the combustion temperatures but also is contacted by the hot exhaust products during the release and exhaust processes. Consequently, the exhaust valve runs hot, and the cooling of this valve is of great importance.

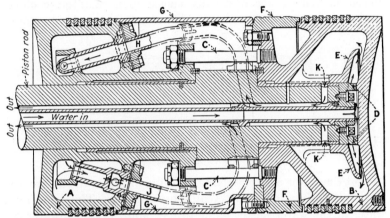

Fig. 308.—Water-cooled piston and rod for a double-acting engine.

The design of the valve and all adjacent parts has a great influence on valve temperature, as indicated in the chapter on Valves and Valve Mechanisms. The hollow salt-cooled valve is partly filled with sodium which melts at about 208°F. The sodium is thrown from one end to the other during engine operation and transfers an appreciable amount of heat from the head to the cooler stem of the valve.

Both water and oil are used to cool valves. A stream of the cooling liquid is directed at the valve head and flows back around the inlet tube which is usually centrally located in the hollow valve stem.

Bearing Cooling.—The high-speed engine transforms considerable work energy into internal energy of the various oil films in the engine as a result of the fluid-friction process. The use of an oil spray to cool pistons also increases the energy and temperature of the oil. Heat is transferred from the oil to the crankcase and then to the surrounding air. However, crankcase areas have not increased appreciably with increase in speed and engine output and in many cases cannot dissipate heat at a rate sufficient to maintain a safe operating oil temperature. In such cases, oil coolers are used that transfer heat from the oil to a cooling fluid which is either water or oil.

Engine Cooling Systems.—Water-cooled engines are supplied with water which flows through the water jackets around the cylinders and heads and absorbs heat from the hot metal it contacts. The water is usually cooled in a radiator, cooling tower, or pond and recirculated. The system may be designed

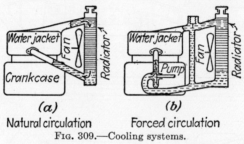

(a) Natural circulation   (b) Forced circulation
FIG. 309.—Cooling systems.

so that the water may circulate naturally, owing to the difference in density of water at different temperatures. Thus, the hot water in the water jacket tends to rise, and the cooler water in the radiator (Fig. 309a) tends to fall. The circulation is at a low rate and is not satisfactory for high-duty engines that require high rates of heat transfer, particularly at some spots.

Forced circulation requires a pump (Fig. 309b), usually of the centrifugal type, which forces water at comparatively high rates at the spots that require the most cooling. A header is usually designed as part of the system to provide equal distribution of water to all cylinders; the header is supplemented by tubes or ducts, which direct water at the exhaust-valve seats.

A quick warm-up after starting an engine is always desirable. This is accomplished in most automotive systems by a thermostat which prevents or restricts the flow of water through the radiator until the jacket water reaches the desired temperature.

Small engines may use "hopper" or evaporative cooling. The hopper is a large water jacket in which the water rises to the boiling temperature. Further absorption of heat evaporates water while the temperature remains at the boiling point.

ENGINE COOLING

Make-up water must be added before evaporation uncovers the surfaces being cooled. The C.F.R. engines use evaporative cooling but condense the vapor formed by contact with a coil through which tap water is flowing.

EXERCISES

1. Compute the heat transfer through a cast-iron cylinder wall of a 3½- by 5-in. engine having a wall thickness of ¼ in. Mean gas temperature is 1200°F. K_1 and K_2 are 30 and 1200 B.t.u. ft.$^{-2}$ hr.$^{-1}$ °F.$^{-1}$ Mean water-jacket temperature is 150°F. Also determine the surface temperatures.

2. Using the same data as in Prob. 1, except $K_1 = 50$ B.t.u. ft.$^{-2}$ hr.$^{-1}$ °F.$^{-1}$, determine the effect on surface temperatures and also the effect of substituting aluminum for cast iron.

3. The conditions of the gases in an engine cylinder at the end of combustion, 20° after top center, are 400 lb. in.$^{-2}$ abs. and 4200°R. (Chart E, Appendix). Determine and plot the temperature variation during adiabatic expansion for an engine with a 6:1 compression ratio and a 4:1 connecting-rod-crank ratio (Table 30). Use the crank angle as the abscissa.

4. Determine and plot the variation in surface exposed for a valve-in-head engine with a flat cylindrical combustion chamber, using the data in Prob. 3.

5. Determine the heat transfer during the expansion stroke after completion of combustion, using the data in Probs. 3 and 4. Engine speed = 4000 r.p.m. Vol. eff. = 0.75. Mean wall temperature = 350°F. Use Janeway's data for K_1.

6. Using Chart E (Appendix), redraw the expansion path allowing for the heat transfer obtained in Prob. 5. Note that the energy equation must be satisfied for any chosen path. Then, plot the temperature variations against crank angle, and determine the heat transfer for this path.

7. Determine and plot the variation in the base surface coefficient K_2 against fin spacing s for a weight of 0.08 lb. of steel fins for each inch of base surface. Air vel. between fins = 100 m.p.h. Air conditions = 80°F. and 14.7 lb. in.$^{-2}$ abs. Outer cylinder dia. = 5.81 in.

8. Solve Prob. 7, using fin thickness as the independent variable.

9. Determine and plot the temperature variation in the piston head, in the example on page 440, from the center to a distance 1.75 in. from the center. Do this problem also with a constant head thickness.

10. Determine and plot the pressure drop for air flow, for complete baffling, against fin spacing for a weight of 0.08 lb. of steel fins for each inch of base surface for an air velocity of 100 m.p.h. Assume that the data in Fig. 304 apply to this problem.

11. Determine and plot the power required to flow the air through the fins in Prob. 10.

CHAPTER XVI

ENGINE PERFORMANCE

Indicated Work.—The maximum work that can be obtained from the internal-combustion-engine process is indicated by the ideal analysis (Chap. V). Various inherent losses (Chap. VI) reduce this performance to the work actually done on the engine pistons. This work may be evaluated by obtaining an indicator

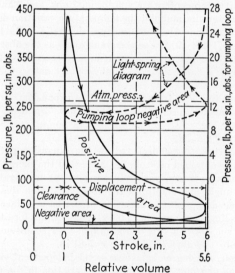

FIG. 310.—Indicator card with light, spring pumping loop. Obtained by the National Bureau of Standards with their balanced-pressure diaphragm indicator. Engine tested in their altitude chamber.

record (Fig. 310) of the pressure-time or pressure-volume variation for a number of engine cycles.

Large slow-speed engines may be indicated very satisfactorily with the usual piston-type indicator. However, the indicator connection changes the clearance space of the small high-compression engine and appreciably affects its performance. Also the piston-type indicator is unable to follow accurately the rapid pressure changes occurring at high speeds. Various other types of indicators[1] have been developed which are satisfactory but usually require considerable additional equipment and refined technique to obtain accurate records.

The common method for approximating the indicated work of high-speed engines is to add the work output of the engine to

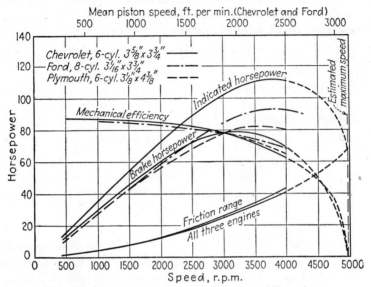

FIG. 311.—Typical power curves, 1939 automobile engines.

the losses obtained by motoring or driving the engine with an electric dynamometer at as near operating conditions as possible.

Thus,   Ind. work = brake work + friction losses.   (1)

This indicated work (Fig. 311) would be equivalent to the positive area of the indicator card (Fig. 310) if the motoring method resulted in the correct friction losses.

Brake or Actual Work.—The actual-work output of an engine is usually determined by a power-absorption device,[2] such as the Prony brake, water brake, fan dynamometer, electric generator,

[1] K. J. DeJuhasz, "Engine Indicator; Its Design, Theory, and Special Applications," Instruments Publishing Co., New York, 1934.
[2] Torque indicators are also used for determining output. See A. L. MacClain and R. S. Buck, Flight-testing with an Engine Torque Indicator, *S.A.E. Trans.*, **33**, 49 (1938).

or electric dynamometer, the electric dynamometer being most satisfactory for testing purposes. In all cases a force F is measured at a given radius r.

Thus, $$\text{Torque} = Fr. \tag{2}$$

The torque measured is that developed by the engine in which case the mean force acts through a distance equal to the circumference of the crank circle each revolution of the engine. Hence, the rate of work output is

$$\text{Work min.}^{-1} = 2\pi rFn \tag{3}$$

where n = speed, r.p.m.

The brake horsepower (b.hp.) (Fig. 311) is obtained by dividing the work in foot-pounds per minute by 33,000 ft.-lb. min.$^{-1}$ hp.$^{-1}$

The approximate brake horsepower for a four-stroke-cycle engine may be estimated by the empirical relationship

$$\text{B. hp.} = \frac{d^2 ns}{2000} \tag{4}$$

where d = cylinder bore, in.,
n = no. of cyl.,
s = mean piston speed, ft. min.$^{-1}$

Equation (4) indicates a straight-line relationship between power and speed. Hence, if the equation fits the low-speed range, it will be too high at the higher speeds. The maximum brake horsepower is attained in automotive engines at piston speeds usually above 2000 ft. min.$^{-1}$ Hence, the factor $d^2 n$ is an approximate measure of the maximum brake horsepower of automotive engines.

The ratio of the brake horsepower to the indicated horsepower is the *mechanical efficiency*.

Friction Losses.—The friction losses in an engine are caused principally by fluid friction since there is very little metal-to-metal contact or bearing surfaces would be quickly destroyed. The losses may be divided into pumping losses and those due to the shearing of the various lubricant films. Pumping losses may be obtained from an indicator record which magnifies this part of the record (Fig. 310). It may be approximated from the mean intake- and exhaust-manifold pressures.

Motoring the engine results in the determination of the total friction losses (Fig. 311). However, the gas pressure on the pistons, its effect on the various bearings, and the temperature conditions of the piston and cylinder walls are changed. The amount of the lubricant film on the cylinder bores is also changed. The pumping work is different under motoring conditions because of the absence of the effect of the release of gases when the exhaust valve opens under actual operating conditions. Some of these effects appear to offset the others, which indicates that the motoring results may be fairly good approximations.

Various methods[1] are used to determine the losses under actual operating conditions. One method is to switch from operating to motoring very quickly without changing speed and to determine the friction losses at various time intervals after switching. Extrapolation of the loss-time relation back to zero time indicates the friction losses under operating conditions provided that there is no abrupt change in the friction losses immediately after switching.

Another method is to operate the engine with and without one of the cylinders firing. With all cylinders firing in a six-cylinder engine,

$$\text{Output}_6 = \text{i.hp.}_6 - \text{engine friction.}$$

With all but one cylinder firing,

$$\text{Output}_5 = \text{i.hp.}_5 - \text{engine friction.}$$

Then, $\quad \text{Output}_6 - \text{output}_5 = \text{i.hp.}_1,$

from which the indicated horsepower for the engine and the engine friction can be determined. This method is subject to the criticism that the friction of the one cylinder is different under the two conditions and that the mixture distribution may be affected.

A method that gives promise of producing the true result is to determine the rate of deceleration of engine speed immediately after opening the ignition circuit. It involves a knowledge of the effective kinetic energy of the moving parts of the engine at various speeds.

Friction Losses of Engine Parts.—The friction losses of engine parts may be determined by progressively removing sets of parts

---

[1] See F. H. Dutcher, The Friction of Reciprocating Engines, *A.S.M.E. Trans.*, **60**, 225 (1938) for a bibliography on engine friction.

and motoring the engine before and after each removal of parts. In attempting to separate piston and connecting-rod friction a difficulty arises that Carson[1] surmounted in the following analysis:

1. Motoring the engine results in total losses (Fig. 312).
2. Removing the tappets and installing check valves in spark-plug holes reduces the valve-mechanism loss, doubles the side-wall friction due to compression and reexpansion, and eliminates pumping.

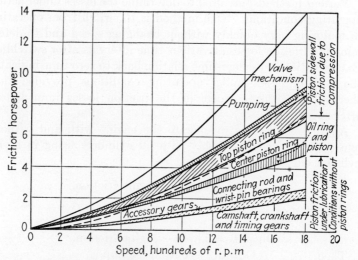

Fig. 312.—Friction losses in White four-cylinder truck engine. Water temperature 180° F.; oil temperature 150° F. (*Carson.*)

3. Removing the cylinder head and replacing it with an open dummy head for cooling purposes eliminates the double side-wall friction due to compression and reexpansion.

4. Replacing the valve tappets adds the part of the valve-mechanism friction that was previously removed and makes possible the determination of pumping loss, since two of the three items forming the difference between losses 1 and 3 are now known.

5. Removing the top compression ring permits the evaluation of its friction, the same procedure being followed with each ring. However, in Carson's tests the removal of the oil ring increased the friction loss apparently because of uncontrolled piston lubrication.

6. Removing the pistons and substituting horizontal oscillating bars having the same inertia effect on connecting-rod bearings eliminates piston friction with uncontrolled lubrication.

[1] G. B. Carson, "Internal-combustion-engine Friction Analysis," Essay for Master of Science in Mechanical Engineering, Yale University, 1932.

7. Removing the connecting rods[1] eliminates connecting-rod bearing losses.

8. Removing the camshaft and the various accessories, one at a time, eliminates the friction of each item.

Carson's tests (Fig. 313) with water-jacket temperature of 60°F. showed the friction losses of the piston assembly in per cent to be almost independent of oil temperature. These losses

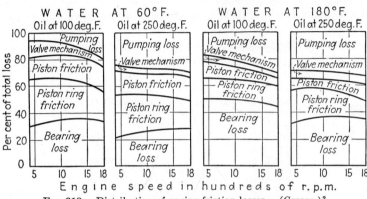

FIG. 313.—Distribution of engine friction losses. (*Carson.*)[2]

amounted to 40 to 50 per cent of the total losses, the ring friction being about 60 per cent of the loss of the piston assembly. With a water-jacket temperature of 180°F. the friction loss of the piston assembly amounted to 20 to 35 per cent of the total losses, the ring friction amounting to about 50 to 70 per cent of the total for the piston assembly.[3] The actual bearing loss decreased with an increase in oil temperature, but the bearing loss percentage increased with an increase in water-jacket temperature.

Mean Effective Pressure.—Mean effective pressure (m.e.p.) is a hypothetical pressure which if acting on the engine piston during a single stroke would result in the actual amount of work. Mean effective pressure may be evaluated for any of the work or power terms.

Thus,   B. hp. $\times$ 33,000 ft.-lb. min.$^{-1}$ hp.$^{-1}$ = $plan$ (5)

---

[1] It would seem desirable also to clamp weights on the crankpins that would produce approximately the same effect on the main bearings as the connecting rods and pistons.

[2] Carson, *op. cit.*

[3] See L. Illmer, Piston-ring Friction in High-speed Engines, *A.S.M.E. Trans.*, **59**, 1 (1936).

where p = b. m.e.p., lb. in.$^{-2}$,
  l = length of stroke, ft.,
  a = piston area, in.^{2},
  n = no. of power strokes or engine cycles min.$^{-1}$

The same procedure may be followed with the indicated and friction horsepower.

Example.—An eight-cylinder 3¼- by 4-in. automobile engine develops 90 b. hp. at 3250 r.p.m. The friction horsepower at this speed is 25. Determine the indicated m.e.p.

$$(90 + 25)33{,}000 = p \times 0.333 \times 8.296 \times 8 \times 1625.$$
$$p = 105.7 \text{ lb. in.}^{-2}$$

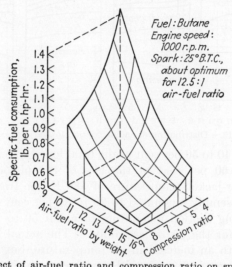

Fig. 314.—Effect of air-fuel ratio and compression ratio on specific fuel consumption. (*Vogt*.)

Fuel Consumption.—The total fuel consumption is determined by measuring the volume or weight of fuel consumed by the engine under given test conditions in a given time. The specific fuel consumption is determined by dividing the total fuel consumption per hour by the power developed, both the brake and indicated power being used.

The fuel consumption of an engine depends principally upon the heating value of the fuel, the air-fuel ratio, and the efficiency of the engine process. Vogt[1] has shown by a family of curves

[1] C. J. Vogt, Tests Show Merits of Butane as an Internal-combustion Engine Fuel, *Auto. Ind.*, **71**, 348 (1934).

(Fig. 314) the variation between specific fuel consumption, air-fuel ratio, and compression ratio for butane.

Gases vary appreciably in heating value. An engine will require about one and one-half times as much blast-furnace gas for a given engine output as compared with what it requires

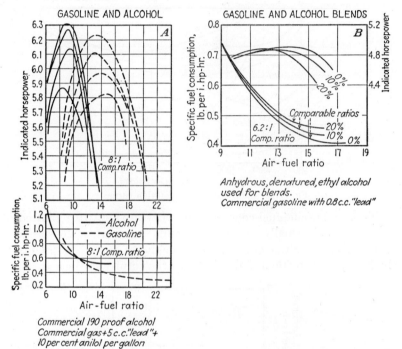

Fig. 315.—Power and specific fuel consumption for gasoline, alcohol, and alcohol blends.[1]

of average natural gas (Table 12). The variation in gasolines may vary the heating value from 1 to 2 per cent so that not much difference can be detected in fuel consumption with different gasolines provided that the volatility characteristics are satisfactory. Ethyl alcohol has 36 and 31 per cent less heating value on a weight and volume basis, respectively, than gasoline and results in larger fuel consumption with practically the same power

---

[1] (A) L. C. Lichty and E. J. Ziurys, Engine Performance with Gasoline and Alcohol, *Ind. Eng. Chem.*, **28**, 1094 (1936).

(B) L. C. Lichty and C. W. Phelps, Gasoline-alcohol Blends in Internal-combustion Engines, *Ind. Eng. Chem.*, **30**, 222 (1938).

output in a given engine (Fig. 315A). The same characteristic is found in comparing alcohol blends with gasoline (Fig. 315B). Alcohol requires less air than gasoline, and comparisons should be made at comparable air-fuel ratios.

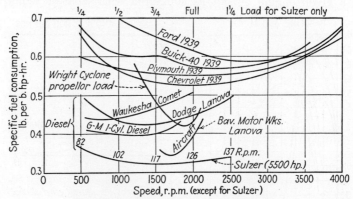

Fig. 316.—Specific-fuel-consumption curves.

Typical specific-fuel-consumption curves (Fig. 316) plotted against speed usually have a minimum value at some speed. The fuel consumption of a carburetor engine at any given speed is high at part throttle and decreases as the throttle is opened.

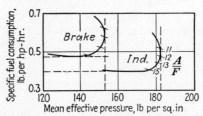

Fig. 317.—Fuel-consumption loops. (Data from test of P. & W. 1340-H cyl.)[1]

The effect of air-fuel ratio on specific fuel consumption and power can be shown by plotting the specific fuel consumption against m.e.p. (Fig. 317). The horizontal and vertical tangents to such plots indicate maximum economy and maximum power values.

Volumetric Efficiency.—The volumetric efficiency of an engine is the ratio of the volume of air or gaseous mixture inducted

[1] O. W. Schey and J. D. Clark, Fuel Consumption of a Carburetor Engine at Various Speeds and Torques, *N.A.C.A. Tech. Note* 654 (1938).

ENGINE PERFORMANCE

in a given time to the total displacement of suction strokes for the same period of time. The volume inducted is determined at the atmospheric pressure and temperature conditions surrounding the engine.

Example.—A gasoline engine has a displacement of 250 in.^{3}. The volume of air induced is 170 ft.^{3} min.$^{-1}$ at atmospheric conditions surrounding the engine and at an engine speed of 3600 r.p.m. Determine the volumetric efficiency.

On the assumption of a four-stroke cycle,

$$\text{Vol. eff.} = \frac{170 \times 1728 \times 2}{250 \times 3600} = 0.65.$$

Example.—A two-cylinder double-acting four-stroke-cycle blast-furnace-gas engine having a bore of 36 in. and a stroke of 50 in. inducts 4060 ft.^{3} of air and 3950 ft.^{3} of gas per minute at an engine speed of 100 r.p.m. The air and gas volumes are determined at atmospheric pressure, the air volume at atmospheric temperature, and the gas volume at its supply temperature. Determine the volumetric efficiency. The piston rod is 6 in. in diameter.

$$\text{Gross piston area} = \pi \times \overline{1.5}^2 = 7.069 \text{ ft.}^2$$
$$\text{Piston-rod area} = \pi \times \overline{0.25}^2 = 0.196 \text{ ft.}^2$$
$$\text{Vol. eff.} = \frac{(4060 + 3950) \times 12}{(7.069 + 7.069 - 0.196) \times 50 \times 2 \times 100} = 0.69$$

The quantity of charge inducted and energy input would vary directly with engine speed if the volumetric efficiency remained

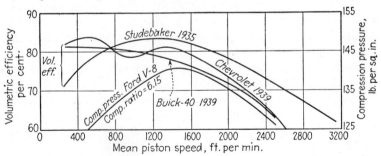

FIG. 318.—Volumetric efficiency and compression pressure vs. piston speed.

constant. An increase in speed increases the charge velocity which requires a higher pressure drop and reduces the charge inducted. This indicates a decrease in volumetric efficiency with an increase in speed. However, the inertia effect of the incoming charge exerts a "ramming" effect, usually causing the volumetric efficiency to have a maximum value at some speed (Fig. 318) that depends on the valve timing.

The product of speed and volumetric efficiency is a measure of the rate at which charge is inducted. This product has a maximum at some speed beyond which the indicated power decreases and approaches the friction loss as the brake output decreases to zero. The maximum desirable speed is below this point since the specific fuel consumption increases very rapidly at the higher speeds.

Geometrically similar engines[1] would have the same ratio for piston, valve, port, and manifold areas. The velocity of flow of the mixture would be the same in such engines at the same piston speed. The pressure drop required for flow is directly proportional to the length of the induction system and inversely proportional to the hydraulic diameter. The length factor increases the pressure drop for the large engine, but this is offset by the larger hydraulic diameter. However, the small hydraulic diameter for the small engine decreases the Reynolds number and indicates a small increase in friction since the velocity of flow is the same and there is little change, if any, in the viscosity of the mixture. Thus, geometrically similar engines will have nearly the same volumetric efficiencies at the same piston speeds, but the volumetric efficiency should decrease slightly as the engine is made appreciably smaller.

The effect of pressure, temperature, fuel, compression ratio, and heat transfer on volumetric efficiency is dealt with in Chap. VII, and the effect of valve timing is dealt with in Chap. XI.

Standard Conditions.—The amount of mixture inducted into an engine and the performance obtained from it vary with the barometric conditions surrounding the engine. Consequently, the engine output should be corrected to standard conditions. The standard conditions adopted by the S.A.E. are 29.92 in. Hg and 60°F. The correction factor is

$$\frac{\text{B. hp., corr.}}{\text{B. hp., test.}} = \frac{29.92}{p}\sqrt{\frac{T}{520}}. \qquad (6)$$

The atmosphere consists principally of air and water vapor. The energy liberation in a combustion process depends on the air. Thus, variations in humidity may result in appreciable variations

---

[1] See Chap. XIX.

in power output. Gardner[1] and Brooks[2] have found that the power output is directly proportional to the dry-air pressure which is equal to the total barometric pressure p, minus the

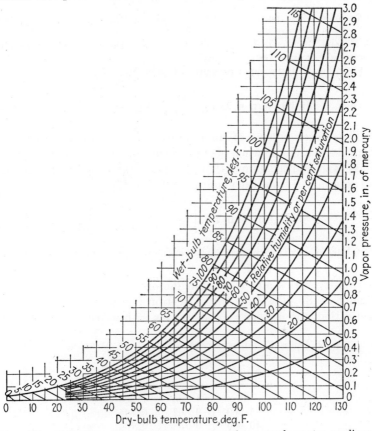

Fig. 319.—Humidity chart for vapor pressure from psychrometer readings. Barometric pressure = 30 in. of Hg. (*Gardner*.)

partial pressure of the water vapor, pp_{wv}. Inserting this correction in Eq. (6), results in

$$\frac{\text{B. hp., corr.}}{\text{B. hp., test.}} = \frac{29.92}{p - pp_{wv}} \sqrt{\frac{T}{520}} \qquad (7)$$

which corrects to a condition of zero humidity.

[1] A. W. Gardner, Atmospheric Humidity and Engine Performance, *S.A.E. Trans.*, **24**, 267 (1929).

[2] D. B. Brooks, Horsepower Correction for Atmospheric Humidity, *S. A. E. Trans.*, **24**, 273 (1929).

460 INTERNAL-COMBUSTION ENGINES

The vapor pressure can be obtained from the wet- and dry-bulb temperatures of a psychrometer and the humidity chart, Fig. 319.

Example.—The power output of an engine is 98.2 hp. The barometer indicates a pressure of 30 in. Hg. The temperature of the atmosphere is 95°F., and the wet-bulb temperature is 90°F. Determine the corrected horsepower at standard conditions.

From Fig. 319, the vapor pressure is 1.35 in. Hg.

$$\text{Then,} \quad \text{B. hp., corr.} = 98.2 \frac{30}{30 - 1.35} \sqrt{\frac{555}{520}} = 108.3 \text{ hp.}$$

The power output of an aircraft engine decreases with an increase in altitude when operating with a fixed throttle opening. Gagg and Farrar[1] found that the power at any altitude could be determined from the empirical relationship

$$\text{B. hp.} = \text{b. hp.}_s \left[\frac{d}{d_s} - \frac{(1 - d/d_s)}{7.55} \right] \quad (8)$$

where d = atmospheric density at any altitude,
s = sea level.

Density ratios and correction factors (Table 27) determined from Eq. (8) show that the correction factor decreases slightly faster than the density ratio decreases with altitude.

TABLE 27.—CORRECTION FACTOR FOR ALTITUDE

Alt., ft.	d/d_s	Factor
0	1.000	1.000
5,000	0.862	0.844
10,000	0.738	0.704
15,000	0.629	0.580
20,000	0.533	0.471

Variation in exhaust back pressure changes the exhaust stroke m.e.p., the pressure of the clearance gases, and the volumetric efficiency of the engine (Chap. VII). The effect of exhaust back pressure on power output is influenced appreciably by the manifold pressure, the compression ratio which determines the

[1] R. F. Gagg and E. V. Farrar, Altitude Performance of Aircraft Engines Equipped with Gear-driven Superchargers, *S.A.E. Trans.*, **29**, 217 (1934).

clearance volume, and valve timing which changes the scavenging effect. The net result for an increase in exhaust pressure of 1 lb. in.$^{-2}$ in a normally aspirated automobile engine having a brake m.e.p. of 100 lb. in.$^{-2}$ is about a $2\frac{1}{2}$ per cent decrease in power output (Chap. X). The exhaust back pressure on aircraft engines decreases with altitude. This results in higher output with altitude exhaust conditions than at sea level, with the same intake conditions at the carburetor in both cases (Fig. 320).

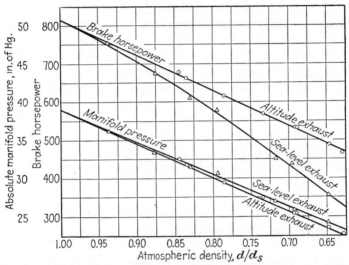

FIG. 320.—Horsepower and manifold pressure with standard altitude conditions at the carburetor. (*Gagg and Farrar.*)[1]

The effect of exhaust back pressure varies with different engines so that test data are required for making corrections to standard conditions (see Fig. 184).

No correction need be applied to the specific fuel consumption unless the mechanical efficiency of the engine is appreciably different under test and standard conditions. Thus, a correction in power output indicates a similar correction in indicated power and fuel supplied, and results in the same specific fuel consumption.

Improving Engine Performance.—The performance of an engine depends primarily on the energy in the mixture that is supplied to the engine and on the efficiency with which the

[1] Gagg and Farrar, *op. cit.*

engine converts the energy supply into work. This indicates two general methods for increasing engine performance.

1. Increasing the energy input.
2. Increasing the efficiency of energy conversion.

The first method requires the induction of more mixture per cycle or per unit of time. The engine displacement may be increased, but this also increases the engine weight. An increase in speed increases the displacement and the mixture input unless the increase in speed results in a greater decrease in volumetric efficiency. Any change in design or operating conditions that increases the volumetric efficiency will increase the engine performance provided that detonation, overheating, or failure of some part to perform its function satisfactorily does not occur.

The second method requires the improvement of the efficiency of the process. This may be accomplished by increasing the compression ratio which necessitates, for the Otto-engine process, the use of fuels with higher antiknock characteristics. Also, a reduction in the deviations from the ideal to the actual process (Chap. VI) and a reduction of the friction losses will increase the output of an engine. However, an increase in compression ratio may increase the friction losses more rapidly than the indicated thermal efficiency is increased, resulting in an optimum compression ratio beyond which the brake thermal efficiency will decrease.

Engine Speed.—The speed of internal-combustion engines is limited by internal stresses, the inability of certain parts to function properly, and the rapid decrease in volumetric efficiency at high speeds. High mean piston speed rather than high r.p.m. is the criterion for a high-speed engine. Practice indicates that the maximum desirable mean piston speeds range from 2500 to 3500 ft. min.$^{-1}$ for aircraft and automobile engines.

High-duty stationary engines, larger than automotive engines and including high-speed Diesel engines, have maximum mean piston speeds that range from 1500 to 2500 ft. min.$^{-1}$ Large gas and Diesel engines have maximum mean piston speeds in the range from 1000 to 1500 ft. min.$^{-1}$

Thus, high piston speeds are used in small-stroke engines, whereas low piston speeds are used in large-stroke engines. The range of engine r.p.m. for any speed classification will depend on the stroke of the engine (Table 28).

ENGINE PERFORMANCE

TABLE 28.—ENGINE SPEED

Classification	High speed				Medium speed				Low speed			
Mean piston speed, ft. min.$^{-1}$	3000				2000				1000			
Stroke, in.	2	3	4	5	7	10	13	16	20	30	40	50
R.p.m.	9000	6000	4500	3600	1714	1200	923	750	300	200	150	120

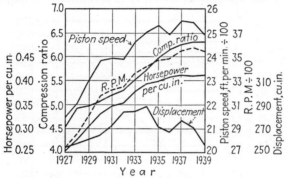

FIG. 321.—Trend in American passenger-car engines. (*Data from Auto. Ind.*, **80**, 218, 1939.)

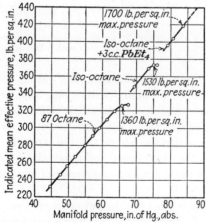

FIG. 322.—Power output with supercharging at constant speed. 6:1 comp. ratio; optimum mixture; man. temp. 200 to 210°F. (*DuBois and Cronstedt.*)[1]

[1] R. N. DuBois and V. Cronstedt, High Output in Aircraft Engines, *S.A.E. Trans.*, **32**, 225 (1937).

464   INTERNAL-COMBUSTION ENGINES

The trend in American automobile engines has been one of increasing compression ratio, r.p.m., and piston speed to obtain increased performance (Fig. 321). The engine displacement has also been increased, but the trend now is downward.

Supercharging.—The charge inducted into an engine may be increased by the use of a supercharger which compresses and forces the air or mixture into the engine at pressures above atmospheric. The output is limited by the detonation characteristics of the fuel (Fig. 322) provided that the engine mechanism can withstand the stresses to which it is subjected. Supercharging

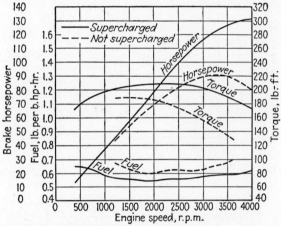

Fig. 323.—Effect of a centrifugal-type supercharger on power and fuel consumption. Graham 8-cyl. 3¼- by 4-in. engine with 6.72:1 comp. ratio. Supercharger speed 5.76 × engine speed. (*Schwitzer.*)[1]

may be used to increase the charging effect at all speeds, but it is particularly desirable at high speeds (Fig. 323) since normally aspirated engines have low volumetric efficiencies under this condition. Supercharging is very desirable for aircraft engines operated at high altitudes where the low specific weight of air reduces the mass of the charge inducted although it does not appreciably affect the volumetric efficiency of the unsupercharged engine.

The ideal indicator card for the supercharged engine is $ABCD$-$EFGA$ (Fig. 324). The work of the ideal compressor for supercharging is represented by the area $JAGFJ$, which makes the

[1] L. Schwitzer, Forced-induction Possibilities for Automotive Vehicles, *S.A.E. Trans.*, **29**, 454 (1934).

net indicated work of the engine equivalent to area $ABCD$ − area $JAEJ$. The ideal normally aspirated engine inducting the same mass of charge and having the same compression pressure as the ideal supercharged engine would have a net indicated work area of $JABCDHJ$, which is larger than the net work area for the ideal supercharged engine by area $JEDHJ$. The same compression ratios being used, the ideal supercharged engine would still be less efficient than the unsupercharged one; for if the work of supercharging is neglected, the thermal efficiency should be practically the same; however, the work of supercharging reduces the ideal work, as indicated in the foregoing, by area $JEAJ$.

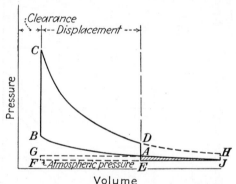

Fig. 324.—Ideal supercharging diagram.

The permissible amount of supercharging depends on the compression ratio, the detonation characteristics of the air-fuel mixture, and the inherent ability of the engine to withstand increased load and heat stresses. An engine operated at a compression ratio that causes incipient detonation cannot be supercharged unless a fuel with better antiknock characteristics is used or unless the compression ratio is lowered. Lowering the compression ratio reduces the thermal efficiency, but supercharging to obtain the same compression pressure increases the output and more than offsets the effect of the loss of thermal officiency on output.

Sparrow[1] developed a theoretical diagram (Fig. 325) for determining the relative effect of various combinations of charge density and compression ratio on power output and thermal

[1] S. W. Sparrow, The Effect of Changes in Compression Ratio upon Engine Performance, *N.A.C.A. Rept.* 205 (1926).

efficiency. This diagram is based on the following assumptions:

   a. At any compression ratio the indicated m.e.p. is directly proportional to the volumetric efficiency.

   b. At any volumetric efficiency the indicated m.e.p. is proportional to the air-standard efficiency as determined by the

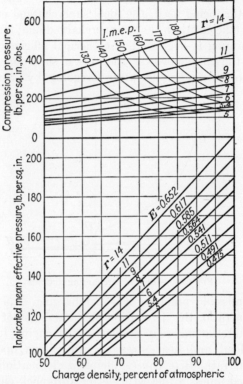

Fig. 325.—Effect of compression ratio and charge density on power output and efficiency. (*Sparrow.*)[1]

compression ratio, one point on the 5.4:1 compression-ratio line being fixed by test data.

Example.—An engine with a 7:1 compression ratio has a relative charge density of 0.65 at a high speed. The compression ratio is as high as can be used with the fuel available. A supercharger is added which increases the relative charge density to 1.0, which is that of the atmosphere at sea level. Determine the desirable compression ratio, the increase in power, and the decrease in thermal efficiency.

[1] Sparrow, *op. cit.*

From Fig. 325, the compression pressure at 7:1 compression ratio and a charge density of 0.65 is 140 lb. in.$^{-2}$ abs. At a charge density of 1.0, the compression ratio should be 5.2:1 to result in the same compression pressure. Thus:

Charge density	Comp. ratio	Ind. m.e.p.	Thermal eff.
0.65	7:1	114	0.541
1.00	5.2:1	156	0.483

This indicates a gain of about 37 per cent in power and 11 per cent loss of efficiency.

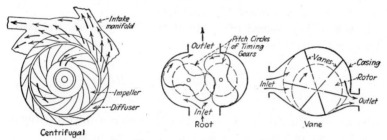

Fig. 326.—Centrifugal and positive types of superchargers.

The foregoing example shows that the maximum efficiency is attained with the unsupercharged engine operating at the optimum compression ratio but that maximum power is attained with the supercharged engine operating with lower compression ratio and efficiency. The Diesel engine is not restricted to lower compression ratios when supercharged, but if the maximum pressure is limited it may be desirable to reduce the compression ratio.

There are three types of superchargers:
1. Piston-cylinder type.
2. Positive rotary blower, Roots and vane types (Fig. 326).
3. Centrifugal type (Fig. 326).

The piston-cylinder type of supercharger is used on large and slow-speed stationary engines, the positive-blower type on medium-size stationary engines as well as automobile and aircraft engines, and the centrifugal type principally on aircraft and automobile engines. The positive-rotary-blower and piston-cylinder types of supercharger are effective at all speeds, whereas the centrifugal type of supercharger is effective at high speeds only.

The ideal indicator diagram for the piston-cylinder type and the centrifugal type of supercharger is $ABCDA$ (Fig. 327). The work area may be evaluated as follows:

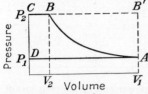

Fig. 327.—Supercharger-compressor diagram.

$$W_{comp} = H_B - H_A$$
$$= \frac{kR}{k-1}(T_B - T_A) \qquad (9)*$$

which assumes that the path AB is described by the equation $PV^k =$ const. and that the process is adiabatic.

If heat is transferred in or out during the process, the path AB may be described by the equation $PV^n =$ const. and Eq. (9) becomes

$$W_{comp} = H_B - H_A + Q_{out} - Q_{in} = \frac{nR}{n-1}(T_B - T_A). \qquad (10)$$

However, when the air is stirred by the rotor of the centrifugal supercharger or leakage occurs past the edges of the rotors or blades of the positive displacement type, transforming available energy by the friction process into internal energy of the air or mixture, the process is irreversible and the part of the equation representing the P-V area is not a measure of the work done.

Example.—A centrifugal compressor increases the pressure of air from 10 to 20 lb. in.$^{-2}$ abs., and the temperature from 500 to 650°R. The net amount of heat transfer is 100 B.t.u. mol^{-1} out of the air. Assume that the path AB (Fig. 327) is described by the equation $PV^n =$ const. Evaluate both parts of Eq. (10).

Obtaining the values of H for air from Table III, Appendix, results in

$$W_{comp} = 1926 - 898 + 100 = 1128 \text{ B.t.u.}$$

Also, $\qquad \dfrac{T_B}{T_A} = \left(\dfrac{P_2}{P_1}\right)^{\frac{n-1}{n}} \qquad$ or $\qquad 1.30 = (2)^{\frac{n-1}{n}}$

from which $\qquad n = 1.608$.

Then, $\quad \dfrac{nR}{n-1}(T_B - T_A) = \dfrac{1.608 \times 1.986}{0.608}(650 - 500) = 788$ B.t.u.

* This relation is obvious when it is noted that $k = c_p/c_v$, $R = c_p - c_v$, and $H_B - H_A = c_p(T_B - T_A)$.

Thus, the *P-V* diagram area is much smaller in value than the actual work when friction is involved. The mechanical work dissipated by friction can be determined from the entropy equation

$$T \Delta S = \Delta F - \Delta Q$$

by evaluating the area under the path AB on the T-S diagram.

Also, from the foregoing data, $\Delta F = 1128 - 788 = 340$ B.t.u. which is dissipated by friction.

The centrifugal supercharger operates at high speeds, the maximum speeds ranging from 10,000 to 30,000 r.p.m. It may be gear driven or exhaust-turbine driven (Fig. 328). The

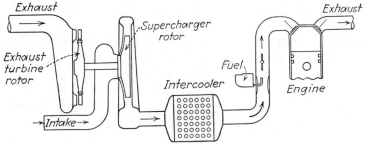

FIG. 328.—Diagrammatic sketch of exhaust-driven centrifugal supercharger. (*Berger and Chenoweth.*)[1]

impeller imparts a high velocity to the inducted medium flowing through the rotor. The velocity is then reduced in the diffuser to increase the pressure. The energy equation for the medium from the tip of the rotor to the intake manifold is

$$KE_t - KE_m = H_m - H_t \qquad (11)$$

where subscripts t and m = rotor tip and intake manifold, resp.

Assuming constant specific heat for the small temperature range involved, results in

$$KE_t - KE_m = c_p(T_m - T_t). \qquad (12)$$

The ideal case would be adiabatic and frictionless, and the term KE_m negligible. Hence, Eq. (12) becomes

$$\frac{mv_t^2}{2g} = c_p T_t \left[\left(\frac{P_m}{P_t} \right)^{\frac{k-1}{k}} - 1 \right] \qquad (13)$$

where m = wt. of a mol of the fluid.

[1] A. L. Berger and O. Chenoweth, Supercharger Installation Problems, *S.A.E. Trans.*, **33**, 472 (1938).

Example.—Air leaves the rotor of a centrifugal supercharger with a velocity of 700 ft. sec.$^{-1}$ and a temperature of 60°F. Determine the ratio of P_m/P_t. $c_p = 6.93$ B.t.u. mol.$^{-1}$ °R.$^{-1}$, and $k = 1.4$.

Then,
$$\frac{28.95 \times 700^2}{64.4 \times 778} = 6.93 \times 520 \left[\left(\frac{P_m}{P_t}\right)^{0.286} - 1 \right],$$

$$\frac{P_m}{P_t} = 1.303.$$

Equation (13) may also be used to determine the theoretical increase in static pressure obtainable from the impact of air at various velocities and altitudes.

The Roots type of supercharger (Fig. 326) traps air or mixture between the impellers and the casing, moves the medium around part of the casing, and forces it out through the outlet. Theoretically, no work is required until the impeller uncovers the outlet port at which time the higher-pressure medium "wiredraws" into the supercharger and almost instantly increases the pressure to that in the discharge line. The theoretical work is indicated by the area $AB'CDA$ (Fig. 327) and is

$$W_{comp} = V_1(P_2 - P_1). \tag{14}$$

A small clearance (0.004 to 0.008 in. depending on size) is provided between the impellers and the casing. This results in leakage past the impellers which will be almost constant irrespective of speed. The capacity of the blower will be the difference between the blower displacement and the leakage.

Thus,
$$Q = \frac{2\pi}{4} d^2 ln \left(1 - \frac{4a_r}{\pi d^2}\right) - cpk(l + d) \tag{15}$$

where Q = actual blower capacity, in.^{3} min.$^{-1}$,
d = rotor dia., in.,
l = length of rotor, in.,
n = rotor speed, r.p.m.,
a_r = rotor cross section, in.^{2},
c = radial cl., in.,
p = pr. difference, in. Hg,
k = a flow coefficient (Fig. 329).

Equation (15) may be written as follows for a given set of conditions:

$$Q = k_1\left(\frac{l}{d}\right)d^3n - k_2\left[\left(\frac{l}{d}\right)d + d\right] \quad (16)$$

where k_1 and k_2 = const.

This equation is based on the assumption of a constant ratio of area of cross section of rotor to area of circle with rotor diameter for all sizes of blowers.

The volumetric efficiency is the ratio of the volume of air discharged [Eq. (16)], measured at entering air conditions, to the blower displacement which is the first term on the right side of Eq. (16). Thus,

$$\text{Vol. eff.} = \frac{\text{air discharged}}{\text{blower displacement}}$$

$$= 1 - \frac{k_2}{k_1} \cdot \frac{\left(\frac{l}{d} + 1\right)}{d^2 n \frac{l}{d}} \quad (17)$$

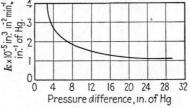

Fig. 329.—Flow coefficients for leakage in Roots-type supercharger. Suction pressure is 1 atm. (*Schopper*.)

Schopper[1] eliminated the factor l/d between Eqs. (16) and (17), differentiated the expression for volumetric efficiency with respect to rotor diameter, and developed an expression for optimum rotor diameter.

Thus,

$$d = \sqrt[3]{\frac{Q}{k_1 n} + \sqrt{\frac{Q^2}{k_1^2 n^2} - \frac{k_2^3}{k_1^3 n^3}}} + \sqrt[3]{\frac{Q}{k_1 n} - \sqrt{\frac{Q^2}{k_1^2 n^2} - \frac{k_2^3}{k_1^3 n^3}}}. \quad (18)$$

With these relationships, the optimum bores and volumetric efficiencies were determined for various blower speeds and capacities (Fig. 330).

The vane-type supercharger (Fig. 326) has a rotor located eccentrically in the casing. Vanes slide back and forth in the rotor and are either mechanically controlled to approach the casing very closely at all times or automatically contact it. The theoretical power required with a small number of vanes approaches that of the Roots blower, whereas with a large number of vanes the power approaches that of the centrifugal

---

[1] K. Schopper, Das Rootsgebläse als Ladungsverdichter an Mercedes-Benz-Motoren, *Automobiltech. Z.*, **10**, 28 (1935).

See also P. M. Heldt, Problems in the Design of Roots Type Superchargers, *Auto. Ind.*, **72**, 450 (1935).

472   INTERNAL-COMBUSTION ENGINES

supercharger. However, the mechanical losses tend to increase with the number of vanes.

Scavenging.—The performance of two-stroke Diesel engines depends on the effectiveness with which the exhaust products

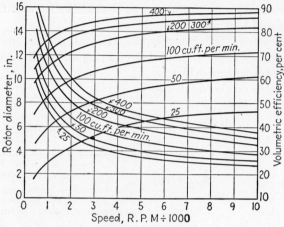

Fig. 330.—Optimum size and volumetric efficiency of Roots-type blowers. Suction pr. = 29.92 in. Hg abs. Discharge pr. = 44.88 in. Hg abs. (*Schopper.*)[1]

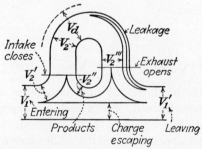

Fig. 331.—Diagram of the scavenging process in a two-stroke engine. (*DeJuhasz.*)[2]

are driven from the cylinder and replaced by the fresh air charge. Ideal scavenging would replace the products in the cylinder and the clearance space and permit the maximum output. Actually, the scavenging blower supplies a volume V_1 (Fig. 331) of fresh air, part of which escapes out the exhaust port during the scaveng-

[1] Schopper, *op. cit.*

[2] K. J. DeJuhasz, Measuring the Scavenging Efficiency of Two-stroke Diesel Engines, *Auto. Ind.*, **73**, 858 (1935).

ing process. A volume V_2' of fresh air is trapped in the cylinder with volume V_2'' of products from the previous process, making a total charge volume of V_2, which at standard conditions is less than the piston displacement V_d. The charge is compressed, fuel injected, and combustion occurs and is followed by expansion to exhaust-port opening. Some leakage occurs, and products are exhausted during release and driven out by the scavenging air, part of which escapes making a total leaving volume of V_1' which is less than the entering volume for the usual hydrocarbon fuel if measured at actual standard conditions.

From the foregoing, $\quad$ Vol. eff. $= \dfrac{V_2}{V_d} = \dfrac{(V_2' + V_2'')}{V_d}$ $\hspace{2em}$ (19)

and, $\quad$ Scav. eff. $= \dfrac{V_2'}{V_2} = \dfrac{V_2'}{(V_2' + V_2'')}.$ $\hspace{2em}$ (20)

Also, $\quad$ Excess air coefficient $= \dfrac{V_1}{V_d} - 1.$ $\hspace{2em}$ (21)

DeJuhasz[1] proposed determining the scavenging efficiency by sampling the cylinder gases immediately after the intake port closes and also immediately before the exhaust port opens. Analyzing for CO_2 gives

$$\frac{V_2''}{V_2' + V_2''} = \frac{CO_2 \text{ in 1st sample}}{CO_2 \text{ in 2d sample}} = 1 - \text{scav. eff.} \hspace{2em} (22)$$

He also proposed determining this efficiency while motoring the engine and artificially introducing CO_2 into the cylinder sometime between the taking of the two samples.

The excess-air coefficient is usually about 0.4 to 0.5. This coefficient depends on the scavenging pressure and the engine speed for a given engine (Fig. 332). Also, the power required to drive the scavenging blower is approximately 50 per cent of the total friction horsepower at rated speeds, this percentage decreasing appreciably at the lower speeds.

Four-stroke-cycle engines may have the clearance space scavenged by using valve overlap and a blower for scavenging or supercharging. This is more desirable for the fuel-injection engine since no fuel will escape into the exhaust with the scavenging medium.

[1] DeJuhasz, *op. cit.*

Blowby.—Gases may leak from the combustion chamber at the valves, the cylinder-head gasket, and at the piston and rings during the compression and expansion strokes. Valve leakage can be remedied by grinding the valves and gasket leakage by replacing the gasket. However, leakage past the piston rings is an inherent loss which should be reduced to an economical minimum.

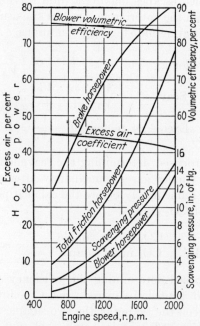

FIG. 332.—Scavenging blower and performance data for General Motors two-stroke Diesel. (*Shoemaker.*)[1]

Piston rings provide a labyrinth passageway for the gases to escape past the piston. The pressure behind the first ring is nearly equal to the pressure in the cylinder (Fig. 333), progressively lower pressures being found behind the second and third compression rings if the rings, grooves, and cylinders are in good condition.

Blowby can be reduced by adding compression rings or increasing the tension of the rings. Adding rings increases the

---

[1] F. G. Shoemaker, Automotive Two-cycle Diesels, *S.A.E. Trans.*, **35**, 485 (1938).

friction loss and is undesirable. However, high-compression pressures require more rings than low-compression pressures, two

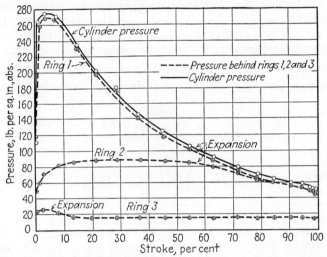

FIG. 333.—Gas pressure behind piston rings. (*Ford and Robertson.*)[1]

rings being standard for automobile engines and four to six rings for Diesel engines.

Blowby is normally low (Fig. 334) up to some critical speed beyond which blowby and oil consumption increase very rapidly with very rapid decrease in power. Apparently, the rings tend to flutter at and above the critical speed which immediately reduces their effectiveness in sealing the piston. Increasing the ring tension by means of an inner-spring ring behind the top compression ring usually eliminates the critical condition.

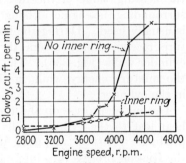

FIG. 334.—Effect of inner ring on blowby. (*Sparrow.*)[2]

Blowby is measured by sealing the crankcase with the exception of one outlet which connects to a gas meter. The outlet

[1] B. J. Robertson, Gas Pressure behind Piston Rings, *Auto. Ind.*, **75**, 118 (1936).

[2] S. W. Sparrow, Problems in the Development of a High-speed Engine, *S.A.E. Trans.*, **30**, 58 (1935).

of the gas meter should be connected to a suction pump which will maintain the crankcase pressure at atmospheric pressure.

A condensing chamber for the oil and water vapors should be installed between the crankcase and the meter. The CO_2, O_2, CO, and N_2, saturated or nearly saturated with water and oil vapor, are measured by the gas meter. Cutter[1] found the blowby gases contained a large percentage of O_2 (Table 29), indicating considerable blowby of mixture which leaks past the top ring or edge of the piston before the flame arrives.

TABLE 29.—BLOWBY AND EXHAUST-GAS ANALYSIS (CUTTER)[1]

Engine speed, r.p.m.	Blowby gas				Exhaust gas			
	CO_2, %	O_2, %	CO, %	N_2* %	CO_2, %	O_2, %	CO, %	N_2, %
1000	3.0	14.0	1.7	81.3	9.5	1.7	4.8	84.0
1500	2.4	15.3	1.2	81.1	10.2	0.6	5.3	83.9
2000	2.7	14.7	1.3	81.3	10.0	1.0	5.0	84.0
2500	3.2	14.5	1.3	81.0	10.7	0.5	4.6	84.2

* By difference between 100 per cent and the sum of the other dry gases.

Energy Distribution.—The net energy supplied an engine is the heating value of the fuel at the atmospheric temperature surrounding the engine, although the value at standard conditions is usually used. During the induction period the entering mixture receives energy from the exhaust system and from the cylinder walls (Fig. 335). The clearance-gas energy is added to the mixture energy while in the cylinder but always remains in the cylinder.

The engine process divides the energy into work on the piston, heat lost to the cylinder walls and pistons, and energy in the products. Part of the work on the piston is transformed by the friction process into internal energy, some of which adds to the energy of the entering mixture and to that in the exhaust and in the cooling medium. The balance of the energy from the friction process is in the lubricating oil and is dissipated through the crankcase and other unjacketed walls to the surrounding medium. High-output engines require oil coolers through

[1] D. C. Cutter, A Study of Blowby in the Internal-combustion Engine, Thesis for Degree of Mechanical Engineer, Yale University, 1933.

which a very appreciable amount of energy is dissipated since the oil serves as a cooling agent for the pistons.

Other obvious exchanges of energy between the energy "streams" for the exhaust, cooling water, and radiation take

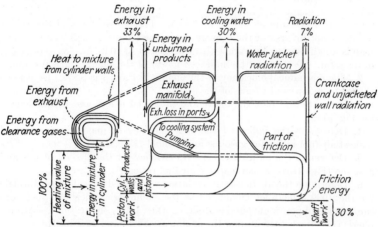

Fig. 335.—Energy distribution within and leaving engine.

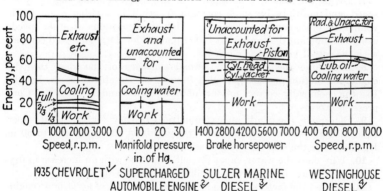

Fig. 336.—Energy distribution of various engines.

place before the energy finally leaves the engine. Eventually, all of the energy entering the engine is divided into four parts

[1] Lichty and Phelps, *op. cit.*

[2] R. B. Sneed, Supercharging, *Auto. Ind.*, **78**, 551 (1938).

[3] G. Eichelberg, Trials of a 5500-B.hp. Sulzer Marine Diesel consuming 0.33 lb./B.hp.-hr. at Full Load, *Sulzer Tech. Review*, No. 2 (1935).

[4] J. Dickson, The Diesel Engine for Rail Transportation, *Mech. Eng.*, **57**, 547 (1935).

(Fig. 336), two of which, exhaust and radiation, may be combined. Any discrepancy is termed "unaccounted-for energy."

EXERCISES

1. The length of an indicator card from a nine-cylinder 5.5-in. bore and 6-in. stroke engine is 3 in. The maximum height of the card is 3.5 in. and represents a pressure of 500 lb. in.$^{-2}$ abs. The positive area of the card is 3.2 in.^{2}, and the area of the negative pumping loop is 1 in.^{2} The pumping loop is a light spring diagram with a spring scale of 10 lb. in.$^{-2}$ in.$^{-1}$ of height of diagram. Determine the indicated power of the engine at 1800 r.p.m.

2. The mechanical efficiency of the engine in Prob. 1 is 0.87. Determine the brake m.e.p. and torque of the engine.

3. Compute and plot the brake and indicated m.e.p. and torque curves for Fig. 311.

4. Determine the indicated thermal efficiency at maximum-power and highest air-fuel ratios for gasoline and alcohol (Fig. 315A).

5. Determine the indicated thermal efficiency for the alcohol blends and gasoline (Fig. 315B) at comparable air-fuel ratios.

6. Determine the air-fuel-ratio variation with speed at wide-open throttle for the 1939 Chevrolet engine. Use the volumetric efficiency, fuel consumption, and power output data found in the various illustrations. Atmospheric conditions surrounding the engine are 80°F. and 29.4 in. Hg.

7. Divide the indicated horsepower of the 1939 Chevrolet engine by the speed, and multiply by some constant to result in a value equal to the volumetric efficiency at 1400 r.p.m. Make computations at other speeds, using the constant determined at 1400 r.p.m. Plot the results, and compare with the actual volumetric efficiency.

8. Determine the work and thermal efficiency of the ideal throttled engine for suction-pressures ranging from 10 in. Hg abs. to 25 in. Hg abs. Use the chemically correct mixture of liquid octane and air at 60°F. Compression ratio is 6:1.

9. Determine the work and volumetric efficiency for the ideal supercharged engine for suction pressures ranging from 30 in. Hg abs. to 60 in. Hg abs. Use the same conditions as in Prob. 8.

10. Why does the water-vapor pressure decrease with an increase in dry-bulb temperature for a given wet-bulb temperature (Fig. 319)?

11. Discuss the factors causing a difference in power of an aircraft engine at sea level and at some altitude with the same intake-manifold pressure.

12. An aircraft engine is to develop 2000 hp. at 10,000 ft. altitude with wide-open throttle. At lower altitudes, detonation occurs and the engine must be throttled. Estimate the power of the engine at sea level and at an altitude of 20,000 ft.

13. The maximum compression pressure desirable for a given fuel is 150 lb. in.$^{-2}$ abs. Determine the desirable volumetric efficiencies, relative power, and relative efficiencies for compression ratios ranging from 5:1 to 9:1. Plot the results.

14. A supercharger compresses air from 1 atm. and a temperature of 40°F. to 2 atm. Determine the work per pound of air for the ideal case,

ENGINE PERFORMANCE

using a Roots blower and a centrifugal supercharger. Assume the process is adiabatic.

15. Assume the process in Prob. 14 is irreversible with a 50°F. temperature rise above that of the adiabatic process. Also, assume a net heat transfer out of the supercharger equal to 3 B.t.u. lb.$^{-1}$ of air. Determine the work required.

16. Determine the possible increases in static pressure due to air speeds of 100 to 400 m.p.h. at sea level, 10,000, 20,000, and 30,000 ft. altitude. Assume a temperature of 80°F. at sea level and a barometer of 29.92 in. Hg.

17. Determine the optimum length of a Roots blower with a rotor diameter of 6 in. and a capacity of 200 ft.^{3} min.$^{-1}$ and having the characteristics indicated in Fig. 330.

CHAPTER XVII

MECHANICS OF PRINCIPAL MOVING PARTS

The forces acting on the principal moving parts of a reciprocating engine are the gas-pressure forces, the inertia forces of the reciprocating parts, and the inertia and centrifugal forces of the rotating parts. The gas-pressure forces are the principal forces at low engine speeds, but the inertia forces may be considerably larger at high speeds. The centrifugal forces also increase rapidly with an increase in speed.

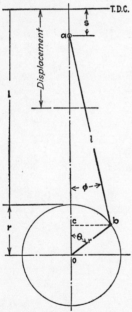

FIG. 337.—Relation of piston to crank.

The gas-pressure and inertia forces result in a net force on the piston which is resisted by the shearing force of the oil film between the piston assembly and the cylinder wall. The net result is a force along the connecting rod that resolves itself into a turning effort or torque on the crankshaft.

Reciprocating Parts.—The principal reciprocating parts of a single-acting engine are included in the piston assembly. The weight of the rings is not included in the weight of the piston assembly, this omission being considered an approximate correction for the friction of the piston assembly which is usually neglected. The upper part of the connecting rod is assumed to have a reciprocating motion, its weight[1] being added to the weight of the piston assembly minus the rings.

---

[1] The weight of the part of the rod which is assumed to reciprocate is determined by weighing the piston end of the rod when the rod is supported in a horizontal position on knife-edges located at the center lines of the bearings.

Crossheads, piston rods, and other parts attached thereto having a reciprocating motion, must be included with the piston, etc., as reciprocating parts in the case of double-acting engines or any engine designed with such parts.

Piston-crank Relationship.—The position of the piston for any crank position depends upon the crank position and the connecting-rod-crank ratio. The distance s (Fig. 337) which the piston has moved from top dead center when the crank has turned θ deg. from the same dead center is

$$s = r + l - (oc + ca)$$
$$= r + l - r \cos \theta - l \cos \phi \qquad (1)$$

where r and l = crank radius and connecting-rod lengths, resp.

From the right triangle abc,

$$l^2 = \overline{bc}^2 + \overline{ca}^2 = r^2 \sin^2 \theta + l^2 \cos^2 \phi$$

from which $l \cos \phi = \sqrt{l^2 - r^2 \sin^2 \theta}$.

Substituting in Eq. (1) and simplifying results in

$$s = r\left[(1 - \cos \theta) + \frac{l}{r}\left(1 - \sqrt{1 - \frac{r^2}{l^2} \sin^2 \theta}\right)\right]. \qquad (2)*$$

Adding $r^4 \sin^4 \theta / 4l^4$ to the terms under the radical to complete the square results in the approximate relation

$$s = r\left[(1 - \cos \theta) + \frac{r}{2l} \sin^2 \theta\right]$$
$$= r\left[(1 - \cos \theta) + \frac{r}{4l}(1 - \cos 2\theta)\right]. \qquad (3)$$

---

* The expression for s may be derived in a slightly different manner and may be expressed in a *Fourier Series*. Thus,

$$s = r\left[k + \sin \theta + k_2 \sin\left(2\theta - \frac{\pi}{2}\right) + k_4 \sin\left(4\theta - \frac{\pi}{2}\right) + \cdots\right]$$

measuring the displacement from the mid-position of the stroke and the angle from the horizontal crank position. The displacement vs. crank-angle curve has two parts that have "mirror" symmetry which eliminates the odd harmonics of the typical Fourier Series and leaves only the fundamental and even harmonics in the foregoing expression.

Values of s in percentage of stroke for every 10 deg. and for various values of l/r are given in Table 30, and are shown graphically in terms of piston position in Fig. 338 for an l/r value of 4.

TABLE 30.—PISTON POSITION IN PERCENTAGE OF STROKE

θ, degrees from top center	Values of l/r				
	3.50	3.75	4.00	4.25	4.50
0	0.0	0.0	0.0	0.0	0.0
10	1.0	1.0	1.0	0.9	0.9
20	3.9	3.8	3.7	3.7	3.7
30	8.5	8.4	8.3	8.2	8.1
40	14.7	14.5	14.2	14.1	14.0
50	22.1	21.8	21.5	21.3	21.1
60	30.3	30.0	29.7	29.4	29.1
70	39.2	38.8	38.4	38.1	37.8
80	48.3	47.8	47.4	47.0	46.7
90	57.2	56.7	56.3	55.9	55.6
100	65.6	65.2	64.7	64.4	64.1
110	73.4	73.0	72.6	72.3	72.0
120	80.4	80.0	79.7	79.4	79.1
130	86.3	86.1	85.8	85.6	85.4
140	91.3	91.1	90.8	90.7	90.6
150	95.1	95.0	94.9	94.8	94.7
160	97.8	97.8	97.7	97.7	97.6
170	99.5	99.4	99.4	99.4	99.4
180	100.0	100.0	100.0	100.0	100.0

Piston Velocity.—The piston velocity is zero at the beginning of a stroke, reaches a maximum near the middle of the stroke, and decreases to zero again at the end of the stroke. The velocity v at any given crank angle θ can be determined by differentiating the expression for s with respect to time t.

Thus,
$$v = \frac{ds}{dt} = \frac{ds}{d\theta} \times \frac{d\theta}{dt},$$
$$\frac{ds}{d\theta} = r \sin \theta + \frac{r^2}{2l} \sin 2\theta,$$
$$\frac{d\theta}{dt} = 2\pi n \text{ rad. min.}^{-1},$$

or
$$v = \frac{2\pi nr}{60 \times 12}\left(\sin\theta + \frac{r}{2l}\sin 2\theta\right)$$
$$= 8.73 \times 10^{-3} nr\left(\sin\theta + \frac{r}{2l}\sin 2\theta\right) \quad (4)$$

where v is ft. sec.$^{-1}$, n is r.p.m., and r and l are in in. A typical piston-velocity curve is plotted in Fig. 338.

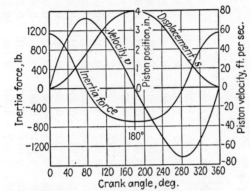

Fig. 338.—Piston displacement, velocity, and inertia curves. $n = 4000$ r.p.m., $r = 2$ in., $l/r = 4$, and $M = 1$ lb.

Piston Acceleration and Inertia Forces.—The acceleration is obtained by differentiating the piston velocity with respect to time.

Thus, $\quad$ Accel. $= \dfrac{dv}{dt} = \dfrac{dv}{d\theta}\dfrac{d\theta}{dt} = \dfrac{2\pi n}{60}\cdot\dfrac{dv}{d\theta}.$

Differentiating Eq. (4) and substituting in the foregoing, results in

$$\text{Accel.} = 9.14 \times 10^{-4} n^2 r\left(\cos\theta + \frac{r}{l}\cos 2\theta\right) \text{ ft. sec.}^{-2} \quad (5)$$

Since force = mass × acceleration,

$$F = \frac{Ma}{g} = 2.84 \times 10^{-5} M n^2 r\left(\cos\theta + \frac{r}{l}\cos 2\theta\right) \text{ lb.} \quad (6)$$

where M = wt. of the reciprocating part, lb.

Values of the term $\left(\cos\theta + \dfrac{r}{l}\cos 2\theta\right)$ are given in Table 31

TABLE 31.—VALUES OF TERMS IN EXPRESSION FOR INERTIA FORCES

θ, degrees from top center	$\cos\theta$	$\cos 2\theta$	$l/r = 3.50$		$l/r = 3.75$		$l/r = 4.00$		$l/r = 4.25$		$l/r = 4.50$	
			$\frac{r}{l}\cos 2\theta$	$\cos\theta + \frac{r}{l}\cos 2\theta$	$\frac{r}{l}\cos 2\theta$	$\cos\theta + \frac{r}{l}\cos 2\theta$	$\frac{r}{l}\cos 2\theta$	$\cos\theta + \frac{r}{l}\cos 2\theta$	$\frac{r}{l}\cos 2\theta$	$\cos\theta + \frac{r}{l}\cos 2\theta$	$\frac{r}{l}\cos 2\theta$	$\cos\theta + \frac{r}{l}\cos 2\theta$
0	1.000	1.000	0.286	1.286	0.266	1.266	0.250	1.250	0.235	1.235	0.222	1.222
10	0.985	0.940	0.269	1.254	0.251	1.236	0.235	1.220	0.221	1.206	0.209	1.194
20	0.940	0.766	0.219	1.159	0.204	1.144	0.192	1.132	0.180	1.120	0.170	1.110
30	0.866	0.500	0.143	1.009	0.133	0.999	0.125	0.991	0.118	0.984	0.111	0.977
40	0.766	0.174	0.050	0.816	0.046	0.812	0.044	0.810	0.041	0.807	0.039	0.805
50	0.643	−0.174	−0.050	0.593	−0.046	0.597	−0.044	0.599	−0.041	0.602	−0.039	0.604
60	0.500	−0.500	−0.143	0.357	−0.133	0.367	−0.125	0.375	−0.118	0.482	−0.111	0.389
70	0.342	−0.766	−0.219	0.123	−0.204	0.138	−0.192	0.150	−0.180	0.162	−0.170	0.172
80	0.174	−0.940	−0.269	−0.095	−0.251	−0.077	−0.235	−0.061	−0.221	−0.047	−0.209	−0.035
90	0.000	−1.000	−0.286	−0.286	−0.266	−0.266	−0.250	−0.250	−0.235	−0.235	−0.222	−0.222
100	−0.174	−0.940	−0.269	−0.443	−0.251	−0.425	−0.235	−0.409	−0.221	−0.395	−0.209	−0.383
110	−0.342	−0.766	−0.219	−0.561	−0.204	−0.546	−0.192	−0.534	−0.180	−0.522	−0.170	−0.512
120	−0.500	−0.500	−0.143	−0.643	−0.133	−0.633	−0.125	−0.625	−0.118	−0.618	−0.111	−0.611
130	−0.643	−0.174	−0.050	−0.693	−0.046	−0.689	−0.044	−0.687	−0.041	−0.684	−0.039	−0.682
140	−0.766	0.174	0.050	−0.716	0.046	−0.720	0.044	−0.722	0.041	−0.725	0.039	−0.727
150	−0.866	0.500	0.143	−0.723	0.133	−0.733	0.125	−0.741	0.118	−0.748	0.111	−0.755
160	−0.940	0.766	0.219	−0.721	0.204	−0.736	0.192	−0.748	0.180	−0.760	0.170	−0.770
170	−0.985	0.940	0.269	−0.716	0.251	−0.734	0.235	−0.750	0.221	−0.764	0.209	−0.776
180	−1.000	1.000	0.286	−0.714	0.266	−0.734	0.250	−0.750	0.235	−0.765	0.222	−0.778

for every 10 deg. The variation of the inertia force per pound of reciprocating weight in an engine having a 4-in. stroke and running at 4000 r.p.m. is shown in Fig. 338.

Gas-pressure Forces.—The gas pressure in the internal-combustion-engine process varies from below atmospheric to above 1000 lb. in.$^{-2}$ in some cases. The gas exerts its pressure against the piston, this force acting along the same line as the inertia force. Consequently, the gas-pressure and inertia forces may be combined algebraically to determine the net force acting along the center line of the cylinder.

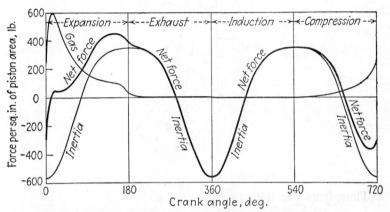

Fig. 339.—Net-force diagram for a one-cylinder single-acting engine. $n = 4000$ r.p.m., $l/r = 4$, $M = 0.5$ lb. in.$^{-2}$ piston area.

Net-force Diagram.—This is a force-crank-angle diagram (Fig. 339) which has a length equivalent to one cycle. The gage gas pressures at various angles are plotted for single-acting engines since the atmospheric pressure acting on the crankcase side of the piston can be eliminated in this manner. Either gage or absolute pressures may be plotted for double-acting engines.

Forces acting toward the crankshaft are plotted as positive values, and vice versa. Thus, only the induction and first part of the compression stroke will have negative gas pressures. The inertia forces are always negative at the top and positive at the bottom of the stroke. The net force is the algebraic summation of the gas pressure and inertia force.

Piston-side-thrust and Connecting-rod Forces.—The net force is exerted in a direction along the cylinder axis. The angularity of the connecting rod causes the net force to be divided into components, one producing piston thrust against the cylinder wall, the other acting along the axis of the connecting rod.

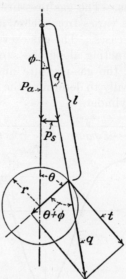

Fig. 340.—Resolution of net force.

Thus, from Fig. 340, $p_s = p_a \tan \phi$

when p_s = side thrust against cyl., lb. in.$^{-2}$ of piston area,
p_a = net force along the cyl., lb. in.$^{-2}$ of piston area.

It can be shown that $r \sin \theta = l \sin \phi$

and since
$$\cos \phi = \sqrt{1 - \sin^2 \phi},$$

$$p_s = \frac{p_a \sin \phi}{\cos \phi} = \frac{p_a \frac{r}{l} \sin \theta}{\sqrt{1 - \frac{r^2}{l^2} \sin^2 \theta}} = \frac{p_a \sin \theta}{\sqrt{\frac{l^2}{r^2} - \sin^2 \theta}}. \quad (7)$$

The force along the connecting-rod axis is

$$q = \frac{p_a}{\cos \phi} = \frac{p_a}{\sqrt{1 - \frac{r^2}{l^2} \sin^2 \theta}} = \frac{p_a l/r}{\sqrt{\frac{l^2}{r^2} - \sin^2 \theta}}. \quad (8)$$

Rotative Force.—The tangential force t at the crankpin is determined by resolving the force q along the rod into two components, one acting tangentially to the crank circle at the crankpin, the other acting radially at the crankpin.

Thus, from Fig. 340, $t = q \sin(\theta + \phi)$. (9)

Also, $\sin(\theta + \phi) = \sin\theta \cos\phi + \cos\theta \sin\phi$
$$= \frac{\sin\theta}{l}\sqrt{l^2 - r^2 \sin^2\theta} + \frac{r}{l}\sin\theta \cos\theta.$$

Substituting this value and the value for q from Eq. (8) into Eq. (9) results in

$$t = \left(\frac{p_a l}{\sqrt{l^2 - r^2 \sin^2\theta}}\right)\left(\frac{\sin\theta}{l}\sqrt{l^2 - r^2 \sin^2\theta} + \frac{r}{l}\sin\theta \cos\theta\right)$$
$$= p_a \sin\theta\left(1 + \frac{\cos\theta}{\sqrt{l^2/r^2 - \sin^2\theta}}\right). \quad (10)$$

Values of the factor $\sin\theta\left(1 + \dfrac{\cos\theta}{\sqrt{l^2/r^2 - \sin^2\theta}}\right)$ for various values of l/r are given in Table 32.

TABLE 32.—VALUES OF $\sin\theta\left(1 + \dfrac{\cos\theta}{\sqrt{l^2/r^2 - \sin^2\theta}}\right)$

Deg.	$l/r = 3.5$	$l/r = 3.75$	$l/r = 4.00$	$l/r = 4.25$	$l/r = 4.50$
0	0.000	0.000	0.000	0.000	0.000
10	0.223	0.219	0.216	0.214	0.212
20	0.434	0.428	0.423	0.418	0.413
30	0.625	0.616	0.609	0.602	0.597
40	0.786	0.777	0.768	0.760	0.754
50	0.910	0.900	0.892	0.884	0.878
60	0.994	0.985	0.977	0.970	0.964
70	1.035	1.028	1.022	1.017	1.013
80	1.036	1.032	1.029	1.026	1.024
90	1.000	1.000	1.000	1.000	1.000
100	0.934	0.938	0.941	0.943	0.945
110	0.845	0.851	0.857	0.862	0.866
120	0.738	0.747	0.755	0.762	0.768
130	0.623	0.632	0.640	0.648	0.655
140	0.499	0.509	0.518	0.526	0.532
150	0.375	0.384	0.391	0.397	0.403
160	0.250	0.256	0.261	0.266	0.271
170	0.124	0.128	0.131	0.134	0.136
180	0.000	0.000	0.000	0.000	0.000

The net-force curve (Fig. 339) provides data for p_a and Table 32 for the balance of Eq. (10) which results in the rotative-force curve (Fig. 341). The areas between the rotative-force curve and the zero-pressure line may be planimetered and added

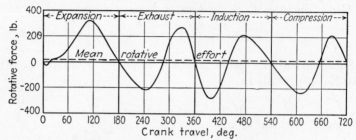

FIG. 341.—Rotative force for a single-cylinder engine; lb. at crankpin in.$^{-2}$ of piston area. (Based on data in Fig. 339.)

algebraically, and the mean rotative force determined. The work done on the crankpin for two revolutions will be

$$\text{Work} = t_{mean} \times 4\pi r. \qquad (11)$$

This should check the work value of the indicator card from which the rotative force was obtained, since the net work of the inertia force for two revolutions is zero.

The torque curve may be obtained by multiplying the various ordinates of the rotative-force curve by the crank radius and the piston area. Obviously, the torque curve will have the same variations as the rotative-force curve.

Torque Variation.—The torque for a single-cylinder engine varies considerably in amount and may reverse in sign eight times per cycle in a high-speed engine. A torque curve for a multicylinder engine is determined by combining the torque curves which are presumably identical for each cylinder. The crank angle between the beginning of expansion or any other common point for each cylinder indicates the phase arrangement for the torque curves. Thus, the torque curve for a four-cylinder engine would require the drawing of four single-cylinder torque curves, starting each one 180 deg. later than the previous one. Obviously, the curve that starts at 180 deg. has the last stroke plotted from 0 to 180 deg., etc.

The greater the number of cylinders, the smoother will be the torque (Fig. 342), six cylinders being required to have no reversals in sign at wide-open throttle with a four-stroke-cycle engine. There is an appreciable variation in torque even with a 16-cylinder engine.

Areas above the mean torque curve (ΔW, Fig. 343) indicate energy in excess of that required, which results in an increase of

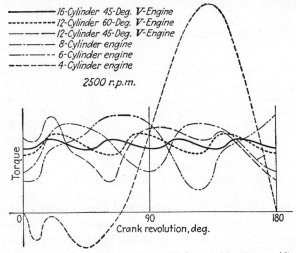

Fig. 342.—Torque curves for various cylinder combinations. (*Crane.*)[1]

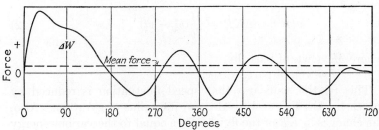

Fig. 343.—Rotative-force diagram for a four-stroke-cycle, single-cylinder, single-acting, slow-speed engine.

speed, and vice versa. The change in speed due to torque variation depends on these excess and deficiency of energy amounts and the inertia of the moving parts. The inertia may be increased by the use of a flywheel that has considerable mass located in the rim. The design of the flywheel depends on the largest excess or deficiency of energy areas which vary with

[1] H. M. Crane, How Versatile Engineering Meets Public Demand, *S.A.E. Trans.*, **28**, 21 (1933).

speed, number of cylinders, and engine design. Thus, a torque curve from a slow-speed single-cylinder engine has one energy area (ΔW, Fig. 343), that is much larger than any other area. This is not so obvious in Fig. 341, which is for a high-speed engine. However, approximate values of the ratio of the energy excess to the indicated work per revolution are given in Table 33.

TABLE 33.—APPROXIMATE VALUES OF ΔW FACTOR

Type of Engine (4-stroke Cycle)	$\Delta W \div$ Indicated Work per Revolution
Single-cylinder	2.40
2-cylinder, opposed, 180-deg. cranks	1.00
3-cylinder, 120-deg. cranks	0.70
4-cylinder, 180-deg. cranks	0.20
6-cylinder, 120-deg. cranks	0.05
8-cylinder, 90-deg. cranks	0.06
12-cylinder V, 120-deg. cranks	0.02
16-cylinder V, 90-deg. cranks	0.01

Flywheels.—The fluctuations in speed due to torque variations are reduced to a minimum by the use of flywheels. A change in speed changes the kinetic energy of the flywheel and other rotating parts according to the relation

$$\Delta KE = \frac{M}{2g}(v_2^2 - v_1^2) \qquad (12)$$

where M = wt. of part, lb.,
v = vel. of part, ft. sec.$^{-1}$

This relation indicates that speed fluctuation is reduced by increasing the flywheel weight and its radius of gyration. Also, the changes in KE of the flywheel are equal to the various energy values indicated by the areas above and below the mean torque curve (Figs. 341 and 343), neglecting the effect of other rotating parts which is negligible in slow-speed engines and those with few cylinders.

Thus, Eq. (12) becomes

$$\Delta W = \Delta KE = \frac{4\pi^2 \rho^2 M}{2g3600}(n_2^2 - n_1^2)$$
$$= \frac{\pi^2 \rho^2 M}{1800g}(n_2 - n_1)(n_2 + n_1) \qquad (13)$$

where n = r.p.m.,
subscripts 1 and 2 = lower and upper limits of speed, resp.,

ρ = radius of gyration, ft.

Also, $$n = \frac{n_1 + n_2}{2} = \text{mean speed}. \tag{14}$$

The ratio of variation in speed to mean speed is defined as the coefficient of speed fluctuation, $k.^{-1}$

Thus, $$\frac{1}{k} = \frac{n_2 - n_1}{n}. \tag{15}$$

Substituting values for $(n_2 - n_1)$ and $(n_2 + n_1)$ from Eqs. (14) and (15) into Eq. (13) results in

$$\Delta W = \frac{\pi^2 \rho^2 M n^2}{900 g k}$$

or $$M = \frac{900 g k \, \Delta W}{\pi^2 \rho^2 n^2}. \tag{16}$$

The value of k varies from about 5 for engines not requiring close speed regulation, to over 200 where close regulation is required.

M is the weight of the wheel in pounds at the radius of gyration, ρ ft. It is usually assumed that all this weight is in the rim, and the balance of the flywheel is neglected so far as flywheel effect is concerned. The radius of gyration may be found by assuming a safe maximum rim velocity. For large stationary engines using cast-iron flywheels, it has long been considered good practice not to exceed a rim speed of 6000 ft. min.$^{-1}$ However, with automotive engines using steel flywheels rim speeds up to 20,000 ft. min.$^{-1}$ are not uncommon.

For automobile engines the flywheel must be heavy enough to take care of unevenness of combined effort of all the cylinders at the lowest speed at which it is desired to operate.

Aircraft engines do not require a flywheel, the counterbalanced crankshaft and propeller providing enough flywheel effect.

Example.—Determine the weight of a flywheel for a four-cylinder 100-hp. engine running at 2000 r.p.m. Assume all the weight of the flywheel is at the radius of gyration, which is 8 in. The flywheel is to be designed to prevent the speed from varying more than 10 r.p.m. from the mean speed.

$$\frac{n_2 - n_1}{2} = 10 \quad \text{and} \quad \frac{1}{k} = \frac{20}{2000} = \frac{1}{100}$$

From Table 33, $\Delta W \div$ indicated work rev.$^{-1}$ = 0.20,

$$\Delta W = \frac{0.2 \times 100 \times 33{,}000}{2000} = 330 \text{ ft.-lb.}$$

From Eq. (16), $M = \dfrac{900 \times 32.2 \times 100 \times 330}{\pi^2 \times (\frac{2}{3})^2 \times 2000^2} = 54.5$ lb.

With this flywheel, it can be shown that at a mean speed of 500 r.p.m. the speed would vary from about 460 to 540 r.p.m. This indicates that the flywheel should be designed for the lowest speed at which control of speed variation is considered important.

Flywheel-velocity and Displacement Diagrams.—The length of the rotative-force diagram represents distance traveled by the crankpin, each increment representing a definite distance. Due to torque variation the time interval between the various

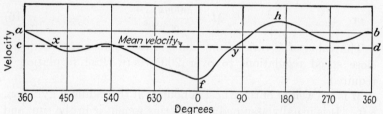

FIG. 344.—Velocity diagram for a four-stroke-cycle, single-cylinder, single-acting, slow-speed engine.

vertical lines is not uniform but may be assumed to be uniform with comparatively little error.

Dividing the rotative-force ordinates by the mass, M/g, which amounts to changing the vertical scale, changes the rotative-force diagram to an acceleration-time diagram. Integrating the area between the curve and the mean-effort line, between any two ordinates on this diagram, determines the change in velocity of the mass, M/g, during the time represented by the distance between the two ordinates. The areas in Fig. 343 have been integrated starting at the 360-deg. ordinate which is point a on the base line ab in Fig. 344. Line cd represents the mean velocity which is displaced from the base line by an amount equal to the area between the curve and the base line divided by the length of the diagram. Points f and h represent the two extreme limits of velocity, and during the time between these points the flywheel absorbs the energy ΔW, Fig. 343.

The product of velocity and time represents distance traversed. Hence, the area between the velocity curve and the mean-velocity line, between any two ordinates, will represent the distance the flywheel has moved from the normal position it would have with the mean velocity. The plot of such integration values is called the "displacement" diagram (Fig. 345). The maximum and minimum points, h' and f', on the displacement diagram will be found at the ordinates x and y (Fig. 344) where the velocity curve crosses the mean velocity line. The normal position of the flywheel, if running with the mean velocity, is indicated by line $c'd'$ drawn so that $gh' = kf'$.

Application of Velocity and Displacement Diagrams.—The excess torque that is exerted on the crankpin, when excess

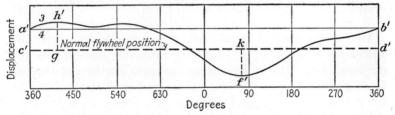

Fig. 345.—Displacement diagram for a four-stroke-cycle, single-cylinder, single-acting, slow-speed engine.

energy ΔW (Fig. 343) is available, is equal to the force that acts on the flywheel multiplied by its radius of gyration ρ. If F represents the value of the total excess force at the crank in pounds, at any point at any crank angle for the area ΔW, and r is the crank radius in feet, the excess torque at the crankpin is Fr.

The mass of the flywheel is M/g, and the acceleration at the radius ρ, in feet per second squared, is $\rho \dfrac{d^2\theta}{dt^2}$. The product of both terms is the force acting on the flywheel at radius ρ.

Then
$$Fr = \left(\frac{M}{g} \cdot \rho \frac{d^2\theta}{dt^2}\right)\rho = \frac{M\rho^2}{g}\frac{d}{dt}\left(\frac{d\theta}{dt}\right). \tag{17}$$

Since the projected length l of the curve in Fig. 343 corresponds to two revolutions, the time dt required for increment of length dl will be

$$dt = \frac{dl}{l/2 \times n/60} = \frac{120}{nl} dl \text{ sec.} \tag{18}$$

where n = r.p.m.

Multiplying the right side of Eq. (17) by dt and the left side (Fr) by the equivalent of dt as given in Eq. (18) results in

$$\frac{120rg}{M\rho^2 nl}\int F\, dl = \int \frac{d}{dt}\left(\frac{d\theta}{dt}\right) dt = \frac{d\theta}{dt} = \omega \qquad (19)$$

which is the angular velocity of the flywheel. Equation (19) indicates the constant by which the $\int F\, dl$ must be multiplied to obtain the value of the ordinates for Fig. 344.

From Eqs. (18) and (19),
$$\int \omega\, dt = \int \frac{d\theta}{dt} dt = \theta \qquad (20)$$

which is the angular displacement.

Also,
$$\int \omega\, dt = \frac{120}{nl}\int \omega\, dl. \qquad (21)$$

Hence, the integral of the velocity curve results in the displacement diagram (Fig. 345) with ordinates representing the displacement. The displacement diagram can be constructed for a given speed, radius of gyration and equivalent weight at the radius of gyration, and the maximum displacement $f'k$ or gh' determined in radians. Increasing the weight decreases the displacement in direct proportion, so that the required flywheel weight for a given maximum displacement can readily be determined, when the displacement has been evaluated for a given weight.

Bearing-load Analysis. *Piston-pin Bearing.*—The net force acting on the piston pin is the algebraic summation of the gas pressure and the inertia force of the piston assembly only. This force is assumed to act along the axis of the cylinder which should be perpendicular to and intersect the axis of the piston pin. This force results in two components, one along the connecting-rod axis and one perpendicular to the cylinder wall similar to that for the total net force (Fig. 340). However, the total side thrust depends upon the total net force which includes the inertia effect of the upper end of the connecting rod.

The load on the piston-pin bearing oscillates through an angle equal to twice the maximum connecting-rod angularity when the piston pin is fastened in either the connecting rod or piston.

Crankpin and Lower Connecting-rod Bearing.—The loads that are impressed upon the crankpin bearing depend on the combination of the net force on the piston, the angularity of the connecting rod, and the centrifugal force of the lower end of the connecting rod. The lower end of the connecting rod is assumed to rotate around the crankshaft center and produce a centrifugal force. The centrifugal force is

$$F = 0.000341 M n^2 r \tag{22}$$

where M = wt. of lower end of rod, lb.,
n = r.p.m.,
r = crank radius, ft.

The force along the rod is combined with the centrifugal force of the crank end of the rod (Fig. 346), and results in the

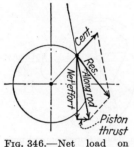

Fig. 346.—Net load on crankpin.

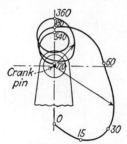

Fig. 347.—Diagram of crankpin forces (in relation to crankpin).

crankpin load and its direction in relation to the crankpin or the connecting-rod bearing.

Plotting the resultant crankpin forces and directions with relation to the crankpin results in a diagram of crankpin pressures which repeats itself for each engine cycle (Fig. 347). A similar diagram can be obtained for the bearing forces with relation to lower connecting-rod bearing.

Where two or more connecting rods connect to one crankpin, there will be a force along each rod that must be combined with the total centrifugal force of the lower ends of both or all connecting rods.

The crankpin-bearing pressures should not exceed 1500 lb. in.$^{-2}$, and the mean pressure should run about half the maximum pressure.

Crankshaft and Main Bearings.—The variation in pressure on main bearings is found in the same manner as for the crankpin. Unbalanced cranks exert a centrifugal force which must be added to that of the lower end of the rod. If each crank is located between bearings, only half the connecting-rod load on each side of the bearing is used. If more than one crank is located between bearings, the forces acting on each crank must be properly distributed between the two adjacent bearings. The various forces are combined (Fig. 348) by starting at the crankshaft center with any one of the forces, the resultant representing the net load which can be plotted with regard to either the main bearing or the crankshaft. The forces acting on an end bearing of a crankshaft for a six-cylinder engine having four main bearings are shown in Fig. 349.

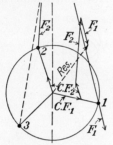

Fig. 348.—Determination of force on end main bearing. Cylinders 1 and 2 between bearings.

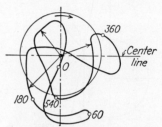

Fig. 349.—Forces on end main bearing. Six cranks—four main bearings.

The bearing pressure should not exceed 600 lb. in.$^{-2}$ for large stationary engines, and for automobile and aircraft engines the pressure should not exceed 2000 lb. in.$^{-2}$ The mean pressure would of course be considerably lower, whereas the maximum pressures might be exceeded in unusual conditions.

The Rubbing Factor, pv.—The mean bearing pressure is determined by averaging the bearing pressures for a cycle. The product of the mean bearing pressure in pounds per square inch of projected bearing area and the velocity of the surface of the shaft in the bearing in feet per second is the rubbing factor. This factor varies from values around 5000 for stationary practice to as high as 50,000 in aircraft-engine practice. In diving an airplane the rubbing factor may go above 200,000.

The satisfactory operation of a bearing depends upon the ability of the oil film to support the load. This necessitates a consideration of the $\mu n/p$ factor which is dealt with in Chap. XIV.

Wear and Oilhole Location.—The wear on a bearing should be proportional to the bearing pressures. Comparative-wear diagrams could then be made from the polar diagrams of bearing pressures. Angle[1] suggested that for each force determined for the polar diagram of forces a 180-deg. arc be drawn from the crankpin center, having a radius that is a measure of the amount of the force, the center of the 180-deg. arc being determined by the direction of the force. The summation of these arcs, to a reduced scale, results in an irregular curve around the crankpin which can be represented by a shaded area (Fig. 350) indicating comparative wear. The oilhole should be located in the low-wear area, preferably somewhat ahead of the center of this area.

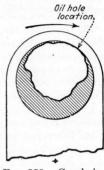

Fig. 350.—Crankpin-wear diagram.

The oil film is thinnest and wear is more apt to occur not in line with the bearing load but in the direction of rotation away from it. Consequently, the oilhole should be located somewhat ahead of the position indicating the center of the low-wear area.

EXERCISES

1. Plot the displacement, velocity, and inertia curves for 360 deg. of crank travel for a 4-in.-stroke engine having l/r ratios of 3 and 3.5. Engine speed is 4000 r.p.m., and M is 1 lb. for the inertia computations. Compare results with the curves in Fig. 338, and state the effect of variation in the l/r ratio on the three factors.

2. Determine the net-force diagram for a speed of 1000 r.p.m. Other data are the same as in Fig. 339.

3. Determine the net-force diagram with the throttle closed. Other data are the same as in Fig. 339.

4. Determine the rotative-force curve for the conditions in Probs. 2 and 3. Compare with Fig. 341.

5. Determine the rotative-force curve due only to gas pressure in Fig. 339.

6. Determine the rotative-force curve due to the inertia force in Fig. 339.

7. Plot the rotative-force curves due only to gas pressure for a 4, 6, 8, 45-deg. V-12, 60-deg. V-12, and 45-deg. V-16 cylinder engines, using the data from Prob. 5.

8. Same as Prob. 7 for the inertia forces, using the data from Prob. 6.

---

[1] G. D. Angle, "Engine Dynamics and Crankshaft Design," Airplane Engine Encylopedia Co., Detroit. Mich., 1925.

9. Same as Prob. 7 for the combined gas and inertia forces.

10. Under what conditions will the torque curves determined in Probs. 7, 8, and 9 be approached or occur?

11. A six-cylinder engine is to develop 1000 b. hp. at 1000 r.p.m. It has a mechanical efficiency of 0.85 under full-load condition at this speed. The indicated torque curve for each cylinder is that obtained in Prob. 2. Design a flywheel for a coefficient of fluctuation in speed of 0.01.

12. Construct the crankshaft-velocity and displacement diagrams for Prob. 11, and determine the displacement of the rotating elements from the mean position due to torque variation.

13. Determine and plot polar diagrams of the forces acting on the piston pin for the conditions in Probs. 2 and 3 and for the net force in Fig. 339. The piston diameter is $3\frac{1}{4}$ in. The total projected piston-pin bearing area is 1.75 in.^{2} The upper end of the connecting rod weighs 1.5 lb. Plot the pressures per square inch of projected bearing area.

14. Determine and plot polar diagrams of the crankpin-bearing pressures for the conditions in Probs. 2 and 3 and for the net force in Fig. 339. The piston diameter is $3\frac{1}{4}$ in. The lower end of the connecting rod weighs 3 lb. The crankpin bearing is $2\frac{1}{2}$ in. in diameter and $1\frac{3}{4}$ in. in length. Plot the diagrams relative to the connecting-rod bearing and the crankpin. $r = 2$ in.

15. Determine and plot polar diagrams of crankshaft-bearing pressures for the conditions in Probs. 2 and 3 and for the net force in Fig. 339. Use the same data as those given in Prob. 14. Do this for the end and the next to the end bearings for a six-cylinder engine with a four-bearing crankshaft. The main bearings are $2\frac{1}{2}$ in. in diameter and 2 in. in length.

16. Determine the mean and maximum bearing pressures per unit area and the mean and maximum pv values for Probs. 13, 14, and 15.

17. Plot wear diagrams for the piston-pin, crankpin, and both main bearings in the previous problems. Indicate desirable oilhole locations.

18. An automobile is traveling at maximum speed on a level road. The throttle is suddenly closed, and all the crankpin bearings are destroyed. Discuss the probable reason for the destruction of these bearings.

CHAPTER XVIII

ENGINE VIBRATION AND BALANCE

Vibration.—A structure or element of a structure may be made to vibrate by applying a force that deflects the structure. No vibration will occur if the force is applied very slowly and is maintained or reduced very slowly. The deflection of the structure by a force application sets up a restoring force in the structure which resists the applied force. The removal of the applied force permits the restoring force to reduce the deflection or cause a motion of the deflected part of the structure. The restoring force approaches zero as the structure approaches its normal position. However, the inertia of the moving part exerts a force that deflects the structure and stresses it in the opposite direction, which results in a vibration.

The deflection of a structure causes a movement of molecules with respect to each other, and some of the energy involved in each vibration is dissipated into internal energy of the material by the friction process.[1] Thus, the deflection of a structure undergoing vibration from a single disturbing force progressively decreases as the energy involved in the initial deflection is dissipated, and the structure gradually comes to rest.

Each structure has a natural frequency of vibration which is shown by the difference in vibrations of various tuning forks. The application of a disturbing force with a frequency equal to the natural frequency of the structure results in a continuous vibration which increases in amplitude with each vibration until the energy involved in each force application is equal to the energy dissipated by the friction process. This is synchronous vibration and may result in the destruction of the vibrating structure if continued for an appreciable time.

The forces acting in an internal-combustion engine that may cause vibration are those due to gas pressure, to the inertia of the reciprocating parts, and to the inertia of the rotating parts.

---

[1] Some energy is dissipated by the agitation of the air around the vibrating structure.

INTERNAL-COMBUSTION ENGINES

Obviously, vibrations can be eliminated if forces are introduced that are equal in amount and opposite in direction to the disturbing forces. Otherwise, material with an inherently high damping characteristic, a design that prevents synchronous vibration, and vibration dampers are used to minimize vibration.

Balance of Rotating Parts.—A rotating part should have both static and dynamic balance. If the bearing sections of a rotating part are placed on parallel knife-edges, set in a horizontal plane, the part will remain in any angular position if it is in static balance. In this case the center of gravity of the rotating part must be at its axis of rotation. However, a body in static balance may not be in "running" or dynamic balance.

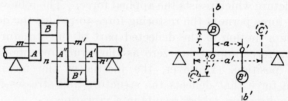

Fig. 351.—Two-throw crankshaft.   Fig. 352.—Dynamic balance.

A two-throw crankshaft (Fig. 351), if carefully machined, has static balance. On dividing the crankshaft into five parts by the lines mm' and nn', it will be seen that the parts A, A', and A'' are symmetrical about the shaft axis, but parts B and B' are not. Parts B and B' can be represented as in Fig. 352. Rotating the shaft sets up a centrifugal force along ob and $o'b'$.

The centrifugal force is equal to

$$F = 3.41 \times 10^{-4} M n^2 r \tag{1}$$

where M = wt. of B or B',
 n = r.p.m.,
 r = radius, ft.

Since the two centrifugal forces are not along the same line, there exists an unbalanced couple equal to $F \times a$, where a is the distance between the two forces. The shaft can be put in dynamic balance by providing an equal but opposite couple lying in the same plane.

Example.—Determine the unbalanced couple if weights B and B' are each 10 lb. Speed = 3600 r.p.m.; a = 6 in.; r = 4 in.

ENGINE VIBRATION AND BALANCE

$$Fa = 3.41 \times 10^{-4} \times 10 \times (3600)^2 \times 0.333 \times 0.5$$
$$= 7363 \text{ lb.-ft.}$$

Example.—For distances of $a' = 10$ in. and $r' = 2$ in., determine the weights necessary to balance the couple in the foregoing example.

$$Fa = F'a' = 3.41 \times 10^{-4} M' n^2 r' a'.$$
$$M = \frac{7363 \times 12 \times 12}{3.41 \times 10^{-4} \times (3600)^2 \times 2 \times 10} = 12 \text{ lb.}$$

There will be no distortion of the shaft if the weights providing the balancing couple can be located so that the centrifugal force due to C' acts along ob, and that due to C acts along $o'b'$. As shown in the diagram the centrifugal forces due to the four weights would tend to deflect the shaft at the points of application of the four forces.

A dynamic balancing machine determines the unbalanced couple which can be balanced by either adding or removing metal at the proper places. Although the crankshaft is the most important of the rotating parts, the others should not be overlooked in endeavoring to determine the cause of vibration.

Balance of Reciprocating Parts.—The motion of the piston of an internal-combustion engine depends on the radius of the crank and the length of the connecting rod (Eq. 3, page 481). With a connecting rod of infinite length, the piston motion would be a simple harmonic and would be indicated by the motion of the crankpin projected on the cylinder axis extended through the crankshaft center.

The motion of the piston with a connecting rod of finite length can be represented by a Fourier Series which has an infinite number of terms, each representing a simple harmonic motion of definite frequency and amplitude. Fortunately, the higher the frequency the smaller is the amplitude of the harmonic, and only a small number of the lower harmonics are required to represent very nearly the true piston motion. Also, the odd harmonics (third, fifth, seventh, etc.) are not present in the series since the piston motion from 0 to 180 deg. is exactly the reverse of the motion from 180 to 360 deg.*

Any of the harmonic motions that constitute the piston motion may be represented by the motion of two equal weights totaling

---

* Curves representing such motions are said to have "looking-glass" or "mirror" symmetry.

the weight of the piston assembly, both of which are coincident at top and bottom centers but rotate in opposite directions at the same speed in a circle the diameter of which represents the amplitude of the harmonic (Fig. 353). Obviously, the movement of the center of gravity of the two rotating weights, which lies on the extended cylinder axis, introduces a force that acts along the cylinder axis and reaches a maximum at top and bottom centers. The frequency of the harmonic fixes the rate of rotation, the first harmonic having crankshaft speed, the second harmonic having twice crankshaft speed, etc., all of the imaginary weights coinciding at top center.

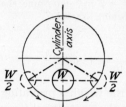

FIG. 353.—Representation of a simple harmonic.

The weight of the upper part of the connecting rod is included as reciprocating weight in the following analyses. The weight of the piston rings is omitted from the weight of the piston assembly on the assumption that piston-ring friction balances the inertia forces of the rings.

Single-cylinder Engine.—The first harmonic for a single-cylinder engine is represented by one weight A (Fig. 354) rotating with the crankpin and another weight B rotating in the opposite direction. The imaginary weight A can be balanced by an equal weight at A', rotating with the crankshaft, whereas it is impossible to balance weight B by either the addition or subtraction of weight from any part of the crankshaft. When half the reciprocating weight is counterbalanced, the unbalanced part of the first harmonic is that due to weight B rotating opposite to crank travel. This force rotates and tends to cause the engine to vibrate equally in all directions in the plane of rotation. In light structures, using single-cylinder engines, such as motor-cycles, it is usually desirable to have more unbalanced force in a fore-and-aft direction rather than up and down. This is accomplished by overbalancing, that is, using a counterbalance heavier than one-half the reciprocating weight.

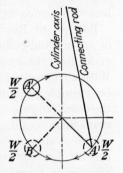

FIG. 354.—First harmonic for a single-cylinder engine.

It is impossible to counterbalance the second and higher harmonics with weights rotating at crankshaft speeds, since their frequency is higher than that of crankshaft rotation.

Four-cylinder Engine.—The first harmonics in a four-cylinder engine are represented by four weights concentrated at the crankpins of the solid-line diameter (Fig. 355b) rotating clockwise and by four weights at the ends of the dashed-line diameter and rotating counterclockwise. The imaginary weights at the crankpins representing half of the reciprocating weights for cylinders 1 and 4 are approaching top dead center, whereas those representing 2 and 3 are approaching bottom dead center. These weights are always diametrically opposite and balance each other in the projection. The other imaginary weights are

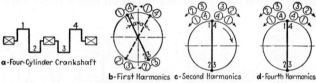

Fig. 355.—Four-cylinder engine balance.

also in balance, so that the first harmonic in a four-cylinder engine is inherently balanced.

These rotating weights set up centrifugal forces acting radially from the various actual and imaginary crankpins. These forces on cranks 1 and 2 set up a couple which is exactly balanced by the couple set up by the centrifugal forces on cranks 3 and 4. The forces at the actual and imaginary crankpins coincide when the crankshaft is in a vertical plane and balance each other when the crankshaft is in a horizontal plane. Thus, although the various forces appear to be balanced when projected on a plane perpendicular to the axis of the crankshaft, couples are introduced that tend to deflect the center bearing of a three-bearing crankshaft up and down when cranks for cylinders 2 and 3 are in the upper and lower quadrants, respectively. A two-bearing four-cylinder-engine crankshaft will be deflected in the same manner.

The weights representing the second harmonic travel twice as fast as the crank. When cranks 1 and 4 are at top dead center (Fig. 355c), all the weights representing the second harmonic traveling in the same direction as well as those traveling in the opposite direction will be at top dead center. Thus, all the weights will be at top dead center and represent an unbalanced force. The center of gravity of all the weights is always on the vertical center line, so that the unbalanced second harmonics cause a vertical vibration with a frequency twice that of engine speed, which is characteristic of all four-cylinder engines with this crank arrangement.

The fourth harmonics (Fig. 356d) and all higher even harmonics act as the second harmonic but have higher frequency and exert less force.

Six-cylinder Engine.—The six-cylinder engine has the following pairs of cylinders connected to cranks in the same plane: 1 and 6 to the end cranks, 3 and 4 to the two middle cranks, and 2 and 5 to the others. With cylinders 1 and 6 on upper dead center the

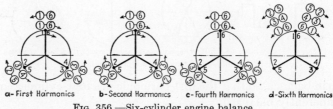

a- First Harmonics   b- Second Harmonics   c- Fourth Harmonics   d- Sixth Harmonics

FIG. 356.—Six-cylinder engine balance.

weights representing the first harmonics are shown in Fig. 356a. When cranks 1 and 6 are on dead center, 3 and 4 have rotated 120 deg. from dead center. This locates one pair of weights for 3 and 4 at cranks 3 and 4 and the other pair at cranks 2 and 5 but rotating counterclockwise. Cranks 2 and 5 have rotated 240 deg. from dead center, which distributes the weights for these cylinders as shown. The projections of these weights on a plane perpendicular to the axis of the crankshaft show all the weights representing the first harmonics to be in balance, for there are three pairs of weights perfectly balanced rotating with the crankshaft, and three pairs also perfectly balanced rotating in the opposite direction.

An analysis for couples on the first three cranks shows that when cranks 1 and 6 are on top center the forces for 2 and 3 can be represented in each case by a vertical force equal in magnitude to one-half of the force on crank 1. The resultant of the downward forces is equivalent to a force downward located halfway between cranks 2 and 3 and of a magnitude equal to the force acting on crank 1. Hence, a couple is set up with an arm equal to the distance between the center of crank 1 and the center line between cranks 1 and 2.

A study of the three other cranks shows that an equal and opposite couple is set up by the forces acting on these cranks. These couples, although perfectly balanced, tend to bend the crankshaft and crankcase and necessitate the use of rigid construction for high-speed operation.

The weights for the second harmonics for cylinders 1 and 6 are shown (Fig. 356b) at top dead center. Cranks 3 and 4 have moved 120 deg. from dead center, and the weights 3 and 4 rotating in the same direction have moved 240 deg. and are found at the position of cranks 2 and 5. Those moving counterclockwise have also moved 240 deg. and are found at cranks 3 and 4. Thus, all the weights are uniformly distributed and are balanced except for couples similar to those of the first harmonics.

The fourth harmonics (Fig. 356c) are also found to be similarly balanced.

The sixth harmonics (Fig. 356d) are completely unbalanced and tend to set up a vibration, in a vertical plane, with a frequency six times that of crankshaft rotation. The magnitude of these forces, however, is very small and practically negligible as a source of vibration.

Radial Engines.—In a single-row three-cylinder radial engine, only one crank is used and all three of the weights rotating with the crankshaft and representing part of the first harmonics are located at the crankpin (Fig. 357a). The crank has moved 120 deg. past dead center for cylinder 2, and weight 2, rotating counterclockwise, is located on the center line of cylinder 3. The crank has moved 240 deg. past dead center for cylinder 3, thus locating weight 3, rotating counterclockwise, on the center line of cylinder 2. Then, for a three-cylinder radial engine the weights moving counterclockwise are perfectly balanced, and the other three can be balanced by a weight equal to 1.5 times the reciprocating weight per cylinder, located opposite the crank and at crank radius.

The second and fourth harmonics are found to be half balanced (Fig. 357b and c), whereas the sixth harmonic is found to be perfectly balanced.

506   INTERNAL-COMBUSTION ENGINES

The unbalanced forces do not produce unbalanced couples in a single-row radial engine. A two-row radial engine with the two cranks 180 deg. apart has an unbalanced couple for each unbalanced harmonic.

Other engines with a different number of cylinders and crank arrangements may be analyzed in a similar manner to determine

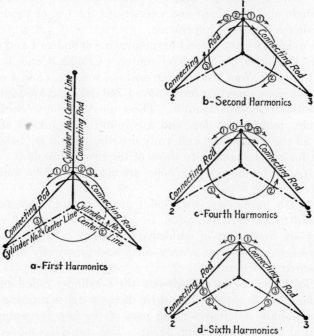

Fig. 357.—Three-cylinder radial engine balance.

the inherent balance of the various harmonics and couples (Table 34). The maximum inertia force for any harmonic in a single cylinder is equivalent to the centrifugal force of the two imaginary rotating weights which coincide at top and bottom centers.

Thus, for the first harmonic,

$$\text{Force}_1 = 3.41 \times 10^{-4} M n^2 r = kM \tag{2}$$

where M = reciprocating wt. lb.

From Eq. (6), page 483, for the second harmonic,

$$\text{Force}_2 = \frac{r}{l} kM. \tag{3}$$

ENGINE VIBRATION AND BALANCE

TABLE 34.—INHERENT BALANCE OF INTERNAL-COMBUSTION ENGINES

Type	Crank arrangement	Inertia forces, max. value, lb.		Couples, max. value, lb. ft.	
		Primary	Secondary	Primary	Secondary
1-cyl.		kM*	$\frac{r}{l}kM(v)$‡	None	None
2-cyl. vertical		Balanced	$2\frac{r}{l}kM(v)$‡	$kMa(v)$‡	None
2-cyl. opposed		Balanced	Balanced	$kMa(h)$‡	$\frac{r}{l}kMa(h)$‡
3-cyl. vertical		Balanced	Balanced	$1.732kMa(v)$‡	$1.732\frac{r}{l}kMa(v)$‡
3-cyl. radial		$1.5kM$†	$1.5\frac{r}{l}kM$	None	None
4-cyl. vertical		Balanced	$4\frac{r}{l}kM(v)$‡	Balanced	Balanced
4-cyl. vertical		Balanced	Balanced	$3.162kMa(v)$‡	Balanced
6-cyl. vertical		Balanced	Balanced	Balanced	Balanced
6-cyl. vertical		Balanced	Balanced	$3.464kMa(v)$‡	Balanced
7-cyl. radial		$3.5kM$†	Balanced	None	None
8-cyl. vertical		Balanced	Balanced	Balanced	Balanced
8-cyl. 90° V		Balanced	Balanced	$3.162kMa$§	Balanced
9-cyl. radial		$4.5kM$†	Balanced	None	None

* Half this force can be eliminated by counterbalancing.
† All this force can be eliminated by counterbalancing.
‡ (v) and (h) indicate that the force or couple is in a vertical or horizontal plane, respectively.
§ May be balanced by counterweights which introduce an equal and opposite couple.

Most of the four-stroke-cycle engines commonly used are inherently well balanced in respect to the lower harmonics and their couples. The typical two-stroke-cycle engines are inherently well balanced except in respect to unbalanced couples in one plane, known as "rocking" couples.

In some cases the maximum value of the rocking couple is not obvious, and the maximum value of the equation representing the rocking couple must be determined.

Example.—Determine the maximum value of the primary rocking couple for a four-cylinder vertical engine with cranks at 90 deg.

In Fig. 358, the crankshaft has moved through an angle θ in a clockwise direction from top-center position for cylinder 1. The inertia forces have vertical components acting upward on cranks 1 and 3 and vertical com-

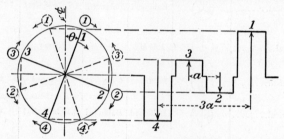

Fig. 358.—Analysis of couples for a four-cylinder 90-deg. crankshaft.

ponents acting downward on cranks 2 and 4. The couples exerted by the forces are $3\,kMa \cos\theta$ on cranks 1 and 4 and $-kMa \sin\theta$ on cranks 2 and 3.

Thus, Net couple $= 3kMa \cos\theta - kMa \sin\theta$.

$$\frac{d(\text{net couple})}{d\theta} = kMa(-3\sin\theta - \cos\theta)$$

or, for the maximum value, $\theta = 180° - 18.4° = 161.6°$.
Then,

Max. net couple $= kMa(-3 \times 0.9489 - 0.3156) = -3.162 kMa$.

The couple will be a positive maximum 18.4 deg. before top center, and negative as found 161.6 deg. after top center position for crank 1.

Barrel or Round Engines.—Another type of engine with cylinders parallel to the shaft (barrel or round engine) has different balance conditions. The ordinary crankshaft cannot be used with this cylinder arrangement, but three other types of mechanism deserve consideration. These are the cylindrical cam, the swash plate, and the wobbler mechanism.

The cylindrical-cam mechanism consists of a drum cam fixed on the shaft and engaging rollers mounted in the piston members. The cam is preferably multiple, with two or more throws, permitting perfect balance without counterweights. With the two-throw cam, the pistons have four strokes per revolution of the shaft, and the shaft may be used as a camshaft for a four-stroke engine.

The swash plate is essentially an oblique slice from a cylinder, the axis being the engine shaft. Connection between the swash plate and the piston is usually made with *slippers* having working faces engaging the swash plate and so pivoted in the piston members as to permit the generation of efficient film lubrication under the working faces. With plane swash-plate faces, the piston motion is simple harmonic and the piston inertia couple can be exactly balanced by the centrifugal couple of the swash plate.

A wobbler is a platelike member mounted on bearings inclined to the shaft. The wobbler is prevented from rotating with the shaft and has arms extending radially into the piston members. These arms act as connecting rods to crossheads with axes perpendicular to cylinder bores. All the engine torque is carried by the piston members if the wobbler mechanism is permitted to have a small motion.

The cylindrical-cam and swash-plate mechanisms necessitate the cutting of each piston member almost in two to permit the cam or swash plate to pass. This is an element of weakness avoided in the wobbler mechanism.

Counterbalancing.—The inertia forces of the reciprocating parts of a single-cylinder engine may be completely counterbalanced by the introduction of equivalent reciprocating parts (dummy pistons) in such a manner (Fig. 359) that the center of gravity of all the reciprocating parts does not change.

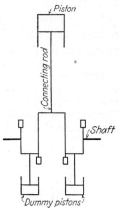

Fig. 359.—Balanced single-cylinder engine.

A smaller crank radius for the dummy pistons necessitates the use of parts heavier than the reciprocating weight of the engine, but the l/r ratio must be the same. Forces that rotate at crankshaft speed and in the same direction of rotation can be counterbalanced by attaching weights to the crankshaft at the proper locations (Fig. 360). There are three forces that may be counterbalanced:

1. The rotating force equivalent of one-half the first harmonic of the reciprocating motion.
2. The force due to rotation of parts of the crank with centers of gravity not at the shaft center.
3. The force due to rotation of the crank end of the connecting rod.

All these forces act radially through the crankpin center line and can be balanced by masses attached diametrically opposite the crankpin at a radius that depends on the total mass attached. The effect of such counterbalancing is to put a single-throw

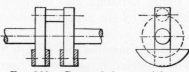

FIG. 360.—Counterbalance weights.

crankshaft out of static balance and a multiple-throw crankshaft out of dynamic balance. However, such counterbalancing not only offsets forces that tend to vibrate the engine but also reduces the main-bearing pressures. This permits the use of smaller main bearings.

Example.—Determine the counterbalance weights to be added to a single-cylinder-engine crankshaft, 3-in. bore and 5-in. stroke, with symmetrical crank cheeks. The crankpin is $2\frac{1}{2}$ in. in diameter and $1\frac{1}{2}$ in. in length. The weight of the reciprocating parts is 1.6 lb. The weight of the lower end of the connecting rod is 0.75 lb.

$$\begin{aligned}
&\text{Wt. of crankpin, } 0.7854 \times \overline{2.5}^2 \times 1.5 \times 0.286 = 2.11 \text{ lb.}\\
&\text{Wt. of lower end of connecting rod} \qquad\qquad = 0.75 \text{ lb.}\\
&\text{Wt. of one-half the reciprocating parts} \qquad\quad = 0.80 \text{ lb.}\\
&\qquad\text{Total wt. to counterbalance} \qquad\qquad\quad = 3.66 \text{ lb.}
\end{aligned}$$

This weight should be added to the crank cheeks, with the center of gravity of the weight located opposite the crankpin and on the crank circle (Fig. 360).

Natural Frequency of Vibration.—Any structure or element thereof has a natural frequency of vibration which depends on the effective mass of the structure and the force required to produce unit deflection.

ENGINE VIBRATION AND BALANCE

Thus, $$f = \frac{1}{2\pi}\sqrt{\frac{gs}{M}} \qquad (4)*$$

where f = natural frequency, vibrations sec.$^{-1}$,
M = equivalent wt. of the deflected mass, lb.,
s = force, lb. ft.$^{-1}$ of deflection,
g = acceleration due to gravity, 32.2 ft. sec.$^{-2}$

The forces occurring in an internal-combustion engine are periodic and may be represented by a series of harmonics. The frequency of the forces and any of their harmonics varies with engine speed. Thus, at some speed the frequency of one of the harmonics will equal the natural frequency of some part of the engine structure on which the force acts and synchronous vibration or resonance occurs. This increases the amplitude (Fig. 361) and stress of the vibrating element very appreciably, which may fatigue and damage the element in a short time.

The amplitudes of the harmonics decrease rapidly with an increase in frequency. Consequently, resonance with one of the high harmonics is not serious, since the inherent damping characteristic of the structure prevents the development of vibration amplitudes of appreciable amounts. This indicates that structures subject to periodic forces should be designed to have high natural frequencies considerably above the frequency of the lower harmonics of the applied force.

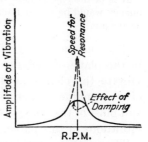

Fig. 361.—Resonance and effect of damping.

Example.—A motorcycle engine caused serious vibration of the foot boards, attached to the engine supports, at an engine speed of 3000 r.p.m. The engine supports consisted of two hollow tubes, 1⅛-in. outside diameter, and No. 12 B.w.g. (0.109-in.) walls.

The frequency of vibration for a steel tube or rod is a function of the radius of gyration and the length of the tube. With a constant tube length, only the tube section can be varied.

$$\text{Radius of gyration, } r = \frac{\sqrt{1.125^2 + 0.907^2}}{4} = 0.361 \text{ in.}$$

* L. S. Marks, "Mechanical Engineers' Handbook," 3d ed., p. 489, McGraw-Hill Book Company, Inc., New York, 1930.

The maximum engine speed was 4000 r.p.m. which is **33** per cent higher than the resonance speed. Increasing the natural frequency of vibration of the supporting tubes by 40 per cent raised the resonance speed above the maximum. To accomplish this, with the same length, a new section of tube was required, such that

$$r_{new} = 1.4 r_{old}$$
$$= 1.4 \times 0.361 = 0.505 \text{ in.}$$

On assuming a tube thickness of $\frac{3}{32}$ in.,

$$0.505 = \frac{\sqrt{d^2 + (d - 0.1875)^2}}{4},$$
$$d = 1.52 \text{ in.}$$

It is of interest to note that with a solid rod instead of a tube the frequency would be lowered and resonance would occur at a speed lower than 3000 r.p.m.

The inertia forces of the principal reciprocating parts act in the plane of cylinder axes and produce couples that tend to vibrate the engine in this plane. The high natural frequency of the engine structure, particularly with cylinders, jackets, and most of the crankcase cast in one piece, practically eliminates vibrations in this plane.

The gas-pressure forces along the connecting rod may be resolved into horizontal and vertical components acting on the main bearings. The horizontal components act perpendicularly toward the plane of cylinder axes and tend to vibrate the crankcase in a transverse plane. The unbalanced half of the first harmonic (Fig. 354), moving opposite to crankshaft rotation, represents a rotating force that also tends to distort the crankcase. Consequently, crankcases should be designed for a high natural frequency of vibration in a plane perpendicular to that of the cylinder axes.

Heldt[1] examined the relation of directions of the various harmonics of the horizontal components of the gas forces in a conventional six-cylinder vertical engine (Fig. 362). On the assumption of a given position for the various harmonics of the horizontal gas-pressure forces on crank 1, the first harmonic for crank 6 is exactly opposite in direction to that of crank 1

[1] P. M. Heldt, Engine Roughness—Its Cause and Cure, *S.A.E. Trans.*, **31**, 47 (1936).

since the frequency of the first harmonic is $n/2$.* Crank 4 lags 120 deg. behind crank 1, which causes the first harmonic for crank 4 to lag 60 deg. behind that of crank 1, etc. The first to the fifth harmonics produce different bending effects on the crankcase, as indicated by their direction and amount. However, the sixth harmonics are in phase with each other and exert a combined effect on the crankcase. Apparently the natural frequency of vibration of the crankcase in a horizontal

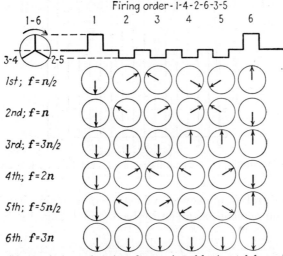

Fig. 362.—Phase relations of various harmonics of horizontal force due to gas pressure in a conventional six-cylinder engine. (*Heldt*.)

plane should be above one-half the product of number of cylinders and maximum speed.

Torsional Vibration.—The variation in torque that occurs with the internal-combustion engine produces torsional vibration of the crankshaft. The torque curve can be expressed as a series of harmonics, all of which increase in frequency directly with engine speed. However, the amplitude of the harmonics decreases with an increase in the order of the harmonic (Fig. 363).[1] A condition of resonance occurs when the frequency of any of

* Since two revolutions are required for each cycle.
[1] K. Lürenbaum, Vibration of Crankshaft-propeller Systems, *S.A.E. Trans.*, **31**, 469 (1936).

the harmonics coincides with the natural frequency of the crankshaft. The vibration will usually be destructive if resonance occurs with a low harmonic that has a high amplitude.

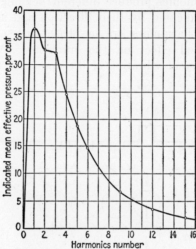

FIG. 363.—Amplitudes of torsional harmonics.[1] Valid for all four-stroke-cycle carbureter engines. (*Lürenbaum.*)

The natural frequency of vibration of a plain shaft with a flywheel on one end, as given by Lack and Jahnke,[1] is

$$f = \frac{1}{2\pi}\sqrt{\frac{I_p G g}{M r^2 l}} \qquad (5)$$

where I_p = polar moment of inertia of shaft, in.^{4},
 G = modulus of elasticity for shear (12×10^6 lb. in.$^{-2}$ for steel),
 g = acceleration due to gravity = 32.2 ft. sec.$^{-2}$,
 M = wt. of flywheel or rotating mass, lb.,
 r = radius of gyration of the wheel, in.,
 l = free length of shaft, in.

This relation indicates that the natural frequency of vibration may be increased by reducing flywheel weight, its radius of gyration, and shaft length, and by increasing the polar moment of inertia.

[1] A. Lack and C. B. Jahnke, Torsion Vibration and Critical Speeds of Shafts. *A.S.M.E. Trans.*, **47**, 493 (1925).

For a solid shaft, of diameter d_o, $I_p = \dfrac{\pi d_o^4}{32}$,

and for a hollow shaft, with inside diameter d_i,

$$I_p = \frac{\pi(d_o^4 - d_i^4)}{32}.$$

A hollow shaft will have a lower moment of inertia and a considerably lower weight.

Example.—Determine the relative natural frequency of torsional vibration of a solid 2-in. shaft and a hollow 2-in. shaft with a 1½-in. hole in it. Both shafts have a heavy flywheel on one end; thus, the effect of the weight of the shaft is negligible compared with that of the flywheel.

For the solid shaft, $I_p = \dfrac{\pi 2^4}{32} = 0.5\pi$.

For the hollow shaft, $I_p = \dfrac{\pi(16 - 5.06)}{32} = 0.342\pi$.

$$I_p \text{ ratio} = \frac{0.5\pi}{0.342\pi} = 1.462.$$

$$\sqrt{1.462} = 1.21.$$

$$\text{Wt. ratio} = \frac{2^2 - 1.5^2}{2^2} = 0.437.$$

The I_p ratio indicates that the hollow shaft would have a lower natural frequency by about 21 per cent but that the shaft would weigh 43.7 per cent less than the solid shaft.

Example.—Determine the outside diameter of a hollow shaft to have the same natural frequency of vibration as the solid shaft in the foregoing example.

$$\frac{\pi(d_o^4 - 5.06)}{32} = 0.5\pi$$

$$d_o = 2.14 \text{ in.}$$

The weight ratio is $\dfrac{2.14^2 - 1.5^2}{2.0^2} = 0.58.$

Thus, increasing the shaft diameter 7 per cent permitted a weight reduction of 42 per cent while the same natural frequency of vibration was maintained as that of the solid shaft.

The natural frequency of vibration for a plain shaft with a flywheel at each end, as given by Lack and Jahnke,[1] is

$$f = \frac{1}{2\pi}\sqrt{\frac{I_p G g (M_1 + M_2)}{M_1 M_2 r^2 l}}. \quad (6)$$

When $M_1 = M_2$, this equation becomes

$$f = \frac{1}{2\pi}\sqrt{\frac{2 I_p G g}{M r^2 l}}. \quad (7)$$

Comparison with Eq. (5) indicates that the addition of a flywheel to a system with only one flywheel on one end increases the natural frequency of vibrations about 41.4 per cent.

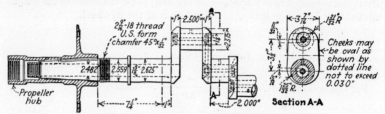

Fig. 364.—Liberty-engine crankshaft.

Natural Frequency of Torsional Vibration of Crankshafts.—A method for determining the natural frequency of torsional vibration of crankshafts has been developed by S. Timoshenko[2] who reduces all cranks to a shaft of definite length having a given torsional rigidity, with the inertia masses reduced to a flywheel system assumed to be located at the longitudinal center of the engine. F. L. Prescott[3] has applied this method to a 12-cylinder Liberty V-type-engine crankshaft (Fig. 364). His computations (pages 518 and 519) indicate a natural frequency of vibration between 90 and 90.5 vibrations sec.$^{-1}$ The observed frequency of the Liberty-engine shaft, as determined by a torsiometer, was found to be 100 vibrations sec.$^{-1}$ which is checked more closely by the expression for the constrained than by that for the unconstrained condition.

[1] Lack and Jahnke, *op. cit.*

[2] S. Timoshenko, "Vibration Problems in Engineering," D. Van Nostrand Company, Inc., New York, 1928.

[3] F. L. Prescott, Vibration Characteristics of Airplane-engine Crankshafts, *Air Corps Information Circ.*, **7**, 664 (1932).

Damping of Vibrations.—Vibrations may be eliminated if forces are introduced that are exactly opposite to the activating forces in direction and amount. The second harmonics of the inertia forces in a four-cylinder engine act in a vertical plane and are in phase. Thus, all these forces act alternately upward and downward in the same plane. Lanchester devised a vibration damper consisting of two unbalanced shafts (Fig. 365) which were mounted below the center main bearing and timed to produce an equal and opposite effect to that of the second harmonics. Obviously, the counterbalance shafts rotate in opposite directions and at twice crankshaft speed.

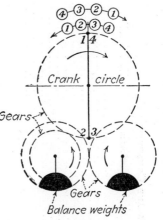

FIG. 365.—Lanchester vibration damper.

The same method is used with the General Motors two-stroke-cycle Diesel engines to balance the unbalanced couples. The camshaft and a shaft parallel to it but on the opposite side of the engine have unbalanced weights attached to both ends of both shafts. The weights are timed to exert a force only in the desired plane and to produce a couple opposite to the inherent couple due to the primary inertia forces of the reciprocating parts.

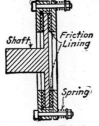

FIG. 366.—Friction vibration damper.

Other types of vibration dampers depend on the vibration producing an offsetting effect. A common type of torsional-vibration damper has a floating mass held in contact with the free end of the crankshaft by means of friction forces (Fig. 366). The occurrence of torsional vibration of the shaft with appreciable amplitude causes relative movement of the shaft with the floating mass, which dissipates the vibration energy with a friction process. This may necessitate the removal of an appreciable amount of heat from the damper to maintain desirable conditions with a resonant speed of the crankshaft.

Resonant-vibration dampers completely eliminate vibrations having a frequency equal to the natural frequency of the damper.

518 INTERNAL-COMBUSTION ENGINES

LIBERTY-ENGINE-CRANKSHAFT DATA AND COMPUTATIONS*

m_1	= rotating wt. per crankpin	= 6.3 lb.
m_2	= reciprocating wt. per crankpin	= 12.4 lb.
L	= length of connecting rod	= ... in.
n	= number of crank throws	= 6
R	= crank radius	= 3.5 in.
w	= width of crank cheek	= 3.4375 in.
h	= thickness of crank cheek	= 1 in.
a	= actual length of crankpin	= 2.5 in.
a_1	= effective length of crankpin	= $a + 0.9h$ in. = 3.4 in.
$2b$	= actual length of main journal	= 2.9 in.
$2b_1$	= effective length of main journal	= $2b + 0.9h$ in. = 2.9 in.
D_1	= outside dia. of main journal	= 2.625 in.
d_1	= inside dia. of main journal	= 1.375 in.
D_2	= outside dia. of crankpin	= 2.375 in.
d_2	= inside dia. of crankpin	= 1.25 in.
E	= Young's modulus of elasticity for steel	= 30×10^6 lb. in.$^{-2}$
G	= modulus of elasticity in shear or torsion	= 12×10^6 lb. in.$^{-2}$
I_1	= moment of inertia of main-journal section	= $\frac{\pi}{64}(D_1{}^4 - d_1{}^4)$ in.^{4}
I_2	= moment of inertia of crankpin section	= $\frac{\pi}{64}(D_2{}^4 - d_2{}^4)$ in.^{4}
I_3	= moment of inertia of crank-cheek section	= $hw^3/12$ in.^{4}
J_1	= polar moment of inertia of main-journal section	= $2I_1$ in.^{4}
J_2	= polar moment of inertia of crankpin section	= $2I_2$ in.^{4}
J_3	= polar moment of inertia of crank-cheek section	= $\dfrac{w^3h^3}{3.6(w^2+h^2)}$ in.^{4}
B_2	= flexural rigidity of crankpin = I_2E	= $1.474 \times 10^6(D_2{}^4 - d_2{}^4) = 1.474 \times 10^6(2.375^4 - 1.25^4) = 43.1 \times 10^6$
B_3	= flexural rigidity of crank cheek = I_3E	= $2.5 \times 10^6 hw^3 = 2.5 \times 10^6 \times 1 \times 3.4375^3 = 101.2 \times 10^6$
C_1	= torsional rigidity of main journal J_1G	= $1.178 \times 10^6(D_1{}^4 - d_1{}^4) = 1.178 \times 10^6(2.625^4 - 1.375^4) = 51.8 \times 10^6$
C_2	= torsional rigidity of crankpin J_2G	= $1.178 \times 10^6(D_2{}^4 - d_2{}^4) = 1.178 \times 10^6(2.375^4 - 1.25^4) = 34.5 \times 10^6$
C_3	= torsional rigidity of crank cheek J_3G	= $3.33 \times 10^6 w^3h^3/(w^2+h^2) = 3.33 \times 10^6 \times 3.4375^3 \times 1^3/(3.4375^2 + 1^2) = 10.6 \times 10^6$
F_2	= cross-sectional area of crankpin	= $\frac{\pi}{4}(D_2{}^2 - d_2{}^2) = \frac{\pi}{4}(2.375^2 - 1.25^2) = 3.2$ in.^{2}
F_3	= cross-sectional area of crank cheek	= $hw = 1 \times 3.4375 = 3.4375$ in.^{2}

ENGINE VIBRATION AND BALANCE

K = factor of complete constraint by main journals (Timoshenko) =

$$\frac{R(a+h)^2}{4C_3} + \frac{aR^2}{2C_2} + \frac{a^3}{24B_2} + \frac{R^3}{3B_3} + \frac{1.2}{12\times 10^6}\left(\frac{a}{2F_2}+\frac{R}{F_3}\right)$$

$$\overline{\frac{aR}{2C_2} + \frac{R^2}{2B_3}}$$

$$= \left[\frac{3.5(2.5+1)^2}{4\times 10.6\times 10^6} + \frac{2.5\times 3.5^2}{2\times 34.5\times 10^6} + \frac{2.5^3}{24\times 43.1\times 10^6} + \frac{3.5^3}{3\times 101.2\times 10^6}\right.$$

$$\left. + \frac{1.2}{12\times 10^6}\left(\frac{2.5}{2\times 3.2}+\frac{3.5}{3.44}\right)\right] \div \left[\frac{2.5\times 3.5}{2\times 34.5\times 10^6}+\frac{3.5^2}{2\times 101.2\times 10^6}\right] = \ldots 9.35 \text{ in.}$$

l_1 = equivalent length of shaft from edge of propeller to first crank throw minus b .... = ... 7.25 in. (actual) or 7.6 in. (equivalent length)

l_2 = equivalent length of shaft, center-to-center of main journals if shaft is unconstrained.. = $2b_1 + a\frac{C_1}{C_2} + 2R\frac{C_1}{B_3} = 2.9 + 3.4\frac{51.8\times 10}{34.5\times 10^6} + 2\times 3.5\frac{51.8\times 10^6}{101.2\times 10^6} = \ldots$ 11.58 in.

l_3 = equivalent length of shaft, center-to-center of main journals if shaft is constrained by main bearings =

$$2b_1 + a\frac{C_1}{C_2}\left(1-\frac{R}{K}\right) + 2R\frac{C_1}{B_3}\left(1-\frac{R}{2K}\right) = 2.9 + 3.4\frac{51.8\times 10^6}{34.5\times 10^6}\left(1-\frac{3.5}{9.35}\right) + 2\times 3.5\frac{51.8\times 10^6}{101.2\times 10^6}\left(1-\frac{3.5}{2\times 9.35}\right) = 9 \text{ in.}$$

$L_f = l_1 + \frac{n}{2}\times l_2$ (unconstrained) .... = $7.6 + 3\times 11.58 = \ldots$ 42.36 in.

$L_c = l_1 + \frac{n}{2}\times l_3$ (constrained) .... = $7.6 + 3\times 9 = \ldots$ 34.6 in.

i_c = moment of inertia of one crank about main-bearing center = ... 92.5 lb. in.$^{-2}$
i = equivalent moment of inertia of one crank system =

$$\left[m_1 + \frac{m_2}{2}\left(1+\frac{R^2}{4L^2}\right)\right]R^2 + i_c = (m_1 + 0.51m_2)R^2 + i_c \text{ for an average value of } R/L = 0.25$$

$I = n[(m_1 + 0.51m_2)R^2 + i_c]$ .... = $6[(6.3 + 0.51\times 12.4)3.5^2 + 92.5] = \ldots$ 1482 lb. in.$^{-2}$

f_f = natural frequency unconstrained .... = $3.125\sqrt{\frac{C_1}{IL_f}} = 3.125\sqrt{\frac{51.8\times 10^6}{1482\times 42.36}} = \ldots$ 90 vibrations sec.$^{-1}$

f_c = natural frequency constrained .... = $3.125\sqrt{\frac{C_1}{IL_c}} = 3.125\sqrt{\frac{51.8\times 10^6}{1482\times 34.6}} = \ldots$ 99.5 vibrations sec.$^{-1}$

* F. L. Prescott, *op. cit.*

Thus, mass M supported by spring S (Fig. 367) cannot vibrate with a frequency equal to the natural frequency of the mass m supported only by spring s. Any tendency to do so sets m into vibration with its natural frequency which completely balances the force tending to make M vibrate with this frequency.

The torque variation of a radial engine is due to the number of power impulses in each engine cycle. The frequency of the torque variation for a four-stroke engine with N cylinders at a speed of n r.p.m. is

$$f = \frac{Nn}{2}. \tag{8}$$

Taylor[1] noted that the harmonic of the torque curve for each cylinder that has the same frequency as the power impulses

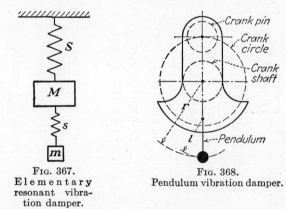

FIG. 367.
Elementary resonant vibration damper.

FIG. 368.
Pendulum vibration damper.

from the engine can cause appreciable torsional vibration and can be eliminated by a pendulum vibration damper attached to the crankshaft (Fig. 368).

The frequency of a simple pendulum is

$$f = \frac{1}{2\pi}\sqrt{\frac{g}{l}} \tag{9}$$

where l = length of the pendulum,
g = acceleration due to gravity.

A pendulum attached to the rotating crank at an appreciable radius r has an angular acceleration impressed on it which makes

[1] E. S. Taylor, Crankshaft Torsional Vibration in Radial Aircraft Engines, *S.A.E. Trans.*, **31**, 81 (1936).

the acceleration due to gravity insignificant. The radial acceleration due to rotation is

$$a_n = \frac{v^2}{r} = 4\pi^2 n^2 r. \qquad (10)$$

Substituting this value for g in Eq. (9) results in the relation for natural frequency of vibration of a rotating pendulum.

Thus,
$$f = n\sqrt{\frac{r}{l}} \qquad (11)$$

which indicates that the natural frequency of vibration of the rotating pendulum varies directly with engine speed and is determined by the design dimensions r and l.

Combining Eqs. (8) and (11) results in

$$\frac{r}{l} = \left(\frac{N}{2}\right)^2 \qquad (12)$$

from which the dimensions for the desired frequency can be determined.

Example.—Determine the length of a simple-pendulum resonant vibration damper attached to the crank cheek at crank radius to have a frequency equal to that of the power impulses of a nine-cylinder radial engine. Crank radius = 3.5 in.

Substituting in Eq. (12),
$$\frac{3.5 + l}{l} = \left(\frac{9}{2}\right)^2$$
or
$$l = 0.182 \text{ in.}$$

The short length of pendulum required for the usual frequencies necessitates the use of equivalent constructions (Figs. 369). Torsiograms taken with and without a resonant damper designed to eliminate the sixth harmonic of the torque curve of a 300-hp. six-cylinder in-line engine show (Fig. 370) the effective elimination of this harmonic. It can be shown[1] to be desirable to have r/l somewhat greater than $(N/2)^2$.

The weight of the pendulum counterbalance does not affect the natural frequency of vibration. However, to obtain the desired balancing force an appreciable weight is used which reduces the required amplitude to a desirable amount.

[1] *Sulzer Technical Review*, No. 1, 6 (1938).

Inherent Damping of Torsional Vibration.—Torsional vibration of a crankshaft imposes a vibratory motion on the connecting rod, piston, and other parts connected to the shaft.

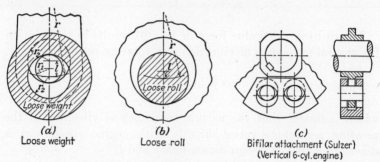

Fig. 369.—Constructions for short pendulums.

All the moving parts are shearing oil films, and the vibratory motion is inherently damped by this friction process. Any part that is deflected dissipates some of the vibratory energy by

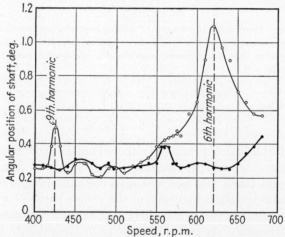

Fig. 370.—Torsiograms with and without dynamic damper. Sulzer, 300-b. hp., four-cycle Diesel engine with six cylinders in line.[1]

internal friction. The damping effect varies with the part and the order of vibration (Table 35). The amount of energy dissipated may be as high as 10 per cent of the power of the engine.

[1] *Sulzer Technical Review*, No. 1, 6 (1938).

TABLE 35.—DAMPING EFFECT OF VARIOUS PARTS IN PERCENTAGE OF TOTAL
ENERGY OF VIBRATION*
(Eight-cylinder single-acting four-cycle engine)

Order of vibration	4	8
Apparent damping by reciprocating parts	7.4	19.8
Cylinder-bore friction	3.1	9.3
Piston-pin friction	0.2	0.7
Crankpin friction	1.8	8.3
Main-bearing friction	2.0	9.1
Additional main-bearing friction due to inclination of crankpins	2.0	8.0
Shocks to crankpins	1.0	2.8
Shocks to cylinder wall	2.5	24.0
Air friction		
Internal friction of shafting	64.0	16.0
Internal friction in crank train	5.0	1.5
Internal friction in engine structure	5.0	1.5

*J. Geiger, Dämpfung bei Drehschwingungen von Motoren, *Automobiltech. Z.*, **38**, 366 (1935).

EXERCISES

1. Plot relative polar-force diagrams for the main bearings of a single-cylinder engine with the primary-inertia forces unbalanced; also, with half of the primary-inertia force balanced, and with one-half of this force overbalanced by 50 per cent. Note that all forces should pass through the center of the bearing and that the direction of the forces at various crank angles is required to plot the various diagrams.

2. Analyze a five-cylinder radial engine for inherent balance as regards the first and second harmonics.

3. Analyze a six-cylinder radial engine for inherent balance as regards the first and second harmonics. The engine consists of two three-cylinder radial engines, one behind the other, with the cylinders of one bank placed between cylinders of the other bank. The crankshaft consists of two cranks 180 deg. apart. Determine the value of the unbalanced primary- and secondary-inertia forces and couples, and discuss the balancing possibilities.

4. A narrow-angle V-8 engine has cranks 1 and 4 at the top of the crank circle. Cranks 2 and 3 follow by 90 deg. Cranks 5 and 8 follow these by 50 deg., and cranks 6 and 7 by 90 deg. more. What should be the angle between the cylinder axes of the two sets of cylinders? Analyze the engine for inherent balance. Determine the unbalanced primary- and secondary-inertia forces and couples, and discuss the balancing possibilities.

5. Prove that the maximum value of the unbalanced primary couple for a six-cylinder engine with cranks at 60° is $3.464kMa$ as given in Table 34.

6. Prove that the value of the unbalanced rotating couple for the eight-cylinder 90-deg. V-type engine is $3.162kMa$ as given in Table 34.

7. The couple in Prob. 6 is to be balanced by a single couple with weights located on the crank cheeks between cranks 1 and 2 and between cranks 3 and 4. Determine the weights and the angular position of the weights.

8. Given an engine with a stroke of 5 in. Crankpins are 2½ in. in diameter and 2 in. in length. Crank cheeks are 1 in. thick, 2½ in. wide, and 5 in. long and do not extend beyond the crankpins or crankshaft at the center line of the cheeks. The crankshaft diameter is 2½ in. Two crank throws are located between bearings (Fig. 351). The reciprocating weight per cylinder is 5 lb., and the lower end of a connecting rod weighs 3 lb. Connecting-rod-crank ratio is 4:1. Design the counterbalance weights to be attached to crank cheeks A and A'.

9. Compute the unbalanced couple acting on the counterbalanced shaft in Prob. 8 when the shaft is rotating alone at 5000 r.p.m.

10. Plot the horizontal force on the crankcase due to gas pressure. Use the data in Fig. 339.

11. Determine the amplitude of the sixth harmonic of the valve-lift curve, Fig. 218.

12. Determine the natural frequency of vibration of a hollow steel shaft with a 75-lb. flywheel. Free length of shaft is 2 ft. Shaft is 2½-in. outside diameter and 1½-in. inside diameter. Assume the flywheel to be a solid disk with a thickness of 1½ in.

13. What effect will decreasing the length of the shaft in Prob. 12 have on the natural frequency of vibration?

14. Determine the natural frequency of vibration of a channel section 10 ft. between supported ends. Height of section is 3 in., width 2 in., and thickness ⅛ in.

Note.—Eq. (4) gives a relationship for natural frequency of vibration. Substitute expression for static deflection (M/s) for this case from any handbook.

15. Add a concentrated load equal to the weight of the beam at the center of the beam in Prob. 14. Determine the frequency of vibration of the beam for this case.

16. In the determination of the natural frequency of torsional vibration of the Liberty-engine shaft in the text, i_c is given as 92.5 lb. in.$^{-2}$ This is the moment of inertia of one crank about the main-bearing center. Check this value, making use of the dimensions in Fig. 364.

17. Plot the torque curve for the gas pressure and other data in Fig. 339.

18. Determine the amplitude of the ninth harmonic for the curve in Prob. 17. See Valve-spring Surge, page 332.

19. Design a vibration damper to balance the unbalanced primary couple for the four-cylinder vertical 90-deg. crank engine in Table 34.

20. Determine the effect of a compound pendulum on the natural frequency of vibration of a vibration damper.

CHAPTER XIX
ENGINE DESIGN[1]

General.—The design of an internal-combustion engine is usually based on current practice with modifications made possible by improvements in fuels, lubricants, materials, and processing of the various parts. The design may be divided into two parts:

1. The determination of the type, speed, size, number of cylinders, cylinder arrangement, etc., required for a definite service demanding a definite power output.

2. The determination of the shape, size, materials, stresses, vibration characteristics, etc., of the various parts.

GENERAL DESIGN

Type of Process.—The Otto-engine process is inherently more efficient than the Diesel-engine process, but due to compression-ratio limitations in the Otto engine there may be little actual difference in economy. The development of better antiknock fuels which permit higher compression ratios for the Otto engine is eliminating the difference in economy.

The Otto engine is particularly desirable when smooth action under throttled conditions is desirable, as in automobiles, and also when the maximum output per unit of weight is desired. However, the two-stroke-cycle Diesel engine has been developed to the stage where it can compete on the latter basis. The carbureted two-stroke-cycle Otto-engine process is wasteful of fuel and has not been developed for high outputs. This may indicate the possibility of development of the two-stroke-cycle Otto-engine process with fuel injection into the cylinder after intake-port closure.

The Diesel process is used in very large engines in marine and stationary service, whereas the Otto process is used in very large gas engines.

[1] Much material relating to the design of various engine parts is located in previous chapters, for which reference to the index should be made. The student should also refer to the "S.A.E. Handbook" which contains much information dealing with materials, specifications, and design standards.

Small engines may use either type of process, but first cost always favors the Otto engine. Fuel cost and fire hazard are factors that also must be considered in deciding on the type of process.

Two- or Four-stroke Cycle.—Two-stroke carbureted spark-ignition engines with crankcase scavenging are used only in low-cost and low-output engines because of poor fuel economy. The two-stroke cycle is well adapted to the Diesel process, particularly with blower scavenging. This combination results in large outputs and low weight per unit output, because of scavenging the clearance space and to one power stroke per cylinder per revolution.

The four-stroke engine has only one power stroke per cylinder in two revolutions. It does not have ports in the cylinder walls or require a separate crankcase for each cylinder or a blower system for scavenging. It results in the most flexible operation and has been developed for automobiles and aircraft until used almost exclusively in these fields.

Single- or Double-acting Engines.—Double-acting engines require crossheads, piston rods, packing glands for piston rods, and a cooling system for the pistons and piston rods. However, double-acting engines develop practically twice the power of single-acting engines with the same bore and piston speed. The added parts and cooling complications limit the use of double-acting engines to the very large power outputs with large cylinders.

Most internal-combustion engines are single-acting, which simplifies the construction and operation and results in the minimum height for a vertical engine.

Number of Cylinders and Arrangement.—A large number of small cylinders rather than a small number of large cylinders is usually desirable from the viewpoint of smoother torque, lower engine weight, and higher efficiency. The smoother torque is obviously due to more power impulses per revolution, which are presumably evenly spaced although this is not required. The lower weight is indicated by the application of the principle of similitude (page 534). Higher compression ratios are permissible and higher thermal efficiencies attainable with the smaller cylinders.

Engines with a large number of small cylinders must operate at high engine speed (r.p.m.) to obtain the foregoing advantages.

ENGINE DESIGN

Large cylinders and consequently a small number are required when very low crankshaft speeds are desirable, as in marine propulsion. Reduction gearing may be used to permit higher engine speed than is desirable for the propeller, in both marine and aircraft engines (Fig. 371), but the advantages of this arrangement must be balanced against the cost, etc., of the additional equipment. Thus, the desirable crankshaft speed is an important factor in deciding on cylinder size and number of cylinders.

The various cylinder arrangements that may be used depend on the number of cylinders and type of engine cycle. The more common arrangements are included in Table 34.[1] The arrangement chosen should result in a desirable inherent balance and degree of compactness, depending on the service.

Stroke-bore Ratio.—This ratio varies from about 1:1 to 2:1 in the various engines built, the majority being near the lower ratio. A reduction in the length of stroke of an engine reduces the over-all dimension of the engine by about three times the reduction in stroke. However, this necessitates an increase in crankshaft speed to obtain the same piston speed and output.

The small stroke-bore ratio results in a thin cylindrical disk for the combustion-chamber shape for a valve-in-head engine. This may be undesirable for a small open-chamber fuel-injection Diesel engine, because of fuel-spray impingement on the walls.

Combustion chambers in engines with small stroke-bore ratios have high surface-volume relationships. This indicates more heat transfer to the combustion-chamber walls with the same engine speed than in engines having large stroke-bore ratios and similar combustion chambers. However, at the same piston speed, less time elapses per cycle for the short-stroke engine, and the heat transfer would be about the same percentage of energy input as in the long-stroke engine.

Displacement per Minute.—The output of an engine is indicated by the relation

$$\text{B. hp.} = k \times \text{disp.} \times Q_p \times \text{vol. eff.} \times \text{ideal eff.} \times \text{rel. eff.} \times \text{mech. eff.} \quad (1)$$

---

[1] A most complete tabulation of cylinder arrangements has been made by L. P. Kalb, Engine Types Adapted to Car Design Trends, *S.A.E. Trans.*, **29**, 13 (1934).

INTERNAL-COMBUSTION ENGINES

The heating value Q_p B.t.u. ft.$^{-3}$ depends on the fuel and the air-fuel ratio. The volumetric efficiency and relative efficiency depend on the engine speed, operating conditions, and design factors. The ideal efficiency depends on the fuel and permissible compression ratio for the carburetor spark-ignition engine and on the compression ratio and combustion process for the compression-ignition engine. The mechanical efficiency depends on engine design and lubrication.

The required displacement per minute is determined from a knowledge of the foregoing fuel and efficiency factors and from the desired output.

Example.—Determine the piston displacement per minute required for a 1000 hp. single-acting single-cylinder four-stroke-cycle blast-furnace gas engine. A low crankshaft speed is required, which indicates large cylinders. A compression ratio of about 6:1 (Table 11, page 149) and multiple ignition would be used.

The heating value of the correct mixture will be about 60 B.t.u. ft.$^{-3}$ (Table 12, page 150).

The volumetric efficiency will be about 0.75 (see example, page 155, and Fig. 81).

A compression ratio of 6:1 indicates an ideal thermal efficiency of 0.35 (Fig. 41).

The relative efficiency will be about 0.75 (page 116).

The mechanical efficiency will be about 0.85 based upon the performance of similar engines.

k will have a value of $\frac{1}{4} \div 33{,}000/778 = 0.25/42.4$ for a four-stroke cycle.

Thus, $$\frac{\text{Disp.}}{\text{Min.}} = \frac{1000 \times 42.4}{0.25 \times 60 \times 0.75 \times 0.35 \times 0.75 \times 0.85}$$

$$= 16{,}850 \text{ ft.}^3 \text{ min.}^{-1}$$

On the assumption of a mean piston speed of 1200 ft. min.$^{-1}$, the engine stroke is 6 ft. for an engine speed of 100 r.p.m. The cylinder area is

$$\frac{16{,}850}{1200} = 14.03 \text{ ft.}^2$$

which indicates a cylinder diameter of about 4.3 ft.

A double-acting cylinder would develop 2000 hp. and a twin-tandem double-acting engine would develop 8000 hp. under these conditions.

Example.—A four-stroke-cycle gasoline engine having a displacement of 300 in.^{3} is to operate at 3600 r.p.m. Estimate the power output.

ENGINE DESIGN

The volumetric efficiency will be about the same as in the preceding example but will be based on the air consumption. The heating value per cubic foot of air is 102 B.t.u. (Table VIII, Appendix). Assuming the same compression ratio and relative and volumetric efficiencies as those used in the preceding example, and substituting in Eq. (1), results in

$$\text{B. hp.} = \frac{0.25}{42.4} \times \frac{300 \times 3600 \times 2}{1728} \times 102 \times 0.75 \times 0.35 \times 0.75 \times 0.85$$

$$= 126 \text{ hp.}$$

A piston speed of 2700 ft. min.$^{-1}$ will be desirable (Fig. 311) and will be attained at 3600 r.p.m. with a stroke of 4.5 in.

$$\text{Total piston area} = 300 \div 4.5 = 66.7 \text{ in.}^2$$

The piston diameter will be 3.26 in. for an eight-cylinder engine.

Example.—Determine the displacement per minute for a six-cylinder two-stroke-cycle Diesel engine to develop 165 hp. The engine is blower scavenged.

The heating value of fuel oil in a correct mixture is about 101 B.t.u. ft.$^{-3}$ of air (Table VIII, Appendix). However, the maximum amount of fuel injected will be about 0.90 of the chemically correct amount.

The volumetric efficiency due to blower scavenging will be about 0.85, the effect of the ports, some heating, the clearance space, and some dilution even with large blower capacity being considered.

A compression ratio of 16:1 would be used. The Otto-engine process would result in an ideal efficiency of about 0.48 with a mixture having 11 per cent excess air, whereas the constant-pressure Diesel process would result in an ideal efficiency of about 0.35. On the assumption that one-third of the fuel is burned at constant volume and two-thirds burned at constant pressure, the ideal efficiency would be about 0.40.

Assuming a relative efficiency of 0.70 and a mechanical efficiency of 0.70, and substituting in Eq. (1), results in

$$\text{Disp. min.}^{-1} = \frac{165 \times 42.4}{0.5 \times 101 \times 0.80 \times 0.85 \times 0.40 \times 0.70 \times 0.70} =$$

$$1040 \text{ ft.}^3 \text{ min.}^{-1}$$

A six-cylinder engine having a bore of 4.25 in. and a stroke of 5 in. would run at a speed of

$$\frac{1040 \times 1728}{0.7854 \times \overline{4.25}^2 \times 5 \times 6 \times 2} = 2110 \text{ r.p.m.}$$

Equation (1) may be written

$$\text{B. hp.} = k' P_m \text{ disp.}_{min.} \tag{2}$$

where P_m = brake m.e.p.

The displacement per minute depends on the number of cylinders, N, the piston area A_p, and the mean piston speed v_p.

Thus, Eq. (2) becomes B. hp. = $k'P_m N A_p v_p$. (3)

A knowledge of the m.e.p. attained in current engines may be used as a basis for the determination of the size of a new engine. Brake m.e.p. ranges from below 100 lb. in.$^{-2}$ to well above this value for normally aspirated engines. High-output aircraft engines have m.e.p. values of 150 to 200 lb. in.$^{-2}$, and experimental values well above this have been obtained (Chap. XVI).

The designer is limited to a piston speed that current practice indicates as being the maximum desirable. However, considerable latitude is permissible in the choice of number and size of cylinders, since Eq. (3) indicates the same output with various combinations of N and A_p. Consequently, the effect of cylinder size on the various factors that influence design and performance should be well understood. It is best approached by the application of the principle of similitude.[1]

THE PRINCIPLE OF SIMILITUDE

General.—Similar engines have their respective parts made of the same materials and have linear dimensions that are proportional. The ratio of the lengths, widths, and thicknesses of similar parts is the same, regardless of the part; consequently similar engines are scale reproductions of each other. Thus, the following relationships exist:

1. The stroke-bore ratios are equal.
2. The compression ratios are equal.
3. The mean flow velocities through the valve ports are equal for equal mean piston speeds. Thus, comparisons will be made assuming the *same piston speed* in all engines.
4. The volumetric efficiencies are equal (page 458).
5. The specific gas pressures are equal.

[1] H. F. P. Purday, "Diesel Engine Design," 4th ed., D. Van Nostrand Company, Inc., New York, 1937.
V. L. Maleev, "Internal-combustion Engines," 1st ed., McGraw-Hill Book Company, Inc., New York, 1933.
E. S. Taylor, Design Limitations of Aircraft Engines, *Aero Digest*, **26**, 26 (1935).
C. F. Taylor, The Next Five Years in Spark-ignition Aviation Engines, *Jour. Aero. Sc.*, **4**, 113 (1937).

The following relationships in terms of the cylinder bore d result from the foregoing:
1. R.p.m. $\propto d^{-1}$.
2. Piston disp. $\propto d^3$.
3. Wt. of any part $\propto d^3$.
4. Wt. $\div$ piston disp. $\propto d^3 \div d^3 = $ const.
5. Bearing areas $\propto d^2$.

Forces, Deflections, and Stresses.—The total gas pressure on a piston is proportional to piston area. Hence, the specific bearing load due to gas pressure is

$$(\text{Unit bearing pressure})_{g.pr.} \propto \frac{d^2}{d^2} = \text{const.}$$

The inertia forces of the reciprocating parts are proportional to the mass of the reciprocating parts, the square of the engine speed, and the length of the crank. Hence, the specific bearing load due to inertia forces is

$$(\text{Unit bearing pressure})_{i.f.} \propto \frac{d^3 d^{-2} d}{d^2} = \text{const.}$$

Unit bearing pressures due to centrifugal forces are found in the same manner as for inertia forces. Hence, the specific bearing load is the same in similar engines having the same piston speeds.

The deflection of a structure is proportional to the load and to the cube of the length and inversely proportional to the moment of inertia. The moment of inertia of a section is proportional to the fourth power of the dimension.

Hence, $\quad\quad$ Deflection $\propto \dfrac{d^2 d^3}{d^4} \propto d$.

The stress is the ratio of the bending moment to the section modulus.

Thus, $\quad\quad$ Stress $\propto \dfrac{d^2 d}{d^3} = $ const.

Thus, deflections are proportional to the bore, and stresses are the same in similar engines having the same piston speed.

Combustion and Detonation.—The rate at which the charge burns in a spark-ignition engine depends on the turbulence of the combustible mixture, and this depends principally on the flow velocity through the intake valve. Similar engines should have the same turbulence with the same piston speed, which indicates the same rate of combustion. The time required for combustion is proportional to the length of flame travel. Then, the crank angle required for combustion is

$$\text{Crank angle} \propto \text{combustion time} \frac{\text{revolutions}}{\text{time}}$$
$$\propto dd^{-1} = \text{const.}$$

Thus, the crank angle required for combustion and the losses due to combustion time should be the same in similar engines at the same piston speed.

Detonation of a combustible mixture in an engine depends on mixture temperature, the rate of burning, the ignition lag of the unburned fraction of the mixture, and the temperature of the surrounding surfaces. The temperature and pressure of the entering charge are assumed to be the same in similar engines. The dilution with clearance gases should be the same. The larger cylinder has longer flame travel and requires more time for combustion which provides more time for the unburned fraction to burst into flame. The piston head runs appreciably hotter and the combustion-chamber walls slightly higher in temperature in the larger engine. Consequently, the permissible compression ratios vary inversely with the cylinder bore in similar engines.

Heat Transfer and Temperature Differences.—The conditions of pressure, temperature, and turbulence in the cylinder are the same in similar engines having the same piston speed. This indicates the same rate of heat transfer from the gases to the walls. The temperature difference between the gases and the walls should be the same. The time for each cycle is $\propto d$, and the cycles per unit of time are $\propto d^{-1}$. Hence, the time for heat transfer is independent of engine size and

$$Q = \text{rate} \times \text{area} \times \Delta T \times \text{time} \propto d^2.$$

The energy supplied the engine in a given time is proportional to engine displacement and engine speed.

Thus, $$Q_{supplied} \propto d^3 d^{-1} \propto d^2$$

which indicates the same percentage loss of energy by heat transfer in similar engines with the same piston speed.

Heat transfer through metals is proportional to temperature difference and area of cross section, and inversely proportional to the length of path. The heat transfer per unit area of cylinder and combustion-chamber walls is the same in similar engines. Consequently, the temperature difference through the walls is proportional to the wall thickness.

Thus, $$\Delta T \propto d$$

which indicates higher inner-wall-surface temperatures for larger engines.

The piston-head thickness is proportional to d in similar engines. The heat transfer into the piston head varies with d^2. The area of cross section through which heat is transferred toward the cylinder walls is proportional to d^2 for any similar position. However, the distance from the piston center to the cylinder walls is proportional to d, and again

$$\Delta T \propto d$$

which indicates higher piston temperatures for the larger engines.

Friction Losses.—Similar bearing operating conditions occur with the same $\mu n/p$ factor. The specific bearing load p is constant in similar engines with the same piston speed. Consequently, the product of viscosity μ and engine speed n is constant.

Thus, $\mu n = $ const., or $\mu d^{-1} = $ const.
or $\mu \propto d$

which indicates that the absolute viscosity of the lubricant should be proportional to d, if the bearing temperatures are the same.

The work of friction in journal bearings is obtained from Eq. (15), page 421. Geometrically similar engines operating at the same piston speed have the same diameter-clearance ratio and the same velocity of the bearing surfaces.

On assuming the same $\mu n/p$ factors,

$$\text{Work of friction} \propto d^2$$

since the bearing diameter and length are proportional to piston diameter.

Piston friction depends on oil viscosity, the shearing area, oil-film thickness, and the piston speed [Eq. (1), page 398].

Thus, Piston friction $\propto dd^2d^{-1}$ const. $\propto d^2$
and, Total engine friction $\propto d^2$.

The centrifugal force throwing oil from the crankpins is proportional to the square of engine speed, n^2, and crank radius r.

Thus, Centrifugal force $\propto d^{-2}d \propto d^{-1}$

which indicates less force throwing oil from the crankpins of the larger engine. Also, the distance between the crank circle and the cylinder increases with d.

Power.—The ideal horsepower depends on the compression ratio and displacement per minute. For the same compression ratio and piston speed

$$\text{Ideal hp.} \propto d^2.$$

Cooling and combustion losses are the same percentage in similar engines.

Hence, I. hp. $\propto d^2$.

If $\mu \propto d$ and all the bearings are at the same temperature in similar engines,

Friction hp. $\propto d^2$
and B. hp. $\propto d^2$

which indicates the same mechanical efficiency for similar engines. However, the larger engine cylinder cannot use the same compression ratio, which gives the advantage to the smaller cylinder.

Weight.—The specific weight is

$$\frac{\text{Wt.}}{\text{B. hp.}} \propto \frac{d^3}{d^2} \propto d$$

which indicates an advantage for the small cylinder.

The weight per cubic inch displacement is

$$\frac{\text{Wt.}}{\text{Disp.}} \propto \frac{d^3}{d^3} = \text{const.}$$

Natural Frequency of Vibration.—The natural frequency of vibration, f, of any engine part is given by the expression

$$f = \frac{1}{2\pi}\sqrt{\frac{gs}{M}} \qquad [\text{Eq. (4), p. 511}]$$

where s = load divided by the deflection,
M = equivalent wt. of the structure.

Thus, $\qquad f \propto \sqrt{\dfrac{d^2/d}{d^3}} \propto d^{-1},$

and since $\qquad$ r.p.m. $\propto d^{-1},$

an increase in cylinder size of similar engines lowers the engine speed at which resonance can occur.

DESIGN OF PRINCIPAL PARTS[1]

Cylinders.—Cylinders require the use of a hard and fine-grained cast iron. It should offer considerable resistance to wear and corrosion. The usual material is a nickel-chromium cast iron, with molybdenum in some cases. The tensile strength varies from 33,000 to 50,000 lb. in.$^{-2}$ depending on the type of service. The Brinell hardness number ranges from about 180 to 270. The stronger and harder of these materials are used in the heavy-duty engines.

Cylinders, cylinder blocks, and liners should be designed to reduce distortion to a minimum. Distortion may be caused by bolting on the cylinder head or any other part. It may also be caused by gas-pressure and temperature stress. Distortion of the cylinder bore is affected by the design of the remainder of the casting of which it is a part. The casting should be symmetrical, but this is impossible with the L-head engine. The anchors for the cylinder-head studs or for studs for any part bolted to the cylinder block should be designed to prevent a bending action being exerted on the cylinder barrel.

Wall Thickness.—The cylinder walls are stressed by the gas pressure and the force due to side thrust of the piston in the usual engine. The gas pressure stresses the cylinder barrel both longitudinally and circumferentially, part of the longitudinal stress being carried by the integral full-length water jacket in the automobile engine.

[1] See footnote, p. 525.

The two stresses due to gas pressure are at right angles to each other, which results in a lower net stress in each direction as indicated by Poisson's ratio.[1]

Thus, $\quad\quad\quad\quad$ Net $s_l = s_l - \dfrac{s_c}{m}$ $\quad\quad\quad\quad$ (4)

and $\quad\quad\quad\quad$ Net $s_c = s_c - \dfrac{s_l}{m}$ $\quad\quad\quad\quad$ (5)

where $\quad s =$ stress, lb. in.$^{-2}$,
$\quad\quad\quad m =$ reciprocal of Poisson's ratio, about 4,
subscripts l and $c =$ longitudinally and circumferentially, resp.

Also, $\quad\quad\quad\quad s = \dfrac{\text{force}}{\text{area}}$ $\quad\quad\quad\quad$ (6)

The thickness of the cylinder wall may be found by using the formula for a thin cylinder, the longitudinal stress being neglected.

Thus, $\quad\quad\quad\quad t = \dfrac{pd}{2s_c} + k$ $\quad\quad\quad\quad$ (7)

where $p =$ max. internal pr., lb. in.$^{-2}$,
$\quad\quad d =$ cylinder bore, in.,
$\quad\quad s_c = 5000$ lb. in.$^{-2}$
$\quad\quad k =$ added thickness for reboring, in.
The value of k depends on the diameter of the cylinder (Table 36).

TABLE 36.—ADDED THICKNESS FOR REBORING

Dia. of cylinder, in	3	4	6	8	10	12	14	16	18	20
k, in	$\frac{1}{16}$	$\frac{3}{32}$	$\frac{5}{32}$	$\frac{1}{4}$	$\frac{5}{16}$	$\frac{3}{8}$	$\frac{7}{16}$	$\frac{1}{2}$	$\frac{1}{2}$	$\frac{1}{2}$

Example.—Determine the thickness of cylinder wall for a 12-in. gas-engine cylinder, with a maximum gas pressure of 450 lb. in.$^{-2}$ Also, determine the stresses.

Substituting in Eq. (7) results in $\quad t = \dfrac{450 \times 12}{2 \times 5000} + 0.38 = 0.92$ in.

[1] L. S. Marks, "Mechanical Engineers' Handbook," 3d ed., p. 412, McGraw-Hill Book Company, Inc., New York, 1930.

ENGINE DESIGN

The apparent longitudinal stress, Eq. (6), is

$$s_l = \frac{\text{force}}{\text{area}} = \frac{0.7854 d_i^2 p}{0.7854(d_o^2 - d_i^2)} = \frac{d_i^2 p}{d_o^2 - d_i^2}$$

where d_i = inside dia. of the cyl., and d_o = outside dia. of the cyl.

Thus, $$s_l = \frac{12^2 \times 450}{13.84^2 - 12^2} = 1363 \text{ lb. in.}^{-2}$$

The apparent circumferential stress, Eq. (6), is

$$s_c = \frac{\text{force}}{\text{area}} = \frac{12 \times 450}{2 \times 0.92} = 2935 \text{ lb. in.}^{-2}$$

From Eq. (4), Net $s_l = 1363 - \dfrac{2935}{4} = 629$ lb. in.$^{-2}$

From Eq. (5), Net $s_c = 2935 - \dfrac{1363}{4} = 2594$ lb. in.$^{-2}$

The foregoing example illustrates the difference between the two stresses and indicates that cylinders need be examined only for circumferential stress. Air-cooled aircraft-engine cylinders are made of forged steel having a tensile strength of 100,000 lb. in.$^{-2}$, and are subjected to higher stresses than are cast-iron cylinders. The circumferential cooling fins (Figs. 371 and 372) provide additional strength against the circumferential stress.

The gas pressure stresses the material near the inner surface more than the material near the outer surface of the cylinder. This inequality of stress increases appreciably with an increase in wall thickness. Thus, wall thickness should be reduced to the minimum desirable, for this reason as well as for cooling purposes (page 433).

Piston Side Thrust.—The angularity of the connecting rod produces a thrust against the cylinder wall that varies throughout the cycle. The thrust can be determined and plotted for the cycle as outlined in Chap. XVII. The section modulus of the cylinder should be large enough to cause the cylinder to resist the side thrust without appreciable deflection. If lower deflections are desired, it is considered preferable to increase the cylinder stiffness by suitable supports or webs rather than to increase cylinder thickness.

The side thrust p_s is usually distributed over the bearing length of the piston, but for analysis it may be considered as

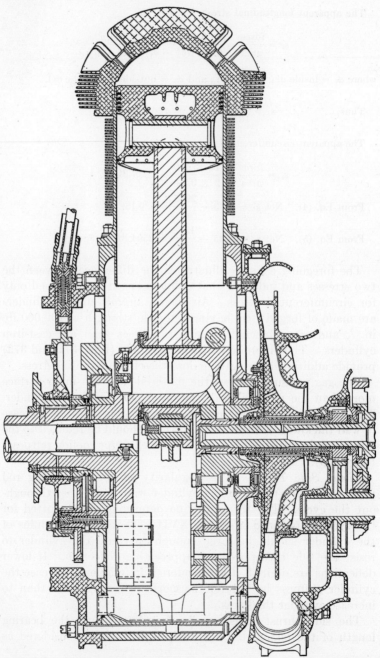

Fig. 371.—Wright "Cyclone" 9-cyl. 1-row radial engine. Model GR1820-G102A, bore 6.125 in., stroke 6-875 in., comp. ratio 6.3:1., supercharger ratio 7.0 to 1, propeller gear ratio 16:11, max. bp. 1100 at 2350 r.p.m., 90 Octane fuel required.

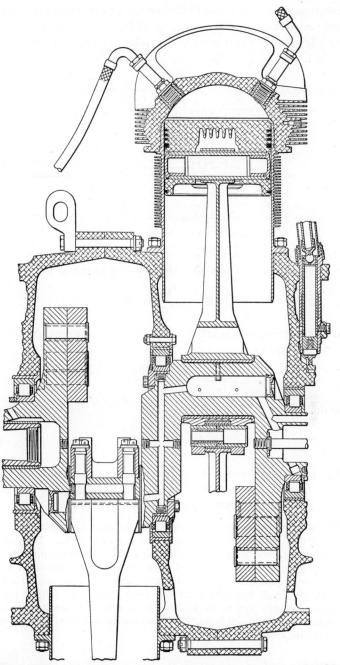

Fig. 372.—Pratt and Whitney 14-cyl. twin-row engine. Model R-1830-5C3-G, bore (5½ in., stroke 5½-in.) comp. ratio 6.7 to 1, blower ratio 7.15 to 1.

concentrated at the point indicated by a perpendicular from the wrist pin to the cylinder wall (Fig. 373). The cylinder sleeve or barrel is usually supported at two places indicated by the two reactions R_l and R_r. Thus, the maximum bending moment will

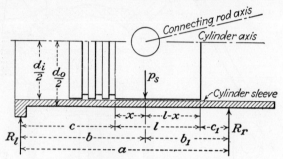

Fig. 373.—Diagram for side-thrust analysis.

occur at the point of application of p_s and will be equal to $R_l b$. But $R_l = p_s b_1/a$. Therefore, the bending moment M is

$$M = p_s \frac{bb_1}{a} \text{ in.-lb.} \tag{8}$$

and on assuming that s_b is the allowable stress the section modulus Z required will be

$$Z = \frac{M}{s_b} = \frac{p_s b b_1}{s_b a} \text{ in.}^3 \tag{9}$$

The deflection of the cylinder will be

$$f = \frac{p_s b^2 b_1^2}{3EIa} \text{ in.} \tag{10}$$

where E = modulus of elasticity which varies from 12×10^6 lb. in.$^{-2}$ for cast iron to 30×10^6 lb. in.$^{-2}$ for steel,
I = the moment of inertia of the cyl., in.^{4}

Example.—A gas-engine cylinder has a cylinder liner with an inside diameter d_i of 12 in. and an outside diameter d_o of 14 in. The distance a between the liner supports (Fig. 373) is 25 in. The wrist pin is located 11 in. from the left point of support when the piston is on top dead center. The connecting-rod-crank ratio is 4.5:1. The stroke of the engine is 15 in.

When the crank has moved 45 deg. from outer dead center, the gas pressure may be as high as 300 lb. in.$^{-2}$, while at dead center the maximum pressure will be about 450 lb. in.$^{-2}$ Determine the deflection of the cylinder liner for both cases.

With 45 deg. of crank travel, the piston will have moved 17.4 per cent of the stroke (Table 30, page 482) or 2.61 in., so that $b = 11 + 2.61 = 13.61$ in.

Thus, $b_1 = a - b = 25 - 13.61 = 11.39$ in.

The side thrust is found from Eq. (7) in Chap. XVII.

Thus,
$$p_s = \frac{p_a r \sin \theta}{\sqrt{l^2 - r^2 \sin^2 \theta}}$$
$$= \frac{300 \times 0.7854 \times 12^2 \times 7.5 \times 0.707}{\sqrt{(4.5 \times 7.5)^2 - (7.5 \times 0.707)^2}} = 5398 \text{ lb.}$$

The section modulus for a cylinder with diameters d_i and d_o is

$$Z = \frac{\pi}{32} \frac{d_o^4 - d_i^4}{d_o} = \frac{\pi}{32}\left(\frac{14^4 - 12^4}{14}\right) = 124 \text{ in.}^3$$

Then, $s_b = \dfrac{p_s b b_1}{Za} = \dfrac{5398 \times 13.61 \times 11.39}{124 \times 25} = 270$ lb. in.$^{-2}$

This stress is low and apparently the cylinder is amply strong, but the maximum deflection should be determined. The moment of inertia of the cross section is

$$I = \frac{\pi}{64}(d_o^4 - d_i^4) = \frac{\pi}{64}(14^4 - 12^4) = 867.8 \text{ in.}^4$$

With a value of $E = 12 \times 10^6$ for cast iron, the deflection is

$$f = \frac{5398 \times 13.61^2 \times 11.39^2}{3 \times 12 \times 10^6 \times 867.9 \times 25} = 0.00017 \text{ in.}$$

The deflection of the cylinder sleeve at dead center will be zero, since the connecting rod does not have any angularity with the axis in this position.

The stress on the water-jacket wall depends entirely on the arrangement of cylinder and jacket walls. When the cylinder and jacket are cast in one piece, the jacket takes part of the axial thrust due to the explosion pressure. The jacket takes all the axial thrust, when cylinder sleeves are used, and should be designed accordingly. In small engines, sheet-iron or copper jackets are sometimes used, and in such cases the cylinder wall is subjected to all loads due to the explosion. In general, the thickness of the jacket wall should be from one-third to three-fourths of the thickness of the cylinder wall, the larger ratio being for the smaller cylinder. The water space between the outer cylinder wall and inner jacket wall should be about ⅜ in. for a 3-in. cylinder to about 3 in. for a 30-in. cylinder.

The cylinder acts as a cantilever beam in the case of the radial, air-cooled engine, and the deflection of the cylinder becomes

$$f = \frac{p_s b_1^3}{3EI} \tag{11}$$

where b_1 = distance from the wrist pin to the cyl. flange.

Example.—Determine the deflection of a radial aircraft-engine cylinder if a maximum pressure of 350 lb. in.$^{-2}$ occurs when the crank is 45 deg. past top dead center. The engine has a 5-in. bore and a 7-in. stroke, with connecting-rod-crank ratio of 4:1. At top dead center the wrist pin is 7 in. above the cylinder flange. The cylinder walls are ⅛ in. thick.

With 45 deg. of crank travel, the piston will have moved 17.8 per cent of the stroke (Table 30, page 482) or 1.25 in. Then $b_1 = 7 - 1.25 = 5.75$ in.

The side thrust from Eq. (7) page 486, is

$$p_s = \frac{p_a r \sin \theta}{\sqrt{l^2 - r^2 \sin^2 \theta}}$$

$$= \frac{350 \times 0.7854 \times 5^2 \times 3.5 \times 0.707}{\sqrt{(4 \times 3.5)^2 - (3.5 \times 0.707)^2}} = 1234 \text{ lb.}$$

The moment of inertia for a cylinder is

$$I = \frac{\pi}{64}(d_o^4 - d_i^4) = \frac{\pi}{64}(5.25^4 - 5^4) = 6.61 \text{ in.}^4$$

The deflection of a steel cylinder at the wrist-pin position is

$$f = \frac{p_s b_1^3}{3EI} = \frac{1234 \times 5.75^3}{3 \times 30 \times 10^6 \times 6.61} = 0.00032 \text{ in.}$$

Cylinder Flanges and Studs.—Cylinders are either integral with the upper half of the crankcase or are attached to the crankcase by means of a cylinder flange, studs, and nuts. The cylinder flange is integral with the cylinder and should be made thicker than the cylinder wall with a generous fillet at the junction of the flange and the cylinder wall.

The studs should be located as near the cylinder as is possible, room being allowed for nuts to turn. The studs are initially under stress due to drawing down the nuts on the cylinder flange. In addition, the force of the explosion pressure is added to this stress, so that the total root area of the studs should be sufficiently large to withstand the total stress.

The studs or bolts for fastening the cylinder to the crankcase should be made of a nickel steel having a yield point of 90,000 to 135,000 lb. in.$^{-2}$ The diameter of the studs may be found by equating the load due to the combustion pressure to the area of all the studs at the root of the thread multiplied by the allowable fiber stress.

Thus,
$$\frac{\pi d_p^2}{4} \times p = \left(\frac{n\pi d_b^2}{4}\right) s_t \qquad (12)$$

and
$$d_b = d_p \sqrt{\frac{p}{n s_t}} \qquad (13)$$

where d_b = dia. of bolt at root of thread,
d_p = cylinder bore,
p = combustion pr., lb. in.$^{-2}$
n = no. of studs,
s_t = allowable fiber stress, 5000 to 10,000 lb. in.$^{-2}$

The number of bolts to use is in the range indicated by

$$n = 0.25 d_p + 4 \text{ to } 0.50 d_p + 4. \qquad (14)$$

A larger number of bolts are used for aircraft-engine cylinders where it is desirable to distribute the stress more uniformly. Small cylinders cast in a block require less studs, the usual number being two more than twice the number of cylinders.

The studs may be relieved between the threaded ends to reduce the weight, since in this section the stud need be no larger than the root diameter of the threads.

The depth to which the studs are set into the crankcase depends principally upon the crankcase material. The shearing stress on the metal around the studs should be determined.

Cylinder Heads.—Cylinder heads are made of alloy cast iron containing at least two of the constituents nickel, chromium, and molybdenum. This alloy is resistant to pressure and temperature and to the action of valves seating directly in the head. The tensile strength varies from 30,000 to 50,000 lb. in.$^{-2}$ while the Brinell hardness number varies from below 200 to about 240.

Cylinder-head walls range from a thickness of ¼ in. for small engines (Fig. 374) to proportionately larger sections for large engines. However, the thickness of the section subjected to combustion pressures depends on the construction of the head.

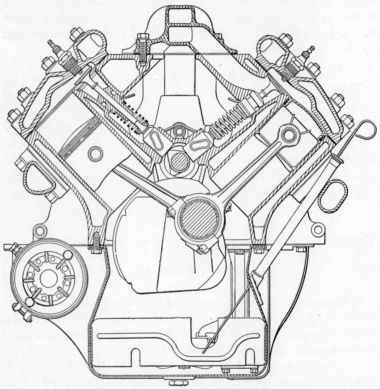

Fig. 374.—Ford V-8, $3\frac{1}{16}$- by $3\frac{3}{4}$-in. engine (1939).

In cases where the cylinder head approximates a flat circular plate, the thickness can be found from

$$t = d\sqrt{\frac{cp}{s}} \tag{15}$$

where t = thickness, in.,
d = dia. of cyl., in.,
p = pr., lb. in.$^{-2}$,
c = a const., 0.1,
s = allowable stress, 5000 to 8000 lb. in.$^{-2}$

Example.—Determine the thickness of a plain cylinder head for a 12-in. cylinder. The gas pressure is 450 lb. in.$^{-2}$

$$t = 12\sqrt{\frac{0.1 \times 450}{5000}} = 0.9 \text{ in.}$$

This is about the same thickness as that of the cylinder walls computed for the same cylinder bore in the example on page 536. In most cases the head and its water jacket are cast integral. The valve-in-head engine has the ports and valve chambers cast integral, which strengthens the head considerably. The metal may be made thinner in such cases and in any case where integral ribs are cast with the metal subjected to the combustion pressure.

Aluminum alloy is used for cylinder heads, owing to high heat-transfer characteristics. The tensile strength of the usual alloys is about 30,000 lb. in.$^{-2}$ Consequently, cylinder walls are usually 50 to 100 per cent thicker than when iron alloys are used. Aluminum heads are used on high-output automobile engines and exclusively on high-output radial aircraft engines (Figs. 371 and 372). In the latter case the heads are screwed and shrunk onto the cylinder barrels.

The cylinder-head bolts or studs are treated in the same manner as the cylinder bolts or studs, with the exception that the area of the head which is subjected to the gas pressure may be considerably larger than the piston area. A sufficient number of bolts should be used to maintain a tight joint between the cylinder and cylinder head.

Crankcases and Engine Frames.—The crankcase should be an extremely rigid structure supporting the crankshaft bearings in such a manner that none of the forces imposed on the crankshaft or crankcase causes misalignment of any of the bearings. It may be more or less boxlike in construction (Fig. 375), transverse webs forming supports for the upper half of the bearings, these transverse webs having other webs that effectively support the bearings. It is rather difficult to analyze the complicated structure that forms the crankcase of the average multicylinder automotive type of engine, its design being based principally upon practical experience and a general application of the principle that a narrow but high section will deflect much less in the vertical direction than a wider section having the same amount of material.

Example.—The deflection of a beam varies inversely with the moment of inertia, which for a rectangular section is equal to $bh^3/12$, where b and h are

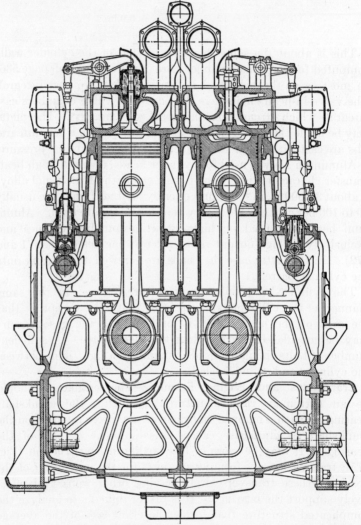

Fig. 375.—Section of a double-bank Diesel engine. (*Sulzer Technical Review*, No. 3, 1938.)

width and height of the section. Determine the relative deflection of two beams having the same rectangular cross-sectional area, one beam having twice the width and half the height of the other.

ENGINE DESIGN

Then, $\qquad b_2 = 2b_1, \qquad h_1 = 2h_2,$
so that $\qquad b_1 h_1 = b_2 h_2.$

Then the ratio of the moments of inertia will be

$$b_1 h_1^3 : b_2 h_2^3 = b_1 h_1^3 : \frac{b_1 h_1^3}{4} = 4:1.$$

Thus, the deflection in the first case will be one-fourth the deflection of the second case, although the weight of material is the same in both cases.

The casting of the cylinders and upper half of the crankcase in one block provides the desirable stiffness in a vertical plane. Full-length water jackets increase this stiffness.

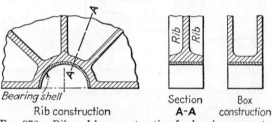

FIG. 376.—Rib and box construction for bearing support.

Stiffness in a horizontal plane may be obtained by the use of comparatively thin longitudinal ribs in the same plane as the crankshaft axis, and also at the junction of the cylinders and the crankcase. These ribs should be wider near the middle of the crankcase than at the ends. The bearing should be amply webbed or of boxlike construction (Fig. 376). Also, all junctions of metal parts should have large radius fillets.

The stiffness of a crankcase can be determined by supporting it in various positions and applying loads at various points. This procedure discloses the weakness of the structure and indicates the locations for the addition of ribs. Proper distribution of the metal should result in the highest stiffness-weight ratio and the highest natural frequency of vibration.

In medium-size and large single-crank engines, a frame construction may be used to maintain cylinder and bearing alignment. In one type (Fig. 377) the neutral axis of the two side beams of the frame is in the same horizontal plane as the center of the shaft. The stress on the two beams is

$$s_t = \frac{P}{2bh}, \qquad (16)$$

where P = total load due to combustion pr. or inertia forces, whichever is greater,
b = section width, in.,
h = section height, in.,
s_t = stress, which should be below 2000 lb. in.$^{-2}$

The side members are subjected to tension and bending when the neutral axis of the members is below the center line of the engine (Fig. 378). The stress in tension due to gas pressure or inertia forces is the same as in the previous illustration.

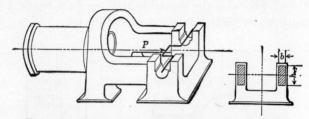

Fig. 377.—Engine frame with side members in tension.

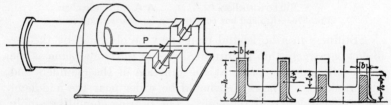

Fig. 378.—Engine frame with side members in tension and bending.

The stress due to bending is

$$s_b = \frac{\text{bending moment}}{\text{section modulus}} = \frac{Pl}{bh^2/3}. \qquad (17)$$

The total stress will be

$$s_{total} = s_t + s_b. \qquad (18)$$

The two side members may be cast integral with a horizontal member (Fig. 379), in which case the distance from the engine center line to the center of gravity of the structure must be determined.

Example.—Determine the relations for total stress in the engine-frame section in Fig. 379.

The center of gravity and the arm l is found by taking moments about an axis at a height h above the base line and adding the distance k.

Thus, $$l = x + k = \frac{2bh \times \frac{h}{2} + fc \times g}{2bh + fc} + k. \qquad (19)$$

The moment of inertia of the section is

$$I = \frac{2bh^3}{12} + 2bh \times e_1^2 + \frac{cf^3}{12} + cf \times e_2^2.$$

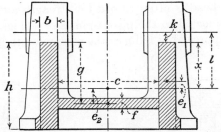

FIG. 379.—Engine frame with horizontal cross member.

The section modulus is I/x. The bending stress in the top fiber is

$$s_b = \frac{Pl}{I/x}$$

The direct tension is $$s_t = \frac{P}{2bh + fc}.$$

The total stress is $$s_{total} = s_b + s_t.$$

It will not be necessary to find the stress in the bottom fibers as these are in compression due to bending, and in tension due to the direct load P unless the inertia force on the exhaust stroke is the major load. The net stress will be the difference of these two and will be less than the total stress on the top fibers.

Pistons. *Gas-pressure Stresses.*—Pistons for single-acting engines are of the trunk type and act as crossheads to guide the upper end of the connecting rod, through which the forces due to gas pressure and inertia are transmitted to the crankshaft. The piston head must withstand the gas-pressure and temperature stresses due to direct contact with the combustion products. The piston head may be considered a flat circular plate with fixed edges, in which case Eq. (15) may be used to determine the head thickness. However, heat-flow considerations usually require thicker sections (page 439) which should be used unless the piston head is liquid-cooled.

Example.—The unsupported diameter of a piston head is 2.5 in. The thickness is 0.2 in. Estimate the stress in the material for a maximum pressure of 600 lb. in.$^{-2}$

Substituting in Eq. (15) results in

$$s = \frac{0.1pd^2}{t^2} = 9375 \text{ lb. in.}^{-2}$$

The piston head is often ribbed for strength and cooling and is directly connected to the piston-pin bosses (Figs. 380 to 383), all of which appreciably reduces the stress due to gas pressure.

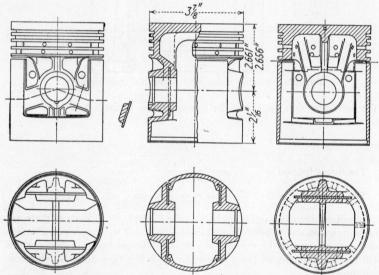

Fig. 380.—Autothermic piston. Low-carbon steel strips in combination with aluminum-alloy control skirt expansion. (*Nelson-Bohnalite, Bohn Aluminum and Brass Corp.*)

Temperature Stresses.—Increasing the piston-head temperature from a uniform temperature of T_2 to a uniform temperature of T_1 increases the diameter.

Thus, $$\Delta d = \alpha d(T_1 - T_2)$$

where α = coefficient of expansion.

Under actual conditions the outside fiber is at T_2 and the center of the head is at T_1, for which condition the deformation will be approximately one-half of the foregoing.

Then, $$E = \frac{\text{unit stress}}{\text{unit deformation}} = \frac{2s}{\alpha(T_1 - T_2)} \qquad (20)$$

when E = modulus of elasticity.

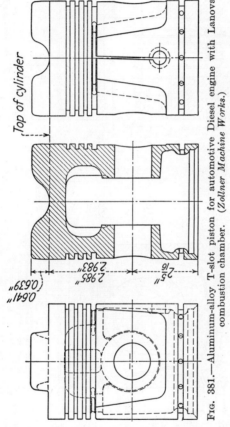

Fig. 381.—Aluminum-alloy T-slot piston for automotive Diesel engine with Lanova combustion chamber. (*Zollner Machine Works.*)

Example.—Determine the stress due to temperature difference of 300°F in the piston head. $\alpha = 5.9 \times 10^{-6}$ and $E = 12 \times 10^6$ for cast iron.

From Eq. (20), $s = \dfrac{12 \times 10^6 \times 5.9 \times 10^{-6} \times 300}{2} = 10{,}620$ lb. in.$^{-2}$

This value is too high since the ring section strengthens the piston head and considerably reduces this stress.

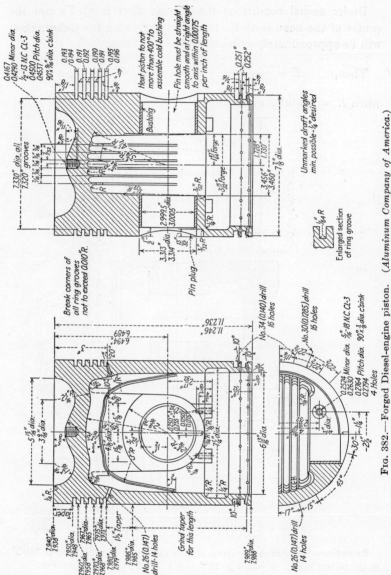

Fig. 382.—Forged Diesel-engine piston. (*Aluminum Company of America.*)

The foregoing example indicates that the temperature stress is higher than the stress due to gas pressure, the temperature stress being about 3500 lb. in.$^{-2}$ per 100°F. temperature difference.

Aluminum alloys have a modulus of elasticity of about 5×10^6 to 6×10^6 and a coefficient of expansion of about 12×10^{-6} to 12.5×10^{-6} which indicates practically the same temperature stress as for cast iron. However, the better conductivity of aluminum alloys compared with iron alloys results in lower temperature differences and lower temperature stresses.

The Ring Section.—The ring section contains the ring grooves with the ring "lands" in between. The width of the top ring land is large in some cases to protect the top ring from the high-temperature condition existing at the top of the piston. The width of the other ring lands varies from 0.75 ring width to ring width. Wide lands increase the length of the piston since no side thrust is carried by the lands which are relieved sufficiently to prevent touching the cylinder walls.

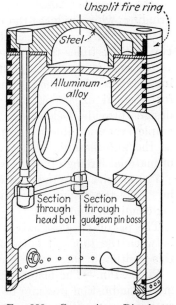

Fig. 383.—Composite Diesel-engine piston. (*Junkers* "*Jumo.*")

The Junkers "Jumo" two-stroke-cycle opposed-piston Diesel aircraft-engine (Fig. 266) uses a steel cylinder head mechanically attached to the aluminum-alloy piston (Fig. 383). A wide unsplit angle "fire-ring," expanded by temperature and pressure protects the lands and rings.

The depth of the ring grooves should be larger than the ring depth; it is given approximately by the following relation:

$$\text{Groove depth} = 0.066 d^{0.775} \tag{21}$$

where d = the cyl. dia., in.

The piston material behind the ring section (Figs. 380 to 383) transfers the heat from the piston head to the ring section and skirt. This section may be tapered since heat is transferred to the walls at the surfaces of the piston and rings that touch the wall or lubricant film.

Piston Skirt.—The piston skirt varies in length from less than piston diameter in aircraft and automotive engines to more than piston diameter in most large Diesel engines. The skirt adjacent to the piston-pin bosses is relieved either in design or by cam grinding, which reduces the diameter along the piston-pin axis. The total skirt area subjected to piston thrust varies from about two to three times the piston-head area.

The skirt is usually tapered, and in some cases the cam grinding is tapered from the bottom to the top of the skirt to provide for unequal expansion due to temperature difference. Some designs provide T-slots which permit less piston clearance and ample expansion due to temperature rise. Other designs, such as the autothermic, control the expansion with combinations of material having different coefficients of expansion.

Piston-skirt clearance varies from about 0.0005 to 0.001 in. in.$^{-1}$ of piston diameter, depending on size, type, material, operating condition, and finish.

The piston-pin-boss axis is located about in the mid-position between the top and bottom of the piston skirt, being above this position in some cases but never below.

Piston Pin.—The piston pin should be designed for the maximum combustion pressure or inertia force of the piston, whichever is larger. The piston pin is usually hollow to reduce the weight and is often tapered on the inside, the smallest inside diameter being at the center of the pin.

A nickel alloy containing chromium, molybdenum, or vanadium or several of these constituents and having a tensile strength of from 100,000 to 130,000 lb. in.$^{-2}$ is used for piston pins.

The piston-pin-bearing area should be about equally divided between the bearing in the connecting rod and in the piston. Thus, the length of the pin in the connecting-rod bearing will be about 0.45 of the piston diameter, allowing for end clearance of the pin, etc. The outside diameter of the pin varies from a value equal to the length in the connecting-rod bearing to a diameter one-third larger. Then, the ratio of the piston area to the projected bearing area in the connecting rod will be

$$\frac{\text{Piston area}}{\text{Bearing area (con. rod)}} = \frac{0.7854 d^2}{(0.45d)^2 \text{ to } 1.33\,(0.45d)^2} = 3.9 \text{ to } 2.9$$

which indicates that the maximum piston-pin-bearing load will be three to four times the maximum combustion pressure or piston-inertia force.

The piston pin is assumed to be uniformly loaded for the distance l (Fig. 384), with supports at the centers of the bosses at both ends, making l' the length of beam between supports.

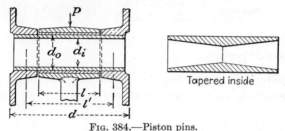

Fig. 384.—Piston pins.

Since l has already been determined, l' may be found from the relation

$$l' = l + \frac{d-l}{2} = \frac{l+d}{2}. \tag{22}$$

Since the load $p_{max} \times$ area $= P$ is evenly distributed over the length l, the maximum bending moment will be at the center of the pin, or

$$M = \frac{P}{2}\left(\frac{l'}{2} - \frac{l}{4}\right). \tag{23}$$

This should be equated to the resisting moment of the pin in bending, which is equal to the stress times the section modulus. For a solid pin the section modulus is $\pi d_o^3/32$, and for a hollow pin $\pi(d_o^4 - d_i^4)/32d$.

Example.—Determine the stress in a piston pin due to a maximum combustion pressure of 1000 lb. in.$^{-2}$ Cylinder diameter is 5 in.

Due to the high combustion pressure the diameter of the piston pin will be assumed to be equal to the length in the connecting-rod bearing.

Thus, $l = 0.45 \times 5 = 2.25$ in.

and $l' = \dfrac{2.25 + 5.0}{2} = 3.63$ in.

Substituting in Eq. (23) results in

$$M = \frac{1000 \times 0.7854 \times 5^2}{2}\left(\frac{3.63}{2} - \frac{2.25}{4}\right) = 12{,}300 \text{ in.-lb.,}$$

$$\text{Stress} = \frac{12{,}300}{\pi 2.25^3/32} = 11{,}000 \text{ lb. in.}^{-2},$$

for a solid pin.

For a hollow pin having an inside diameter of 1.5 in.,

$$\text{Stress} = \frac{32 \times 2.25 \times 12{,}300}{\pi(2.25^4 - 1.50^4)} = 13{,}700 \text{ lb. in.}^{-2}$$

The hollow pin increased the stress about 28 per cent but reduced the weight about 56 per cent.

The piston pin may "float" in both the connecting rod and piston bosses, being retained by spring rings or soft plugs at the

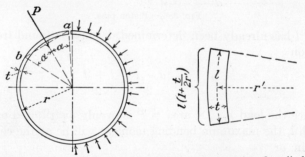

Fig. 385.—Concentric ring and ring section compressed to fit cylinder.

ends of the piston pin, or be clamped or fastened in either the upper end of the connecting rod or the piston.

Piston Rings.—Piston rings are made of cast iron or of cast-iron alloy because of the comparatively good wearing qualities of this material and also because the spring characteristics are retained at temperatures attained during engine operation. The tensile strength of the material varies from about 25,000 to 45,000 lb. in.$^{-2}$ depending on its composition. A nickel constituent is responsible for the higher strengths.

The width of piston rings is comparatively small, varying from $\frac{1}{8}$ in. in small sizes to about $\frac{1}{2}$ in. in sizes about 30 in. in diameter. Wide rings tend to reduce the wear, but narrow rings seat more rapidly. (See page 413.)

ENGINE DESIGN

The radial thickness of the ring depends on the wall pressure desired. The total pressure normal to the chord ab of an elementary section (Fig. 385) is equal to $pw \times$ length of ab.

Thus, $\qquad$ Total pr. $P = 2\,pwr \sin \alpha \qquad (24)$

where p = mean specific wall pr. for the given section,
$\qquad w$ = face width of the ring.

The bending moment of this force about b is

$$2\,pwr^2 \sin^2 \alpha.$$

The bending moment is

$$M = \frac{sI}{c} = EI\left(\frac{\text{unit strain}}{c}\right) \qquad (25)$$

where s = unit stress, lb. in.$^{-2}$,
$\qquad E$ = modulus of elasticity, lb. in.$^{-2}$,
$\qquad I$ = moment of inertia of ring section, $wt^3/12$, in.^{4},
$\qquad c$ = distance from outer fiber to neutral axis, $t/2$, in.

Then, substituting values for M and I in Eq. (25),

$$\frac{\text{Unit strain}}{c} = \frac{M}{EI} = \frac{24\,pr^2 \sin^2 \alpha}{Et^3}. \qquad (26)$$

The mean pressure $2wrp_m$ for one-half the ring, which is known as the "ring tension," is acting when $\alpha = 90$ deg. For this angle, Eq. (26) becomes

$$\frac{\text{Unit strain}}{c} = \frac{24 p_m r^2}{Et^3}. \qquad (27)$$

Assuming a constant unit strain, Eqs. (26) and (27) may be combined and

$$p = \frac{p_m}{\sin^2 \alpha}. \qquad (28)$$

This indicates an infinite pressure at the ends and a minimum pressure opposite the gap.

The unit stress at any section of the ring is

$$s = \frac{M}{I}c = \frac{12pr^2 \sin^2 \alpha}{t^2}. \qquad (29)$$

If a uniform wall pressure, which is desirable, is assumed, the stress in the ring should vary from zero at the gap to a maximum opposite the gap.

Example.—Determine the maximum stress in a 3- by ⅛-in. piston ring with a thickness of 0.135 in. The ring tension is 14.5 lb.

$$p_m = \frac{2wrp_m}{2wr} = \frac{14.5}{2 \times ⅛ \times 1.5} = 38.7 \text{ lb. in.}^{-2}$$

With $\alpha = 90°$, $\sin \alpha = 1$, and

$$s = \frac{12 \times 38.7 \times 1.5^2 \times 1^2}{(0.135)^2} = 57,300 \text{ lb. in.}^{-2}$$

This is higher than the tensile strength of the unworked material and indicates that a larger ring thickness is required to reduce the stress.

High stresses are used in piston-ring design to obtain high wall pressure which is desirable for high engine speed. Low-tension rings have low frequencies of vibration and owing either to cylinder-wall irregularities or to lateral piston movement tend to vibrate at some engine speed. Increasing the ring tension increases the natural frequency of vibration, and ring "flutter" will occur only at higher engine speeds.

F I G. 386.— Ring section with soft metal inserts. (*American Hammered Ring Co.*)

Increasing ring tension increases the tendency of the ring to break through the oil film at the lower speeds and particularly at the upper end of the stroke. Thus, the higher the ring tension, the greater the metal-to-metal friction, particularly in the lower range of speeds.

This has resulted in the development of composition rings with the soft bearing-metal inserts in the outer surface (Fig. 386) and of surface treatments such as oxide coating and tin-plating[1] to prevent "scuffing" during the running-in period of the ring.

Stress in Installing Ring on Piston.—An elementary length of piston ring (Fig. 385) having a neutral axis of length l and a radius of $r' = r - \dfrac{t}{2}$ will have an outside-surface length of $l[1 + (t/2r')]$. The outside-surface length in its free state will be $l[1 + (t/2\rho')]$ where ρ' is the radius of the neutral axis of the ring in the free state. Substituting these values in Eq. (25), results in

---

[1] B. A. Yates, Recent Developments in Piston Ring Materials, *S.A.E. Trans.*, **34**, 49 (1939).

$$\frac{Mc}{I} = \frac{Et}{2}\left(\frac{1}{r'} - \frac{1}{\rho'}\right) \tag{30}$$

or
$$s = \frac{Et}{2}\left(\frac{1}{r'} - \frac{1}{\rho'}\right) \tag{31}$$

and,
$$\rho' = \frac{Etr'}{Et - 2sr'}. \tag{32}$$

The ring is expanded until the neutral axis has a radius of ρ'' in installing the ring on the piston. The stress will be, from Eq. (31),

$$s' = \frac{Et}{2}\left(\frac{1}{\rho'} - \frac{1}{\rho''}\right) \tag{33}$$

where $\rho'' = r + t/2$.

Example.—Determine the ring stress due to slipping the ring in the previous example over the piston. $E = 17 \times 10^6$ lb. in.$^{-2}$ and $s = 30{,}000$ lb. in.$^{-2}$

From Eq. (32),

$$\rho' = \frac{17 \times 10^6 \times 0.135 \times (1.5 - 0.063)}{17 \times 10^6 \times 0.135 - 2 \times 30{,}000(1.5 - 0.063)} = 1.493 \text{ in.}$$

From Eq. (33),

$$s' = \frac{17 \times 10^6 \times 0.135}{2}\left(\frac{1}{1.493} - \frac{1}{1.5 + 0.063}\right) = 34{,}400 \text{ lb. in.}^{-2}$$

The foregoing illustrations indicate the difficulty of obtaining high radial pressure and account partly for the trend to thinner rings with spring-steel inner rings which provide the desired pressure uniformly around the entire ring.

Oil-control Rings.—Oil-control rings are slotted to provide an escape for the oil that the slot edges cut from the cylinder wall. These rings usually have the same thickness and consequently the same wall pressure as the compression rings. Spring-steel inner rings are used behind oil-control rings to increase the wall pressure. The subject of oil control and the combinations of rings is discussed in Chap. XIV.

Ring-gap and -groove Clearance.[1]—Piston rings operate at temperatures about 150 to 200°F. above wall temperatures,

---

[1] "Engineers' Piston Ring Handbook," Koppers Co., The American Hammered Piston Ring Div., Baltimore, 1939.

depending upon engine conditions. On the assumption of a maximum difference of 200°F. and a coefficient of expansion of 6×10^{-6} in. °F.$^{-1}$, the ring circumference will expand about 0.004 in. in.$^{-1}$ in diameter more than the cylinder circumference. Thus, a ring gap of about 0.005 in. in.$^{-1}$ of cylinder bore should be provided to permit the ring to expand. If too small an end clearance is provided, the ends will butt and the ring will break.

Enough clearance must be provided between the ring and the sides of the ring groove so that the movement of the ring in the groove is unrestricted. This implies about 0.001 to 0.002 in. for small-bore automotive type of engines. This clearance should be increased with an increase in cylinder diameter and also with heavy-duty and high-speed engines. The top ring requires more clearance than the others.

Connecting Rods.—The material used for connecting rods ranges from plain carbon steel to the high-grade nickel alloys which are used for heavy-duty engines or where low weight is of importance. The tensile strength varies from about 60,000 lb. in.$^{-2}$ for the carbon steel to 135,000 lb. in.$^{-2}$ for the nickel alloys. Nonferrous alloys are also used.

Connecting rods should be designed for direct compression of the small or piston end of the rod, as a pin-ended column in the plane of rotation, and finally for stresses due to "whipping" or bending because of the inertia of the rod. It is quite common practice to make the connecting rod for large engines rectangular in section, the larger dimension being that in the plane of rotation. In this case the rod should be investigated as a fixed-ended column in a plane normal to the plane of rotation.

The cross-sectional area A of the small end of the connecting rod is determined from the relation

$$A = \frac{P}{s} \tag{34}$$

where P = max. load on piston pin due to gas pr. or inertia, whichever is greater,

s = allowable stress for alternate compression and tension.

The allowable stresses are 15,000 lb. in.$^{-2}$ for carbon steel, 25,000 lb. in.$^{-2}$ for nickel alloys, and 10,000 lb. in.$^{-2}$ for the stronger nonferrous alloys. Lower stresses should be used where large factors of safety are desirable.

Example.—The maximum combustion pressure in an automotive Diesel engine is 1000 lb. in.$^{-2}$ Cylinder bore is 5 in. Determine the desirable area of cross section (Fig. 387) based on combustion pressure only, of the small end of the nickel-steel connecting rod.

$$A = \frac{P}{s_c} = \frac{0.7854 \times 5^2 \times 1000}{25,000} = 0.785 \text{ in.}^2$$

Various shapes of cross section may be used for connecting rods, such as rectangular, tubular, I-section (Fig. 387), and H-section (Fig. 388), the latter two being the most desirable. The

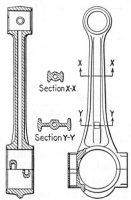

FIG. 387.—I-section connecting rod. (*Waukesha Motor Co.*)

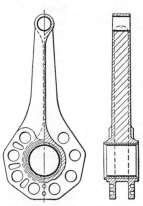

FIG. 388.—H-section master connecting rod. (*Wright Aeronautical Corp.*)

rods are forged and the outer surfaces usually unfinished except in high-duty aircraft engines which have highly finished rod surfaces.

The *Rankine*[1] formula may be used to design the rod as a column in the plane of rotation:

$$s = \frac{P}{A}\left(1 + k\frac{l^2}{\rho^2}\right) \qquad (35)$$

where s = stress in material, lb. in.$^{-2}$
P = total load on column, lb.,
A = sectional area, in.^{2},
l = length of connecting rod, in.,
ρ = radius of gyration at the center of the rod, in.,
k = 1.6 × 10^{-4} for a pin-ended column
   = 0.4 × 10^{-4} for a fixed-ended column.

[1] Marks, *op. cit.*, p. 479.

Example.—Determine a desirable I-section (Fig. 389) for the middle of a connecting rod 12.5 in. long for the data given in the previous example. Substituting in Eq. (35) results in

$$25{,}000 = \frac{0.7854 \times 5^2 \times 1000}{A}\left(1 + 1.6 \times 10^{-4}\frac{12.5^2}{\rho^2}\right)$$

or
$$\rho^2 = \frac{1.965 \times 10^{-2}}{A - 0.7854}. \qquad (A)$$

The value of ρ^2 for an I-section with an over-all width of b and over-all height of h with a constant thickness of t is

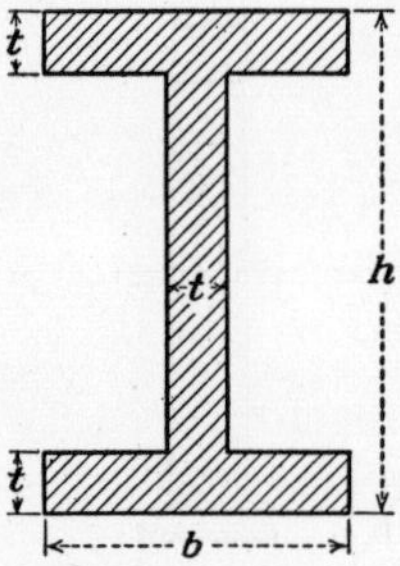

$$\rho^2 = \frac{bh^3 - (b - t)(h - 2t)^3}{12[bh - (b - t)(h - 2t)]}. \qquad (B)$$

Equation (A) may be written

$$\rho^2 = \frac{1.965 \times 10^{-2}}{2bt + (h - 2t)t - 0.7854}. \qquad (C)$$

Fig. 389.—I-section.

Assuming a value for the thickness of the metal results in various values of b and h that will make Eqs. (B) and (C) equal. Thus, with $t = 0.25$ in., a value of $b = 1.0$ in. will require a value of $h = 1.82$ in.

The rod should then be examined as a column with fixed ends, the plane of bending being perpendicular to the plane of rotation.

In this plane,
$$\rho^2 = \frac{2tb^3 + (h - 2t)t^3}{12[2tb + (h - 2t)t]}$$
$$= \frac{2 \times 0.25 \times 1^3 + (1.82 - 0.50)0.25^3}{12[2 \times 0.25 \times 1 + (1.82 - 0.50)0.25]} = 0.0523 \text{ in.}$$
$$A = 2bt + (h - 2t)t = 0.83 \text{ in.}$$

Substituting in Eq. (35), results in

$$s = \frac{0.7854 \times 5^2 \times 1000}{0.83}\left(1 + 0.4 \times 10^{-4}\frac{12.5^2}{0.0523}\right)$$
$$= 26{,}600 \text{ lb. in.}^{-2}.$$

This is slightly larger than the stress used, which indicates that the dimension b might be increased slightly.

The stress in a connecting rod of uniform cross section due to whipping, or, in other words, due to the inertia of the rod in the plane of rotation, may be found by the formula derived by Bach:

ENGINE DESIGN

$$s_b = 2 \times 10^{-6} n^2 r A d \frac{l^2}{Z} \tag{36}$$

where s = stress in material, lb. in.$^{-2}$,
n = r.p.m.,
r = radius of crank, in.,
A = area mean section of rod, in.^{2},
d = density of rod material, lb. in.$^{-3}$,
l = length of rod, in.,
Z = section modulus of rod, in.^{3}

The total stress in the rod is the sum of the stress as a column and that due to the whipping action. The first is a maximum at top center and the other a maximum at about 90 deg. crank travel from top center. Consequently, only a portion of the column stress is acting when the stress due to whipping is a maximum.

Example.—Determine the whipping stress in the connecting rod in the previous example. The engine stroke is 5 in., and the maximum speed is 2000 r.p.m.

$$Z = \frac{bh^3 - (b - t)(h - 2t)^3}{6h} = \frac{1 \times 1.82^3 - 0.75(1.32)^3}{6 \times 1.82} = 0.394 \text{ in.}$$

Substituting in Eq. (36) results in

$$s_b = 2 \times 10^{-6} \times 2000^2 \times 2.5 \times 0.83 \times 0.28 \times \frac{12.5^2}{0.394} = 1840 \text{ lb. in.}^{-2}$$

The stress due to whipping is of little consequence in this case.

The big-end connecting-rod-bearing diameter should be 0.50 to 0.65 of the cylinder bore, while the length of the bearing may vary from 0.40 to 0.60 of the cylinder bore. This indicates a projected bearing area that varies from $0.2d^2$ to $0.4d^2$. High values are used with high-speed and heavy-duty engines and low values with low-speed and light-duty engines.

The piston-pin bearing is usually a bronze bushing, although in some heavy-duty engines an antifriction needle bearing is used. This consists of an inner and an outer race (Fig. 390) between which small-diameter rollers are located. The piston pin is used as the inner race.

For years, babbitt has been the almost universal bearing metal for the crankpin bearing. This has probably been due to its low melting point which causes high spots, which break through the oil film, to soften and be wiped away without damaging the bearing. The increase in severity of service that has accompanied engine improvement has led to the development of various bearing metals with improved performance, such as copper-lead and cadmium-silver compositions.[1]

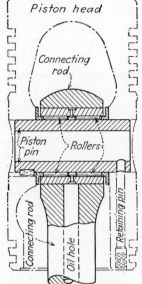

Fig. 390.—Roller connecting-rod bearing. (Bantam "Quill.")

The various bearing materials (Table 37) are attached to thin bronze or steel shells which fit into the big ends of the connecting rods. The bearings are replaceable and usually require no fitting, having been built to size. The bearing clearance depends principally on the crankpin or crankshaft diameter (Table 38).

Connecting-rod bolts should be made of a nickel-alloy steel having a tensile strength that ranges from 100,000 to 150,000 lb. in.$^{-2}$ The design of the big end of the connecting rod for the connecting-rod bolts should not introduce weak sections.

Crankshafts.—Crankshafts are either forged (Fig. 391) or cast. The forged crankshafts are made of plain carbon steel or a nickel alloy. The tensile strength may vary from 70,000 to 140,000 lb. in.$^{-2}$ The cast crankshafts are made from a nickel-alloy iron having a tensile strength that ranges from 50,000 to 75,000 lb. in.$^{-2}$

Crankshafts consist of the shaft parts which rotate in the main bearings, the crankpins which fit in the big-end connecting-rod bearings, the crank cheeks or webs which connect the crankpins and the shaft parts, and any extensions which carry the flywheel or other required parts. All parts should be designed to provide

[1] S. W. Sparrow, Recent Developments in Main and Connecting-rod Bearings, *S.A.E. Trans.*, **29**, 229 (1934).

H. C. Mougey, The Newer Bearing Materials and Their Lubrication, *Ind Eng. Chem.*, **14**, 425 (1936).

P. M. Heldt, Bearing Materials, *Auto. Ind.*, **78**, 412 (1938).

ENGINE DESIGN

TABLE 37.—BEARING MATERIALS AND PERFORMANCE FACTORS*

Description	Analysis, %	Permissible bearing pressure, lb. in.$^{-2}$	Min. permissible $\mu n/p_{max}$†	Max. $p_{max}v$†	Oil reservoir temp., °F.	Min. shaft hardness	Affected by corrosion
Tin-base babbitt (standard)	Copper, 3.50 Antimony, 7.50 Tin, 88.75 Lead (max.), 0.25	1000	20	35,000	235	Not important	No
Tin-base babbitt (Alpha process)	Same as above	1500	15	42,500	235	Not important	No
High-lead babbitt	Tin, 5 to 7 Antimony, 9 to 11 Lead, 82 to 86 Copper (max.), 0.25	1800	10	40,000	225	Not important	No
Cadmium silver	Silver, 0.75 Copper, 0.50 Cadmium, 98.75	1800 to 3850	3.75	90,000 and up	260	250 Brinell	Not likely if temp. is maintained as specified and proper lubricating oil is used
Copper lead	Copper, 0.60 Lead, 0.40	>1800	3.75	90,000 and up	260	300 Brinell	

* Albert B. Willi, Chief Engineer, Federal-Mogul Corporation.
† Note.—p_{max} is used.

TABLE 38.—BEARING CLEARANCE*

Crankshaft or crankpin dia., in.	Diametral cl., in.	End cl., in.
2 to 2¾	0.0015	0.004 to 0.006
2 13/16 to 3½	0.0025	0.006 to 0.008
3 9/16 to 3¾	0.003	0.008 to 0.010
3⅞	0.0035	0.008 to 0.010
4	0.004	0.008 to 0.010

* Federal-Mogul Corporation.

a maximum of rigidity with regard to the stresses involved. The design of an elementary single-throw crankshaft illustrates the principles involved.

Example.—Design a plain carbon-steel crankshaft for a 16- by 24-in. single-acting four-cycle single-cylinder engine to operate at 200 r.p.m.

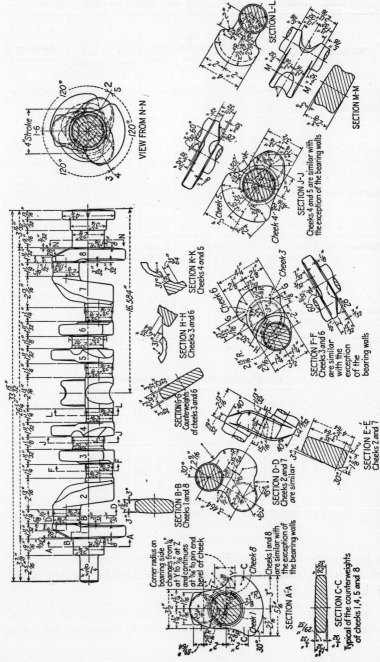

Fig. 391.—Unfinished forged crankshaft for a six-cylinder 3 7/16- by 4-in. engine. (*Wyman-Gordon Co.*)

ENGINE DESIGN

The m.e.p. is 70 lb. in.$^{-2}$, and maximum combustion pressure is 375 lb. in.$^{-2}$ The width of belt required will be about 11 in. with a thickness of $\frac{7}{32}$ in., making the face of the flywheel 12 in. The diameter of the flywheel will be assumed to be 8.6 ft. The tension in the tight side of the belt will be

$$T_1 = \frac{7}{32} \times 11 \times 400 = 960 \text{ lb.} \tag{1}$$

where 400 = working tension of belt, lb. in.$^{-2}$ The difference in tensions will be

$$T_1 - T_2 = \frac{72 \times 550}{90} = 440 \text{ lb.} \tag{2}$$

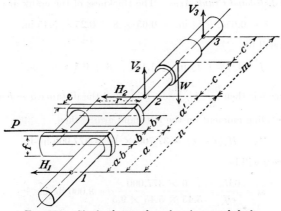

FIG. 392.—Single-throw, three-bearing crankshaft.

where 90 = vel. of the belt, ft. sec.$^{-1}$ From Eqs. (1) and (2), $T_2 = 520$ lb. The sum of the tensions will then be

$$T_1 + T_2 = 960 + 520 = 1480 \text{ lb.}$$

The total weight of the flywheel W will be assuumed to be 12,000 lb.

Crankpin.—The distance n (Fig. 392) may be assumed to be twice the cylinder diameter or 32 in. The total piston pressure with the crank on dead center is

$$P = \frac{\pi D^2}{4} \times p_{max} = \frac{\pi 16^2}{4} \times 375 = 75,400 \text{ lb.} \tag{3}$$

If a and a' are assumed equal, the reactions H_1 and H_2 will be equal, 37,700 lb. each. The bending moment at the center of the crankpin will then be

$$M_b = H_1 a = 37,700 \times 16 = 603,200 \text{ in.-lb.} \tag{4}$$

The assumed stress (12,000 lb. in.$^{-2}$) due to this bending moment will be

$$s_b = \frac{M}{Z} = \frac{32M}{\pi d^3} \tag{5}$$

from which $\quad d = \sqrt[3]{\dfrac{603{,}200 \times 32}{12{,}000 \times \pi}} = 8$ in.

Allowing 1450 lb. pr. in.$^{-2}$ of projected area, based on the explosion pressure, the length of the crankpin is found to be

$$l = \frac{75{,}400}{8 \times 1450} = 6.5 \text{ in.} \tag{6}$$

Stress in Left-hand Crank Arm.—The thickness of the crank arms will be

$$e = 0.65d + \tfrac{1}{4} \text{ in.} = 0.65 \times 8 + 0.25 = 5.45 \text{ in.} \tag{7}$$

The width will be

$$f = 1\tfrac{1}{8}d + \tfrac{1}{2} \text{ in.} = 1.125 \times 8 + 0.5 = 9.5 \text{ in.} \tag{8}$$

The distance b is then $\dfrac{6.5}{2} + \dfrac{5.45}{2} = 6$ in., and the distance $a - b$ will be 10 in. The bending moment on the left-hand arm is then

$$M_b = H_1(a - b) = 37{,}700 \times 10 = 377{,}000 \text{ in.-lb.} \tag{9}$$

and the stress will be

$$s_b = \frac{6M_b}{e^2 f} = \frac{6 \times 377{,}000}{5.45 \times 5.45 \times 9.5} = 8{,}000 \text{ lb. in.}^{-2} \tag{10}$$

The stress due to direct compression will be

$$s_c = \frac{H_1}{ef} = \frac{37{,}700}{5.45 \times 9.5} = 730 \text{ lb. in.}^{-2} \tag{11}$$

The total stress will be

$$s_{total} = s_b + s_c = 8000 + 730 = 8730 \text{ lb. in.}^{-2} \tag{12}$$

which is well below the limit.

Stress in Right-hand Crank Arm.—The bending moment on the right-hand crank arm will be

$$M_b = H_1(a + b) - Pb \text{ in.-lb.,} \tag{13}$$

which is the same as that on the left-hand arm, and since the size of the arm is the same the stress will be the same. This stress is on the face of the broad side of the arm.

Shaft under Flywheel.—Before the moments under the flywheel may be computed, it is necessary to find c, c', the length l_2 of bearing 2, and the length l_3 of bearing 3. In this type of shaft, the length of all the main bearings may be assumed to be the same.

Thus, $\quad l_2 = 2\left(a' - \dfrac{l}{2} - e\right) = 2\left(16 - \dfrac{6.5}{2} - 5.45\right) = 14.6$ in. $\quad$ (14)

The face of the flywheel is 12 in. The bending moment due to the weight of the flywheel will be

$$M_w = \frac{W}{2}c' = \frac{12{,}000}{2} \times 21 = 126{,}000 \text{ in.-lb.} \quad (15)$$

where $c = c' = 21$ in., allowing space for gearing and clearance.

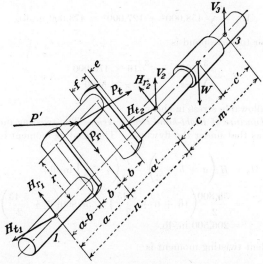

Fig. 393.—Crank at angle of maximum twisting moment.

The bending moment due to the belt pull will be

$$M_b = \frac{T_1 + T_2}{2}c' = \frac{1480}{2} \times 21 = 15{,}120 \text{ in.-lb.} \quad (16)$$

The combined bending moment according to *Guest's Law*, will be

$$M_{total} = \sqrt{M_w^2 + M_b^2} = \sqrt{126{,}000^2 + 15{,}120^2} = 127{,}000 \text{ in.-lb.} \quad (17)$$

The diameter to correspond will be

$$d_w = \sqrt[3]{\frac{32 M_{total}}{\pi s}} = \sqrt[3]{\frac{32 \times 127{,}000}{\pi \times 6000}} = 6 \text{ in.} \quad (18)$$

Crank at Angle of Maximum Twisting Moment.—The force P' along the rod at the time of maximum twisting moment is 56,800 lb. Resolving this force parallel and normal to the crank produces $P_r = 45,400$ lb. and $P_t = 38,200$ lb. With reference to Fig. 393, it is found that

$$H_{t_1} = H_{t_2} = 19,100 \text{ lb.},$$
$$H_{r_1} = H_{r_2} = 22,700 \text{ lb.},$$
$$V_2 = V_3 = 6,000 \text{ lb.}$$

Shaft under Flywheel.—The combined bending moment under the flywheel is the same as before, 127,000 in.-lb. The twisting moment is

$$M_t = P_t r = 38,200 \times 12 = 458,000 \text{ in.-lb.} \tag{19}$$

If this is combined with the bending moment by Guest's Law, the equivalent twisting moment will be

$$M_{t_e} = \sqrt{458,000^2 + 127,000^2} = 473,000 \text{ in.-lb.}$$

The diameter to correspond is

$$d_w = \sqrt[3]{\frac{16 M_{t_e}}{\pi s_s}} = \sqrt[3]{\frac{16 \times 473,000}{\pi \times 5000}} = 8 \text{ in.} \tag{20}$$

when s_s = allowable stress in shear.

Shaft at Juncture of Right-hand Crank Arm.—The twisting moment here is the same as that under the flywheel. The bending moment is

$$M_b = H_1\left(a + b + \frac{e}{2}\right) - \frac{P'}{2}\left(b + \frac{e}{2}\right) \text{ in.-lb.} \tag{21}$$
$$= \frac{56,800}{2}\left(16 + 6 + \frac{5.45}{2}\right) - 56,800\left(6 + \frac{5.45}{2}\right)$$
$$= 206,500 \text{ in.-lb.}$$

The equivalent twisting moment is

$$M_{t_e} = \sqrt{458,000^2 + 206,500^2} = 502,000 \text{ in.-lb.}$$

The diameter to correspond is

$$d_2 = \sqrt[3]{\frac{16 \times 502,000}{\pi \times 6000}} = 7.5 \text{ in.}$$

The stress used here is higher than that used under the flywheel for two reasons: the condition of loading which was assumed at this point is more severe than the actual condition, and the deflection under the flywheel must be kept small.

Crankpin.—The bending moment is

$$M_b = H_{r_1} a = \frac{56,800}{2} \times 16 = 455,000 \text{ in.-lb.,} \tag{22}$$

and the twisting moment is

$$M_t = H_{t_1}r = 19{,}100 \times 12 = 229{,}000 \text{ in.-lb.} \tag{23}$$

The equivalent twisting moment is

$$M_{t_e} = \sqrt{455{,}000^2 + 229{,}000^2} = 511{,}000 \text{ in.-lb.}$$

The diameter to correspond is

$$d = \sqrt{\frac{511{,}000 \times 16}{\pi \times 5000}} = 8 \text{ in.}$$

Left-hand Crank Arm at Juncture of Pin.—The bending moment due to the radial component is

$$M_{b_r} = H_{r_1}(a - b) = 22{,}700 \times 10 = 227{,}000 \text{ in.-lb.} \tag{24}$$

The stress to correspond is

$$s_{b_r} = \frac{6M_{b_r}}{e^2 f} = \frac{6 \times 227{,}000}{5.45 \times 5.45 \times 9.5} = 4830 \text{ lb. in.}^{-2} \tag{25}$$

The bending moment due to the tangential component is

$$M_{b_t} = H_{t_1}r' = 19{,}100(12 - 4) = 152{,}800 \text{ in.-lb.} \tag{26}$$

where $r' = r - \tfrac{1}{2}$ dia. of crankpin.

The stress to correspond is

$$s_{b_t} = \frac{6 \times 152{,}800}{5.45 \times 9.5 \times 9.5} = 1860 \text{ lb. in.}^{-2}$$

The stress in direct compression is

$$s_c = \frac{22{,}700}{5.45 \times 9.5} = 440 \text{ lb. in.}^{-2}$$

The total compressive stress is

$$s_{c_{total}} = 4830 + 1860 + 440 = 7130 \text{ lb. in.}^{-2}$$

The twisting moment on the arm is

$$M_t = H_{t_1}(a - b) = 19{,}100 \times 10 = 191{,}000 \text{ in.-lb.,} \tag{27}$$

and the stress to correspond is

$$s_s = \frac{9M_t}{2fe^2} = \frac{9 \times 191{,}000}{2 \times 9.5 \times 5.45 \times 5.45} = 3050 \text{ lb. in.}^{-2}$$

The total combined stress in the arm will be

$$s_{c_{max}} = \frac{s_{c_{total}}}{2} + \sqrt{\frac{s_{c_{total}}^2}{4} + s_s^2} \qquad (28)$$

$$= \frac{7130}{2} + \sqrt{\frac{7130^2}{4} + 3050^2} = 7280 \text{ lb. in.}^{-2}$$

Right-hand Crank Arm.—The bending moment due to the radial component is

$$M_{b_r} = H_{r_1}(a + b) - P_r b = 22{,}700 \times 22 - 45{,}400 \times 6 = 27{,}000 \text{ in.-lb.}$$

The stress to correspond is

$$s_{b_r} = \frac{6 \times 227{,}000}{5.45 \times 5.45 \times 9.5} = 4830 \text{ lb. in.}^{-2}$$

The bending moment due to the tangential component is a maximum where the arm joins the shaft and is

$$M_{b_t} = P_t\left(r - \frac{d_2}{2}\right) + H_{t_1}\frac{d_2}{2} \qquad (29)$$

$$= 38{,}200(12 - 3.75) + 19{,}100 \times 3.75 = 386{,}600 \text{ in.-lb.}$$

The stress to correspond is

$$s_{b_t} = \frac{6 \times 386{,}600}{5.45 \times 9.5 \times 9.5} = 4720 \text{ lb. in.}^{-2}$$

The direct stress in compression is the same as in the other arm, 440 lb. per in.$^{-2}$ The total compressive stress on one corner is

$$s_{c_{total}} = 4830 + 4720 + 440 = 9990 \text{ lb. in.}^{-2}$$

The twisting moment on the arm is

$$M_t = H_{t_1}(a + b) - P_t b = 19{,}100 \times 22 - 38{,}200 \times 6 = 191{,}000 \text{ in.-lb.} \qquad (30)$$

The twisting stress is

$$s_s = \frac{9 \times 191{,}000}{2 \times 9.5 \times 5.45 \times 5.45} = 3050 \text{ lb. in.}^{-2}$$

The total combined stress in the arm will be

$$s_{max} = \frac{s_{c_{total}}}{2} + \sqrt{\frac{s_{c_{total}}^2}{2} + s_s^2} \qquad (31)$$

$$s_{max} = \frac{9990}{2} + \sqrt{\frac{9990^2}{4} + 3050^2} = 10{,}850 \text{ lb. in.}^{-2}$$

Length of Bearing at 2.—The bearing at 2 has the largest total reactions; hence this particular one should be designed for bearing pressure. The mean reactions there are due to the mean pressure exerted by the pressure in the cylinder and the weight of reciprocating parts, one-half the flywheel weight and one-half the belt pull. The mean reaction due to the piston pressure and reciprocating parts should be found from the indicator card and the inertia diagram for each stroke and the mean of the four used. In this case the mean pressure of the four strokes of the cycle will be assumed to be 40 lb. in.$^{-2}$ of piston. The total reaction at 2 will then be

$$Q_2 = \frac{40 \times 201 + 1480 + 12{,}000}{2} = 10{,}760 \text{ lb.}$$

The diameter of the shaft at this point, bearing 2, was found to be 7½ in. From the length of crankpin and thickness of arm, the length of bearing 2 can be found.

Thus, $\quad \dfrac{l_2}{2} = a' - \dfrac{l}{2} - e = 16 - \dfrac{6.5}{2} - 5.45 = 7.3 \text{ in.},$

from which the total length of l_2 is 14.6 in. The work of friction per square inch of projected area per second is

$$F_2 = \frac{fQ_2 \pi d_2 N}{12 \times 60 d_2 l_2} \qquad (32)*$$

$$= \frac{0.05 \times 10{,}760 \times \pi \times 7.5 \times 200}{12 \times 60 \times 7.5 \times 14.6} = 32.2 \text{ ft.-lb.}$$

As this value is far below the maximum value of 70, the total pressure on bearing 2 will be investigated. This is

$$R_2 = \frac{201 \times 375 + 1480 + 12{,}000}{2 \times 7.5 \times 14.6} = 406 \text{ lb. in.}^{-2}$$

Since this value may run up to 600 lb. in.$^{-2}$ and since the work of friction is low, the length l_2 may be reduced. If $l_2 = 12$ in., then a' (Fig. 393) will be

$$a' = \frac{6.5}{2} + 5.25 + 6 = 14.5 \text{ in.}$$

instead of 16 in. as was first assumed. The calculations should be gone over again in exactly the same manner and new diameters and new stresses found as before. In order to avoid tiresome repetition, this recalculation is omitted here.

---

* The same method can be used for the crankpin by substituting P_n for Q_2.

574   INTERNAL-COMBUSTION ENGINES

Bending Moment in Bearing 2.—The dimensions being used as in the previous calculations the bending at bearing 2 may be found by the theorem of three moments. Referring to Fig. 394, the values of K' and K'' will each be 0.5. The load P' between 1 and 2 will be 56,800 which may be assumed to be in a horizontal direction. The load W between bearings 2 and 3 is

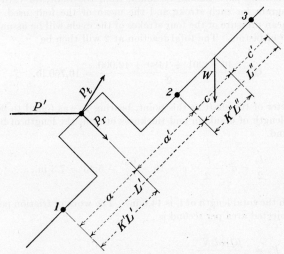

Fig. 394.—Diagram of shaft at maximum twisting moment.

12,000 lb., the belt pulls being neglected. The bending moment at bearing 2 due to P' is

$$M_{p2} = -\frac{P'L'^2(K' - K'^3)}{2(L' + L'')} \text{ in.-lb.} \tag{33}$$

$$= -\frac{56,800 \times 32^2 \times (0.5 - 0.5^3)}{2 \times (32 + 42)} = -147,500 \text{ in.-lb.}$$

$$R_{p1} = \frac{M_{p2}}{L'} + P'(1 - K') \tag{34}$$

$$= \frac{147,500}{32} + 56,800(1 - 0.5) = 23,780 \text{ lb.}$$

$$R_{p3} = \frac{M_2}{L''} \tag{35}$$

$$= -\frac{147,500}{42} = -3510 \text{ lb.,}$$

$$R_{p2} = P' - R_{p1} - R_{p2} \tag{36}$$
$$= 56,800 - 23,780 + 3510 = 36,530 \text{ lb.}$$

where R represents the reaction at the point indicated.

ENGINE DESIGN

The bending moment at bearing 2 due to W is

$$M_{w_2} = -\frac{WL''^2(2K'' - 3K''^2 + K''^3)}{2(L' + L'')} \quad (37)$$

$$= -\frac{12,000 \times \overline{42}^2 \times (2 \times 0.5 - 3 \times 0.25 + 0.125)}{2 \times (32 + 42)}$$

$$= -53,700 \text{ in.-lb.}$$

$$R_{w_1} = \frac{M_{w_2}}{L'} = -\frac{53,700}{32} = -1675 \text{ lb.} \quad (38)$$

$$R_{w_3} = WK'' - \frac{M''_{w_2}}{L''} \quad (39)$$

$$= 12,000 \times 0.5 - \left(-\frac{53,700}{42}\right) = 7280 \text{ lb.}$$

$$R_{w_2} = W - R_{w_1} - R_{w_3} \quad (40)$$
$$= 12,000 - 7280 + 1675 = 6395 \text{ lb.}$$

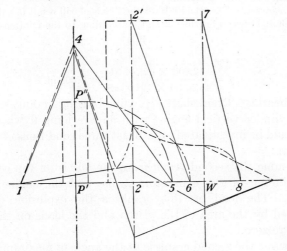

FIG. 395.—Combination of moments.

The twisting moment on the shaft is

$$M_t = 38,200 \times 12 = 458,000 \text{ in.-lb.}$$

The bending moment at the center of the crankpin due to P' is

$$M_p = R_{p_1} \times a = 23,780 \times 16 = 380,000 \text{ in.-lb.} \quad (41)$$

The bending moment under W due to W is

$$M_w = R_{w_3} \times c_1 = 7280 \times 21 = 153,000 \text{ in.-lb.} \quad (42)$$

The bending-, twisting-, and combined-moment diagrams are shown in Fig. 395. The moments due to the piston pressure and flywheel being at right angles to each other are combined by finding the square root of the sum of their squares. The combined bending moments are shown by the dotted broken line. At the center of the crankpin P', at bearing 2, and at the center of the flywheel W, the combined bending moments are combined with the twisting moments at those points, resulting in equivalent twisting moments 4–5 at the crankpin, 2'–6 at the center of bearing 2, and 7–8 at the center of the flywheel. These moments have the following values:

$$4\text{–}5 = 469,000 \text{ in.-lb.,}$$
$$2'\text{–}6 = 483,000 \text{ in.-lb.,}$$
$$7\text{–}8 = 466,000 \text{ in.-lb.}$$

In the computations for this shaft as a simple beam, it was found that the combined bending and twisting moment at the center of the crankpin was 511,000 in.-lb., that at the place where bearing 2 joins the crank arm it was 502,000 in.-lb., and that under the flywheel it was 473,000 in.-lb. Thus the shaft is safe as already designed.

Shaft Deflection.—The weight W of the flywheel will result in a deflection f of the shaft between bearings 2 and 3 according to the relationship

$$f = \frac{Wm^3}{48\ EI} \tag{43}*$$

Thus, $$f = \frac{12,000 \times 42^3 \times 64}{48 \times 30 \times 10^6 \times \pi 8^4} = 0.003 \text{ in.}$$

Five-bearing Crankshaft.—A five-bearing crankshaft for a four-cylinder vertical engine with an overhung flywheel (Fig. 396) should be investigated for at least three conditions of loading as follows:

1. Assume the second crank from the left is on top center at the beginning of the expansion stroke and the engine is just starting. The load on this crank is the explosion pressure multiplied by the area of the piston and the loads on the other cranks are zero.

2. Assume the second crank is at the angle of maximum twisting moment. In this position the load on this crank should be found as in the foregoing illustrative example. The loads on the other cranks will be practically zero.

3. Assume the cranks are in the first condition but the engine up to speed. The load P'' (Fig. 397) will be the explosion pressure minus the force of inertia multiplied by the area of the piston, and may be acting upward or downward. The other loads will be the inertia forces in each case multiplied by the area of the piston. P''' will be upward, P' and P'''' will be downward.

* Marks. *op. cit.*, p. 437.

In all these cases the bending moments should be found by the theorem of the three moments and plotted as was done for the three-bearing shaft. In the second case there will be twisting in addition to the bending.

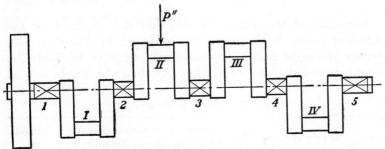

Fig. 396.—Four-throw, five-bearing crankshaft.

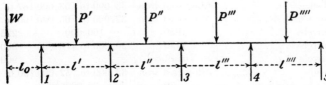

Fig. 397.—Diagram of loads on a four-throw crankshaft.

The moments may be found by the following equations in which the spans are all the same length except l_o and the loads are in the middle of the spans in every case except the flywheel which is an overhung load. The subscripts of the bending moments refer to the bearing numbers. Thus,

$$M_1 = -Wl_o, \tag{44}$$

$$M_1 l' + 2(l' + l'')M_2 + M_3 l'' = -P'l'^2(\tfrac{1}{2} - \tfrac{1}{8})$$
$$- P''l''^2(1 - \tfrac{3}{4} + \tfrac{1}{8}) \tag{45}$$

$$M_2 l'' + 2(l'' + l''')M_3 + M_4 l''' = -P''l''^2(\tfrac{1}{2} - \tfrac{1}{8})$$
$$- P'''l'''^2(1 - \tfrac{3}{4} + \tfrac{1}{8}) \tag{46}$$

$$M_3 l''' + 2(l''' + l'''')M_4 + M_5 l'''' = -P'''l'''^2(\tfrac{1}{2} - \tfrac{1}{8})$$
$$- P''''l''''^2(1 - \tfrac{3}{4} + \tfrac{1}{8}) \tag{47}$$

$$M_5 = 0. \tag{48}$$

The shear at the right of the various supports will be as follows:

$$V_1 = \frac{M_2 - M_1}{l'} + \frac{P'}{2}. \tag{49} \qquad V_2 = \frac{M_3 - M_2}{l''} + \frac{P''}{2}. \tag{50}$$

$$V_3 = \frac{M_4 - M_3}{l'''} + \frac{P'''}{2}. \tag{51} \qquad V_4 = \frac{M_5 - M_4}{l''''} + \frac{P''''}{2}. \tag{52}$$

The moments under the loads will then be

$$M' = M_1 + \frac{V_1 l'}{2}. \quad (53) \qquad M'' = M_2 + \frac{V_2 l''}{2}. \quad (54)$$

$$M''' = M_3 + \frac{V_3 l'''}{2}. \quad (55) \qquad M'''' = M_4 + \frac{V_4 l''''}{2}. \quad (56)$$

Summary of Materials.—Various materials are used for the different parts, depending on the service. Typical materials, tensile strengths and hardness values are given in Table 39. Highly stressed parts should have gradual changes in section

TABLE 39.—TYPICAL MATERIALS*

Engine part	Material or S.A.E. no.	Tensile strength, lb. in.$^{-2}$	Brinell no.
Cyl. blocks	Alloy cast iron	33,000 to 53,000	187 to 269
Cyl. liners and sleeves	Alloy cast iron	32,000 to 65,000	190 to 286
Cyl. barrels	1050	100,000	225
Cyl. barrels	4140	120,000 to 170,000	300
Cyl. barrels	Nitralloy		900
Cyl. heads	Alloy cast iron	32,000 to 52,000	190 to 240
Cyl. heads	39	32,000	65
Pistons	Alloy cast iron	30,000	200
Pistons	34	26,000 to 34,000	85 to 150
Pistons	39	32,000	65
Pistons	321		100
Pistons	N.F.-2	55,000	95
Piston rings	Alloy cast iron	20,000 to 50,000	190 to 220
Piston pins	2330	130,000	260
Piston pins	3140	145,000	285
Piston pins	6150	220,000	440
Connecting rods	2340	130,000	250
Connecting rods	3140	145,000	285
Connecting rods	4340	160,000	330
Crankshafts	Alloy cast iron	50,000 to 70,000	280 to 320
Crankshafts	3240	160,000	330
Crankshafts	4340	180,000	360
Bolts and studs	2330	130,000	250
Bolts and studs	3130	145,000	280

* Data obtained from the following sources:
"S.A.E. Handbook," S.A.E., 1939.
G. P. Phillips, Use of Alloy Cast Iron Grows, *Auto. Ind.*, **74**, 732 (1936).
J. B. Johnson, Aircraft Engine Materials, *S.A.E. Trans.*, **32**, 153 (1937).
H. E. Blank, Jr., Nickel—A Versatile Automotive Metal, *Auto. Ind.*, **78**, 538 (1938).
Automotive Uses of Nickel Alloy Steel, *Bull. U*-1, Int. Nickel Co., Inc.

and should be highly finished to remove any tool marks or finish scratches.

EXERCISES

1. A nine-cylinder radial aircraft engine has a brake m.e.p. of 175 lb. in.$^{-2}$ at 2000 r.p.m. It develops 1500 hp. and weighs 1600 lb. Determine the output, the weight per horsepower, the engine speed, and the cylinder bore for similar engines having the same total displacement and piston speed with 18, 27, and 36 cylinders. Plot the results against number of cylinders.

2. Determine the effect on similar-engine performance of a given oil viscosity and the same bearing temperatures. Assume a mechanical efficiency of 0.90 with the largest bore engine and plot the indicated horsepower, brake horsepower, and mechanical efficiency of the various engines. Discuss results.

3. Assume all air-cooled cylinders have the same cooling-fin surface per square inch of outside-cylinder surface. Discuss the possible advantage of small or large cylinders with regard to inside-wall temperature.

4. Assuming the same stroke and piston speed, estimate the effect of variation in cylinder bore on the various performance and design factors.

5. A single-cylinder blast-furnace gas engine develops 1000 b. hp. at a piston speed of 1000 ft. min.$^{-1}$ Estimate the relative weight, size, etc., for a V-type engine having a piston speed of 3000 ft. min.$^{-1}$

6. Estimate the stress in the cylinder head of the P. and W. aircraft engine (Fig. 372), assuming a maximum combustion pressure of 500 lb. in.$^{-2}$ Neglect the effect of the fins. Scale the illustration for dimensions.

7. Estimate the effect of the fins on the stress in Prob. 6.

8. Assume the gas pressure during expansion in the P. and W. aircraft engine is described by the relation $PV^{1.3}$ = const. Estimate the cylinder deflection at the point of maximum side thrust at a speed of 2000 r.p.m.

9. Design the studs and nuts to hold down the cylinder in Prob. 6. The studs are screwed into an aluminum crankcase. Estimate the pull on a wrench to be used to tighten the nuts on the studs.

10. Estimate the effect of the side thrust in Prob. 8 on the stress in the cylinder hold-down studs. Make a reasonable assumption regarding the distribution of the added stress due to side thrust.

11. Design a piston pin for the example on page 555; it should have an inside taper (Fig. 384) such that the pin will be equally stressed to 13,800 lb. in.$^{-2}$ theoretically at all sections. Compare with weight of nontapered pin.

12. Determine how much thinner an alloy cast-iron piston head can be for equal stress compared with an aluminum-alloy piston. Estimate relative piston weights.

13. Design a compression piston ring to have a ring tension of 28 lb. The ring is for an 8-in. cylinder bore and the face of the ring is $\frac{1}{4}$ in. wide.

14. Estimate the stress in the ring in Prob. 13 during installation on the piston. Assume a diametral piston clearance at the ring section of 0.015 in.

15. Assume $\frac{1}{8}$, $\frac{1}{4}$, and $\frac{3}{8}$ in. for the metal thickness of the I-section of the connecting rod in the example on page 562. Determine the desirable width and height of the section for the same column and whipping stress in both directions. Compare the areas of cross sections.

16. Referring to Fig. 397, assume $l' = l'' = l''' = l'''' = 15$ in., $l_o = 11$ in., $W = 3000$ lb., $P' = 1000$, $P'' = 18,000$, $P''' = 1200$, and $P'''' = 2000$ lb. Find the bending moments at bearings 1, 2, 3, 4, and 5 and at each load. Draw the bending-moment diagram.

17. Examine the latest published data, such as appear in *Automotive Industries*, and obtain such design data as brake m.e.p., piston speed, stroke-bore relation, cylinder-bore and compression-ratio relation, and the other design-dimension relationships compared with cylinder bore for the various types and classifications.

18. Design a single-cylinder four-stroke-cycle completely-balanced engine to operate at 5000 r.p.m. with a brake m.e.p. of 300 lb. in.$^{-2}$ Incorporate a Roots-type blower in the design.

APPENDIX

TABLE I.—Relation of Energy Units

Watt-second (w.-sec.)	Foot-pound (ft.-lb.)	British thermal unit (B.t.u.)	Horsepower-hour (hp.-hr.)	Kilowatt-hour (kw.-hr.)
1	0.738			
1.355	1	1.285×10^{-3}		
1.055	778	1	0.393×10^{-3}	
2.684×10^6	1.980×10^6	2544	1	0.745
3.600×10^6	2.656×10^6	3413	1.341	1

TABLE II.—Specific Heat Equations

Gas or vapor	Symbol	Equation for C_p in B.t.u. mol^{-1} °R^{-1}.	Range, °R.	Source
Oxygen	O_2	$c_p = 11.515 - \dfrac{172}{\sqrt{T}} + \dfrac{1530}{T}$	540 to 5000	*
		$= 11.515 - \dfrac{172}{\sqrt{T}} + \dfrac{1530}{T} + \dfrac{0.05(T-4000)}{1000}$	5000 to 9000	*
Nitrogen	N_2	$c_p = 9.47 - \dfrac{3.47 \times 10^3}{T} + \dfrac{1.16 \times 10^6}{T^2}$	540 to 5000	*
Carbon monoxide	CO	$c_p = 9.46 - \dfrac{3.29 \times 10^3}{T} + \dfrac{1.07 \times 10^6}{T^2}$	540 to 5000	*
Hydrogen	H_2	$c_p = 5.76 + \dfrac{0.578}{1000}T + \dfrac{20}{\sqrt{T}}$	540 to 4000	*
		$= 5.76 + \dfrac{0.578}{1000}T + \dfrac{20}{\sqrt{T}} - \dfrac{0.33(T-4000)}{1000}$	4000 to 9000	*
Water	H_2O	$c_p = 19.86 - \dfrac{597}{\sqrt{T}} + \dfrac{7500}{T}$	540 to 5000	*
Carbon dioxide	CO_2	$c_p = 16.2 - \dfrac{6.53 \times 10^3}{T} + \dfrac{1.41 \times 10^6}{T^2}$	540 to 6300	*
Hydroxide	OH	$c_p = 7.401 - 0.7306 \times 10^{-3}T + 0.386 \times 10^{-6}T^2$	491.6 to 1620	†
		$c_p = 10.151 - \dfrac{8217}{T} + \dfrac{5.67 \times 10^6}{T^2}$	1620 to 5940	†
Nitric oxide	NO	$c_p = 6.550 + 0.889 \times 10^{-3}T$	491.6 to 1620	†
		$c_p = 9.406 - \dfrac{2293}{T}$	1620 to 5940	†
Carbon	C	$c_p = 0.7041 + 3.488 \times 10^{-3}T - 0.556 \times 10^{-6}T^2$	491.6 to 1440	†
		$c_p = 6.512 - \dfrac{540}{T} - \dfrac{3.24 \times 10^6}{T^2}$	1440 to 2700	†
Methane	CH_4	$c_p = 4.22 + 8.211 \times 10^{-3}T$	491.6 to 1800	†
		$c_p = 27.0 - \dfrac{14,400}{T}$	1800 to 5940	†
Ethylene	C_2H_4	$c_p = 6.0 + 8.33 \times 10^{-3}T$	720 to 1440	‡
Ethane	C_2H_6	$c_p = 6.6 + 13.33 \times 10^{-3}T$	720 to 1440	‡
Ethyl alcohol	C_2H_6O	$c_p = 4.5 + 21.1 \times 10^{-3}T$	680 to 1120	‡
Methyl alcohol	CH_4O	$c_p = 2.0 + 16.67 \times 10^{-3}T$	680 to 1100	‡
Benzene	C_6H_6	$c_p = 6.5 + 28.9 \times 10^{-3}T$	520 to 1120	‡
Octane	C_8H_{18}	$c_p = 14.4 + 53.3 \times 10^{-3}T$	720 to 1440	‡
Dodecane	$C_{12}H_{26}$	$c_p = 19.6 + 80.0 \times 10^{-3}T$	720 to 1440	‡

* R. L. Sweigert and M. W. Beardsley, Empirical Specific Heat Equations Based upon Spectroscopic Data, *Ga. School Technology, State Eng. Expt. Sta., Bull.* 2 (1938).

† Carl Schwarz, Die Spezifischen Wärmen der Gase als Hilfswerte zur Beruchung von Gleichgewichten, *Arch. Eisenhüttenwesen,* **9**, 389 (1936).

‡ G. S. Parks and H. M. Huffman, *A.C.S., Mon.* 60 (1932).

582 INTERNAL-COMBUSTION ENGINES

TABLE III.—INTERNAL ENERGY OF GASES*
(B.t.u. mol⁻¹ above 520°R.)

Temp., °R.	O_2	N_2	Air	CO_2	H_2O	H_2	CO	OH	NO	Monatomic O, H, A, He, etc.	C_8H_{18}	$C_{12}H_{26}$	$C_8H_{18} + 59.5$ air 60.5	$8CO_2 + 9H_2O + 47N_2$ 64	$PV = RT$	T, °R.
520	0	0	0	0	0	0	0	0	0	0	0	0	0	0	1,033	520
536.6†	83	81	81	115	101	80	81	84	85	49	674	994	90	88	1,066	536.6†
540	100	97	97	139	122	96	97	101	103	59	814	1,201	109	106	1,072	540
560	200	196	196	280	244	193	196	202	207	119	1,648	2,433	220	213	1,112	560
580	301	295	295	424	367	291	295	304	312	179	2,504	3,697	332	320	1,152	580
600	402	395	395	570	490	390	396	407	418	238	3,381	4,993	443	430	1,192	600
700	920	896	897	1,320	1,110	887	896	914	943	537	8,088	11,955	1,016	979	1,390	700
800	1,449	1,399	1,403	2,120	1,734	1,386	1,402	1,422	1,468	834	13,326	19,716	1,600	1,536	1,589	800
900	1,989	1,905	1,915	2,965	2,366	1,886	1,913	1,930	1,994	1,132	19,099	28,278	2,198	2,102	1,787	900
1000	2,539	2,416	2,431	3,852	3,009	2,387	2,430	2,439	2,522	1,430	25,403	37,639	2,811	2,678	1,986	1000
1100	3,101	2,934	2,957	4,778	3,666	2,889	2,954	2,948	3,053	1,728	32,241	47,800	3,441	3,267	2,185	1100
1200	3,675	3,461	3,492	5,736	4,339	3,393	3,485	3,458	3,596	2,026	39,612	58,762	4,089	3,869	2,383	1200
1300	4,262	3,996	4,036	6,721	5,030	3,899	4,026	3,968	4,154	2,324	47,516	70,523	4,754	4,452	2,582	1300
1400	4,861	4,539	4,587	7,731	5,740	4,406	4,580	4,480	4,730	2,622	55,953	83,085	5,437	5,107	2,780	1400
1500	5,472	5,091	5,149	8,764	6,468	4,916	5,145	4,996	5,321	2,920	64,923	96,446	6,138	5,743	2,979	1500
1600	6,092	5,652	5,720	9,819	7,212	5,429	5,720	5,516	5,922	3,218	74,425	110,607	6,856	6,392	3,178	1600
1700	6,718	6,224	6,301	10,896	7,970	5,945	6,305	6,041	6,532	3,516	84,461	125,569	7,592	7,054	3,376	1700
1800	7,349	6,805	6,889	11,993	8,741	6,464	6,899	6,571	7,150	3,813	95,030	141,330	8,346	7,726	3,575	1800
1900	7,985	7,393	7,485	13,105	9,526	6,988	7,501	7,106	7,773	4,111	106,132	157,892	9,116	8,407	3,773	1900
2000	8,629	7,989	8,087	14,230	10,327	7,517	8,109	7,646	8,399	4,409	117,767	175,253	9,900	9,098	3,972	2000
2100	9,279	8,592	8,698	15,368	11,146	8,053	8,722	8,191	9,029	4,707				9,798	4,171	2100
2200	9,934	9,203	9,314	16,518	11,983	8,597	9,339	8,741	9,663	5,005				10,507	4,369	2200
2300	10,592	9,817	9,934	17,680	12,835	9,147	9,961	9,299	10,302	5,303				11,224	4,568	2300
2400	11,252	10,435	10,558	18,852	13,700	9,703	10,558	9,867	10,946	5,601				11,946	4,766	2400
2500	11,916	11,056	11,185	20,033	14,578	10,263	11,220	10,445	11,595	5,899				12,673	4,965	2500

APPENDIX

2600	12,584	11,682	11,817	21,222	15,469	10,827	11,857	11,039	12,249	6,197			13,407	5,164	2600
2700	13,257	12,313	12,453	22,419	16,372	11,396	12,499	11,628	12,906	6,495			14,417	5,362	2700
2800	13,937	12,949	13,095	23,624	17,288	11,970	13,144	12,223	13,566	6,792			14,894	5,561	2800
2900	14,622	13,590	13,742	24,836	18,217	12,549	13,792	12,823	14,228	7,090			15,646	5,759	2900
3000	15,309	14,236	14,394	26,055	19,160	13,133	14,443	13,428	14,892	7,388			16,406	5,958	3000
3100	16,001	14,888	15,051	27,281	20,117	13,723	15,097	14,037	15,559	7,686			17,173	6,157	3100
3200	16,693	15,543	15,710	28,513	21,086	14,319	15,754	14,649	16,229	7,984			17,944	6,355	3200
3300	17,386	16,199	16,369	29,750	22,066	14,921	16,414	15,264	16,902	8,282			18,718	6,554	3300
3400	18,080	16,855	17,030	30,991	23,057	15,529	17,078	15,882	17,577	8,580			19,494	6,752	3400
3500	18,776	17,512	17,692	32,237	24,057	16,143	17,744	16,504	18,254	8,878			20,273	6,951	3500
3600	19,475	18,171	18,356	33,487	25,067	16,762	18,412	17,131	18,931	9,176			21,055	7,150	3600
3700	20,179	18,833	19,022	34,741	26,085	17,385	19,082	17,762	19,610	9,474				7,348	3700
3800	20,887	19,496	19,691	35,998	27,110	18,011	19,755	18,397	20,291	9,771				7,547	3800
3900	21,598	20,162	20,363	37,258	28,141	18,641	20,430	19,036	20,973	10,069				7,745	3900
4000	22,314	20,830	21,037	38,522	29,178	19,274	21,107	19,678	21,657	10,367				7,944	4000
4100	23,034	21,500	21,714	39,791	30,221	19,911	21,784	20,324	22,343	10,665				8,143	4100
4200	23,757	22,172	22,393	41,064	31,270	20,552	22,462	20,975	23,030	10,963				8,341	4200
4300	24,482	22,845	23,073	42,341	32,326	21,197	23,140	21,631	23,719	11,261				8,540	4300
4400	25,209	23,519	23,755	43,622	33,389	21,845	23,819	22,292	24,409	11,558				8,738	4400
4500	25,938	24,194	24,437	44,906	34,459	22,497	24,499	22,957	25,100	11,856				8,937	4500
4600	26,668	24,869	25,120	46,193	35,535	23,154	25,179	23,626	25,792	12,154				9,136	4600
4700	27,401	25,546	25,805	47,483	36,616	23,816	25,860	24,297	26,485	12,452				9,334	4700
4800	28,136	26,224	26,491	48,775	37,701	24,480	26,542	24,969	27,179	12,750				9,533	4800
4900	28,874	26,905	27,180	50,069	38,791	25,148	27,226	25,643	27,874	13,048				9,731	4900
5000	29,616	27,589	27,872	51,365	39,885	25,819	27,912	26,319	28,570	13,346				9,930	5000
5100	30,361	28,275	28,566	52,663	40,983	26,492	28,600	26,997	29,267	13,644				10,129	5100
5200	31,108	28,961	29,262	53,963	42,084	27,166	29,289	27,677	29,966	13,942				10,327	5200
5300	31,857	29,648	29,958	55,265	43,187	27,842	29,980	28,360	30,666	14,240				10,526	5300
5400	32,607	30,337	30,655	56,569	44,293	28,519	30,674	29,045	31,367	14,537				10,724	5400

* Based upon data of R. L. Hersey, J. E. Eberhardt, and H. C. Hottel, Thermodynamic Properties of the Working Fluid in Internal Combustion Engines, *S.A.E. Trans.*, **31**, 409 (1936).

† 536.6°R. is equivalent to 25°C., the standard temperature for calorimeter tests.

TABLE IV.—ENTROPY OF GASES*
(B.t.u. mol^{-1} °R.$^{-1}$ at 1 atm.)

Temp., °R.	O_2	N_2	Air	CO_2	H_2O	H_2	CO	OH	NO	O	H	$8CO_2 + 9H_2O + 47N_2$ / 64	$C_8H_{18} + 59.5$ air / 60.5	Temp., °R.
520	0	0	0	0	0	10.54	20.70	9.07	2.97	13.52	15.83	0	0	520
540	0.29	0.27	0.27	0.34	0.32	10.82	20.98	9.34	3.23	13.71	16.02	0.28	0.29	540
560	0.54	0.52	0.52	0.67	0.63	11.07	21.24					0.55	0.56	560
580	0.78	0.76	0.76	0.99	0.93	11.31	21.48					0.81	0.83	580
600	1.03	1.00	1.00	1.30	1.22	11.55	21.71					1.07	1.09	600
700	2.16	2.08	2.09	2.76	2.43	12.64	22.79					2.21	2.29	700
800	3.12	3.01	3.02	4.10	3.48	13.55	23.73					3.21	3.32	800
900	3.98	3.84	3.86	5.34	4.46	14.37	24.57					4.11	4.26	900
1000	4.76	4.59	4.61	6.50	5.36	15.11	25.33					4.94	5.11	1000
1100	5.48	5.27	5.30	7.57	6.20	15.78	26.02					5.69	5.90	1100
1200	6.16	5.91	5.95	8.57	6.97	16.39	26.66					6.39	6.65	1200
1300	6.80	6.50	6.54	9.51	7.67	16.95	27.27					7.04	7.34	1300
1400	7.39	7.05	7.10	10.41	8.34	17.47	27.82					7.65	7.99	1400
1500	7.94	7.56	7.62	11.26	8.97	17.96	28.31					8.22	8.61	1500
1600	8.47	8.05	8.11	12.06	9.57	18.43	28.79					8.77	9.20	1600
1700	8.97	8.52	8.58	12.83	10.15	18.87	29.27					9.29	9.76	1700
1800	9.44	8.97	9.03	13.57	10.71	19.29	29.75					9.79	10.30	1800
1900	9.89	9.40	9.47	14.28	11.23	19.69	30.19					10.27	10.84	1900
2000	10.32	9.81	9.88	14.95	11.75	20.06	30.60					10.73	11.34	2000
2100	10.73	10.20	10.28	15.59	12.25	20.41	31.00					11.16	.....	2100
2200	11.12	10.57	10.65	16.21	12.75	20.74	31.38					11.58	.....	2200
2300	11.50	10.92	11.01	16.81	13.22	21.07	31.74					11.98	.....	2300
2400	11.86	11.26	11.36	17.39	13.67	21.39	32.09					12.37	.....	2400
2500	12.21	11.60	11.70	17.95	14.11	21.69	32.43					12.75	.....	2500

Absolute Entropy† at 536.6°R.

Substance (Gaseous)	Entropy, B.t.u. Mol.$^{-1}$ °R.$^{-1}$
O_2	49.03
N_2	45.79
H_2	31.23
CO	47.32
CO_2	51.07
H_2O	45.17
CH_4	44.35

2600	12.55	11.03	12.03	18.49	14.53	21.99	32.76				....	2600
2700	12.88	12.24	12.35	19.02	14.94	22.29	33.07				....	2700
2800	13.21	12.54	12.65	19.53	15.35	22.57	33.37				....	2800
2900	13.53	12.83	12.94	20.03	15.74	22.83	33.66				....	2900
3000	13.84	13.11	13.22	20.51	16.13	23.09	33.95				....	3000
3100	14.14	13.39	....	20.98	16.51	23.35	34.23	21.81	16.50	22.22	....	3100
3200	14.43	13.67	....	21.44	16.88	23.61	34.51	22.08	16.79	22.39	....	3200
3300	14.70	13.94	....	21.89	17.25	23.87	34.78	22.34	17.07	22.54	....	3300
3400	14.95	14.20	....	22.32	17.61	24.12	35.04	22.59	17.34	22.69	....	3400
3500	15.21	14.45	....	22.74	17.96	24.36	35.29	22.84	17.61	22.84	25.00	3500
3600	15.46	14.69	....	23.15	18.30	24.58	35.54	23.08	17.87	22.99	25.15	3600
3700	15.71	14.92	....	23.54	18.64	24.80	35.77	23.31	18.11	23.13	25.30	3700
3800	15.96	15.15	....	23.91	18.97	25.02	35.99	23.53	18.34	23.26	25.44	3800
3900	16.20	15.38	....	24.28	19.29	25.23	36.21	23.75	18.57	23.39	25.57	3900
4000	16.43	15.60	....	24.65	19.51	25.44	36.43	23.97	18.79	23.51	25.70	4000
4100	16.65	15.79	....	25.02	19.92	25.64	36.65	24.19	19.02	23.63	25.82	4100
4200	16.87	16.00	....	25.38	20.22	25.84	36.87	24.40	19.24	23.75	25.94	4200
4300	17.08	16.21	....	25.73	20.51	26.04	37.08	24.61	19.46	23.87	26.06	4300
4400	17.29	16.42	....	26.08	20.79	26.24	37.29	24.81	19.68	23.99	26.18	4400
4500	17.50	16.62	....	26.42	21.07	26.44	37.49	25.01	19.89	24.11	26.30	4500
4600	17.70	16.81	....	26.74	21.35	26.63	37.69	25.20	20.09	24.23	26.42	4600
4700	17.90	17.00	....	27.06	21.62	26.82	37.88	25.39	20.28	24.34	26.54	4700
4800	18.10	17.18	....	27.37	21.89	27.00	38.06	25.58	20.47	24.45	26.65	4800
4900	18.30	17.36	....	27.68	22.16	27.18	38.24	25.76	20.66	24.56	26.76	4900
5000	18.49	17.54	....	27.98	22.42	27.35	38.42	25.94	20.85	24.67	26.87	5000
5100	18.68	17.72	....	28.28	22.68	27.52	38.59	26.12	21.03	24.78	26.98	5100
5200	18.86	17.89	....	28.57	22.93	27.69	38.76	26.29	21.21	24.88	27.09	5200
5300	19.04	18.06	....	28.86	23.18	27.86	38.93	26.46	21.38	24.97	27.19	5300
5400	19.22	18.23	....	29.14	23.42	28.02	39.10	26.63	21.55	25.06	27.28	5400

* Hersey, Eberhardt, and Hottel, *op. cit.* (Table III).

The entropy change during reaction is assigned to the fuel except in the case of the values for the octane-air mixture. The values for mixtures of gases do not include the terms $R \log p.p.$ for each constituent when compared with the values for each constituent alone.

† E. D. Eastman, *Chem. Reviews*, **18**, 257 (1936).

TABLE V.—HEATING VALUES AND CHEMICAL ENERGIES
(For gaseous fuels)

Fuel	Symbol	Reaction	Latent heat, B.t.u. mol^{-1} at 77°F.	Heating value, B.t.u. mol^{-1} at 77°F. (Water formed is condensed)		Chemical energy, B.t.u. mol^{-1} for the reaction indicated (use only with Tables III and VII)
				Q_p (const. pr.)	Q_v (const. vol.)	
Oxygen	O	$O = 0.5O_2$	.....	.....	.....	105,833
Hydroxide	OH	$OH = 0.5H_2 + 0.5O_2$	.....	.....	.....	10,700
Nitric oxide	NO	$NO = 0.5N_2 + 0.5O_2$	.....	.....	.....	38,746
Hydrogen	H	$H = 0.5H_2$	.....	.....	.....	92,822
	H_2	$H_2 + 0.5O_2 = H_2O$	.....	122,963[a]	121,365	103,486
Carbon monoxide	CO	$CO + 0.5O_2 = CO_2$	.....	121,721[b]	121,188	121,181
Methane	CH_4	$CH_4 + 2O_2 = CO_2 + 2H_2O$	.....	383,022[b]	380,891	345,214
Ethylene	C_2H_4	$C_2H_4 + 3O_2 = 2CO_2 + 2H_2O$	.....	607,104[c]	604,973	569,298
Ethane	C_2H_6	$C_2H_6 + 3.5O_2 = 2CO_2 + 3H_2O$	.....	671,058[d]	668,394	614,866
Ethyl Alcohol	C_2H_5OH	$C_2H_5OH + 3O_2 = 2CO_2 + 3H_2O$	18,216[e]	606,204[e]	604,072	550,553
Methyl Alcohol	CH_3OH	$CH_3OH + 1.5O_2 = CO_2 + 2H_2O$	16,092[e]	328,644[e]	327,045	291,374
Benzene	C_6H_6	$C_6H_6 + 7.5O_2 = 6CO_2 + 3H_2O$	14,495[f]	1,417,041[g]	1,414,377	1,360,843
n-Heptane	C_7H_{16}	$C_7H_{16} + 11O_2 = 7CO_2 + 8H_2O$	15,750[h]	2,085,751[i]	2,080,421	1,937,660
n-Octane	C_8H_{18}	$C_8H_{18} + 12.5O_2 = 8CO_2 + 9H_2O$	17,730[h]	2,368,089[i]	2,362,226	2,201,618
n-Dodecane	$C_{12}H_{26}$	$C_{12}H_{26} + 18.5O_2 = 12CO_2 + 13H_2O$	23,000[i]	3,498,659[i]	3,490,664	3,258,667

[a] *Nat. Bur. Standards, Jour. Research*, **6**, 1 (1931).
[b] *Nat. Bur. Standards, Jour. Research*, **6**, 37 (1931).
[c] *Nat. Bur. Standards, Jour. Research*, **13**, 249 (1937).
[d] *Nat. Bur. Standards, Jour. Research*, **12**, 735 (1933).
[e] *Nat. Bur. Standards, Jour. Research*, **13**, 189 (1934).
[f] *Nat. Bur. Standards, Jour. Research*, **6**, 881 (1931).
[g] *Nat. Bur. Standards, Jour. Research*, **2**, 375 (1929).
[h] *Nat. Bur. Standards, Jour. Research*, **13**, 21 (1934).
[i] *Nat. Bur. Standards, Jour. Research*, **18**, 115 (1937).
[j] *Nat. Bur. Standards, Misc. Pub.*, 97 (1929).

TABLE VI.—Logarithms of Equilibrium Constants*

$$\log_{10} K_p = \log \frac{[C]^c[D]^d}{[A]^a[B^b]} \text{ for reaction } aA + bB = cC + dD \text{ (see Table VII)}$$

[A] = partial pr. of A constituent in equilibrium mixture, atm.

Temp., °R.	(1) $H_2 = 2H$	(2) $O_2 = 2O$	(3) $N_2 = 2N$	(4) $N_2 + O_2 = 2NO$	(5) $H_2 + O_2 = 2OH$ $(8) - (7)$	(6) $2CO_2 = 2CO + O_2$	(7) $2H_2O = 2H_2 + O_2$	(8) $2H_2O = 2OH + H_2$	(9) $2H_2O + O_2 = 4OH$ $2(5) + (7)$	(10) $H_2 + CO_2 = H_2O + CO$ $\frac{(6) - (7)}{2}$	(11) $2H_2O + N_2 = 2H_2 + 2NO$ $(7) + (4)$
3000	−7.89	−8.93	−15.68	−4.30	+0.06	−8.68	−9.72	−9.66	−9.60	0.520	−14.02
3100	−7.44	−8.41	−14.97	−4.13	0.11	−8.12	−9.21	−9.10	−8.99	0.545	−13.34
3200	−7.01	−7.93	−14.28	−3.97	0.14	−7.58	−8.72	−8.58	−8.44	0.570	−12.69
3300	−6.61	−7.48	−13.63	−3.81	0.16	−7.08	−8.26	−8.10	−7.94	0.593	−12.07
3400	−6.23	−7.06	−13.01	−3.66	0.19	−6.60	−7.83	−7.64	−7.45	0.616	−11.49
3500	−5.87	−6.67	−12.43	−3.52	0.21	−6.15	−7.42	−7.21	−7.00	0.638	−10.94
3600	−5.53	−6.30	−11.89	−3.39	0.24	−5.72	−7.04	−6.80	−6.76	0.658	−10.43
3700	−5.21	−5.94	−11.37	−3.27	0.27	−5.32	−6.68	−6.41	−6.14	0.676	−9.95
3800	−4.91	−5.60	−10.88	−3.15	0.29	−4.95	−6.34	−6.05	−5.76	0.693	−9.49
3900	−4.62	−5.27	−10.41	−3.03	0.31	−4.59	−6.01	−5.70	−5.39	0.709	−9.04
4000	−4.34	−4.96	−9.97	−2.92	0.33	−4.25	−5.70	−5.37	−5.04	0.723	−8.62
4100	−4.08	−4.66	−9.55	−2.82	0.37	−3.93	−5.41	−5.04	−4.67	0.736	−8.23
4200	−3.83	−4.38	−9.16	−2.72	0.40	−3.63	−5.13	−4.73	−4.33	0.748	−7.85
4300	−3.60	−4.11	−8.79	−2.62	0.43	−3.35	−4.87	−4.44	−4.01	0.760	−7.49
4400	−3.38	−3.86	−8.44	−2.53	0.45	−3.08	−4.62	−4.17	−3.72	0.770	−7.15
4500	−3.17	−3.62	−8.10	−2.44	0.48	−2.83	−4.39	−3.91	−3.43	0.779	−6.83
4600	−2.97	−3.39	−7.77	−2.36	0.50	−2.59	−4.17	−3.67	−3.17	0.787	−6.53
4700	−2.77	−3.17	−7.45	−2.28	0.52	−2.36	−3.95	−3.43	−2.91	0.795	−6.23
4800	−2.58	−2.96	−7.15	−2.20	0.53	−2.12	−3.73	−3.20	−2.67	0.803	−5.93
4900	−2.40	−2.76	−6.86	−2.13	0.55	−1.90	−3.52	−2.97	−2.41	0.810	−5.65
5000	−2.22	−2.56	−6.68	−2.06	0.57	−1.70	−3.33	−2.76	−2.19	0.816	−5.39
5100	−2.05	−2.37	−6.31	−2.00	0.59	−1.51	−3.15	−2.56	−1.97	0.823	−5.15
5200	−1.88	−2.19	−6.05	−1.94	0.61	−1.32	−2.98	−2.37	−1.76	0.829	−4.92
5300	−1.72	−2.02	−5.80	−1.88	0.62	−1.13	−2.80	−2.18	−1.56	0.835	−4.68
5400	−1.58	−1.86	−5.56	−1.82	0.62	−0.94	−2.62	−2.00	−1.38	0.840	−4.44

* B. Lewis and G. von Elbe, Heat Capacities and Dissociation Equilibria of Gases, *A.C.S. Jour.*, **57**, 612 (1935).

TABLE VII.—REACTION DATA

$Q_p + H_{left\ side} = C_{right\ side} - {left\ side} + H_{right\ side}$. See text, pages 40 to 53

Reaction no.	Reaction	Equilibrium relationship		ΔS, B.t.u. °R.$^{-1}$ at 540°R.	Q_p, B.t.u. at 540°R. (gaseous H_2O)	C, B.t.u. (based on Table III)
		In terms of x and y, indicating part dissociated	In terms of partial pressures			
1	$H_2 = 2H$	$K_p = \dfrac{4x^2}{1-x^2}P$	$K_1 = \dfrac{[H]^2}{[H_2]}$	21.22	186,738*	185,644
2	$O_2 = 2O$		$K_2 = \dfrac{[O]^2}{[O_2]}$	27.13	212,756*	211,666
3	$N_2 = 2N$		$K_3 = \dfrac{[N]^2}{[N_2]}$			
4	$N_2 + O_2 = 2NO$	$K_p = \dfrac{4x^2}{(1-x)^2}$	$K_4 = \dfrac{[NO]^2}{[N_2][O_2]}$	5.90	77,501*	77,492
5	$H_2 + O_2 = 2OH$		$K_5 = \dfrac{[OH]^2}{[H_2][O_2]}$	7.57	21,407	21,401
6	$2CO_2 = 2CO + O_2$	$K_p = \dfrac{x^3 P}{(1-x)^2(2+x)}$	$K_6 = \dfrac{[CO]^2[O_2]}{[CO_2]^2}$	41.57	243,450†	242,362
7	$2H_2O = 2H_2 + O_2$		$K_7 = \dfrac{[H_2]^2[O_2]}{[H_2O]^2}$	21.29	208,092†	206,972
8	$2H_2O = 2OH + H_2$		$K_8 = \dfrac{[OH]^2[H_2]}{[H_2O]^2}$	28.86	229,499*	228,373
9	$2H_2O + O_2 = 4OH$	$K_p = \dfrac{64x^4 P}{(1-x)^3(x+3)}$	$K_9 = \dfrac{[OH]^4}{[H_2O]^2[O_2]}$	36.43	262,104	249,774
10	$CO_2 + H_2 = CO + H_2O$	$K_p = \dfrac{x(1-y)}{y(1-x)}$	$K_{10} = \dfrac{[CO][H_2O]}{[CO_2][H_2]}$	10.14	17,679	17,695
11	$2H_2O + N_2 = 2H_2 + 2NO$	$K_p = \dfrac{4x^2 y^2 P}{(1-y)(1-x)^2(3+y)}$	$K_{11} = \dfrac{[H_2]^2[NO]^2}{[H_2O]^2[N_2]}$	27.19	285,496	284,464

* Hersey, Eberhardt, and Hottel, *op. cit.* (Table III).
† *Nat. Bur. Standards, Jour. Research,* **6**, 1, 37 (1931).

TABLE VIII.—COMBUSTION DATA

Fuel	Chemical symbol	Ft.³ O_2 ft.$^{-3}$ fuel[a]	Ft.³ of air ft.$^{-3}$ fuel[a]	Lb. of air lb.$^{-1}$ fuel[a]	% CO_2 in products (H_2O cond.)	Products Mixture ratio by volume (H_2O not cond.)	Heating value at const pr., B.t.u. (fuel in gaseous state and H_2O cond.)			
							Lb.$^{-1}$ of fuel[c]	Ft.$^{-3}$ of fuel[d]	Ft.$^{-3}$ of air[d]	Ft.$^{-3}$ of corr. mix.[d]
Hydrogen	H_2	0.5	2.38	34.45	....	0.85	60,990	314	132	93
Carbon to CO	C	....	....	5.74	....	0.85	4,340			
Carbon to CO_2	C	....	....	11.48	21.0	0.83	14,520			
Carbon monoxide	CO	0.5	2.38	2.46	34.7	0.85	4,350	311	131	92
Methane	CH_4	2.0	9.52	17.23	11.7	1.00	23,890	978	103	93
Acetylene	C_2H_2	2.5	11.90	13.25	17.5	0.96	21,510	1429	120	111
Ethylene	C_2H_4	3.0	14.28	14.76	15.1	1.00	21,660	1550	109	109
Ethane	C_2H_6	3.5	16.66	16.08	13.2	1.03	22,330	1713	103	97
Propane	C_3H_8	5.0	23.80	15.66	13.8	1.04	21,670	2438	102	98
Butane	C_4H_{10}	6.5	30.94	15.44	14.1	1.05	21,320	3161	102	99
Pentane	C_5H_{12}	8.0	38.08	15.31	14.3	1.05	21,100	3884	102	99
Hexane	C_6H_{14}	9.5	45.22	15.22	14.4	1.05	20,950	4606	102	100
Methyl alcohol	CH_4O	1.5	7.14	6.46	15.1	1.06	10,260	839	118	103
Ethyl alcohol	C_2H_6O	3.0	14.28	8.99	15.1	1.07	13,160	1547	108	101
Benzene	C_6H_6	7.5	35.70	13.25	17.5	1.01	18,160	3618	101	98
n-Octane (gasoline)	C_8H_{18}	12.5	59.50	15.11	14.5	1.06	20,750	6047	102	100
Alcohol blend	$\{0.1 C_2H_6O, 0.9 C_8H_{18}\}$[b]	10.22	48.65	14.42	14.6	1.06	19,900	4969	102	100
n-Dodecane (kerosene)	$C_{12}H_{26}$	18.5	88.06	15.00	14.7	1.06	20,560	8934	101	100

[a] Obtained from the chemically correct reaction equation.
[b] Proportions by liquid volume. This is equivalent to about 0.24 mol of C_2H_6O and 0.76 mol of C_8H_{18}.
[c] Using a molecular weight of H_2 as 2.016. Fuels in gaseous or vapor state.
[d] At a pressure of 14.7 lb. in.$^{-2}$ and a temperature of 77°F. At these conditions 1 mol = 391.7 ft.³

INDEX

A

Abbott, 400
Acceleration, 132, 220–221, 483
Acetylene, 146–148, 177, 589
Adiabatic, 21
Air, 32
Air cell, 394
Air cleaner, 224
Air filter, 224
Air heater, 277
Air injection, 260
Air service, 284
Air standard, 9, 55
Air swirl, 389
Air-fuel ratio, 90, 172, 173, 175–176, 208, 289, 291, 294, 295, 297
  desirable, 204
A.C. Spark-plug Co., 297, 350
Alcohol, 140, 148, 155, 181–182, 455
  blends, 156, 455
Aldrich, 138, 157
Aliphatic, 122, 169
Allen, 228
Alleva, d', 160, 162
Almen, 407
Altitude, 225
  correction factors, 460
  effect on carburetion, 225
Aluminum Co. of America, 552
American Hammered Ring Co., 558–559
American Petroleum Inst., (A.P.I.), 124, 176
American Society of Mechanical Engineers (A.S.M.E.), 251, 263, 285, 289, 306
American Society for Testing Materials (A.S.T.M.), 124–128, 143, 190, 191, 200
Anderson, D. E., 386

Anderson, G. V., 225
Angle, 497
Aniline, 181–182
Argon, 32, 582
Aromatics, 118
Ash, 144
Aske, 372, 386
Atomic weight, 33
Atomization, 204, 229, 232
Autoignition, 171
Availability, 23, 110

B

Bach, 562
Backfiring, 228
Baffling, 441
Bahlke, 416
Baker, H. W., 441
Baker, J. B., 73, 82
Balance, 499, 507
  dynamic, 500
  four-cylinder engine, 503
  radial engine, 505
  reciprocating parts, 501
  rotating parts, 500
  single-cylinder engine, 502
  six-cylinder engine, 504
  static, 500
  Table, 507
Barber, 134
Barnard, D. P., 133, 407, 416
Barnard, E. R., 407
Barnett, 2
Barrel engine, 508
Bartholomew, 173–174, 176, 293
Baster, 281
Battery, 346
Beall, 351
Beardsley, E. G., 233
Beardsley, G. E., 227

Beardsley, M. W., 581
Bearing load, 494
　crankpin, 494, 495
　crankshaft, 495
　connecting rod, 494
　main, 495
　piston pin, 494
Bearings, 417, 534
　cooling of, 421–422
　materials for, 565
　roller, 534
　supports for, 547
　temperature of, 420
Beau de Rochas, 3–5
Becker, 176, 198
Benzene, 120, 139, 146, 148, 155, 170, 181–182, 197, 581, 586, 589
Benzol, 197, 199
Berger, 469
Best, H. W., 191–192
Best, R. D., 103, 165, 368–369
Biermann, 178, 437–439
Blackwood, 295, 354
Blank, 578
Blast-furnace gas, 146, 148–149
Blends, 156, 455
Blowback, 283
Blowby, 474–476
Blower (see Superchargers)
　horsepower, 474
Boerlage, 388
Bonds, 118, 169
Bone, 100
Bore-stroke ratio, 527
Bosch, 246, 248, 255
Bouchard, 105
Bouncing pin, 190
Bouvy, 314–315, 319, 320
Boyd, 105, 162, 165–169, 177, 181, 187, 368
Brake horsepower, 450, 461, 464, 474
Brake work, 449
Breaker mechanism, 347
Brevoort, 435–436, 441
Brewster, 173–174, 293
Bridgeman, 127–131, 134, 136, 138, 157, 419, 421
British thermal unit (B.t.u.), 581

Broeze, 388
Brooks, 228, 459
Brown, G. G., 138, 182
Buck, 449
Buick, 385, 456–457
Burned fraction, 357, 365
　temperature of, 357, 361
Burstall, 104, 148
Butane, 144, 148–149, 177, 454, 589
Butylene, 123, 144

C

Calorimeter, 35, 583
Cam followers, 313–314
Campbell, J. F., 238
Campbell, J. M., 167–169, 177, 181, 187, 191–192
Cams, 313, 509
Camshafts, 312
Carbon, 33, 581, 589
　formation, 180, 416
　residue, 143
Carbon dioxide, 32, 581–582, 584, 587–588
Carbon monoxide, 146–149, 581, 582, 584, 586–589
Carburetion, 204
Carburetors, auxiliary air valve, 213
　Carter, 220
　commercial, 223
　design and construction, 301
　elementary, 210
　idling, 212
　restricted air bled, 217
　unrestricted air bled, 215
　variable fuel orifice, 219
Carnot, 7, 24, 55
Carson, 452–453
Castleman, 230, 425
Centrifugal force, 410, 500
Cetane number, 201, 242
Chalk, 173–174, 293
Chapman, 147
Chard, de, 351
Charts (see **Appendix**)
　combustion, 48
　volatility, 129–130

Charts, volatility, corrections for, 131
  warming up, 132
Chemical energy, 18, 34, 586, 588
Chemical equilibrium, 40, 588
Chemical reaction, 33, 586–588
  entropy of, 25
Chenoweth, 469
Chevrolet, 324, 385, 449, 456–457, 477
Circuits (electrical), 346, 349
Clark, 456
Clearance, 27, 112, 380
  bearing, 408
  gases, 73, 175, 206
Clerk, 5, 101, 361
Coats, 199
Coefficients, compressibility, 247
  discharge, 209, 256
  excess air, 473–474
  flow, 471
  friction, 251, 400, 418–419
  heat transfer, 428, 431, 435, 437
  speed fluctuation, 491
Coke-oven gas, 145
Colwell, 308
Combustion, 32, 72, 166, 531
  charts, 48, Appendix
  data, 589
  incomplete, 115
  rate, 107
  spontaneous, 170
  temperatures, 38, 44
  time, 102, 106, 300
Combustion chambers, 200, 355
  antechamber, 392, 394
  air cell, 394
  design, 178, 387
  efficiency, 387
  fuel injection, 388
  material, 180, 386
  maximum output, 383–384
  open, 390–391
  precombustion, 391
  shapes, 179
  spark ignition, 355
Combustion process, 355, 361, 367, 388

Combustion process, effect of piston motion, 363
Combustion roughness, **377**
Combustion shock, 379
Common rail, 239
Compressibility, 247
Compression, 69, 88
Compression ignition, 7, 142
Compression pressure, 194, 196, 457
Compression ratio, 58, 65, 148, 175–176, 194, 197
  critical, 177, 199–200
  optimum, 466
Compression temperature, 194–195
Condenser (electrical), 346
Conductivity, 433–434
Conradson, 143
Constant, equilibrium, 25, 42, 44, 587–588
  universal gas, 11
Constant-pressure process, 21
Constant-temperature process, 21
Constant-volume process, 21
Cooling, 428
  bearing, 445
  cylinder, 433, 435
  oil, 410
  piston, 444
  valve, 445
Cooling systems, 446
Cooperative Fuel Research (C.F.R.) 126, 139, 170–171, 173, 176, 187–188, 193, 197, 200–201
Corrosion, 126
Counterbalancing, 500, 505, 507, 509–510
Couples, 500, 505–507
Coward, 146, 148
Cracking, 122
Cragoe, 131
Crane, 489
Crankpin, 567
  bearing loads, 494
Crankshaft, 495, 564
  bearing loads, 495
  deflection, 576
Critchfield, 352–353
Critical velocity, **237, 251**

Cronstedt, 463
Cummins, 259
Cutter, 476
Cycles, 22
  air standard, 9, 55
  Carnot, 9, 24, 55
  comparison, 9, 63
  Diesel, 7, 59, 61, 63
  four stroke, 5, 526
  Otto, 5, 57, 63
Cyclobutane, 120
Cyclopropane, 120
Cylinder, 535

D

D'Alleva, 160, 162
Dalton, 12
Davis, 405–406
De Chard, 351
De Juhasz, 263, 449, 472–473
Dean, 405–406
Dennison, 285, 286, 306–307, 423
Design, 525
  cam, 313
  connecting rod, 560
  crankcase, 545
  crankshaft, 564
  cylinder, 535
  cylinder head, 543
  frame, 545
  piston, 549
  piston pin, 554
  piston ring, 556
  valve, 309
    spring, 327
Detonation, 165, 531
  factors, 166
  incipient, 170, 176
  pressure, 182
  suppressors, 180
Deviations, 97
Dew point, 136
Dickinson, 190, 225, 419, 421
Dicksee, 197, 238
Dickson, 477
Diesel, 7, 59, 61, 83, 86

Diesel index, 198
Dilution, 90, 175, 206
  crankcase, 136, 417
  effect of, 360–361
Diolefin, 123
Displacement, 65, 91, 112, 528–529
Dissociation, 32
Distillation, 120, 126, 173
  A.S.T.M., 125
Distribution, 109, 173
  effect on temperature, 295
  energy, 476–477
Distributor (electrical), 346
Dispersion, 237
Dixon, 146
Dodge, 399
Dodge Lanova, 456
Dopes, 201
Du Bois, 296, 463
Duchene, 373
Dutcher, 451

E

Eastman, 585
Eberhardt, 48, 583, 585, 588
Edgar, 187, 193
Efficiency, combustion, 163
  chamber, 387
  time, 106
  mechanical, 450
  relative, 98
  thermal, 88, 100, 107
  volumetric, 92
Egloff, 125, 142–143, 201
Eichelberg, 477
Eisinger, 131, 133, 416
Energy, 14, 18
  absolute, 18, 36
  chemical, 18, 34, 588
  electrical, 18
  internal, 17, 582
  kinetic, 17
  potential, 16
  relative, 18, 36
  values, 19, 581
  work, 14–15

INDEX 595

Engine, Carnot, 7
  Clerk, 5
  C.F.R., 187–188, 197
  Diesel, 7
  Ford, 544
  free piston, 2, 3
  general analysis, 87
  Langen, 2–3
  Manly, 6
  Otto, 2–5, 7
  Pratt and Whitney, 539
  Plymouth, 381
  Radial, 6
  Round, 508
  Street, 2
  Sulzer, 546
  Two-cycle, 114, 339, 342
  Wright, 538
Enthalpy, 17–18
Entropy, 15, 24, 584, 588
Equation, characteristic, 10–11
  energy, 19
  entropy, 24
  equilibrium, 42, 44, 49
  reaction, 32, 586, 588
  specific-heat, 15
Equilibrium, 25
  constants, 25, 42, 44, 162, 587–588
Equilibrium air distillation (E.A.D.), 128
Ethane, 144, 146–148, 177, 581, 586, 589
Ether, 146, 148
Ethyl gasoline, 181
Ethyl Gasoline Corp., 176
Ethyl iodide, 182
Ethylene, 123, 146–148, 177, 369, 581, 586, 589
Ethylene dibromide, 181
Ethylene dichloride, 181
Excello, 245
Excess air, 241
Exhaust, 81
Exhaust gases, 158
  analysis 292, 294, 299–300, 476
Expansion, 73
Expansion ratio, 61, 63, 106

F

Faller, 357
Farrar, 460–461
Fedden, 336–337
Federal-Mogul Corp., 565
Fenney, 196–197, 243
Finn, 108
Fins, 435
  optimum dimensions, 438
Fiock, 370
Fire point, 402
Firestone, 400
Firing order, 354
Fischer, 198
Fitzsimons, 416
Flame front, 357, 365
  area, 370, 372
Flame photographs, 166, 370
Flame progress, 368, 370
Flame speed, 147, 172, 373
Flash point, 402
Float chamber, 210
Flow, 207, 423
  manifold liquid, 283
  port and valve, 306, 324
Flywheel, 490
  displacement diagram, 492–493
  velocity diagram, 492–493
Forces, 531
  centrifugal, 410, 500, 506, 534
  connecting rod, 485
  gas pressure, 485, 512
  inertia, 483, 512
  net, 485–486
  piston side thrust, 485
  rotative, 486, 488–489
Foot-pound, 581
Ford, 224, 324, 449, 456–457, 544
Fourier series, 481, 501
Frames, 545, 548
Frederick, 336
Friction, 28–29, 251, 285, 400, 420, 474, 533
Fuel, 118
  consumption, 289, 296, 298–299, 454–456, 464
  loops, 456

Fuel, gaseous, 144
  injection, 204, 229
    air, 238
    duration, 243
    hydraulics of, 262
    nozzles, 253
      multiple, 267
    pilot, 244
    pumps, 245
    quantity, 240
    rate, 241
    solid, 238
    systems, 238
    timing, 242
    valves, 253
  lines, 249
    resistance of, 250
  liquid, 118, 155
  loss, 294
  oil, 142
  rating, 190, 198

G

Gagg, 460–461
Gardner, 459
Gary, 134
Gases, 144, 150
  blast furnace, 145, 148, 150
  blue water, 145, 148, 150
  carbureted water, 145, 150
  cleaning of, 146
  coal, 145, 150
  coke oven, 145, 150
  exhaust, 158
  illuminating, 145, 148, 150
  monatomic, 16
  natural, 144, 148, 150
  oil, 145, 150
Gasoline, 124, 148, 155, 199
  aviation, 124
  freezing point, 126
  grades, 124
  latent heat, 124
  motor, 124
  specifications, 125
Geiger, 523
Gellales, 234, 256

General Motors Corporation, 258, 271, 276, 390, 456, 474
Gerrish, 158, 160, 162, 293, 294
Geschelin, 308
Gillette, 386
Glauert, 441
Gleason, 158, 162
Goodenough, 73, 82
Graf, 158, 162
Gum, 126, 138

H

Harmonics, 501, 513, 522
  amplitudes, 333, 514
Hawley, 176
Heat, 15
  specific, 15
    equations, 581
Heat engine, 1, 8
Heat loss, 109
  distribution, 111
Heat transfer, 28, 153, 428, 532
  cooling medium, 434
  cylinder wall, 433
  integration areas, 430
  piston head, 439
Heating surface, 277
Heating value, 34, 37
  of fuels, 586, 588
  of mixture, 144, 155, 588
Hebl, 178
Heldt, 471, 512, 564
Heptane, 118–119, 122, 177, 181, 187, 199, 586
Heptiadene, 123
Heptylene, 122
Hercules, 394
Hersey, M. D., 251–252, 407, 423
Hersey, R. L., 48, 583, 585, 588
Hesselman, 254, 260
Hetzel, 198, 200
Hexane, 118, 119, 177, 373, 589
Hiel, 199
Highest useful compression ratio (H.U.C.R.), 186
Holaday, 192
Holley, 219

Honaman, 102, 368
Hopf, 251
Hopkinson, 361
Horning, 308
Horsepower-hour, 581
Hottel, 48, 583, 585, 588
Hubner, 125, 142–143, 198–199, 201–202
Huffman, 581
Humidity, 459
Hydraulic valve lifter, 315
Hydraulics, 262
Hydrocarbons, 41, 118
  normal, 118
Hydrogen, 25, 33, 146–149, 183, 581–582, 584, 586–589
Hydrogenation, 123

I

Ice formation, 227
Idling, 207
  jet, 212
Ignition, 345
  delay, 193, 196, 200
  lag, 193, 197, 242
  methods of, 345–346
  quality, 193
  temperature, 146, 169, 195
  timing, 107, 176, 185, 352
  (*See also* Spark timing)
Illmer, 453
Illuminating gas, 145, 148–149
Indicated horsepower, 449, 455
Indicated work, 449
Indicator card, 448
Induction, 75
Induction coil, electrical, 346–347
Inflammability, 147
Injection (*see* Fuel injection)
Injection nozzles, 253, 257, 259
Injection pressure, 259
Intake temperatures, 175
Ionization, 102–103
Irreversibility, 26
Isobutane, 144
Isomers, 119
Isoöctane, 177, 181, 187

J

Jäfar, 244
Johnke, 514, 516
James, 225
Janeway, 377, 382, 430–432
Jardine, F., 380, 382
Jardine, R., 308
Jardine, R. S., 308
Jehle, 334, 336
Jessup, 124
Johnson, 578
Jones, 148
Journal bearing, 417
Juhasz, de, 263, 449, 472–473
Junkers, 389, 553

K

Kalb, 527
Kalmar, 104, 105, 298–300, 351, 368, 370, 387
Kass, 295, 354
Kaye, 199
Keenan, 199
Kent's, 176
Kerosene, 197, 199
Kettering, 165
Keyes, 13
Kilowatt-hour, 581
Kinematic viscosity, 199, 402–403
Kishline, 386
Knock intensity, 174, 185, 192
Knock testing, 165, 185
  accuracy, 191
  methods, 185
  procedure, 190
Krupp, 260
Kuchler, 306–307
Kulason, 134
Kuring, 182

L

Lack, 514, 516
Langley, 5–6
Lanova, 394, 395

Laws, Dalton's 12, 49
  first, of thermodynamics, 20
  second, of thermodynamics, 29
Leakage, 248
Lee, 230–233, 237, 267
Lewis, B., 26, 101, 587
Lewis, G. H., 25
Lewis, O. G., 295, 354
Lichty, 73, 156, 158, 162, 182, 204, 266, 357, 455, 477
List, 286
Losses, friction, 27, 449–450, 533
  heat, 109
  incomplete combustion, 115
  pumping, 100, 112
  summation, 116
  two stroke, 113
Lovell, 105, 160, 162, 165, 167–169, 177, 181, 187, 191, 368
Lubrication, 398
  piston ring, 424
  thick film, 418
  thin film, 407
  systems, 408
Lubricants, properties, 402
Lürenbaum, K., 513

M

Macauley, 301
MacClain, 449
MacCoull, 134
MacKenzie, 102, 368
McDavid, 146
McDermott, 170
McKechnie, 230
McKee, 419, 421–422
Magneto, 348
Maleev, 530
M.A.N., 394
Manifolds, 270
  dual, 283
  eight cylinder, 281
  exhaust, 286–287
  four cylinder, 279
  four port, 282
  multiple, 281
  pressure, 272, 282, 284, 289

Manifolds, resistance, 272
  six cylinder, 280
  shape, 273
  size, 270, 286
  supercharging, 285
  two cylinder, 279
Manly, 6
Marks, 511, 536, 561
Marsh, 248, 267
Marvel, 239
Marvin, 103–104, 107, 165, 368–369
Materials, 578
  bearing, 565
  combustion chamber, 180, 386
  summary, 578
  valve, 309, 312
  valve seat, 309
Mean effective pressure (m.e.p.), 65, 91, 300, 453, 529
  supercharging, 463
Mechanics, 480
Media, 10
Methane, 11, 21, 118, 146–149, 177, 369, 581, 586, 589
Methylnaphthalene, 201
Midgley, 181
Minter, 108
Miscibility, 157
Mixture, 12, 68, 98
  accelerating, 220
  actual, 224
  altitude, 225
  condition, 172, 270
  distribution, 270, 289, 300
  formation, 204
  fractions, 291
  heating, 137, 275
  maximum economy, 206, 289, 455–456
  maximum power, 205, 221, 289, 455
  starting, 131
Molecular structure, 167
Molecular weight, 32
Monatomic, 16
Moore, 199
Mougey, 564

INDEX

Murdock, 197–200
Murphy, 125, 198–202

N

Naphthenes, 118, 120, 169
Natural gas, 144, 148–149
Nelson, 550
Nitrogen, 32–33
Nonflow, 20
Nonreaction, 20
Normal (n-), 118
Nusselt's number, 436, 443

O

Octane, 177, 181–182, 187–188, 199, 356, 463, 581–582, 584, 586, 589
Octane number (O.N.), 126, 176, 187, 191, 356
Oil, consumption, 411
  viscosity effect, 412
Oil film, 399, 418
Oil flow, 423
Oil-hole location, 497
Olefin, 122
Operating conditions, 400
Oregon State Agricultural College, 158
Otto, 2–5, 7, 57, 68, 82
Oxygen, 32, 33

P

Paraffins, 118, 167
Parks, 581
Partial pressure, 12
Paul, 158, 162
Payman, 147
Peletier, 172–173, 175–176
Pendulum, 520, 522
Penetration, 233
Pennsylvania State College, 198, 234
Pentane, 118–119, 144, 148, 177, 181, 589
Perfect gases, 10

Performance, 448
  improving, 461
  multicylinder, 290
Phelps, 156, 162, 182, 455, **477**
Phillips, 578
Pinkel, 255
Pintle, 255
Piston, 549
  acceleration, 483
  inertia, 483
  position, 482
  velocity, 324, 482
Piston-crank relation, 481
Piston pin, 554
Piston ring, 413, 559
  clearance, 559
  grooves, 553
  slotted oil, 413, 559
  tension, 557
Plunger displacement, 248
Plunger velocity, 248
Plymouth, 324, 381, 449, 456
Poise, 398
Polymerization, 123
Pope, 197–200
Ports, 275, 338–339, 342
  size, 303
Pour point, 402
Power, 289, 296, 298–299, 528, 534
  air cooling, 439
  curves, 449
  trend, 463
Pratt and Whitney, 539
Preignition, 184
Prescott, 516
Pressure, 10
  atmospheric, 225
  mean effective, 65, 91
  rate of rise, 374, 377
Process, adiabatic, 21
  analysis, 9, 20, 29, 68, 87
  combustion, 72
  compression, 69
  constant pressure, 21
  constant volume, 21
  cracking, 122
  Diesel engine, 83, 86
  exhaust, 81

Process, expansion, 73
  friction, 28–29
  heat engine, 1
  heat transfer, 28–29, 428
  hydrogenation, 123
  induction, 75, 324
  internal-combustion engine, 1, 8, 68
  isothermal, 21
  nonflow, 20
  nonreaction, 20
  Otto engine, 68
  polymerization, 123
  refining, 120
  release, 77
  reversible, 15
Producer gas, 145, 149
Properties, 10, 13
Propane, 144, 146, 148–149, 177, 369
Propylene, 123, 144, 177
Pumping, loss, 100, 112
  work, 99
Pumps, 245, 247
Purday, 530
Purdue University, 210

R

Rabezzana, 104–105, 298–300, 350–351, 368, 370, 387
Randall, 25
Rankine, 11, 561
Rassweiler, 105, 165, 364, 366–367, 370
Rating, 190, 198
  Diesel fuel, 202
  road, 192
Reaction, 33, 586–588
  velocity, 101, 373
Reboring, 536
Reciprocating parts, 480
Reference fuels, 187, 201
  Shell, 197, 200
Refining, 120
Refrigerator, 8
Reid, 126, 134
Relative efficiency, 98
Release, 77

Rendel, 132, 178, 201
Reversibility, 15, 26
Reynolds number, 231, 237, 250, 256, 436, 442–443
Ricardo, 165–166, 175, 179, 186–187, 333, 379–380, 384, 388, 415
Ring section, 553
Ristine, 386
Road rating, 192
Robertson, 475
Robie, 138
Rocking couples, 508
Roeder, 165, 370
Roensch, 414
Rogers, E. C., 228
Rogers, T. H., 407
Rose, 242
Rosecrans, 100, 373
Rothrock, 165, 248, 250, 263, 266
Roughness, 512
Round engines, 508
Rubbing factor (pv), 421, 496

S

Saturation, 122
Saybolt, 142, 403
Scavenging, 114, 389, 472
  pressure, 474
Scherger, 414–415
Schey, 456
Schnauffer, 102–103, 368
Schopper, 471–472
Schwarz, 581
Schweitzer, 198, 200, 231, 234–237, 243–245
Scintilla, 348
Secondary reference fuels, 187
  calibration, 188
Selden, 242
Self-ignition, 169–170, 196
Shoemaker, B. H., 407
Shoemaker, F. G., 258–259, 474
Siamese, 280
Side thrust, 485, 537, 540
Siemens and Halske, 103
Signaigo, 169

Silent Knight, 336
Similitude, 530
Singer, 182
Sleeve valves, 336
Sligh, 128
Smith, D. W., 306–307
Smith, M. F., 357
Smoke, 243
Sneed, 477
Socony-Vacuum Company, 200
Spark advance, 107, 174, 189, 352
  incipient detonation, 354
  maximum power, 353
  optimum, 109
  road load, 352–353
Spark ignition, 5, 345
Spark plug, 297, 350
  location, 179, 369, 382
  thermocouple, 297
Spark voltage, 349, 351
Sparrow, 414–415, 465–466, 475, 564
Specific fuel consumption (*see* Fuel conservation)
Specific gravity, 124
Spencer, 165
Spiller, 334, 336
Splash system, 408
Spray form, 230
Spontaneous combustion (*see* Self-ignition)
Spring design, 328
Spring vibration, 332
Stability, 406
Standard conditions, 458
Standard Oil Development Company, 187
Starting, 131
State, 10
Steam engine, 8
Stevens, 100, 370
Stewart, 412
Stoke, 402
Straight-chain structures, 118–119
Street, 2
Stroke-bore ratio, 527
Stromberg, 224
Studebaker, 457
Sulphur, 126, 139

Sulzer, 456, 477, 521–522, 546
Superchargers, 467
  capacity, 472
  centrifugal, 467
  exhaust driven, 469
  Roots, 467, 470
  vane, 467
Supercharging, 285, 464
  indicator card, 465
Surface variation, 430
Swan, 333
Swash plate, 509
Sweigert, 581

T

Tappets, 315
Taub, 224, 272, 301, 351
Taylor, C. F., 105, 530
Taylor, E. S., 105, 520, 530
Tee, 274
Temperature, 10
  chemical equilibrium, 44
  combustion, 38
  cylinder head, 296
  exhaust, 79, 278
  gas, 429, 433
  piston, 441
  spark plug, 298–300
  standard, 458, 583
  stresses, 550
Tessmann, 158
Tetraethyllead, 126, 170, 181–182, 188
Texas Company, The, 412
Thermodynamics, 10, 25
Thompson, 399
Thomsen, 400
Thornycroft, 187
Throttle, 210
Throttling, 99, 174
Tice, 206, 221–222, 281–283, 327
Timken, 407
Timoshenko, 516
Tognola, 351
Toluene, 120, 148, 177, 181–182
Torque, 450, 464, 489
  variation, 488

Torsiograms, 522
Torsional vibration, 513, 522
Townend, 100
Townsend, 351
Transfer passage, 380
Trimethylpentane, 177, 187
Turbulence, 103, 375, 389

U

Unavailability, 24
Unburned fraction, 358
    cooling, 360
Universal Oil Products, 142
University, of Illinois, 73, 100, 373
    of Michigan, 138
Unsaturated, 122
Upton, 108

V

Vacuum engines, 2
Vail, 322
Valve, acceleration, 315, 319
    diagrams, 338
    harmonic analysis, 334
    inserts, 305, 308
    lift, 315, 319, 324
    mechanisms, 303, 321
    overlap, 283, 287
    seats, 304, 310
    springs, 327
        design, 328
        surge, 332
    stems, 311
        bosses, 312
        guides, 312
    timing, 111, 154, 175, 323
    velocity, 315, 319
Valves, 303
    disk, 336
    multiple, 307
    rotary, 336
    shrouded, 389
    sleeve, 336
Vapor lock, 134
Vapor pressure, 126, 134
Vapor venting, 135

Vauxhall, 224
Veal, 192
Vibration, 499
    dampers, 517, 520, 522
    damping, 511, 517, 520, 522–523
    natural frequency, 510, 514, 516, 521, 535
    shaft, 514, 516
    torsional, 513
Viscosity, 142, 398, 402, 422
    absolute, 398, 402
    chart, 404, 406
    classification, 403, 405
    index, 405
Viscosity-gravity number, 199
Volatility, 125, 136
    curves, 127
    effective, 137
Volumetric efficiency, 92, 151, 154–155, 240, 456–457, 471, 472, 474
von Elbe, 26, 101, 587
Vogt, 454
Voss, 160, 162, 293–294

W

Wahl, 328–331
Walls, 322
Warming up, 132
Water, 12–13, 25
    tolerances, 158
Watt-second, 581
Waukesha Comet, 456
Waukesha Motor Company, 189, 561
Wear, 413, 497
    cylinder, 414–416
Weight, 534
Westinghouse, 477
Wharton, 165
Whatmough, 379, 384
White, 134
Whitefield, 289
Wilcox-Rich, 315
Wilkin, 407
Willi, 565
Wilson, 242
Withrow, 105, 165, 166, 364, 368

Wobbler, 509
Work, flow, 15
  mechanical, 14
  pumping, 99
World Power Conference, 187
Wright Aeronautical Corporation, 561
Wright "Cyclone," 456, 538
Wyman-Gordon Company, 566

X

Xylene, 120, 177, 182

Y

Yale University, 170, 196
Yates, 558
Young, 226

Z

$\dfrac{\mu n}{p}$ factor, 407, 418, 421
Zenith, 220, 224–225
Ziurys, 156, 204, 455
Zuchrow, 210